HILLSLOPE MATERIALS AND PROCESSES

HILLSLOPE MATERIALS AND PROCESSES

Second Edition

M. J. Selby

with a contribution by
A. P. W. Hodder

Oxford University Press · Oxford
1993

Oxford University Press, Walton Street, Oxford OX2 6DP

Oxford New York Toronto
Delhi Bombay Calcutta Madras Karachi
Kuala Lumpur Singapore Hong Kong Tokyo
Nairobi Dar es Salaam Cape Town
Melbourne Auckland Madrid
and associated companies in
Berlin Ibadan

Oxford is a trade mark of Oxford University Press

Published in the United States
by Oxford University Press Inc., New York

British Library Cataloguing in Publication Data

Data available

Library of Congress Cataloging-in-Publication Data
Selby, Michael John.
   Hillslope materials and processes / M.J. Selby.—2nd ed.
   Includes bibliographical references.
   1. Slopes (Physical geography) 2. Weathering. I. Title.
GB448.S44 1993   551.4'36—dc20   92-35120
ISBN 0-19-874165-0
ISBN 0-19-874183-9 (pbk.)

1 3 5 7 9 10 8 6 4 2

Typeset by Best-set Typesetter Ltd., Hong Kong

Printed in Great Britain
on acid-free paper by
Butler & Tanner Ltd.
Frome, Somerset

Preface to the Second Edition

In the period between publication of the first and second editions of this book there have been many advances in knowledge of the properties of rock and soil and of the processes acting on and in the materials of which hillslopes are formed. Of equal significance for earth scientists concerned with hillslopes is the strengthening of the interchanges amongst scientists and engineers; symposia concerned with hillslopes commonly include contributions from geologists, engineers, hydrologists, soil scientists, and geomorphologists. Those who wish to study and contribute to advances in knowledge of hillslope materials, processes, and development require a comprehensive understanding of the relevant disciplines. The objective of this book is to provide an integrated review of basic knowledge which forms the foundation for advanced study and a contribution to developments in understanding through the appropriate disciplines.

In my own experience, students have varied degrees of prior knowledge which they bring to undergraduate studies. I have therefore included material, which for some may be unnecessary, that I believe to be essential for understanding major concepts. The inclusion of chapters on chemical bonds, and much of the material on particles and fabrics, and on rheology, is part of an attempt to ensure that students have ready access to the basic knowledge required for an understanding of weathering, cohesion, and the behaviour of rock and soil under stress.

Preparation of this book was undertaken during a period of study leave from the University of Waikato and I am greatly indebted to its Council both for that leave and others which have allowed me to undertake fieldwork in some of Earth's most beautiful and interesting landscapes. My colleagues in the Department of Earth Sciences have supported the concept of integration of disciplines and provided a congenial and stimulating atmosphere which we have all found productive. I am particularly grateful to Dr A. P. W. Hodder for agreeing to contribute parts of Chapters 2, 8, and 9 and for his helpful comments on several other chapters. Dr V. G. Moon kindly read part of the text and Dr D. I. Campbell commented on Chapter 11; any blemishes which remain are entirely my responsibility. I owe much to Mrs E. Norton, who did all of the word-processing with unfailing care and helpful good humour; to Mr F. E. Bailey for preparing the figures and to Mr G. M. Oulton for making prints of these figures. Staff of the photographic unit prepared prints of the photographs. All photographs not otherwise acknowledged are my own.

M. J. Selby

Hamilton, New Zealand
August 1991

Acknowledgements

Thanks are due to the following authors, publishers, and learned bodies for permission to reproduce copyright material.

American Association for the Advancement of Science: Figs. 13.12, 8.4a, 13.19a,b,c

American Society of Civil Engineers: Figs. 5.26a, 7.4c, 7.5h

American Society for Testing and Materials: Figs. 6.3, 7.6c. Copyright ASTM.

Branz: Figs. 17.1, 17.2, 17.3

E Brown: Fig. 6.6d after E. Brown (1981), *Rock Characterization Testing and Monitoring*, © E. Brown 1981.

Butterworth-Heinemann Ltd: Fig. 2.3b

Cambridge University Press: Figs. 2.3a, 8.10, 12.8

Chapman & Hall: Fig. 4.9c

The University of Chicago Press: Fig. 9.4a

K. E. Easterling: Fig. 3.15, *Mechanisms of Deformation and Fracture* (1979). © K. E. Easterling.

Elsevier Science Publishers BV: Figs. 4.8e, 4.10, 7.6b, 7.7a,b,c, 8.5d, 9.5, 13.2a

Department of Energy, Mines and Research, Mines Branch, Ottawa: Fig. 4.13

W. H. Freeman: Fig. 12.11

Geological Society of America: Figs. 4.12, 8.7c, 8.8b,c, 14.2, 14.3, 14.11, 15.17, 16.8. All © the authors.

Gebruder Borntraeger: Figs. 1.1, 1.2, 8.5b,c, 9.8b,c, 10.2, 16.1, 19.6

University of Guelph, Department of Geography/Geo Abstracts: Fig. 6.2

HarperCollins (Allen & Unwin): Figs. 19.2b, 19.3

Institute of Mining and Metallurgy: Figs. 5.23, 5.27

Institution of Engineers, Australia: Fig. 13.23a

International Association of Engineering Geology: Fig. 5.29

James Cook University of North Queensland: Fig. 5.20i

Japanese Geomorphological Union: Figs. 13.2b,e

McGraw-Hill, Inc.: Figs. 3.2, 3.3, 3.5, 5.20b,c,d, 12.1

National Research Council of Canada: Figs. 5.26b, 7.3c, 17.7

New Zealand Hydrological Society: Figs. 11.7c, 11.8, 15.26

Oliver & Boyd: Fig. 8.1

Oxford University Press: Fig. 8.7b

Oxford University Press, Inc.: Fig. 8.12

Pergamon Press Ltd: Fig. 8.11b. © 1982 Pergamon Press Ltd.

Pergamon Press, Inc: Figs. 4.9a,b, 5.21, 5.30a,b, 6.7

Prentice-Hall, Englewood Cliffs, NJ: Fig. 8.11a, from *The Geochemistry of Natural Waters*, 2 edn., © 1988

A. Roberts: Figs. 4.3h–k, 4.4 and 5.22 after A. Roberts, *Geotechnology* (1977). © A. Roberts 1977.

Scandinavian University Press, Oslo, Norway: Figs. 4.11, 13.9, 19.1, 19.2a, 19.4. Copyright is held by Scandinavian University Press.

Science Press, Beijing: Fig. 5.19a,b,c,d

Scientific American Books: Fig. 4.7

SIR Publishing: Figs. 10.5, 11.6, 13.2d

Society of Economic Petrologists and Mineralogists: Fig. 8.10b

Society for Mining, Metallurgy, and Exploration, Inc., Littleton, Colo.: Fig. 6.1g

Soil and Water: Fig. 12.16

Soil and Water Conservation Society of America: Fig. 12.9

Springer-Verlag: Fig. 6.6c.ii

Thomas Telford Services Ltd: Figs. 3.8, 5.15, 5.26c, 13.20

US Army Corps of Engineers: Fig. 15.20

United States Department of Agriculture: Figs. 12.7, 12.8

United States Geological Survey: Figs. 7.5g, 13.5

Van Nostrand Reinhold: Fig. 16.10

John Wiley & Sons Ltd.: Figs. 3.10, 4.3b, 4.8g, 5.10, 6.13, 8.6, 14.5, 14.7, 16.3, 16.9, 16.11, 17.6

John Wiley & Sons Inc.: Fig. 13.16

Contents

x   Contents

Symbols Used

Note: Variables denoted by italic type. Acronyms and indices are denoted by roman type.

A	area		g	acceleration due to gravity
A_j	total area of joint surface within a failure plane		H	height
			H_c	critical height
a	ground acceleration in earthquakes		h	thickness, height
B	sorbtivity of soil, width, rainfall obliquity, Bubnoff unit		h_c	height of capillary rise, depth of crack
			h_w	depth of water below the phreatic surface
b	length along a specified axis			
C	transport rate, solubility		I	intensity of rainfall
C_c	compression index		I_B	brittleness index
CEC	cation exchange capacity		i	angle of inclination of asperities
CSR	cyclic stress ratio		JCS	joint wall compressive strength
c	cohesive strength		JRC	joint roughness coefficient
c'	cohesive strength with respect to effective stresses		K	bulk modulus, a constant
			K_o	coefficient of geostatic stress
c_e	effective cohesion of a rock mass		KE	index of erosivity
c_r	residual cohesive strength		k	hydraulic conductivity, coefficient of permeability in saturated soils
c_u	cohesive strength as indicated in an undrained shear test		L	length
			LL	liquid limit
D	depth, thickness, diameter, duration, displacement		LS	Linear shrinkage
			l	length
D_{50}	median grain size of particles		M	mass of substance, Richter magnitude, bending moment, rank
d	depth, diameter			
E	Young's modulus of elasticity, erosion, energy		M_w	mass of water
			M_s	mass of solids
E_d	Young's modulus derived from dynamic tests		MM	modified Mercalli scale
E_{st}	Young's modulus derived from static tests		m	mass, fractional part of
			N	number of units
E_k	kinetic energy		NMC	natural moisture content
E^o	reduction potential		n	porosity, Manning's roughness coefficient
ESP	exchangeable sodium percentage			
e	void ratio		OCR	over-consolidation ratio
F	factor of safety		P	dispersive stress
f	a function of, infiltration rate, coefficient of friction		PI	plasticity index
			PL	plastic limit
G	shear modulus		p	pressure
G_s	specific gravity		p_o	overburden pressure

Q	total discharge per unit of time
q	flow rate
q_s	splash transport
R	hydraulic radius, resistance
RMS	rock-mass strength
r	radius, rating
r_c	radius of curve
r_u	pore-pressure ratio
S	slope, gradient, sediment transport rate
SPT	standard penetration test
S_f	softening factor
S_t	sensitivity
S_u	undrained strength (shear vane)
T	surface tension, weight of trees
T_c	critical thickness
t	time
U	velocity
UCS	unconfined compressive strength test
USLE	universal soil loss equation
u	pore-water pressure
V	velocity, hydrostatic pressure, volume
V_a	volume of air
V_p	velocity of compression waves
V_s	velocity of shear waves, volume of solids
V_v	volume of voids
V_w	volume of water
v	velocity
W	weight, width, degree of weathering
w	weight, width
x	distance
y	elevation of a point
Z	depth, modulus of section
Z_c	critical depth
z	thickness, height
α (alpha)	angle of tilt, inclination, dip, energy, coefficient of linear thermal expansion
β (beta)	angle of slope
γ (gamma)	unit weight of substance
γ_b	unit weight of boulder
γ_D	unit weight of debris
γ_d	unit weight of dry soil
γ_{sat}	unit weight of saturated soil
γ_{sub}	unit weight of submerged soil or rock
γ_w	unit weight of water
ε (epsilon)	strain
$\dot{\varepsilon}$	strain rate
ε_{ax}	axial strain
ε_d	diametral strain
ε_e	elastic strain
ε_t	total strain

ε_v	viscous strain
ε_D	Duncan free-swelling coefficient
η (eta)	coefficient of viscosity
η_p	coefficient of plastic viscosity
η_B	coefficient of Bingham viscosity
η_N	coefficient of Newtonian viscosity
θ (theta)	angle
λ (lambda)	linear grain concentration, energy
μ (mu)	micro (millionth part of), friction
v (nu)	Poisson's ratio
ξ (xi)	coefficient of turbulent friction
ρ (rho)	bulk density
ρ_b	bulk density of boulder
ρ_f	bulk density of fluid
ρ_w	bulk density of water
σ (sigma)	stress
σ_1	major principal stress
σ_2, σ_3	minor principal stresses
σ_{ax}	axial stress
σ_b	bending stress
σ_c	compressive stress at failure (= compressive strength)
σ_d	transverse stress (diametral)
σ_h	horizontal stress
σ_ℓ	longitudinal stress
σ_n	normal stress
σ_o	yield stress, stress due to overburden
σ_t	tensile stress, thermal stress
σ_v	vertical stress
σ'	effective stress
τ (tau)	shear stress
τ_f	shear resistance at failure (= shear strength)
τ_o	yield shear resistance
τ_p	peak shear resistance
τ_r	residual shear resistance
ϕ (phi)	friction angle
ϕ_{cv}	friction angle at constant sample volume
ϕ_D	dynamic angle of internal friction
ϕ_d	internal friction angle as determined in a drained test
ϕ_j	joint friction angle
ϕ_{jr}	residual joint friction angle
ϕ_p	peak friction angle
ϕ_r	residual friction angle
ϕ_μ	friction angle for sliding between two macroscopically smooth surfaces
ϕ'	friction angle with respect to effective stresses
$\tan \phi$	coefficient of plane sliding friction
ψ (psi)	soil-water potential, axial modulus of

elasticity (in confinement)

ψ_g that part of soil-water potential that is due to gravity

ψ_m matric potential

ψ_o osmotic potential

ψ_p pressure potential

Notes:

1. The superscript prime (') is used with c, σ, and ϕ to denote strength and stress as modified by buoyancy in water, and therefore indicates effective stresses (in speech, the word 'dash' is used instead of 'primo');

2. Subscripts are used with c and ϕ for the various shear strength test conditions: thus c_u, ϕ_u for undrained tests; c_{cu}, ϕ_{cu} for consolidated-undrained tests, and c_d, ϕ_d for drained tests.

1 Introduction

Hillslopes occupy most of the landsurface with the exception of terraces and plains formed by river deposits. Even extensive surfaces, like the Great Plains of North America and the plateaux of Africa, are largely formed of hillslopes of low angle between crests of interfluves and valley floors. The most spectacular hillslopes are the great rock cliffs of the high mountain chains and the extensive valley slopes of deep gorges in uplifted plateaux.

The scientific study of hillslopes has passed through a number of recognizable stages. (1) Most geomorphological work carried out before the middle of the twentieth century was related to the classical analytical studies of G. K. Gilbert (1904) and the explanatory concepts of W. M. Davis (1899) whose theories were concerned particularly with long-term decline in hillslope gradients; with the ideas of Walther Penck (1924) who advocated long-term hillslope replacement; and those of L. C. King (1953) who stressed the importance of parallel retreat of rock cliffs. (2) In the 1950s geomorphologists paid particular attention to measuring and recording the nature of hillslope forms (Fair 1947, 1948a,b; Savigear 1952). (3) From the mid-1960s to the present much geomorphological work on hillslopes has concentrated on the processes of hillslope denudation. During the 1970s also, some engineers, hydrologists, and soil scientists became interested in natural hillslopes, and their influence has encouraged a greater attention to the geomechanical properties, and resistance to erosion, of slope materials. (4) Work on processes has rekindled an interest in models of slope evolution, but modern modelling has assumed a mathematical form, unlike the approach of W. M. Davis, but advancing from the approach of Fisher (1866) who predicted changes in the form of cliffs subject to weathering and rockfall,

to involve the action of several processes acting together to change hillslope forms. Attempts to account for the morphology of actual hillslopes through the actions of processes have, however, been few in number and of limited validity because of the complexity of the problem.

Complexity in hillslope evolution

The rate at which hillslope forms change is extremely varied and affected by many influences of variable intensity. One steep hillslope unit in a mountainous terrain subject to extreme climatic and seismic conditions, and formed on rock with closely spaced joints, may change the detail of its form in a few years and retreat at average rates exceeding 1 m/ky; another hillslope unit of low gradient, on strong bedrock in a cratonic fragment of one of the southern hemisphere arid zones, may be subject to a change only at the rate of weathering of resistant rock minerals, and suffer average lowering rates of a few mm/My. As most geomorphic studies of processes are carried out over fewer than three years, the validity of short-term measurements of processes is at best uncertain and at worst they are irrelevant (see Conacher 1988). Furthermore, a single catastrophic storm, flood, or earthquake may do more geomorphic work, and of a different kind, than the processes studied in an observational programme. Add to such problems a consideration of the variability of climate, and therefore vegetation cover, over the last three million years, and recognize the importance of etch forms developed at the weathering front and revealed by surface erosion of the regolith (Twidale 1990), and it is not difficult to question the usefulness of many studies of hillslopes and the validity of many models of hillslope evolution.

Practical value of hillslope studies

Geomorphology began as, and continues to be, a study of the Earth's surface worth undertaking because of its inherent interest to many people but, since about 1960, it has become of increasing importance as a science which has application to the solution of many problems associated with human settlement on the land and the use of natural resources. Geomorphologists have increasingly found employment as practitioners who can serve a range of needs. Knowledge of hillslope materials and processes has been found useful in many activities: for anticipating the consequences of making deep cuts for quarries, mines, and transport routes; creating artificial slopes such as spoil heaps; erecting buildings and structures on natural and artificial hillslopes; and cultivating sloping ground. Less directly, the study of hillslopes has a contribution to make to studies of water and soil conservation, environmental planning, and resource use. Geomorphology is also a contributing discipline to advances in related earth sciences (Henkel 1982). The short time-scale of most engineered projects ensures the relevance of studies of modern processes and the response of soils and rock to them.

The sharing of interest in hillslope stability and processes has produced a huge increase in knowledge but also a great diversity of methods of research and investigation. The student of hillslopes now needs to have a basic understanding: of the rock and soil materials on which hillslopes have developed; of the processes operating within and at the surface of the materials; of the evidence for past formative events and processes; of the methods of investigation developed within rock and soil mechanics, soil physics, geochemistry, and geophysics; of the field investigative techniques of engineering geologists and geomorphologists; and of the information needed by engineers and conservators who plan and design land-use schemes, conservation, remedial and protective structures. It is inevitable that no single investigator or student will have all of the knowledge, skill, and capability to work alone and find solutions to major problems; team work is already the common form of investigation, but to it must be taken a basic understanding of the work and contribution that other scientists and engineers offer.

The objectives of this book are to assist students in the acquisition of a basic understanding of hillslope materials, processes, and development so that they will be able to: appreciate the complexity of hillslope forms and origins as major components of the natural environment; prepare themselves for more advanced study of some aspect of the science of hillslopes; and, if they come to have a professional interest in this science, to participate in team investigations.

Hillslope systems

In many engineering projects of small magnitude, such as a quarry or single road cutting, the factors involved can be identified, assessed, and accommodated in a design. In larger projects, such as construction of a large open-pit mine or a motorway through and over a mountain range, the number and complexity of interaction of factors increases rapidly. The longer the projected life of the project the greater is the complexity. In research projects to investigate the development of natural hillslopes over hundreds of thousands of years the unknowns are so many that they swamp the accessible information. Most geomorphologists recognize that they may never achieve a complete understanding of the development of natural hillslopes. There is, however, a much greater chance of achieving an understanding of current processes of change and, by recognition and interpretation of relict deposits left by formerly active processes, of realizing the magnitude and relative significance of past and present processes in creating the modern form.

Rocks and soils have properties which are the result of their origins and alteration by weathering, erosion, and deposition. Rock is not only formed by igneous, sedimentary, and metamorphic processes, but contains locked in stresses as strain energy. Release of that stress, tectonism, and weathering create joints and other planes of weakness along which weathering can act preferentially. Weathering creates new materials either as fragments and grains of unaltered original rock or, by chemical alteration, as new species, especially of clay minerals. Rock and soil respond to applied stresses in ways which are controlled partly by their inherent properties and partly by the nature, magnitude, and frequency of the stress application. The behaviour of rock and soil under stress has a major influence upon the effects of processes of weathering and erosion.

Weathering and removal of rock and soil on

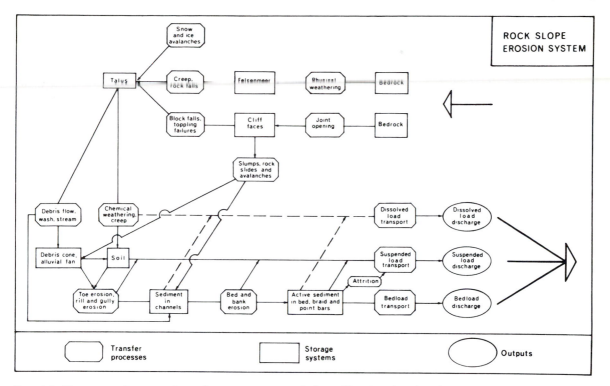

FIG. 1.1. The system of stores and transfer processes on a rock slope. (Developed on the scheme of Dietrich and Dunne used in Fig. 1.2.)

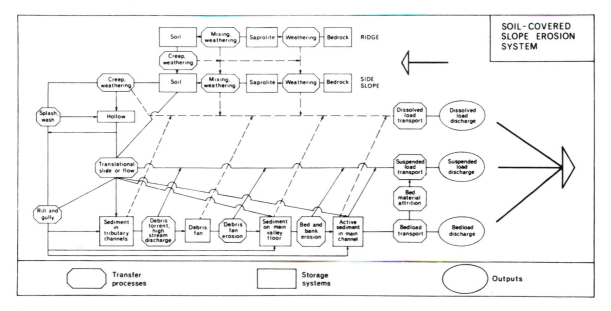

FIG. 1.2. The system of stores and transfer processes on a soil-covered hillslope. (After Dietrich and Dunne 1978.)

hillslopes is not a uniform process in either time or space: it is episodic and depends upon the availability of energy and a transporting medium. As a result hillslopes can be regarded as a system of stores (Figs. 1.1, 1.2) which are periodically unlocked by processes. Very resistant stores, such as are provided by massive hard rock outcrops, may only yield material at very infrequent intervals. Soil slopes in a humid tropical climate may yield solutes almost continuously, but solids by landslide processes much less frequently. Each process, therefore, has its own magnitude and frequency of operation which is controlled by the resistance of the hillslope rock and soil, and by the intensity of the denudational processes.

By tracing and measuring the movement of material from hillslopes by different processes, and by measuring the modification of hillslope form produced by them, it is possible to evaluate short-term changes of slope profiles. In theory this should eventually provide an understanding of how hillslopes evolve. In most environments hillslopes change too slowly for the progression from long steep slopes to slopes of lesser angle to be observed. That such changes occur is indicated from geological sections, in which erosional surfaces have been cut across complex structures to produce unconformities which may be preserved in the depositional record. Attempts to compensate for the lack of observations of slope change usually involve one of two possible procedures. In a few rather rare situations space may be substituted for time, as where Savigear (1952) was able to measure slope profiles along a cliff which had been protected from wave attack at its base for varying periods, so that a sequential development of hillslope forms was assumed to have been produced. Less secure methods involve the measurement of different slope profiles in one area and the assembly of these profiles into a sequence. Such methods are extremely uncertain because the underlying assumption of sequential change cannot be verified.

The second method—that of the development of process-response models—forms the basis of most modern hillslope geomorphology (Fig. 1.3). Analysis involves the measurement of the resistance of rock and soil to change, the force and mode of action of a process causing change, and variation in the rate of change through time and in space. A model seeks to link variations in the rate of change over the hillslope to development of slope profiles (frequently by using differential

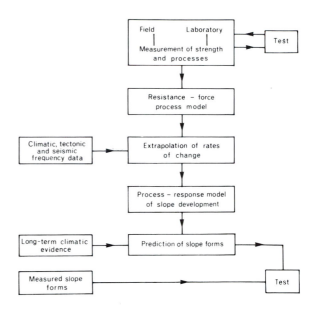

FIG. 1.3. The pattern of hillslope studies. (Modified from Carson and Kirkby 1972.)

equations). Field or laboratory methods often involve the establishment of a statistical relationship between a slope change and measurements of soil or rock properties and a process such as rainfall. The primary difficulty in such methods is that of adequately sampling slopes, rock and soil types, and process types and rates. A second difficulty arises from the uncertain degree of inherited influences upon current hillslope forms, so that when a certain suite of slope forms is predicted from a sequence of processes the prediction may not be tested against natural slopes with a known history. There is at present little hope of developing predictive slope-form models which incorporate the multitudinous climatic, and hence process intensity, changes of the Quaternary. The emphasis of this book is therefore upon those components of hillslope studies which can be studied effectively—the strength of slope materials, the effect and intensity of processes acting on those materials and the resulting short-term changes on slopes.

Energy available for hillslope processes

The energy available for slope processes is derived from three sources: solar radiation, gravity, and endogenetic forces. Solar radiation directly promotes weathering processes but much of it is effective by driving the circulation of water in a hydrological cycle between the atmosphere, pedosphere, lithosphere, and ocean. Raindrops striking the ground, water flowing, and boulders falling or rolling downslope, have energy provided by the gravitational force that attracts them towards the centre of the Earth. Endogenetic forces are generated by radioactive decay of natural isotopes producing heat. Geothermal heat drives volcanic activity and creates the stresses which are released in earthquakes. Over most of the land surface the energy available from solar and gravitational sources for geomorphic work is several thousand times greater than that available endogenetically. Only where volcanic activity, or sudden release of seismic energy, concentrates power can internal energy produce distinctive landforms or land-forming processes.

Plan of the book

The book begins with six chapters devoted to the properties of rock and soil, especially to their behaviour under stress, and to their strength. The next two chapters are concerned with the alteration of rock and soil by weathering processes and with the minor landforms created by weathering. In Chapters 10 and 11 are discussed the soils and deposits occurring on hillslopes, the evidence derived from them of past environments, and the action of water within soils. Chapters 12 to 15 are devoted to study of the processes acting to modify slopes, and Chapter 16 to the development of hillslopes and use of models to simulate and study that development. Chapter 17 is both a brief comment on hazards arising from the presence of people in hill country and of investigations undertaken to reduce hazards. The book concludes with two chapters in which are discussed both the magnitude and frequency of hillslope erosional events and the rates at which separate processes, and combinations of processes, change hillslopes.

2 Bonds

States of matter

Matter exists as gases, liquids, and solids. Distinctions between these states are dependent on the interactions between constituent atoms and molecules. The atoms or molecules in gases are in constant motion, diffusing past each other with minimal interaction, to be dispersed throughout the vessel that contains them; they behave as independent particles, interacting with one another only during collisions. This indicates that the inter-molecular forces are too weak, as a result of distance between molecules, to force them to aggregate. In air at atmospheric pressure the distance of separation of molecules is about 20 nm; in liquids and solids it is 0.3–0.5 nm. It is clear, however, that repulsive forces must exist between molecules, otherwise they would go on condensing to ever greater densities. In liquids diffusion still occurs, but there are stronger interactions between the molecules. This interaction does not rigidly confine the molecules, they can slide past each other: this is why liquids can change their shape to fit a confining vessel. In the solid state the constituent atoms or molecules form such strong short-range bonds between each other that they tend to stay in constant positions with respect to one another, consequently the shape of a solid is maintained (Fig. 2.1).

A given material may not always be in the same state: under particular temperature and pressure conditions, for example, solid ice may be transformed to liquid water; or water may change to steam. In this example ice, liquid water, and steam are three phases of water. The transformation from one phase to another is usually distinguishable, but there can be exceptions as in some solid-solid transformations, and between a very viscous liquid and a weakly bonded solid.

Phase changes are achieved by diffusion. Atoms move relative to each other by an adjustment of geometry as the total available energy, usually heat, is absorbed or released. An excellent review of modern understandings of the nature of matter is given in Cotterill (1985).

Types of bonds

Atoms are held together by bonds which can be classified into six types in two groups. The three strong primary bonds are (1) ionic, (2) covalent, and (3) metallic. Two of the weak secondary bonds are related to the action of dipoles and are known as van der Waals attractive forces; they are caused by (4) electron dispersion, and (5) molecular polarization. The third kind of weak secondary bond is (6) hydrogen bonding. At least two of the six bonds operate in all materials, with the types of bonding largely controlling the properties of the materials. The relative strength of the six classes of bond are given in Table 5.1.

The nature of all six forms of bond is controlled by the structure of atoms and their constituent charged particles. An atom consists of small electrons orbiting a far more massive nucleus. The hydrogen nucleus consists of one positively charged proton. The nuclei of all other elements are heavier and contain both protons and neutrons. Protons carry a positive charge and neutrons are electrically neutral. Electrons have a negative charge which is equal in magnitude to that of the protons. In an electrically neutral atom, then, the number of protons equals the number of electrons.

Electrons orbit the nucleus of the atom, with those furthest from the nuclei being the most loosely bound and thus the most likely to interact with electrons of neighbouring atoms. These

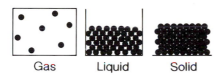

Gas   Liquid   Solid

FIG. 2.1. Schematic distribution of molecules or atoms in the three states of matter.

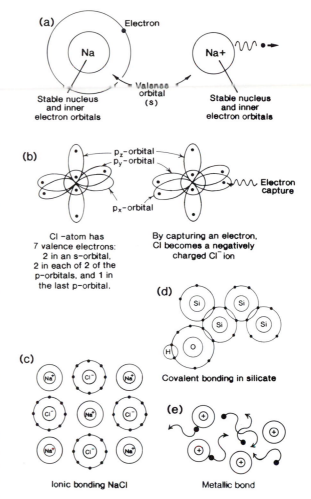

FIG. 2.2. (a) Formation of a positively charged ion by loss of an electron. (b) Formation of a negatively charged ion by capture of an electron. (c) NaCl molecules with bonding through electrical attraction and formation of dipoles; consequently oppositely charged ends of adjacent molecules attract and similarly charged ends repel. (d) Covalent bonding between molecules through shared pairs of valence electrons. (e) Representation of metallic bonding.

outermost electrons are generally responsible for the chemical behaviour of the elements, and the distribution of their charge in space defines orbitals. These comprise the spherically symmetrical 's' orbital and three 'p' orbitals—dumb-bells aligned along each of the three cartesian axes. The approach of other atoms may cause these orbitals to combine, as is shown in Fig. 2.3c.

Each orbital may contain up to two electrons: thus the outer or valence orbitals optimally contain six electrons in the p-orbitals and two in the s-orbital: a total of eight. This is the most stable electron arrangement: bonding between atoms serves either to reduce or increase the number of outer electrons to this number. This can be best achieved either by 'donating' or 'receiving' electrons—the ionic bond, or by 'sharing' electrons—the covalent bond.

Ionic bonding

The electronic configurations of sodium and chlorine atoms are shown in Fig. 2.2a and b. It is immediately apparent that if the sodium atom were to lose an electron to become a positively charged ion, and the chloride atom to gain that electron to become a negatively charged ion, each would have their outermost orbitals filled. Sodium is electropositive, and readily ionizes in this way. Similarly chlorine is electronegative. A very strong bond is made by this transfer of electrons from electropositive to electronegative elements, to form ionic bonds, as exist in common salt, NaCl (Fig. 2.2c).

Covalent bonding

When two elements of comparable electronegativity come together each may transfer electrons to the other: the outer electrons are mutually shared. This covalent bond is strong and is the form of bonding between silicon and oxygen in the silicate minerals (Fig. 2.2d).

Metallic bonding

In an extreme case, when only electropositive elements are involved, the outermost electrons become detached, producing a lattice of positive ions in a sea of conducting shared electrons moving in a complex joint orbital system. This type of bonding is the principal form occurring in metals (Fig. 2.2e).

van der Waals forces

The time-averaged electron distribution within atoms defines the electron orbitals. However, the random movement of electrons means that there

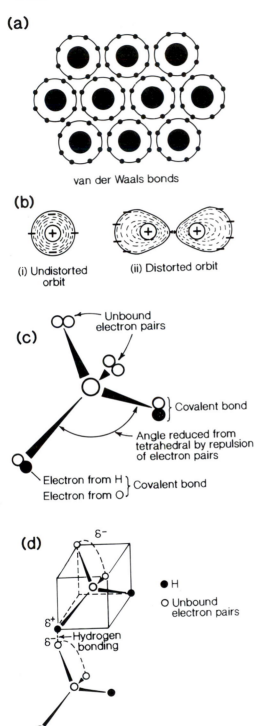

(a)

van der Waals bonds

(b)

(i) Undistorted orbit

(ii) Distorted orbit

(c)

Unbound electron pairs

Covalent bond

Angle reduced from tetrahedral by repulsion of electron pairs

Electron from H ⎱ Covalent bond
Electron from O ⎰

(d)

δ^-

● H

○ Unbound electron pairs

δ^+

δ^-

Hydrogen bonding

are instantaneously more electrons on one side of the atom than the other. This produces an electric dipole: a separation of charges which constantly changes its direction and strength. Overall, this generates a weak attractive force between each nucleus and the electrons of adjacent atoms. The magnitude of these forces, named for their discoverer, J. D. van der Waals, varies between materials but they are weak in comparison with the repulsive forces between atoms. Materials which owe their strength to van der Waals forces are likely to be mechanically weak.

Molecular polarization is the result of permanent dipole interaction, of the atoms of a molecule, created when the electric field of one atom distorts the charge distribution of an adjacent atom. This permanent dipole bond is directional (Fig. 2.3b). Hydrogen bonding is a particular type of permanent dipole interaction.

Hydrogen bonding and the structure of water

The valence electrons in oxygen occupy the p and s orbitals, but in this—and other—cases the orbitals are hybridized to give four tetrahedrally oriented orbitals. Two of the orbitals are occupied by one electron from each of the hydrogen and oxygen atoms, the other two pairs are located in the remaining orbitals. While the angle between the two hydrogen and oxygen atoms (H-O-H) might be expected to be that of the tetrahedron 109.5°, the pair of electrons repel each other and reduce the bond angle to about 104°.

The pairs of unbonded electrons give the molecule a permanent charge imbalance: a permanent dipole, negatively charged at the oxygen 'end' and positively charged at the hydrogen 'ends'. This electric dipole adds sufficiently to the van der Waals forces to make the melting and boiling temperatures of water higher than would be expected for molecules of this type, and also

FIG. 2.3. (a) Weak bonding through van der Waals forces. (b) Distortion of the electron orbitals within adjacent atoms in close proximity, creating polarization of their charges. (c) The dipolar nature of a water molecule resulting from the combination of the outer orbitals (one s-orbital and 3 p-orbitals) to give four tetrahedrally distributed orbitals (sp^3 hybridization) of the oxygen atoms covalently bonded to hydrogen atoms. (d) Hydrogen bonds linking separate water molecules; the hydrogen 'ends' having a permanent positive charge (δ^+), the unpaired electrons on the oxygen 'end' having a permanent negative charge (δ^-). (a, c, d, based on Cotterill 1985.)

enables water to interact with ionic species. Water molecules, for example, can rotate and the negative (oxygen) end of the dipole be attracted to cations, the positive (hydrogen) end to anions and other water molecules. This explains the strong solvent action water has for ionic compounds, and its role in the hydration and dehydration of crystalline solids.

This permanent dipole, hydrogen bonding, is found in other compounds with hydrogen where there is an asymmetric charge distribution, particularly organic molecules.

Crystalline solids

Many minerals, including all rock primary minerals and most clay minerals, have that systematic, repetitive three-dimensional geometric pattern, called a lattice, of their constituent atoms and molecules, which is the characteristic of crystalline solids. In pure crystals each atom or molecule is surrounded by exactly the same relative orientation and spacing of its neighbours as every other unit within the structure (Fig. 2.4). There need be no limit to the number of atoms or molecules making up a crystal so crystals do not have a fixed maximum size.

Imperfections in the geometry of atomic patterns, or impurities in the form of foreign materials, will affect the properties of the solid. Even in nearly pure solids, crystals will develop boundaries which separate the mass into discrete units, along breaks in the atomic arrangements.

Chemical reactions

Because of their bonding properties, atoms will react with each other in certain limited and predictable ways. The less stable the atomic electron configuration of an atom, the more likely is it to be chemically active. Thus, covalent bonds will form spontaneously with groups of atoms such as hydrogen and oxygen to form molecules which are relatively more stable than the atomic state.

Similarly, ionic bonds may be formed or disrupted depending on the relative stability of atomic or ionic electron configurations, and the characteristics of the environment in which chemical changes occur.

As atoms, ions, and molecules are rearranged into more stable arrays heat may be absorbed or released. Where this rearrangement involves no external phases, this heat is described as 'latent heat', but where other chemical species are involved it is the 'heat of reaction'. In a chemical reaction molecules are not necessarily broken

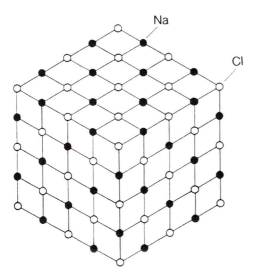

FIG. 2.4. The regular organization of sodium (black dots) and chlorine atoms (open circles) in a crystal of NaCl, which has perfect cubic symmetry.

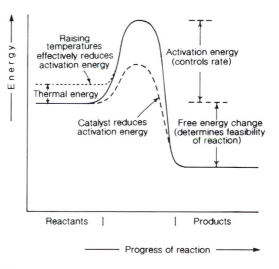

FIG. 2.5. Energy relationships in chemical reaction. Reaction is energetically feasible if there is an overall decrease in free energy, but whether it will proceed depends on whether the activation energy barrier can be surmounted.

down into their smallest constituent atoms: only one type of atom may be lost to form a new compound leaving the remaining atoms either as a separated chemical species or a new compound. The dissociation of feldspars, to give ions in solution and a clay mineral is one such example.

The difference in energy between reactants and products of chemical reactions is not the only controlling influence on their occurrence. Chemical reactions show much variation in their rate or speed. Oxygen, for example, in combination with atoms and molecules may react very quickly—as in the combustion of say hydrogen and oxygen to give water, or it may react rather more slowly at lower temperatures—as in the rusting of iron.

The relative ease of initiating and propagating a chemical reaction depends on (1) the types of bonds present between the reacting substances, (2) proximity of particles of substances as influenced by particle size, shape, degree of mixing, and type of contact, and (3) availability of sufficient energy to initiate the reaction. This process of 'activating' a reaction and its relationship to the overall energetics of the reaction is often shown schematically as in Fig. 2.5. Clearly, if a reaction is to be speeded up, the best strategy is either to lower the activation energy, generally by catalysis, or to decrease the energy difference between the reactants and the activated complex. Raising the temperature or increasing the contact between reacting molecules, by increasing the pressure or by physical stirring, are ways of increasing slow natural rates of reaction to rates more appropriate for industrial processes.

3

Particles and Fabrics of Soil and Weak Sediments

The strength, behaviour under stress, void space, permeability, capacity to retain water, chemical reactivity, and other properties of soils, which influence the nature and development of hillslopes, are due to the minerals and particles of the soil. The relative proportions of gravel, sand, silt, and clay define the texture of the soil. The patterns in which the particles are arranged define the fabric or structure of soils. The term 'fabric' is normally used in reference to geological deposits and 'structure' with reference to organic soil.

Particle size

The terms 'clay', 'silt', 'sand', and 'gravel' are used to denote the size of particles as indicated by their effective diameter, as in 'sand size'; they may also refer to texture, where 'sand' may refer to a 'sandy soil' in which the sand fraction dominates the soil.

Several classifications of particle size are in use (Fig. 3.1); most use the same boundary between clay and silt ($2\,\mu m = 0.002\,mm$) and between sand and gravel ($2000\,\mu m = 2\,mm$). The chosen boundaries between silt and sand, and within silt and sand, are varied. It is therefore necessary to specify the system being used.

Particle size determination is useful for some agricultural purposes because it is an indication of such soil properties as water storage capacity, drainage, and nutrient availability and it is not likely to change with agricultural practices. For many other purposes, particle size is most usefully indicated for the fine fraction, silt plus clay, according to the characteristic behavioural properties. This procedure is followed in the 'Unified Soil Classification' of the United States Bureau of Reclamation (Earth Manual, 1963). The Unified

Classification is followed by many engineers, and its main elements are given below.

Clay soils (usually having more than 20 per cent clay size particles), when moist, have a smooth greasy feel and can be moulded or rolled into threads (a test of plasticity) over a considerable range of water contents. The thread will support its own weight if held at one end and thus exhibits tensile strength caused by its own cohesion (a silt will not do this). On drying, a thread of clay will crack and break off in flakes, the thread may shrink and residual lumps may be too hard to be broken readily in the fingers.

Silty soils (usually more than 50 per cent silt) exhibit the properties of cohesion and plasticity to a limited extent. They can be rolled into threads between the fingers, but the threads tend to crumble as they dry. If a wet sample is shaken in the hand its surface becomes wet and shiny. If it is then compressed between the fingers the water disappears into the sample and the surface becomes dull and dry, at the same time the soil becomes firmer and gives increasing resistance to compression. This change, as pressure causes rearrangement of grains into a larger volume, with greater void space into which water is drawn, is called dilatancy.

Fine-grained soils vary in their behaviour, depending upon the proportions of clay and silt they contain and on the species of the clay minerals present. A full classification of such soils therefore involves a number of tests which define specific forms of behaviour (see Chapter 7).

Sands and gravels of the coarse fraction are not cohesive and consequently retain single grain forms and behaviour. They do not exhibit plasticity, but they do undergo volume change on being shaken and so are dilatant. Table 3.1 summarizes the major properties of the size fractions in soils and

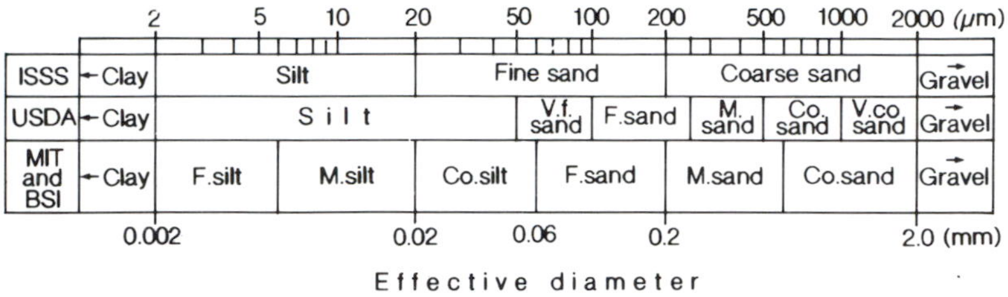

FIG. 3.1. The most commonly used classifications of soil particle sizes: International Society of Soil Science (ISSS), United States Department of Agriculture (USDA), Massachusetts Institute of Technology (MIT), and British Standards Institute (BSI). (F. = fine, M. = medium, Co. = coarse, V. = very.)

TABLE 3.1. *Physical properties of the major size fractions in soil*

Property	Gravel	Sand	Silt	Clay
Volume change from dry to wet conditions	None	None	Slight	Large (marked swelling)
Tensile strength when wet	Low	Low	Intermediate	High
when dry	Low	Lower than when wet	Higher than when wet	Very high
Compressibility when wet	Very low	Very low unless particles separated by water when change is high	Intermediate unless saturated with water when change is high	Very high
when dry	Very low	Very low	Low	Intermediate to low when very dry
Plasticity when wet	None	Slight	Intermediate	Very high
when dry	None	None	None	None, partial cementation
Porosity	Very high	High	High	Very high
Permeability	Very high	High	Intermediate to low	Very low
Size of voids	Large	Intermediate	Capillary	Subcapillary
Shape of particles	Rounded	Round to angular	Angular	Thin tabular
Water retention	Very low	Low	High	Very high

Sources: Dapples (1959), *Basic geology for science and engineering* (New York: Wiley); US Department of the Interior, Bureau of Reclamation (1963), *Earth Manual* (Washington, DC: Government Printing Office).

thus illustrates the importance of particle size. However, it should be remembered that the behaviour of a soil is dependent on its dominant fraction; a soil which is composed of, say, 20 per cent gravels by volume, 20 per cent sand, and 60 per cent clay may have properties controlled entirely by the clay, if the gravel and sand grains are dispersed throughout the clay and are not in contact with each other.

Origins of particles

Gravels, sands, silts, and clays are nearly all derived ultimately from the weathering of rocks, primary rock minerals, and transport by water, ice, and air to sites of deposition on land, in lakes, and in seas. In the processes of transportation weathered particles may undergo comminution and sorting, with the result that bodies of sediment may have textures ranging from extreme mixtures, as in some glacial tills, to extremely well-sorted uniform grain sizes such as those of some marine clays, loess, and desert sand dunes.

Sands and silts are commonly composed of quartz grains because quartz is one of the most abundant and resistant rock minerals to chemical attack and to abrasion. Clays, by contrast, have been classified into many mineral species, each of which has distinctive crystal structures and behaviour. The properties which make clays so important to studies of hillslopes and for engineering purposes result from: the small size of their particles which result in very large surface areas per volume of soil and hence high chemical reactivity; the effects of surface electrostatic forces; water absorbed to the crystal units which therefore separates the units from each other and allows them to move relatively to each other; and the small diameter of the individual voids within the clay body, which permits retention of water against gravity and causes low permeability (Table 3.1).

Clay minerals

Most clay minerals occur as clay-size particles. Their species can be grouped into three main classes: the layer silicates (or phyllosilicates); the weakly crystalline short-range order aluminosilicates; and the hydrous oxides of iron, aluminium, and manganese (see Grim 1968; Brindley and Brown 1980).

Phyllosilicate mineral structures

The phyllosilicates include primary minerals such as mica and many secondary clay minerals which have been formed as crystalline materials usually in weathering profiles and more rarely in the high temperature conditions of hydrothermal activity. In soils the most important species of phyllosilicates are: kaolinite, halloysite, montmorillonite, illite (a hydrous mica), vermiculite, and chlorite.

The layer silicates are composed of two types of structural unit arranged in various combinations: tetrahedral silica sheets and octahedral alumina sheets.

Tetrahedral sheets consist of silicon and oxygen atoms arranged with a silicon ion at the centre of each tetrahedron, while oxygen anions form the four corners (Fig. 3.2). The individual tetrahedra are connected with adjacent tetrahedra by sharing

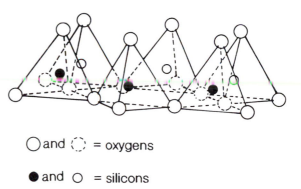

$\bigcirc$ and $\langle\ \rangle$ = oxygens

● and $\bigcirc$ = silicons

Fig. 3.2. Structure of a tetrahedral silica sheet. (After Grim 1968.)

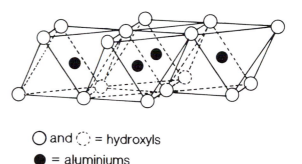

$\bigcirc$ and $\langle\ \rangle$ = hydroxyls

● = aluminiums

Fig. 3.3. Structure of an octahedral alumina sheet. (After Grim 1968.)

the three basal corner oxygens which thereby constitute a hexagonal mesh arrangement. The fourth oxygen anion, at the apex of each tetrahedron, forms part of the neighbouring octahedral sheet.

Octahedral sheets (Fig. 3.3) comprise medium-size aluminium cations at the centres and oxygens at the eight corners of each octahedron. Individual octahedra are linked laterally with neighbouring octahedra and across the thickness of the sheet with tetrahedra, by sharing oxygens. The plane of junction between tetrahedral and octahedral sheets comprises the shared oxygens, together with unshared hydroxyls. These OH^- groups are located at the centre of each hexagon at the same level as the apical oxygens.

Octahedral sheets have the characteristic structure of gibbsite, $Al(OH)_3$, and are sometimes referred to as gibbsite sheets. In some cases Al^{3+} is replaced by divalent cations such as Mg^{2+} or Fe^{2+}, a process called isomorphous (i.e. equal form) substitution. Silicon, Si^{4+}, may also be sub-

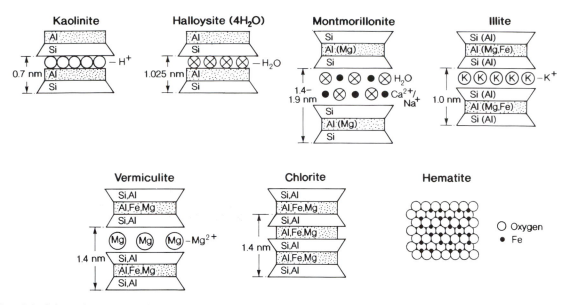

FIG. 3.4. Schematic representations of the layers forming the common phyllosilicate clay minerals, and of the iron oxide mineral, hematite.

stituted by Al^{3+} in the tetrahedral layer. There are consequently two ways in which various species of clay minerals may be created: (1) by isomorphous substitution of cations, and (2) by variations in the assemblage of tetrahedral and octahedral sheets (also called layers).

Two main types of layer are recognized. (1) The 1:1 layer minerals consist of the assemblage of one tetrahedral sheet with one octahedral sheet; such an assemblage is typical of the kaolinite group. (2) The 2:1 layer minerals consist of an octahedral sheet between two tetrahedral sheets. The apical oxygens of each tetrahedral sheet face the octahedral sheet and can be shared with it in all 2:1 minerals. Most clay minerals such as montmorillonite, vermiculite, and illite, as well as primary mica and talc, are of the 2:1 layer type.

The clay mineral chlorite has a 2:1:1 assemblage which includes an additional octahedral sheet (Fig. 3.4).

Structural units

Layers of 1:1, 2:1, and 2:1:1 sheets are separated from each other by a space called an interlayer (Fig. 3.5). The interlayers are devoid of any chemical elements if the layers are electrostatically neutral, as when all structural cations are compensated by oxygen or hydroxyl anions. Most clay mineral species have an excess of layer negative charge which is neutralized by cations such as Ca^{2+}, Mg^{2+}, Na^+, K^+, and H^+ (Fig. 3.6).

The sum of layers held together through shared ions with an interlayer is called a structural unit (Fig. 3.5). Its thickness is between 0.7 and 1.9 nm depending on the type of layer and the interlayer content of each unit. The thickness for 1:1 kaolinite is 0.7 nm, for 2:1 montmorillonite is 1.4–1.9 nm, and for 2:1:1 chlorite is 1.4 nm. The accumulation of many structural units constitutes a clay particle. Some kaolinites may have as many as 100 units in a clay particle. Particle sizes (maximum thicknesses) vary from about 2 nm to 2 μm, but most are in the range of 10 nm to 0.2 μm. Clay particles represent the most common components of the clay fraction of soils and of many sedimentary rocks, in which they occur as aggregates of various sizes, and in combination with such components as organic matter, aluminium and iron oxides, carbonates, phosphates, free silica, and various aluminosilicates.

Characteristics

Kaolinite (1:1) has a well-developed hexagonal crystal form which can grow about all three of its axes. Its tetrahedral silicon sheet is firmly attached by hydrogen bonding to its octahedral aluminium

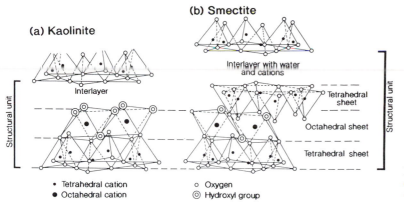

(a) Kaolinite

(b) Smectite

Interlayer with water and cations

Interlayer

Structural unit

Tetrahedral sheet

Octahedral sheet

Tetrahedral sheet

Structural unit

• Tetrahedral cation
● Octahedral cation

○ Oxygen
◎ Hydroxyl group

Fig. 3.5. Three-dimensional structure of two common clay minerals. (After Grim 1968.)

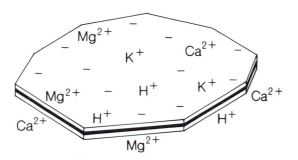

Fig. 3.6. A clay structural unit with adsorbed ions attracted by the negative charge of its surface.

Plate 3.1. A stack of kaolinite platelets, each formed of several structural units. Such stacks are said to have an 'open-book' structure. (Scanning-electron micrograph by A. G. Beattie.)

Plate 3.2. Kaolinite platelets filling pore spaces in a mudrock. The crystals have a hexagonal form and loose arrangement of the platelets. (Scanning-electron micrograph by A. G. Beattie.)

hydroxide sheet, producing a rigid structure which cannot expand.

Halloysite (1:1) has a similar composition to kaolinite but contains interlayer water and sheets rolled into a tubular form. The water is firmly attached by hydrogen bonding resulting in a rigid structure. However, dehydration can occur readily, causing collapse of the structure to a form which is similar to that of kaolinite. In some situations, halloysite may be a precursor to kaolinite.

Montmorillonites (which are major members of the class of smectites (2:1)), have a structure in which one-sixth of the aluminium ions have been replaced by magnesium in the octahedral layer. This causes an imbalance of charge which is usually satisfied by ions such as Ca^{2+} or Na^+, substitution of iron may also occur. Smectites have a small size and a great affinity for water which is drawn into the interlayer spaces. This looseness in the structure causes a great capacity for swelling and shrinkage as water is gained and lost.

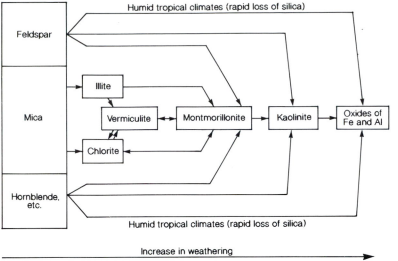

<figure>**FIG. 3.7.** Possible pathways for formation of clay minerals by soil weathering. (Based on data in Fieldes and Swindale 1954; Fieldes 1968; Bisdom *et al.* 1982.)</figure>

Illite (2:1) is an alteration product of mica. Illite is hydrated but is incapable of expanding and contracting because potassium ions maintain the interlayer spacing.

Vermiculite (2:1) is a hydrated mica in which potassium has been replaced by calcium and magnesium, and in which there is increased substitution into the structural units.

Chlorites (2:1:1) are closely related to the micas but have additional positively charged octahedral sheets containing aluminium and/or magnesium ions. Chlorites may be formed both as primary minerals or as a result of weathering of primary micas.

Formation of phyllosilicate clays

Clay minerals are formed by two mechanisms: (1) neoformation which is the result of precipitation forming clay crystals from solution or reaction of non-crystalline material; and (2) transformation in which clay has inherited some of its structure from an earlier mineral form while undergoing ion exchange and/or layer transformation in which the structure of the layers is modified. Clay in soils may also be inherited directly from a rock, such as a marine mudrock (Millot 1970; Eberl 1984) and from eolian dust.

Kaolinite is commonly formed by direct precipitation, with $Al(OH)_3$ and SiO_2 separately forming the sheet structures, but several of the common phyllosilicates clays are closely related and may develop from each other. In Fig. 3.7

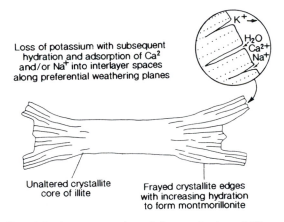

<figure>**FIG. 3.8.** A representation of the weathering of illite to montmorillonite. (After Hawkins *et al.* 1988.)</figure>

some possible pathways for clay formation are illustrated. Primary mica has an octahedral alumina sheet sandwiched between two tetrahedral silica sheets. Potassium ions (K^+) are too large to fit into the sheets but lie in the interlayers and weakly bond them. Because of the weakness of the bond micas cleave readily along this plane. As micas weather the K^+ is lost and water and hydrated cations enter the interlayers, increasing their width. The result is the formation of clay minerals differentiated from micas by the ions isomorphously replaced in the crystal structures and by the bonding and spacing across the interlayers. In Fig.

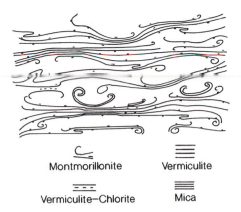

Montmorillonite   Vermiculite

Vermiculite–Chlorite   Mica

FIG. 3.9. Features of a predominantly montmorillonitic phyllosilicate clay complex formed by weathering of mica. (Based on Jackson and Sherman 1953.)

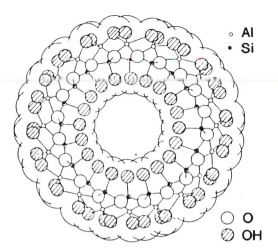

○ Al
• Si

○ O
◎ OH

FIG. 3.10. Schematic cross-section through an imogolite tube. (After Brown *et al.* 1978.)

3.8 the weathering of illite to montmorillonite is illustrated schematically.

The solutions passing through a soil have variable compositions and the processes of transformation in a soil are neither constant in rate nor uniform over space. The type of clay mineral formed at any place will be influenced by the minerals of the original rock or soil, climate, drainage, vegetation, and weathering time. Consequently, in any soil, aggregates of clays may form with several species and with intergrades between distinct species (Fig. 3.9).

Short-range order aluminosilicates

Early studies of aluminosilicate clays recognized a group of materials which showed no sharp peaks on the graphs resulting from X-ray diffraction analyses. The lack of peaks seemed to imply that these clays have random or non-crystalline structures and they were often described as being X-ray amorphous. Other techniques of analysis, particularly electron microscopy, have shown that they do have structural order, but this order holds only over short ranges so is irregular, also the particle sizes are too small to show sharp peaks on X-ray diffractograms.

The clays which fall into this group are imogolite, proto-imogolite, and allophane. Imogolite is the best ordered of the group and consists of bundles of long tubes. Each tube has an internal diameter of ~1 nm and an external diameter of 2.1–2.7 nm. A cross-section through a tube (Fig. 3.10) indicates that the external layer of the tube is an octahedral gibbsite sheet and the internal layer is basically a tetrahedral silica sheet with hydroxyl groups attached to silicon atoms.

The term proto-imogolite is used to refer to clays which are almost like imogolite but generally have poorly shaped or incomplete tube forms.

Allophane is a material consisting chemically of O^{2-}, OH^-, Al^{3+}, and Si^{4+}, and is characterized by short-range order and a predominance of Si-O-Al bonds. It occurs as small, hollow, porous spheres which tend to aggregate when soils become dry.

The short-range aluminosilicate clays develop most commonly in soils derived from volcanic ash (tephra) and so are very important in countries such as Japan, Indonesia, and New Zealand. Allophane forms in the soil where Si abundance is less than a critical level; above that level halloysite forms. Free drainage and high rainfall tend to encourage formation of allophane (Lowe 1986). These clays have very large surface areas—5 g of allophane has the area of a football field—can adsorb large quantities of water but after drying are very slow to become wet again. As a result allophanic clays are prone to flow and liquefy when saturated and shrink and crack when dry. The over-dried aggregates tend to behave as sand grains, lacking both cohesion and plasticity.

Allophane is also now recognized as being formed in many environments where tephras are not present and is much more common than was once thought.

Oxide clays

Aluminium in soils is usually in the form of a hydroxide, and iron in the form of an oxide. Precipitates of both aluminium hydroxides and iron oxides develop crystalline forms of clay-size with structures based on hexagonal (α) and cubic (γ) close-packed oxygen anion frameworks, either as an independent mineral or, in the case of aluminium hydroxide, within interlayers of vermiculites and smectites.

The most important oxide clays are: gibbsite, γ–Al(OH)$_3$; hematite, α–Fe$_2$O$_3$; maghemite, γ–Fe$_2$O$_3$; goethite, α–FeOOH; lepidocrite, γ–FeOOH; ferrihydrite, 5Fe$_2$O$_3$, 9H$_2$O; and magnetite, Fe$_3$O$_4$. Gibbsite and goethite are the most common of these minerals in soils (Taylor 1987).

Oxide clays may form in many climatic environments but are most abundant where extreme leaching of silicates has occurred so are best represented in soils of the humid tropics and in long-stable landscapes.

Clays, climate, and drainage

The Hawaiian Islands are thought to have had relatively minor variations in climate over middle and late Quaternary times and they have a single rock type, basalt. Clay neoformation can therefore be assumed to vary as a result of rainfall and drainage. It was found, by Bates (1962), that smectite forms on the dry, leeward sides of the islands where rainfall in less than 500 mm/y, whereas gibbsite forms on the windward sides and in mountains where rainfall is more than 10 000 mm/y. Halloysite is predominant in zones with intermediate rainfall. This study illustrates the simple principle that clays, such as smectites, with more soluble elements, form in dry sites and climates where their ions can accumulate, but clays with the least soluble elements, such as gibbsite, form under severe leaching conditions such as well-drained sites in the humid tropics where silicates are removed but aluminium and ferric iron can remain. The kaolinite minerals form in intermediate zones where silicon, as well as aluminium can be retained. Similar results were obtained in studies by Barshad (1966) and Clemency (1975) in the mountains of California and near São Paulo, Brazil, respectively. Barshad's results for two rock types are illustrated in Fig. 3.11.

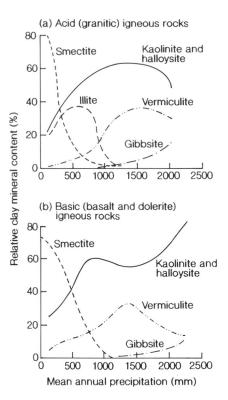

FIG. 3.11. The effect of precipitation on the frequency of clay minerals of given species in soils in California, for acid rocks of high silica content and basic rocks of lower silica but higher ferromagnesian content. (After Barshad 1966.)

Adsorbed water

Clay particles in soils are always surrounded by layers of water molecules, bonded to the clays as adsorbed water. These adsorbed molecules should be considered as part of the clay surface when behaviour of clay is being considered, because they influence the physical properties of soil such as plasticity and strength, they also influence water movement within the soil.

Water adsorption to clay surfaces at early stages of wetting of dry soils may occur by any of three processes: (1) hydration, or incorporation into the crystal lattice, is a major mechanism; (2) intermolecular attraction between the solid surface and water over a short range may occur as a result of van der Waals forces; and (3) by hydrogen bonding to exposed oxygen atoms of the solid surface.

Water adsorbed at the clay surface has been variously described as dense, viscous, strong, non-liquid, solid, quasi-crystalline, and strongly

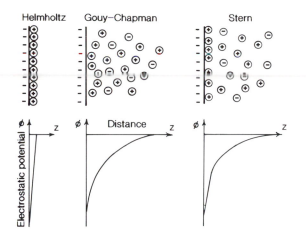

FIG. 3.12. Models of the electric double layer formed by an ionic solution in contact with a negatively charged surface. The graphs indicate dependence of the electrostatic potential on the distance from the surface. (After Bedzyk *et al*. 1990.)

bonded. Such diversity of terms is unhelpful and some of them are misleading. Direct measurements of the density and viscosity of water at the clay surface are extremely difficult and have produced density values in the range of 1.7 to 0.7 Mg/m³; thicknesses of the adsorbed water films have been variously estimated as being in the range of 1 to 4 nm indicating an equivalence to 4 to 16 molecular layers (Grim 1968). However, while the existence of any long range structure in water at interfaces is still open to question (Conway 1977), the existence of a layer of adsorbed water at clay surfaces is not questioned, nor is the concept of reduction of the strength of bonding beyond a distance of a few molecular diameters towards a zone of free water. Bonding at the surface does not restrict the relatively free movement of water molecules from one bonded position to another; such movement is extremely important to the processes of realignment of clay crystals as a result of stress.

Electric double layer

The term electric double layer describes the distribution of ions in the solution surrounding negatively charged colloids. The simplest model, known as the Helmholtz double layer, is of a layer of cations adsorbed to, and balancing, the negatively charged colloid surface (Fig. 3.12).

A more realistic model of distribution of ions around colloid surfaces is known as a diffuse double layer, or as the Gouy–Chapman model. This model describes a film of water, around the clay particle, in which cations grade continuously from a high concentration at the mineral surface to a lower concentration further away. Positions of cations are governed by the opposing effects of electrostatic attraction towards the surface and diffusion away from it in the direction of decreasing concentration. The diffuse double layer is consequently the clay surface and the swarm of cations close to it. Any anions in the water will then occur at some distance from the clay surface until the ratio of cations to anions equals that of the bulk solution. The thickness of the diffuse layer can be calculated for different ions and for different ionic concentrations, it is in the range of 1 nm to >20 nm. In general, the higher the valency of the cations, and the greater the concentration of the solution, the thinner is the diffuse layer.

The Stern model incorporates a Helmholtz layer of specifically adsorbed ions at the colloid surface and beyond this is a Gouy–Chapman diffuse layer of electrostatically retained ions. This model has been tested experimentally (Bedzyk *et al.* 1990) using a biological membrane as the negatively-charged surface and a long period X-ray standing wave for measuring the ion distribution profile in a solution layer. The diffuse layer was found to vary between 0.3 and 5.8 nm in thickness and the results agreed qualitatively with the Stern model. Electrostatic potential decreases linearly across the layer of adsorbed ions and exponentially across the diffuse layer. Natural soil solutions contain highly variable quantities of many species of ions, consequently the thickness of an electric double layer is expected to be variable but the general nature of the bonding strength appears to be well established. A clay particle together with its electric double layer is called a 'micelle'.

Because of the differences between clay species in the size of clay crystals, there is considerable variation in the number of crystals and the amounts of adsorbed and double layer water in equal volumes of clay of different species (Fig. 3.13). Montmorillonitic soils, for example, can hold far more water than kaolinitic soils and this fact, together with the much greater swelling capacity of montmorillonites, has a major effect upon the behaviour of such soils (see Chapter 7).

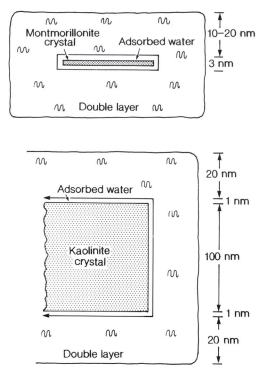

Note: The Kaolinite crystal is shown smaller than it is in comparison with the montmorillonite.

Fig. 3.13. Schematic representation of the relative sizes of kaolinite and montmorillonite crystals with their adsorbed water and electric double layers. Note that the kaolinite crystal is composed of many layers.

The number of clay particles per unit volume of soil is very large, of the order of 10^{15} to 10^{16} per cm^3. The total particle surface area is about 50 to 100 m^2 per cm^3 (Hansbo 1979). The amount of clay in a soil consequently has a major effect on soil properties.

Electric charges

Electric charges in soils can be associated with organic and inorganic colloids as permanent charge and variable charge.

Permanent charge results from isomorphous substitution within structural units of clay minerals. Cations within tetrahedral and octahedral sheets can be substituted by ions of a smaller charge (e.g. Al^{3+} for Si^{4+}; Mg^{2+}, Fe^{2+} for Fe^{3+}) producing a deficit of positive charge and an excess of negative charge within the layers. This negative charge is neutralized by the presence of cations and water in the interlayers. The ionic substitutions in clay minerals are primarily responsible for their physical properties, such as capacity to adsorb and lose bonded water and to form organo-mineral complexes. In many clay mineral species, particularly the vermiculites and smectites, these cations are not held strongly between layers but are in equilibrium with, and can be exchanged by, cations in the soil solution. It is the weak bonding strength in such species which permits their interlayer spaces to widen or shrink, especially if water is drawn into or expelled from the interlayers. Vermiculite and montmorillonite are consequently described as swelling or expanding clays.

Micas and illites have potassium ions (K$^+$) held tightly between the layers and are described as low-swelling or low-expanding clays. Chlorites have layer charges balanced by tightly held gibbsite interlayers and the 1:1 clay minerals have little isomorphous substitution and therefore have only limited permanent negative charge. Chlorites and kaolinites are therefore also low-swelling.

The low-swelling clays have a limited capacity for cation exchange and the swelling clays a high cation exchange capacity. Cation exchange capacity (CEC) is expressed in units of milliequivalents, of which one equivalent of an ion is defined as the quantity (in grams) of an ion that has the same number of charges as one equivalent of any other ion; thus, for monovalent ions, 1 mole equals one equivalent; for divalent ions, 1 mole equals 2 equivalents. An example may be written as:

$$Na^+ \text{ Clay} + CaCl \rightleftharpoons Ca^{2+} \text{ Clay} + 2NaCl.$$

On silicate clay surfaces, the net negative charges resulting from ionic substitutions and unsatisfied valencies are not evenly distributed across mineral surfaces. They are thought to be concentrated at edges and corners of clay layers and to be more common on exposed surfaces than in interlayer positions. Concentrations on edges may account for edge to face arrangements of clay crystal units in soil fabrics (see below).

Variable charge is dependent upon soil pH and therefore OH$^-$ ion concentration. As soil pH decreases (i.e. H$^+$ ion concentrations increase) the amount of negative charge decreases, and as soil pH increases (i.e. the OH$^-$ ion concentration increases) so does the negative charge on colloid surfaces. Most soils, and particularly those in temperate regions have an overall net negative charge as a result of the permanent negative charge of 2:1 clay minerals, even at low soil pH values.

Soils dominated by highly weathered 1:1 clays and aluminosilicates may have an overall positive charge. Such soils are most common in the humid tropics.

Fabric of clays in sedimentary rocks

Clays in weak rocks

Clay minerals are major constituents of marine muds, mudrocks, shales, and argillites and are present in many siltstones and sandstones. The most important sources of these minerals are terrestrial weathered rocks and soils. Transport is largely by rivers to the sea but eolian dusts are increasingly recognized as significant contributors, especially in the central Atlantic where dust is derived from the Sahara, and in the northeastern Pacific where dust comes from central Asia (Pye 1987).

Clays are commonly flocculated at the freshwater–saline water boundary and so estuarine and deltaic sediments have a very open structure with many and large pores separating nets of adhering clay, silt, and organic particles. Sediment carried into deeper water may accumulate to form very thick bodies of mudrocks, rich in both clay and silt together with organic particles such as the hard skeletal fragments from zooplankton (Chamley 1989).

Mudrock fabrics

Fabric is a term used to indicate the arrangement of grains in a sediment or soil and the pattern of partings and units of soil. Fabric is influenced by (1) the distribution of grains within the material, (2) the orientation of the grains, (3) the packing of the grains, and (4) opening of fissures.

Early workers postulated a number of models of clay fabrics which were believed to develop as a result of distinct electrochemical environments in which three types of contact (Fig. 3.14) were visualized as a result of considering the distribution and effect of electrical charges on clay surfaces: (1) edge–edge (E–E); (2) edge–face (E–F); and (3) face–face (F–F) contacts (Van Olphen 1977). E–F contacts are thought to develop preferentially in open fabric clays because there is a concentra-

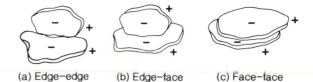

(a) Edge–edge   (b) Edge–face   (c) Face–face

Fɪɢ. 3.14. Orientation of clay crystals in contact.

Pʟᴀᴛᴇ 3.3. Kaolinite platelets with predominantly F–F contacts but with some E–F contacts. (Scanning-electron micrograph by A. G. Beattie.)

tion of unsatisfied valencies at edges and corners of clay crystals.

Since about 1970, electron microscopy has made possible direct observation of natural microfabrics and forced a recognition that many factors influence fabrics. Clay content and mineralogy, silt content, grain shape, ion concentration, rate of sedimentation and depth of burial act in complex ways to produce a wide range of microfabrics.

Initial deposition of floccules may involve E–E, E–F, and F–F contacts and random orientation. Flocculation is accounted for by a balance between attractive and repulsive forces. When the double layers of two clay particles come in contact repulsion occurs between their attached cations, and perhaps also between water molecules. At the same time van der Waals forces provide a general long-range attractive force which causes the adhesion needed to form low-density floccules. Van der Waals forces are not effective at ranges beyond about 20–30 nm (Sridharan and Jayadeva 1982; Nagaraj and Srinivasa Murthy 1986); these forces are therefore most effective where double layer thicknesses are limited, as they are in the

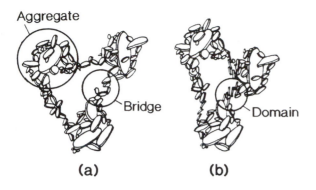

Aggregate

Bridge

Domain

(a)   (b)

Fig. 3.15. Schematic picture of clay frameworks; (a) a common microstructural pattern, and (b) formation of face–face alignments in domains. (After Hansbo 1979.)

Plate 3.4. Clay aggregates making up a mudrock which has obvious voids. Note the kaolinite stack at top-centre. (Scanning-electron micrograph by A. G. Beattie.)

high ion concentrations of sea water. Increasing consolidation of sediment underneath an overburden of fresher sediment leads to reorientation and closer spacing of particles in aggregates held together by electrostatic bonds and linked by bridges of very small grains (Fig. 3.15). As consolidation increases the voids between aggregates diminish and the free water in the pores is squeezed out, leaving adsorbed and double layer water, with contained ions, to act as the agent which continues to give the aggregate structure some rigidity (Hansbo 1979). The randomly oriented bridges become oriented progressively towards E–F then F–F contacts (when they are sometimes called domains), until F–F contacts at pressures >40 MPa (which, theoretically, may represent about 2 km thickness of overburden) may be so close that there is dry contact between clay crystals; bonding in the dry state is valent (i.e.

through shared ions) (Osipov 1978), as it is within crystal layers. The results of such orientation are turbostratic and laminar fabrics (Fig. 3.16).

Valent bonding is virtually irreversible and is an important part of lithogenesis, the transformation of soil to rock, together with precipitation of cementing materials.

Laminar fabrics cause marked anisotropy (i.e. variation in properties in different directions) and the development of short partings, creating the overall fissile structure which is the characteristic of clay shales. Such features reduce the strength of the material along planes parallel to the partings but have little effect on strength across them. Consequently, recognition of fabric is important for developing an understanding of the behaviour of fine-grained, sedimentary rocks.

The presence of silt grains in various propor-

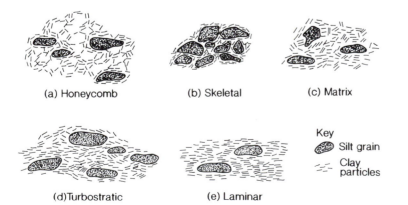

(a) Honeycomb   (b) Skeletal   (c) Matrix

(d) Turbostratic   (e) Laminar

Key

Silt grain

Clay particles

Fig. 3.16. Models of principal microstructural types of sedimentary clays with included silt grains. (Based on Grabowska-Olszewska *et al.* 1984.)

PLATE 3.5. Clay platelets forming a turbostratic fabric with elongated voids and clay units wrapping around silt grains. (Scanning-electron micrograph by A. G. Beattie.)

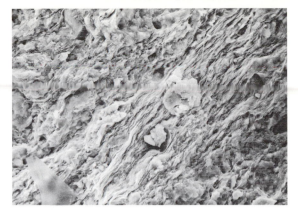

PLATE 3.6. A band of clay platelets with laminar fabric stretches from bottom-left to top-right through a body of material with turbostratic fabric. (Scanning-electron micrograph by A. G. Beattie.)

tions, disturbance by organisms (bioturbation), and large non-platy grains of inorganic or organic origin give rise to more complex structures. A classification which recognizes these complexities is that of Sergeyev *et al.* (1980) (Fig. 3.16). This classification is noteworthy because it has been possible to link to it general statements about the physical properties of the materials which contain such fabrics (Huppert 1986, 1988). Table 3.2 summarizes some of these relationships (see also Grabowska-Olszewska *et al.* 1984).

The presence of silt in sediments has the effect of creating a loose, open structure in freshly deposited sediment in fresh and salt water. Under

consolidation, the silt increases the frictional strength of the materials and reduces the degree of preferred orientation of platy minerals.

Unloading

Mudrocks which have been deeply buried have densities and fabrics which are in accord with their thickness of overburdens. If they are subsequently raised above sea-level and the upper strata are removed by erosion, the once deeply buried rocks will be over-consolidated with respect to the current overburden. Over periods of tens to hundreds of years joints may develop, water may penetrate,

TABLE 3.2. *Relationships between mudrock fabrics and geomechanical properties: a New Zealand example*

Microstructural	Sediment type	Clay species	Porosity (%)	Behaviour
Honeycomb	Weakly compacted, recent	>25% Illite-smectite	60–90	Low strength UCS <0.01 MPa
Skeletal	Weakly compacted, recent	>40% Silt, illitic clay	~50	Compressible
Matrix	Moderately compacted	>20% Illite or mixed clays	30–50	Moderate strength UCS <2 MPa
Turbostratic	Compacted from honeycomb or matrix	Illitic, silts, mixed clays	30–50	1–5 MPa (strength is anisotropic)
Laminar	Clay crystallized in place or sheared sedimentary clays	>40% Clays, various species	Large range of properties	

Notes: 1. Many microstructures are transitional between those specified and several types may exist in one rock body;
2. UCS is Unconfined Compressive Strength.

Sources: Huppert (1986, 1988).

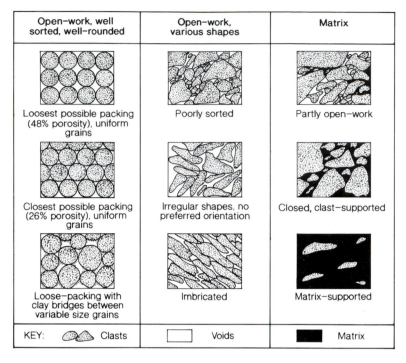

FIG. 3.17. Models of fabrics occurring in gravels, sands, and coarse silts.

weathering occur, and some of the laminar fabrics may be converted into brecciated skeletal fabrics (see Fig. 5.4). The large-scale result will be a swelling of the rock body. This process is called unloading and it may lead to pressures on walls of tunnels, release of plates of rock on cliff faces, and weakening of mudrock masses leading to development of landslides (see Chapter 13).

Structure in pedogenetic soils

Pedogenetic soils develop structure or fabric by distinct processes which occur in soil profiles. These are discussed in Chapter 10. The major effect of structure is the opening of fissures which become pathways for the movement of water, zones of softening of the soil and therefore zones of possible weakness. The overall effect in dense fine-grained soils is the development of linked macropores which greatly increase permeability and which may become linked to create planes of failure of landslides (see Rowe 1972).

Fabrics of coarse-grained soils

Gravels, sands, and, to a lesser extent, silts have properties which are different from those of clays. The coarse particles are non-cohesive, do not have films of adsorbed water, are comparatively inert chemically, and do not have electrostatic bonding operating between them. They may, however, be held together by the intermolecular bonds of cementing precipitates such as silica, calcium carbonate, and iron oxides, and by bridges of clays. Non-cemented coarse particles in direct contact with each other owe their mass strength entirely to friction at their points of contact. This frictional strength increases in proportion to the number of point contacts between grains and the area of contact.

Single-grain structures are common in sediments produced by settlement of grains, through air or water, under gravity. Each grain rolls to a position of equilibrium among its neighbours. Single-grain sediments may be deposited in a loose state with a high void ratio or in a dense state with a low void ratio (void ratio is defined as volume of pores/volume of solids). Uniform particle sizes give rise to high void ratios but mixed particle sizes result in voids between large particles being filled with small particles. To sedimentologists uniform sizes are an indication of well-developed sorting; to engineers uniform particles give rise to low strength and high permeability, and therefore are usually unfavourable fabrics and are referred to as poorly graded. The terms 'poorly sorted' and 'well-graded' are

TABLE 3.3. *Sedimentary soils: origin, fabrics, and properties*

Origin	Agency of transportation	Fabrics	Properties of deposits
Littoral and marine	Waves, currents, and tides	Open-work to matrix-supported	Variable degrees of cementation with $CaCO_3$ derived from marine organisms. Fabrics may collapse if saline pore water replaced by fresh; shocks or vibrations may induce landslides in both marine and subaerial soils.
Estuarine and deltaic	Tidal rivers depositing in saline waters	Partly open-work to matrix-supported	Low strength, if fabrics are disturbed may suffer extreme loss of strength (i.e. they are sensitive); variable properties; compressible.
Talus (coarse colluvium)	Rockfall and other mass wasting	Open-work to closed clast-supported with lenses of matrix-supported from sheet wash	Irregular, angular particles; variable texture and fabric, sometimes lens-shaped structures, variable permeability; coarsest grains near source or in foot-slope mounds.
Colluvium, fine-grained	Sheet wash, mass wasting	Closed clast-supported to matrix-supported	Irregular particle sizes and shapes give variable properties. Generally low strength; may be dispersive leading to piping; variable pore-water pressures.
Eolian deposits	Wind	Open-work (sand dunes) Open-work with bridges of clay and sometimes $CaCO_3$ cements (loess)	Dunes may have collapsible fabrics; high permeability; subject to remobilization. Loess fabric may collapse if saturated; strength lost on wetting; highly erodible by water.
Reworked sands	Sheet wash, wind, termites	Open-work to partly open-work	Collapsible fabric; compressible.
Alluvium	River	Well-bedded with dune and ripple forms; partly open-work to closed open-work	Variable texture and bedding in a vertical sequence; lenses of unbedded fines (over-bank deposits); variable pore-water pressures and perched water-tables.
Lacustrine	Rivers flowing into lakes or pans, organic deposits	Closed clast-supported to matrix-supported; lenses of fines and organic sediments	Highly compressible; low strength; may be expansive; saline in coastal and arid environments.
Tephra	Air-fall volcanic ash	Well-bedded; well-sorted, open-work	Coarsest grains near to source; low density particles easily eroded; collapsible and compressible; dispersive and subject to piping; allophanic clays may have extreme swelling and shrinking; extreme loss of strength on saturation.
Pyroclastic flows	Air-fall to ground-hugging dense flows of volcanic material	Open-work to partly open-work; variable texture and bedding	Variable texture; little to moderate sorting; angular particles; low density pumice easily eroded; piping; collapsible.

TABLE 3.3. (*cont'd.*)

Origin	Agency of transportation	Fabrics	Properties of deposits
Glacial till	Ice	Closed clast-supported to matrix-supported	Variable texture; basal tills overconsolidated by overburden of ice and therefore strong; lenses of flow and lacustrine deposits of low strength and sometimes high plasticity; lenses of water-bearing fluvial sands.
Glacial outwash	Rivers	Open-work	Coarse deposits near source have least particle rounding and sorting; also least bedding. Finer grains, better sorting, stronger bedding away from source.

used to describe the opposite situation of mixed grain sizes (Fig. 3.17).

Two other general conditions of coarse particles are recognized. (1) Voids may become filled with fines such as silt and clay by inwashing after deposition, in which case the strength of the material is still largely from the frictional contact between coarse grains. (2) In some deposits, such as landslide, glacial, and colluvial materials, the coarse particles may be separated by and enveloped in a matrix of fine grained material, in which case the strength of the deposit is derived from the matrix. A classification of fabrics of coarse-grained materials is illustrated in Fig. 3.17, and Table 3.3 indicates some of the properties of sedimentary soil deposits. As a result of tectonism, sea-level change and environmental change, any of these deposits may become the base materials in which hillslopes are formed.

4 Stress, Strain, and Rheology of Materials

Definitions

Force and stress

There are two types of forces which can act upon a body composed of continuous matter: body forces and surface forces. Body forces act throughout the body and are produced without physical contact with other bodies; examples are gravitational and magnetic forces. Surface forces are impressed on external surfaces of a body and result from physical contact with other bodies.

When impressed forces are applied to the external surface of a body, they set up internal forces within that body, which is then said to be in a state of stress, with a balanced internal reaction to the external action. Stress (σ), is specified as the intensity of these internal reactions as the average force (P) acting on a unit area (A) in any chosen direction ($\sigma = P/A$); a stress therefore has units of N/m^2. Stresses may be tensile, shearing, or compressive, and produce, or tend to produce, strain.

Strain

Strain (ε) is deformation; measured as a ratio of the change in dimensions of the stressed body to its original dimensions. Strain may occur as an increase or decrease in dimensions and may involve changes in length (linear strain, $\delta l/l$), volume (volumetric strain, $\delta v/v$) displacement or distortion (shear strain, $d/l = \tan\theta$, see Fig. 4.1). Strain may be recorded as a percentage change in dimensions, in units per unit (e.g. mm/mm) or, where strains are very small, in millistrains (10^{-3}) or microstrains (10^{-6}, e.g. μm/m).

Strength

Crystalline solids respond to imposed loads elastically, that is the deflection is proportional to the load and the original dimensions are recoverable on removal of the load. This elastic behaviour is caused by changes in the interatomic spacing within the material. For most engineering materials and strong rocks the change in dimensions lie between 0.1 and 1.0 per cent of the original dimensions. However, if the applied load exceeds the capacity of the solid to deform elastically then the solid may undergo ductile, irrecoverable, change of form and finally fracture. The stress required to cause fracture is a measure of the ultimate strength of the solid, but the rate of stress application, the presence of voids or planes of weakness in the solid, and inclusions of water can all influence the flow and deformations occurring within materials. The science of such behaviour is rheology.

Modern concepts of force, stress and strain, and strength are derived from the works of Galileo Galilei (1564–1642), Isaac Newton (1642–1727), Robert Hooke (1635–1703), Siméon-Denis Poisson (1781–1840), and Augustin Cauchy (1789–1857). An account of the development of materials science is given by Gordon (1988). The rheological properties of rocks are discussed in such texts as those by Jaeger and Cook (1976), and Lama and Vutukuri (1978). The rheology of soils is discussed by Roberts (1977) and Vyalov (1986).

Rheology

The object of most forms of science is the prediction of consequences of actions, or behaviour of phenomena. Rocks and soils vary greatly in their responses to stress because of their great range of properties. In order to predict the behaviour, for example, of one rock mass under load of a thick overburden of rock, of a mass of soil saturated with water, or a body of sand compared with a body of clay, it is desirable to express the behaviour

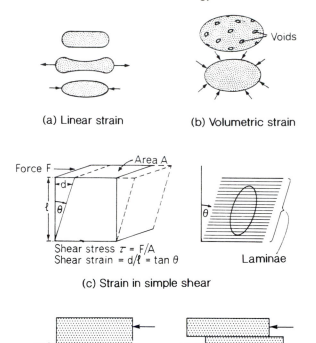

(a) Linear strain

(b) Volumetric strain

(c) Strain in simple shear

(d) Strain in direct shear

FIG. 4.1. Forms of strain.

mathematically. Developing appropriate equations is done most readily by using idealizations, or models, which can be understood in simplified terms. Simple models are then evaluated against laboratory tests of simulated or equivalent materials which are simplifications of complex materials, and then against complex materials such as rocks and soils.

Mechanical testing of complex and anisotropic rock and soil is based on samples which are thought to be representative of a large body. The results depend partly upon the inherent properties of the natural material, partly upon the characteristics of the testing equipment, and partly upon the methods followed in applying the test. The value of the interpretation of the test results, consequently, depends upon the appropriateness of the sample and upon recognition of the contributions to the test result.

The value of mathematical idealizations for prediction depends entirely upon the extent to which the model describes the behaviour of the natural materials. Many of the physical properties of rock and soil are, as yet, imperfectly understood.

Ideal models are commonly considered by reference to mechanical models consisting of springs, dashpots, and sliding blocks (Fig. 4.2a, b, c) and the corresponding ideal behaviours are elastic, viscous, and plastic. Combinations of these simple models and behaviours describe elastoplastic, elasticoviscous, viscoelastic, and plasticoviscous materials among many other possible types. Elastoplastic behaviour is modelled by a spring and block in series, elasticoviscous by a spring and dashpot in series, viscoelastic by a spring and block in parallel, and plasticoviscous by a block and dashpot in parallel.

Elasticity is a property of solids and viscous behaviour is a characteristic primarily of fluids near the surface of the Earth. Plasticity is displayed by moist soils. There are no ideally elastic, viscous, or plastic materials in nature.

Elastic behaviour

A spring is a mechanical model for an elastic material. Such a material responds immediately to an applied load and its change in dimensions is directly proportional to the applied stress. When the load is removed the entire strain is recoverable. The magnitude of the stress (σ) required to produce a given strain (ε) is a characteristic of the material and is known as Young's modulus (E) (a modulus is a constant factor):

$$\sigma = E\varepsilon, \quad E = \sigma/\varepsilon, \quad \varepsilon = \sigma/E. \tag{1}$$

This relationship between stress and strain is termed 'Hooke's Law' and a perfectly elastic substance is said to be Hookean (Fig. 4.2d). The graphical representation in the stress–strain plane for such a solid is a straight line that passes through the origin and has a slope describing E for that material. In brittle materials the behaviour is elastic up to a point, known as the yield point, at which there is a change from elastic to plastic behaviour. The corresponding stress is the yield stress. For a brittle material, however, plastic yielding is very limited and fracture occurs under continuing strain.

The greater the value of E for a material, the less will be the deformation produced by a given value of stress, and the stronger the material will be. In an ideal elastic material E will be the same both in tension and compression, but in a real material this may not necessarily be so. Real materials may also display non-linear elasticity, in which, although the stress–strain path may be the same both for

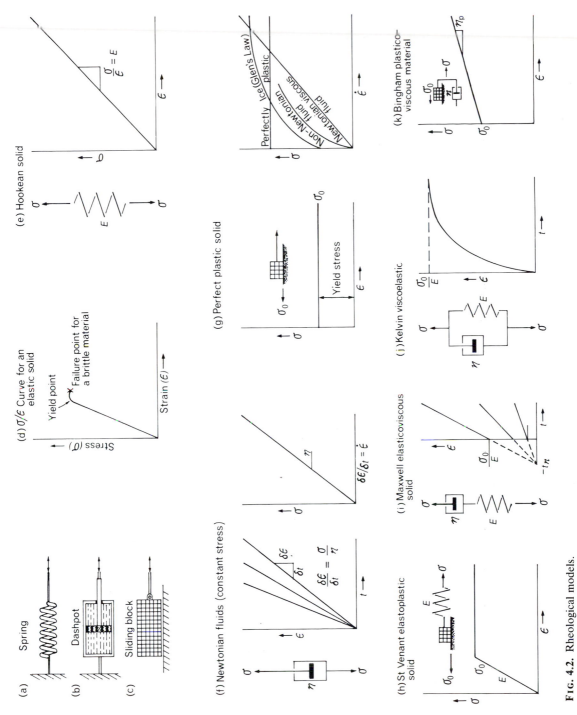

Fig. 4.2. Rheological models.

unloading and loading, the path is a curvilinear one, with different values of E at different levels of stress. An isotropic material displays the same value of E in all directions, at a given value of stress, but in an anisotropic material E differs in different directions through the material.

Viscosity

In viscous materials deformation develops over time. A dashpot is a mechanical model for a perfectly viscous fluid, also known as a Newtonian substance. A dashpot is a tube, filled with oil, in which a loosely fitting piston can be pushed back and forth. Moving the piston causes oil to move from one end to the other of the cylinder by flowing round the edges of the piston. The movement of oil (ε) is proportional to the force applied (σ) to the piston and the time it acts (t), and inversely proportional to the coefficient of viscosity of the oil (η):

$$\varepsilon = \frac{\sigma t}{\eta} \quad \text{and} \quad \sigma = \eta\dot{\varepsilon} \qquad (2)$$

where $\dot{\varepsilon}$ is the strain rate.

The graphical representation in the strain–time plane is a straight line passing through the origin and having a slope of σ/η. On a stress–strain rate diagram the gradient of the curve is a measure of dynamic viscosity of the fluid and is related to its molecular composition. It is a measure of the internal frictional resistance to flow and has units of $N\,s/m^2$. An alternative measure is the kinematic viscosity, which is the dynamic viscosity divided by the density of the fluid, the units being m^2/s.

Over a geological time-scale, and at high temperatures, all rocks can flow. Pitch, salt, and lead are well known solids which flow in human time-scales. In the shorter time-scale of geological investigations only molten rock and soils are usually capable of viscous behaviour, and the latter only as a result of a high water content. Viscosity, however, is a concept that is applied to all fluids and solids and links the behaviour of the two types of material.

Plasticity

A sliding block is a mechanical model for a perfectly plastic material. For the block to be moved an initial stress of sufficient magnitude has to be applied to overcome the friction between the block and the surface on which it lies. For any perfectly plastic material the initial shear stress has to be of

TABLE 4.1. *Dynamic viscosities (N s/m²)*

Earth's mantle	10^{20}
Shale	10^{16}
Anhydrite (evaporites)	10^{13}–10^{16}
Glacier ice	10^{12}–10^{13}
Magma	10^{2}–10^{3}
Flowing lava	10^{1}–10^{5}
Debris flows	7.5×10^{2}
Mudflows	2×10^{2}–6×10^{2}
Solifluction	10^{2}
Water at 20 °C	10^{-3}

sufficient magnitude to equal the yield strength. Once this threshold value is exceeded, deformation occurs at a constant rate as long as the stress is uniform. Once the stress is removed the strain is permanent. In a stress–strain plane the graphical representation of a perfectly plastic material is a straight line parallel to the strain axis and intersecting the stress axis at the yield strength (= yield stress, σ_o). Clays commonly exhibit plastic behaviour and ice approximates to it even though ice yields at very low stresses, and then at an increasing rate as the stress increases, before approaching perfect plasticity.

Mixed behaviour

A perfectly elastoplastic, or St Venant material is perfectly elastic for stresses less than the yield stress and perfectly plastic for stresses equal to σ_o. The appropriate mechanical model is a spring in series with a sliding block and the behaviour is described by:

$$\sigma = E\varepsilon + \sigma_o. \qquad (3)$$

An elasticoviscous, or Maxwell material behaves as though it has an elastic element in series with a viscous element and the same stress must act on both elements. The total strain (ε_t), therefore, is the sum of the elastic strain (ε_e) and the viscous strain (ε_v), thus:

$$\varepsilon_t = \varepsilon_e + \varepsilon_v = \frac{\sigma}{E} + \frac{\sigma t}{\eta}. \qquad (4)$$

Assuming that a constant stress is applied suddenly, then the graphical representation will be a straight line intersecting the strain axis at a value of σ_o, thus:

$$\varepsilon_t = \frac{\sigma_o}{E} + \frac{\sigma_o t}{\eta}, \qquad (5)$$

with the rate of strain being controlled by the viscous element (i.e. the fluid in the dashpot).

The graphs in Fig. 4.2i show plots for three values of σ_o. All the graphical lines emanate from a single point on the time axis given by $t_n = -\eta/E$: this point is the negative relaxation time when stress is removed. A Maxwell solid exhibits instantaneous elastic strain.

A viscoelastic, or Kelvin material behaves as though it has viscous and elastic elements in parallel. For this model the strain in the dashpot must equal the strain in the spring. The total stress (σ_t) is the sum of the viscous stress (σ_v) and the elastic stress (σ_e):

$$\sigma_t = \sigma_v + \sigma_e = E\varepsilon + \eta\dot{\varepsilon}. \qquad (6)$$

The elastic strain, σ_o/E, which would be obtained instantly for an elastic material is approached exponentially for a viscoelastic material.

The combination of a Maxwell model and a Kelvin model in series is a Burger model. If this model is subjected to a constant stress σ_o, maintained for a time t, then the strain in Burger's model ε_B will be the sum of the strains in the Kelvin model ε_K and the Maxwell model ε_M, that is:

$$\varepsilon_B = \varepsilon_K + \varepsilon_M. \qquad (7)$$

When the stress σ_o is removed, the elastic strain will be recovered immediately, thereafter gradual creep recovery will occur but will never be complete. Burger's model is the closest representation of behaviour of many rocks. The graphical form is similar to that illustrated in Fig. 4.4b.

A plasticoviscous, or Bingham material behaves as though it has an initial plastic behaviour but, as the level of stress is increased, the yield stress is reached and the material behaves like a fluid. After release of applied stress, the strain remains permanent. Such a material may be represented by a dashpot and block in parallel. The total stress is the sum of the yield stress and the plastic viscosity (η_p):

$$\sigma_t = \sigma_o + \eta_p \dot{\varepsilon}. \qquad (8)$$

In a true Bingham material the rate of shear is linear; toothpaste, oil paints, and drilling muds of smectite clays exhibit such behaviour. There are other materials which also exhibit plastic properties and a yield stress before viscous deformation occurs, but the deformation thereafter has rates of shear which are not linear. Such materials are generally referred to as generalized Bingham plastics.

In the discussion so far it has been assumed that applied stresses can be maintained permanently and that strain will be continuous. This is clearly unrealistic for an elastic material: a spring can only be stretched to a limited value before losing its form and capacity to recover, and beyond this proportional limit the strain is no longer proportional to the load. At higher increments of load, behaviour ceases to be elastic (the elastic limit) and may become essentially plastic with none of the strain being recoverable. Continued loading will eventually cause the spring to fracture. The extent to which a material can sustain plastic deformation is known as ductility (Fig. 4.3a). For most rocks the proportional limit and elastic limit are taken as representing the same value, and the same value may also represent the yield strength in brittle rocks.

Rheology of rock

Under static loading of a rock sample only the hardest, densest, crystalline, and non-porous igneous and metamorphic rocks, such as basalt (Fig. 4.3b), approach ideal elastic behaviour. The generalized stress–strain relationship for most rocks takes a more curvilinear form. With increase of load from a zero condition the pores and interstices of most rocks will be closed and this may show up as approximating an initial plastic behaviour (marble and schist in Fig. 4.3d, e). The consolidated rock may then behave elastically until the continued loading causes internal slip, dislocation, and microfracturing, which shows as plastic behaviour. In a hard, dense rock rupture will result from internal microfracturing, but in rocks with platey, aligned, silicate minerals, or in evaporites and rock salt, shearing within crystals followed by creep may permit large plastic deformations before rupture occurs. The significance of mineral alignment in a rock is shown by the different behaviours of schist cored parallel and perpendicular to the foliation (Fig. 4.3d, f).

On relief of load from an elastic rock which has not reached rupture the stress–strain relationship will follow that produced during loading. Any de-

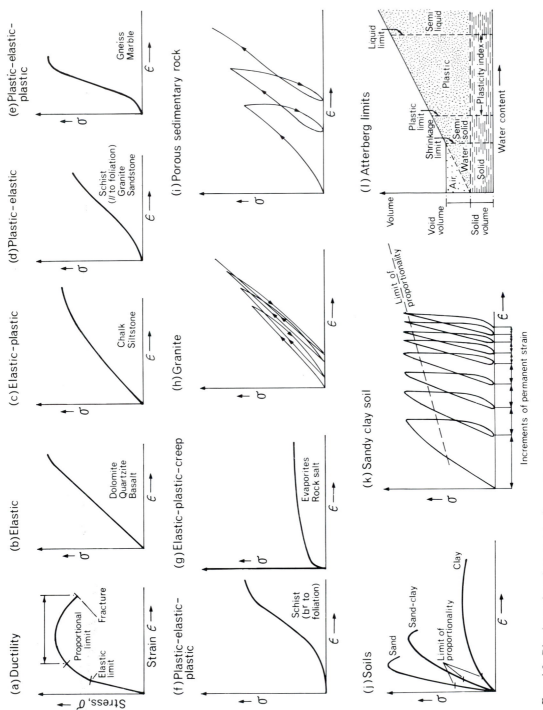

FIG. 4.3. Rheology of earth materials. (b–g, after Hendron 1968; h–k, after Roberts 1977.)

parture from linearity and ideal elasticity will show up as a hysteresis loop as the load is alternately applied and removed during load-cycling. If the elastic recovery is not instantaneous some residual strain will remain on relief of load. That part of the residual elastic strain recovered in the course of time is termed the delayed elastic recovery. Any plastic deformation resulting from increased load will not be recovered on unloading and will be evident in an increased width of the hysteresis loop. The effects of plastic components in rheology of a granite and a porous sedimentary rock are evident in Fig. 4.3h, i. Load-cycling is a valuable method of studying rock behaviour.

Rheology of soils and weak sediments

Soil is commonly treated as an elastoplastic material, allowing relatively simple mathematical representation. Most soils, however, are viscoelastic-plastic materials with non-linear behaviour (Vyalov 1986). Elastic recovery results from reversible displacement of mineral particles and bending of flaky minerals; at very high stresses, the water of double layers may be squeezed out and recovered on removal of the load, but this would require overburdens of several kilometres and millions of years. Plastic deformation is the result of irreversible sliding of particles against one another, especially of hydrated clay particles. Viscous behaviour is derived from free water which is not adsorbed to clay minerals.

Behaviour of soils is greatly influenced by grain size, packing, fabric, and water content. Under fluctuating load and vibration, gravels and sands will increasingly compact and an approximate elastic behaviour can be identified in many granular soils for stresses well below their maximum strength. The effect of load-cycling is displayed in Fig. 4.3k. The limit of proportionality increases with each progressive cycle, while the increment of permanent strain decreases. Eventually, either the hysteresis loop will close and the material will behave elastically over that range of stress, or the increment of residual strain per cycle will attain a constant value as the material behaves plastically.

In fine-grained soils, water is the major control on behaviour. Depending on water content, soil may behave as a solid, semi-solid, plastic, or liquid. The amount of water needed to change the behaviour depends largely upon the species of clay minerals present. For each soild it is possible to define limits to the type of behaviour in terms of water quantity (by weight) in a sample (Fig.

4.3j, l). Because fabric and structure vary from one soil sample to another these are eliminated, for the purposes of measurement, by crushing the fabric units and thoroughly remoulding the soil until all structure is lost. The behaviour of soils under stress at various water contents is expressed as Atterberg Limits; these are discussed in Chapter 7.

Most cohesive soils, such as marine clays, display elements of elastic and viscous behaviour and are appropriately modelled as Kelvin bodies. The shape of the Kelvin curve (Fig. 4.2j), when compared with that for clay in Fig. 4.3j, illustrates this point. Well-compacted sands are predominantly elastic in behaviour unless the grains are crushed under the applied stress, when behaviour becomes increasingly plastic. Earthflows, debris flows, and some lahars may have behaviours largely influenced by their fine-grained soil content and high water content which separate the coarse sands and gravels in the flow. The combination of behaviours is often most appropriately modelled as Bingham behaviour.

Viscous behaviour in soils is almost invariably due to a high water content. The influence of the solids in the fluid makes most flowing soils non-Newtonian; that is, their viscosity is not constant for all rates of strain, usually because of the change of structure in the fluid during flow; an example is that of smectite muds which will flow under very low stresses if they are first subjected to high sudden stresses. Removal of the stress allows the muds to recover some of their original structure and strength. Materials with such behaviour are said to be thixotropic; they may become highly unstable when subjected to earthquake shocks or vibrations from traffic.

Viscosity measurements of muds are relatively simple in the laboratory: a steel ball dropped through columns of fluids of different density, for example, will descend at a rate proportional to the resistance and provide a relative measure of viscosity; cylinders rotated at known rates within the fluid provide another method. Measuring the viscosity of molten lavas and mudflows containing boulders is not so easy and is usually a field problem. If the flow is laminar its viscosity may be estimated from:

$$\eta = \frac{g \, \rho \, \sin\beta \, d^2}{3V}, \qquad (9)$$

where ρ is the density, β is the angle of slope, d is the thickness, and V is the mean velocity. For flow in restricted channels, the coefficient in the denominator is replaced by 4. For non-Newtonian flow materials the effective viscosity will be greater at low velocities and shear rates than at high velocities of flow (see Fig. 4.2g). Viscosities of common earth materials are given in Table 4.1.

Time-dependent behaviour of earth materials

Time-dependent behaviour in earth materials is commonly referred to as 'creep'. It is usually promoted by factors, in addition to time, such as temperature and temperature variations, water-content and especially alternating hydration and dehydration, pore-water pressure, and ambient stresses such as the loads of overburdens.

Creep processes are usually pictured as a strain–time curve representing the deformation of material (Fig. 4.4a). After an initial component of instantaneous elastic response to load, the strain–time curve represents the deformation of material at constant stress (Fig. 2a). After an initial elastic response to load, the strain–time graph is curvilinear with a decreasing strain rate until a time t_1. This period is called the primary or transient creep. If the stress is removed at any instant during primary creep the elastic component of strain is recovered immediately in cohesive rock—but not in a non-cohesive soil: the creep component of strain is recovered by a time-dependent process in cohesive materials at a decreasing strain rate (Fig. 4.4b).

The primary creep phase is followed by a secondary stage (t_1 to t_2) of constant strain rate and permanent deformation. At high stress levels a tertiary phase of accelerating deformation may lead to rupture (t_2 to t_3).

Creep is an important process in cohesive soils on slopes. There is still much debate about its cause, but water, temperature, stress, and biotic variations are usually regarded as important, with evidence that rearrangement of soil particles, following solution of some materials (Young 1978), could be a major factor. In rock, creep may be of major importance and magnitude in evaporites and porous sedimentary formations, particularly where fluids are present in the pore space. Sedimentary rocks display creep which is evident in closures of tunnels and shafts and swelling of walls of rock cuts and in natural cliffs. Fine-grained igneous and metamorphic rocks have such low rates and magnitudes of creep (Fig. 4.4c) at the temperature and pressure conditions of engineering and geomorphic interest that tunnels

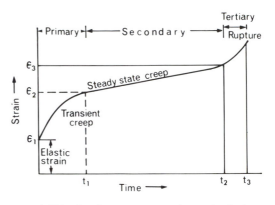

(a) Idealized creep curve, at constant stress

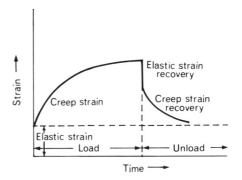

(b) Creep and recovery curve

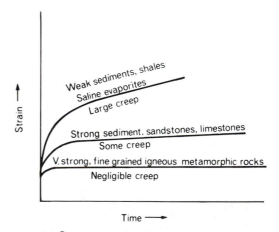

(c) Creep in rocks, for same stress

FIG. 4.4. Creep in rocks. (After Roberts 1977.)

TABLE 4.2. *Elastic properties of rocks and selected materials*

Rock type	Density (Mg/m³)	Modulus of elasticity, E (GPa)	Shear modulus, G (GPa)	Poisson's ratio, v
Andesite	2.4	35–50	5–10	0.2
Basalt	2.6–2.8	45–100	10–30	0.28
Conglomerate	2.5–2.7	20–30	6–20	0.1–0.29
Diorite	2.7	35–80	15–25	0.1–0.28
Dolomite	2.4–2.7	50–80	20–30	0.2–0.4
Gabbro	3.0	55–90	20–40	0.2–0.38
Gneiss	2.7–2.8	55	—	0.21
Granite	2.64	35–70	15–50	0.21–0.28
Greywacke	2.5	20–60	—	0.2–2.5
Limestone	2.3–2.7	26–63	3–30	0.2–0.23
Marble	2.7	17–100	17–30	0.1–0.28
Phyllite	2.8	25–40	—	0.3
Quartzite	2.5–2.8	50–70	1–40	0.1–0.4
Salt	2.2	28	—	0.3
Sandstone	2.2–2.5	10–70	1–40	0.1–0.45
Schist	2.4–2.8	20–60	8–25	0.1–0.3
Shale	1.8–2.5	20–50	1–15	0.1–0.3
Siltstone	2.0–2.7	20–40	1–22	0.1–0.4
Tuff	1.9–2.4	3–60	—	—
Aluminium	2.7	73	49	0.34
Concrete	2.7–3.2	14	—	—
Diamond	—	1200	—	—
Kevlar 49	—	62	—	—
Steel (mild)	7.8	210	135	0.29

Sources: Wuerker (1955); D'Andrea *et al*. (1965); Tennent (1971); Lama and Vutukuri (1978); Rahn (1986); Imazu (1986); Gordon (1988). The values given above have been selected as being the most characteristic values from samples which did not fracture or open along cleavage planes during testing.

and shafts in them do not have to be lined to resist creep, but may need bracing to prevent falls of fractured rock.

Moduli of elasticity

Young's recognition that the relationship between applied load and deformation (the stiffness) is a constant for any elastic material (that is, it is a modulus) was first expressed in the terms of stress and strain by Navier (1785–1836) as:

$$\textit{Young's modulus} = \frac{\text{longitudinal applied force/}{\text{cross-sectional area}}}{\text{change in length/}{\text{original length}}}$$

$$= \frac{\text{axial stress}}{\text{axial strain}}.$$

The range of values of E for natural materials is enormous (Table 4.2): from about $1\,200\,000\,\text{MN/m}^2$ for diamond, through $1\,100\,000–2000\,\text{MN/m}^2$ for strong to weak rocks, to $0.2\,\text{MN/m}^2$ for soft biological materials.

Young's modulus for rocks and soils is more easily measured in compression than in tension by placing a carefully machined core of the rock in a compression machine and measuring the vertical height of the core as stress is applied. The resulting stresses and strains are plotted graphically (Fig. 4.5b).

The curvilinear relationship shows that the magnitude of the modulus of elasticity, E, varies from a minimum at zero load to a maximum somewhere between zero load and the yield point. We must, therefore, specify to what point on the stress–strain curve we are referring, and this choice should reflect the nature of the problem

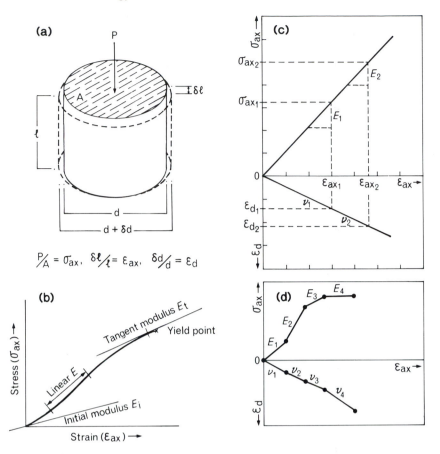

FIG. 4.5. (a) Deformation in uniaxial compression. (b) A generalized stress–strain curve for a rock, up to the yield point, with definitions of the forms of Young's modulus. (c) An idealization of a linear elastic body showing constant values for Young's modulus, E, and Poisson's ratio, v. (d) An elastic body with variations of the moduli as a result of variation in the strain response to increasing stress.

being investigated. The linear modulus is usually quoted for the stress–strain at half the ultimate compressive strength. For conditions requiring a particular representative value the secant modulus is used (the secant modulus is defined as: the slope of a straight line joining the origin of the axial stress–strain curve to a point on the curve at some fixed percentage of the peak strength); for known stress levels a tangent modulus at the given stress is required; and at very low stresses an initial modulus is appropriate. The tangent modulus, E_t is defined by:

$$E_t = \frac{\delta\sigma_{ax}}{\delta\varepsilon_{ax}}. \qquad (10)$$

Stretching or compression along the longitudinal axis of a specimen is accompanied by transverse thinning or thickening—as seen most obviously with a stretched rubber band. It was Poisson who first expressed this relationship mathematically and the relationship between longitudinal (axial) and lateral (diametral) strain is a constant for a material, known as 'Poisson's ratio, v':

$$\frac{Poisson's}{ratio\ (v)} = \frac{\text{change in diameter/original diameter}}{\text{change in length/original length}}$$

or

$$v = \frac{\varepsilon_d}{\varepsilon_{ax}}. \qquad (11)$$

The stronger the rock the smaller will be the value of v. Although Poisson's ratio is usually regarded as a constant this is not a reliable statement, for near the ultimate strength the stress–strain curves indicate that the lateral strain increases more than the axial strain resulting in an increased value of *v*. Most published values of Poisson's ratio refer to values at, or below, 50 per cent ultimate strength.

For most common engineering materials the values of Poisson's ratio are in the range of 0.25–0.33. For most rocks they are 0.1–0.4. Values lower than 0.5 indicate that a material under tension will increase its volume by increasing the spaces within crystal lattices or by opening of microfractures. Under compression the reverse will occur and the material will decrease its volume. A value of 0.5 indicates no volume change. Some biological materials, such as muscle material, can have Poisson's ratios higher than 0.5 and close to 1.0. In rocks, values higher than about 0.45 usually indicate that specimens are suffering internal fracturing while under test; values as high as 0.8 have been recorded but they indicate irrecoverable strain and not elastic properties.

Where a specimen is confined, longitudinal extension or contraction is limited by the surrounding material and Young's modulus cannot apply. In such cases another constant has to be used:

$$Axial\ modulus\ (\psi) = \frac{\text{longitudinal applied force/}\ \text{cross section area}}{\text{(in confinement) change of length/}\ \text{original length}}.$$

Shear stresses tend to induce sliding of atoms or particles within a body of material. Shear stress (τ) is the shear force per unit area of the cross section of the material on which the force is acting. The units of shear stress are thus N/m^2. Shear strain is an angle θ (see Fig. 2.1c), usually expressed in radians. The shear modulus G is the stress that would distort an element of the material through an angle equal to one radian. The units of G are those of a shear stress:

$$Shear\ modulus,\ G = \frac{\text{shear force/unit area}}{\text{angular deformation}}.$$

The shear modulus is also known as the rigidity or torsion modulus. G is related to the other moduli of elasticity by:

$$G = \frac{E}{2(1 + v)}. \tag{12}$$

The changes of volume in response to stress are expressed by the bulk modulus K, otherwise known as the modulus of incompressibility:

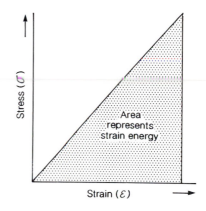

FIG. 4.6. Strain energy is equal to the area under the stress strain curve ($\frac{1}{2}\sigma\varepsilon$).

$$Bulk\ modulus,\ K = \frac{\text{applied force/unit area}}{\text{change in volume/unit volume}}.$$

The reciprocal of K $(1/K)$ is called the 'compressibility', and the bulk modulus is related to E and v by:

$$K = \frac{E}{3(1 - 2v)}. \tag{13}$$

Strain energy and fractures

Stretching or compressing a spring stores energy in it which is released when the applied load is removed. All elastic materials store energy when they are stressed. The energy (λ) stored in an elastic, Hookean, material subjected to an axial stress (σ), causing a strain (ε) is given by:

$$\lambda = \tfrac{1}{2}\varepsilon\,\sigma \quad \text{(Fig. 4.6).} \tag{14}$$

In a uniform material, subject to axial tension or compression, the stress may be envisaged as passing from one atom or molecule to another in a series of parallel lines which are oriented in the stress direction. If the cross-section of the material reduces gradually (Fig. 4.7b) and smoothly, the stress will be increased in the narrowed zone in proportion to the reduction in cross-sectional area but will still be evenly distributed. If the constriction is abrupt, as at the tip of a crack or sharp bend, the stress trajectories will be crowded together and the local stress concentration may be very large (Fig. 4.7c, d). Notches, angular pits, cavities, and breaks all give rise to stress con-

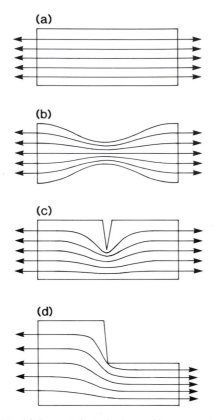

FIG. 4.7. (a) Stress trajectories in a uniform material under axial tension; (b) with a smooth constriction; (c) at a notch; (d) at a step. c and d demonstrate stress concentrations. (After Gordon 1988.)

centrations at all scales from molecules to those of cliffs.

Strain energy is diffused throughout a uniform body of material which has no cracks or variations in form. If a tensile stress (σ_t) is applied to that body, the body will contain $\sigma_t^2/2E$ units of energy per unit volume. If the body is held rigidly by surrounding bodies no mechanical energy can pass into or out of the system. However, if a crack is introduced into the system the material on

either side of the crack will relax, that is, it will release much of its stored strain energy and this energy becomes available to operate the fracture mechanism and cause the crack to spread.

The total energy needed to propagate a crack in rock may be very large because, not only do the atoms and molecules in the plane of fracture have to be separated, the inter-atomic bonds of the material a few millimetres (a thickness of half a million atoms) on either side of the crack may also be broken.

Fractures in rock

Modern study of crack propagation began with the work of A. A. Griffith (1893–1963) who studied fracturing of glass. Rocks are far more anisotropic than glass, but the mechanisms observed in glass can be recognized in some fractures in rocks.

Stress systems in nature can be resolved into three stress directions σ_1, σ_2, σ_3 (Fig. 4.8a), for the sake of simplicity the following discussion will assume that $\sigma_2 = \sigma_3$ and that we are therefore dealing with a two dimensional problem involving only σ_1 and σ_3. The forms of stress in the crust fall into four classes.

(1) Pure tensile stress results in pulling apart of rock bodies with the formation of joints in planes perpendicular to the principal (largest) stress, σ_1 (Fig. 4.8b).

(2) A state of compression in one direction and tension in the other may result in tensile failure despite the presence of the compressive stress—many geological processes, such as stress relief as overburdens are eroded, involve such stress directions (Fig. 4.8c).

(3) Under pure compression tensile cracks can also develop and align themselves with the direction of the major principal stress. Such cracks are often called wing cracks or Griffith's cracks (Fig. 4.8e).

(4) Shear failure may result from the cases of 2 and 3 above (Fig. 4.8d).

FIG. 4.8. In all parts of this figure, relative stress magnitudes are indicated by lengths of stems of arrows and crosses. (a) A three-dimensional stress system. (b) Development of a joint in a block under pure tension. (c) Development of a joint parallel to the principal stress, σ_1, under a σ_3 of low magnitude. (d) Shear failure occurring under high magnitude compressive (above) and tensile (below) stresses with low magnitude minor stresses. (e) Development of wing, or Griffith's, cracks by tensile failure in a far-field compressive stress regime (after Einstein and Dershowitz 1990). (f) The stress directions and magnitudes around a crack (after Hudson and Cooling 1988). (g) Mechanisms of fracture propagation caused by displacement on a shear plane (after Hoek 1968). (h) Crack initiation mechanisms observed in granular marble (after Olsson and Peng 1976). (i) Development of a major shear fracture by linking of extended cracks (after Johnson 1970).

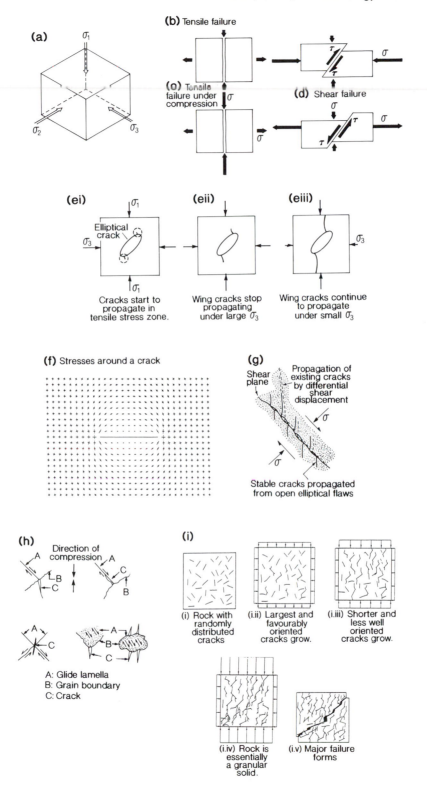

(a)

(b) Tensile failure

(o) Tensile failure under compression

(d) Shear failure

(ei) Cracks start to propagate in tensile stress zone.

(eii) Wing cracks stop propagating under large σ_3

(eiii) Wing cracks continue to propagate under small σ_3

Elliptical crack

(f) Stresses around a crack

(g) Propagation of existing cracks by differential shear displacement

Shear plane

Stable cracks propagated from open elliptical flaws

(h) Direction of compression

A: Glide lamella
B: Grain boundary
C: Crack

(i)

(i) Rock with randomly distributed cracks

(i.ii) Largest and favourably oriented cracks grow.

(i.iii) Shorter and less well oriented cracks grow.

(i.iv) Rock is essentially a granular solid.

(i.v) Major failure forms

The development of wing cracks can be envisaged by considering the stresses around an elliptical opening (Hoek 1968; Einstein and Dershowitz 1990). Stresses which are overall (i.e. in the far-field) compressive, are tensile where they are tangential to the tip of the ellipse and at such sites a wing crack may form. If the minor stress σ_3 is close in magnitude to the stress level of the principal stress σ_1 the crack will be held closed and not extend. If σ_3 has a low magnitude it will not prevent crack propagation and the wing crack will extend in the direction of the major stress axis, σ_1, and create joints (Fig. 4.8e).

Once a crack has formed it may be propagated by its own effect on the surrounding stress field (Fig. 4.8f) as stresses become concentrated at crack tips, and even where the far-field stresses are equal in magnitude ($\sigma_1 = \sigma_3$) the local stresses may exceed the tensile strength of the rock.

Crack propagation may stop at the site of a change in rock properties such as another crack, an inclusion of a plastic material such as clay, cementing material, or a strong crystal. Another possibility for crack formation is the pressure of one grain or crystal on another (Fig. 4.8g). A crystal may shear against its neighbours and then transmit a compressive stress to either a grain aligned in the stress direction, or fail itself. Internal shearing in a grain or crystal may also result in crack development with cracks always tending to propagate in the axial plane of the maximum stress.

Shearing within a rock body with confining (compressive stresses) large enough to cause 'tearing' of the sliding rock may also be responsible for opening cracks at large angles with the shear plane (Fig. 4.8h).

Whatever the influence of anisotropies in the rock and different magnitudes of stresses, it is the linking of Griffith's cracks which is the major cause of joint formation in a compressive or tensile stress system (Fig. 4.8i). The cracks may link to form a shear fracture or an axial fracture depending on the tangential stress to pre-formed cracks in the rock.

The importance of natural planes of separation in rocks as controls on the orientation of propagating cracks is illustrated by consideration of crystal units of the mineral muscovite. This form of mica has a tensile strength in the plane of its structural units, or laminae, of about $3000\,MN/m^2$, but the laminae can be pulled apart, in a direction across the planes, by finger pressure of a few N/m^2.

Rocks, such as schist, with high mica concentrations arranged in oriented fabrics have, therefore, large differences in strength along and across the planes of schistosity, and the micas also act as planes of weakness in shear parallel to the laminae of the mineral crystals.

The strain energy needed for propagation of cracks is dependent upon the strength of the crystals, grains, and cementing materials between them, as well as the void ratio of the material. Strong, dense rocks suffer small elastic strains under stress; in comparison, weaker rocks suffer greater elastic strains. Strong rocks therefore store less strain energy than the weaker ones, and for the same stress strong rocks have fewer cracks which link to form fewer joints. Very weak rocks, such as some mudrocks, and soils have higher void ratios than denser, stronger rocks, as well as a greater density of short partings, within their fabrics, which limit the propagation of fractures. In addition, clay-rich rocks and soils yield plastically and this further limits fracture propagation. In general, then, continuous joint systems do not develop in very weak rocks and soils such as marine clays, weakly cemented sandstones, unwelded ignimbrites, and Quaternary terrestrial sediments. The structural units and cracks in such materials are usually the result of hydration and dehydration causing swelling and shrinking, and are essentially superficial with the partings having limited dimensions and continuity.

Stresses in the upper crust

Four major sources of stress may be active in rock bodies at shallow depths: residual, gravitational, tectonic, and thermal.

Residual stresses

Residual stresses are induced in a rock body by plastic deformation resulting from original crystallization, overburden weight, thermal gradients, tectonic pressures, or other processes which create a meta-stable stress condition in the absence of external stresses (Haxby and Turcotte 1976). Removal of overburden or confining stresses disturbs the meta-stable state. Rock near the ground surface will contain residual stresses which may then be relieved by strains which may cause fracture propagation, buckling, slab failure, and other forms of instability.

Gravitational stresses

Gravitational stresses at a point within a rock body are induced by the weight of the column of rock above that point: thus the total vertical stress (σ_z), which is also known as the lithostatic stress, is given by:

$$\sigma_z = \gamma z, \tag{15}$$

where γ is the rock unit weight (units of N/m^3), and z is the depth (in metres) of the point below the surface of the ground. Over a wide range of measurements this relationship has been found to be approximately true (Fig. 4.9a). Scatter in the data is caused by insensitivity in the measuring techniques and by proximity to major geological and topographic features.

In porous or fissured rock containing water, the relationship should be expressed in terms of effective geostatic stresses (σ_z') which take into account the thickness of the body of rock saturated by water and the pressure of water in the column of rock being studied:

$$\sigma_z' = \gamma z_1 + \gamma_{sub} z_2 + \gamma_w h_w, \tag{16}$$

where: γ_{sub} is the submerged unit weight of rock equal to $\gamma - \gamma_w$,

γ_w is the unit weight of water,

h_w is the depth of water below the phreatic surface, or water table (for saturated rock $z_2 = h_w$),

z_1 and z_2 are the thicknesses of dry and saturated rock bodies respectively.

Gravitationally induced vertical stresses are related to horizontal stresses because rocks tend to expand horizontally in response to a vertical load, this is, in directions perpendicular or transverse to the applied compressive stress. The transverse expansion is described by Poisson's ratio (v).

If a rock is not free to expand transversely a transverse stress (σ_d) is created. The magnitude of this transverse stress is related to the coefficient of lateral earth pressure or coefficient of geostatic stress (K_o) and to σ_z:

$$K_o = \frac{v}{1 - v}, \tag{17}$$

and

$$\sigma_d = \frac{v}{1 - v} \sigma_z', \quad \text{or} \quad \sigma_d = K_o \sigma_z'. \tag{18}$$

This transverse component (σ_d) of the total horizontal stress (σ_h) has a value of about one-third of

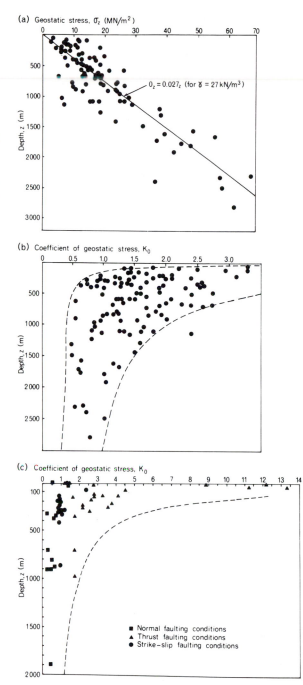

FIG. 4.9. (a) Relationship between total geostatic stress and depth. Data from many sites. (b) Relationship between coefficient of geostatic stress and depth. (c) Effect of tectonic processes on the coefficient of geostatic stress. (a, b, after Brown and Hoek 1978; c, after Farmer 1983, from an unpublished paper by Jamison and Cook.)

the vertical stress (σ_z), if the value of v is 0.25, which it is for many rocks.

Weak rocks are less effective at preventing transverse expansion than strong rocks, consequently they have higher Poisson's ratios than strong rocks, and the magnitude of σ_d is greater in weak rocks for any value of the vertical stress. Weak rocks therefore tend to fracture readily with closely spaced joints, when confining rock bodies are removed by erosion, and strong rocks fracture less readily with more widely spaced joints in response to removal of confining rock.

Tectonic stresses

Tectonic stresses result from earth movements and are especially important in areas of converging and diverging lithospheric plates. The compressive forces commonly active in zones of convergence induce stress fields in which horizontal stresses exceed vertical stresses. Very high horizontal stresses are common in zones of thrust faulting and give rise to high values for the coefficient of geostatic stress. Most K_o values are <4 and are rarely >7 (McGarr and Gay 1978) (Fig. 4.9b, c). Zones of normal faulting are associated with crustal extension and values of K_o are usually <1, if all horizontal stresses are relieved. Where strike-slip faulting is present, the stress field may be weakly compressional, but values of K_o approximate to unity. Field evidence for tectonic stresses exists not only in the presence of faulting but also of jointing.

In all stress fields the values of horizontal and vertical stresses become increasingly similar at depth and K_o approximates to unity. It has been suggested by Turcotte (1974) that only the upper 25 km of the lithosphere, at temperatures below 300 °C, behaves elastically. At greater depths or temperatures, stresses are relieved by plastic flow.

Thermal stresses

Thermal stresses result from the prevention of expansion or contraction of a solid during heating or cooling. In a homogeneous rock the relationship between strain (ε) and temperature change (ΔT) is:

$$\varepsilon = \alpha \, \Delta T \qquad (19)$$

where α is the linear coefficient of thermal expansion of the rock.

If the rock is not free to expand or contract,

stress is generated in the rock with a magnitude described by the elastic stress–strain relationship:

$$\varepsilon_d = \frac{1}{E}\left[\sigma_t - v(\sigma_t + \sigma) \right] + \alpha \, \Delta T \qquad (20)$$

where E is Young's modulus of elasticity.

If a confined rock ($\varepsilon_d = 0$) is cooled through a temperature interval ΔT, the induced thermal stress, σ_t, ignoring the applied longitudinal stress (σ_ℓ), is given by (Voight and St Pierre 1974):

$$\sigma_t = \frac{\alpha E \, \Delta T}{1 - v}. \qquad (21)$$

Consider, for example, a confined rock that cools 100 °C but may not contract ($\alpha = 10^{-6}/°C$, $E = 10^5$ MPa, $v = 0.25$). The induced tensile stress is then −13 MPa, which is similar in magnitude to the tensile strength of many rocks, and joints are likely to form.

Joints created by thermal contraction are most readily formed in heterogeneous rocks composed of minerals and grains with variable values for α, v, and E, and variable rates of cooling. Thermal stresses created at the scale of individual and interlocking crystals are major contributors to residual stresses.

Finite-element stress analyses

Analyses of the effects of stresses in the upper crust and in individual bodies of rock and soil making up hillslopes have been made possible by the use of a method first developed as a way of studying stresses set up in aircraft bodies (Turner *et al.* 1956). Known as the finite-element method, it has been applied to geomorphological problems at a range of scales from small cliffs and individual slopes (Yu and Coates 1970; Lee 1978; Kohlbeck *et al.* 1979; Valliapan and Evans 1980; McTigue and Mei 1981; Savage *et al.* 1985), to isolated mountain peaks (Augustinus and Selby 1990) and mountain massifs (Sturgul *et al.* 1976).

The finite-element method uses the assumption that the response of a body to loading can be approximated by considering a cross-section of the body to be made up of deformable elements, taken as either triangular or quadrilateral in shape, each connected at the corners or nodes. The amount of deformation will depend on the amount of load and how close the element is to the applied load.

The chosen elements must be small enough for the straight lines which make up the sides of the elements to remain straight after deformation. The calculation involves determining the nodal displacement resulting from the total loading of the system, then the strains induced and hence the stresses within each element (Zienkiewicz 1971).

In a study of a rock mass, a true scale cross-section through it is drawn and the input data needed are the distribution of rock densities and the elastic parameters, v and E. Consequently, the stratigraphy and structure must be known for a realistic determination of the stresses. It is common practice to plot the data showing the resulting direction of action and magnitude of the maximum and minimum principal stresses. The length of each bar indicates magnitude of the stress: compressive stresses are indicated by plain bars or converging arrows; tensile stresses are indicated by diverging arrows (see Fig. 4.13 as an example). Predicted strains at an open face, such as a cliff may be displayed as predicted profiles after displacement by the yielding to stress (see Fig. 4.11).

Examples of finite-element analyses

(1) In a study of a 60 m high cliff behind a hydro-electric power station at Niagara Falls, Lee (1978) showed that the cliff was formed by recession of the Horseshoe Falls over a period of the order of

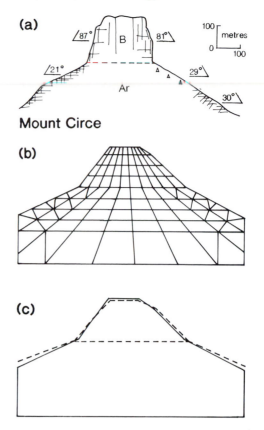

Mount Circe

FIG. 4.11. (a) A profile of Mt. Circe, Antarctica, showing slope angles: B—the Beacon Heights Orthoquartzite, and Ar—the Arena Sandstone. (b) The finite-element mesh, and (c) the calculated displacements (pecked line) compared with a simplified measured profile. (After Augustinus and Selby 1990.)

400 years. As a result, the rock of the cliff had become unconfined and had swelled progressively with tensile development of vertical joints behind the cliff face. The horizontally bedded sedimentary rocks of the cliff have a range of elastic properties and tensile strengths (Fig. 4.10), with the lowest strengths existing in the shales. Drilling horizontally into the rock has shown that vertical joints are very closely spaced near the cliff face (50–100 mm apart) and more widely spaced in the interior of the rock mass. Joint development has occurred up to 24 m into the rock. The opening of joints causes slabs of rock to separate from the face and the cliff to recede by successive rockfalls. Calculated strains resulting from residual stresses being relieved are in conformity with the processes observed in the field.

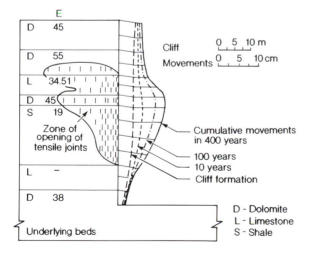

FIG. 4.10. A simplified vertical section of the cliff at Niagara showing cumulative rock movements since cliff formation and the zone of tension in the cliff. Elastic modulus values (E in GPa) are given for each rock unit. Values of Poisson's ratio were taken as 0.3 for all units. (After Lee 1978.)

PLATE 4.1. Mt. Circe, in the centre, with a talus, on the right, largely derived from fallen joint blocks freed by stress releases on the cliffs above.

(2) Augustinus and Selby (1990) studied the gravitational stresses and predicted strains in an isolated mountain peak formed on strong orthoquartzite. The finite-element mesh (Fig. 4.11) is based on a simplification of the natural relief, and the predicted strains indicate that there is a tendency for the peak to settle vertically as lateral extension occurs in the sandstone underlying the peak. It is assumed that the sandstone deforms plastically in the interior of the peak, but field observations also show that the base of the upper cliff suffers deformation by opening of tensile joints parallel to the cliff face. Such joints are the major cause of slope failure and recession (Plates 4.1, 4.2).

There is evidence from some sites of the Colorado Plateau, USA, that where massive sandstones overlie shales, the weight of the sandstone may cause lateral ductile deformation of the shale. This lateral spreading may, in turn, promote a tensile stress within the base of the sandstone and propagation upwards of joints. The large vertical joints in the sandstone may, consequently, be partly a result of tensile stresses developed at the contact of the sandstone and shale, and partly a result of stress relief expansion of the sandstone itself. The resulting joints control the development of the sandstone cliffs and the formation of rock spires at the margins of the buttes and mesas (Plate 4.3a, b).

(3) In their study of the Hochkönig massif, Austria, Sturgul *et al.* (1976) set out to test the hypothesis that the stresses in the massif rocks are due to gravitational loading and not to tectonic forces. Measured stress orientations were available from the Mitterberg copper mine situated in the Palaeozoic basement beneath the calcareous rocks of the massif (Fig. 4.12). The finite-element analysis confirmed that the measured and calculated stresses have a similar orientation and hence that

graben known as 'sackung', and small scarps aligned parallel with mountain ridges (e.g. Gerber and Scheidegger 1969; Savage and Varnes 1987).

The dominant vertical stress direction in the basement beneath the mountain is progressively transformed into a dominant horizontal stress pattern below the lowlands to the north and south of the massif. If deep and rapid incision by streams or ice were to occur in such localities then it is probable that the stresses imposed by the gravitational body forces of the massif would control joint opening and the details of valley landform development. This study shows also an example of the effect of a particular rock unit on stress patterns. The strata shown on the geological cross-section by parallel-line shading are of greywacke with a low value of Young's modulus and a Poisson's ratio of 0.34. These rocks have a tendency to swell transversely to a higher degree than those above and below them, and show this in their stress pattern.

(4) Studies by Yu and Coates (1970) and Stacey (1973) of stress induced by excavating open-pit mines with a range of depths, shapes, and volumes and with rocks with a range of physical properties, indicate the value of the finite-element model when used as a simulator of a large range of slope conditions. In Fig. 4.13, the effect of varying the coefficient of lateral earth pressure alone is shown to be very great. The dominant compressive stresses are nearly vertical in (a), although parallel to the hillslope; but in (b) are more nearly parallel to the ground surface at shallow depths, and are of large magnitude at the toe of the slope. In (c) it is shown that a tensile zone has formed near the face of the slope. The depth of this zone into the rock is proportional to the value of K_o. The stress pattern and magnitude has a major effect, through joint propagation and opening, on the stability of rock slopes, especially the development of slope-parallel joints (stress-release joints).

(5) Several studies have shown that stress-release joints develop parallel to the ground surface in the floors of quarries and open-pit mines and result in heaving of the surface rocks into small anticlines (e.g. Adams 1982). Similar joint development has been recognized beneath valley floors (Matheson and Thomson 1973; Ferguson and Hamel 1981). The opening of joints at the foot of valley-side slopes by relief of tension and up-arching of valley floors with subsequent joint opening may be of much greater importance as a

PLATE 4.2. Dominant joints on a cliff face are nearly vertical stress-release joints parallel to the cliff face. The horizontal bedding joints are tightly closed and have no influence on the stability of the cliff.

the stress pattern is that which would develop from gravitational loading alone.

The figure also indicates that below the mountain the largest, or compressive, stresses are aligned close to the vertical, but near the flanks of the massif they trend parallel to the surface of the ground. The significance of the latter observation is that where the stresses become tensional and of such magnitude that they exceed the tensile strength of the rock, then joints will open parallel with the rock face and give rise to slab failures. The figure also shows that tensile stresses are concentrated near the peaks of the mountain: tensile failures in such areas have been described in several papers as being responsible for opening of joints, the formation of elongated miniature

PLATE 4.3a, b. Mesas and buttes of Monument Valley, near the Arizona–Utah border. Detail of the joints, formed at the contact between the sandstone and underlying shales, is shown in the enlargement to the right. Some of the vertical joints have propagated from the contact.

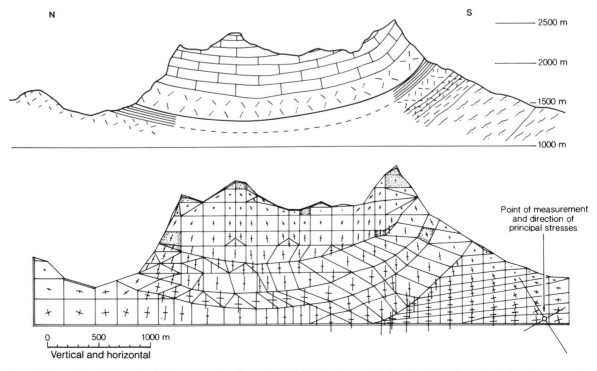

FIG. 4.12. A simplified geological cross-section through the Hochkönig massif, Austria (above), and a finite-element mesh (below). The calculated magnitudes and orientations of the principal stresses are shown by crosses. Zones of tension are stippled. (After Sturgul *et al.* 1976.)

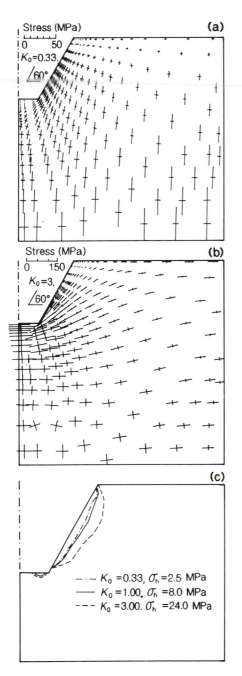

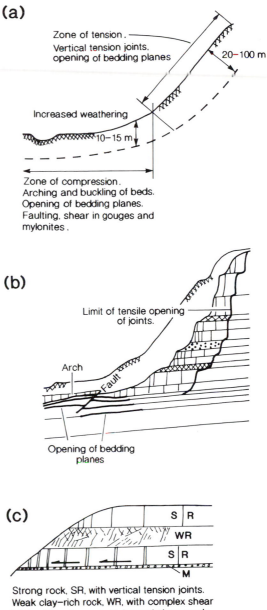

FIG. 4.13. (a, b) Direction and magnitude of principal stresses in an open pit with 60° slope and indicated values of the coefficient of geostatic stress. (c) Development of tensile zones around a 60° slope under compressive horizontal field stress. (After Yu and Coates 1970.)

FIG. 4.14. Schematic cross-sections of a valley; (a) zones of tension and compression with possible results; (b) zones of joint opening and propagation; (c) intensity and styles of joint formation. (Based, in part, on Ferguson and Hamel 1981.)

weathering process, and aid to river erosion, than has been commonly recognized.

Tension joints in strong, horizontally bedded sedimentary rocks are commonly nearly vertical; in weak rocks and clay-rich shales and mudrocks they are commonly curved and discontinuous, and show evidence of shearing (Ferguson and Hamel 1981). Joint spacing commonly increases into the slope and away from the crest of valley slopes.

Where stream incision has been rapid, up-warping of the valley sides, slope crests, and valley bottoms can occur, with the elastic rebound being up to 5 per cent of the depth of the valley (Matheson and Thomson 1973). The upwarping influences the dip of local rock units and is likely to be accompanied by the development of shear zones with weakened, brecciated, and finely crumbled rock called gouge, or mylonite (Fig. 4.14) (Cabrera 1986).

(6) Glacial troughs are usually regarded as resulting from the erosive action of the glaciers occupying them. It has been postulated by Gerber (1980) that many features of mountains own their form to stress patterns in their rocks, rather than to the glaciers, which are primarily removers of rock pre-fractured by tensile failure and stress-release. In studies of mountain valleys in New Zealand, Augustinus (1988) has demonstrated by finite-element modelling that tensile failures of jointed rock are common at the foot of trough walls, and has supported similar conclusions in respect of Norwegian fjords (Bjerrum and Jorstad 1968; Myrvang and Grimstad 1984). The overall form of troughs is, by this hypothesis, due primarily to stress-release jointing at the bases of trough walls and valley floors and to the strength of the rock of the upper valley walls.

5

Strength of Earth Materials

Definitions

Strength

The term 'strength' is used in three senses in the earth sciences: it may be used for the ability of material to resist deformation by compressive, tensile, or shear stresses; (2) it may be used for the ability of a rock or soil to resist abrasion; and (3) it may be used to indicate the resistance of loose or unconsolidated mineral grains to being transported by a fluid. In this chapter strength is being used in the first of these senses.

Rock and soil

Earth materials forming hillslopes are rocks and soils, but there is not a universally accepted definition of what is meant by the terms 'rock' and 'soil', definition is dependent upon the interest of the user. To most engineers, rock is a hard, elastic substance not significantly weakened by immersion in water. For a geologist rock is that material which is below the depth of modern weathering and showing some degree of strength due to burial, infilling of pores, or crystallization of minerals. To the geologist it is the process of formation and the composition of the rock that is the focus of interest rather than the strength or behaviour. To an engineer soils are naturally occurring loose or soft deposits which either disintegrate or soften on immersion in water. To a soil scientist, soils are a complex of mineral and organic material formed at the surface of the Earth in response to physical, chemical, and biological processes acting on mineral and organic materials to produce distinct profiles.

Difficulty arises in the use of all of the above definitions because there is a group of materials which have some of the properties commonly regarded as characteristic of both rocks and soils.

In the following discussion, therefore, distinctions will be made in four categories: (1) rock, (2) weak rock, (3) soil, and (4) the solum.

Rock is a material composed of mineral grains and crystals tightly bound together by cements and interlocking of crystals; it is a hard, elastic substance which does not significantly soften on immersion in water; in a mass it is a discontinuous material with joints and other partings which divide the mass into discrete blocks; it is the partings which largely control the resistance of a rock mass, to the forces acting upon it, and not the strength of grains, crystals, or cementing materials holding blocks together.

Weak rock (also called 'soft') is distinct from 'rock' in that its strength under compression, tension, and shear is greatly reduced by immersion in water (up to 60 per cent reduction for some soft rock containing clay), is likely to swell after immersion, and, most importantly, soft rock does not develop continuous systems of joints, although it may display close fissuring on exposure to weathering processes. The major differences between rock and soft rock therefore relate to jointing and softening on immersion in water, not to strength of a small specimen. The distinction between soil and soft rock is not easily made, is arbitrary and has no universally accepted definition. The simplest distinction is that of strength, as in Table 5.3 where a strength in compression of 1 MPa has been adopted as the boundary.

Soil is a naturally occurring loose or soft deposit formed at the surface of the Earth; it is weakened or softened by immersion in water; it may be the result of physical, chemical, and biological processes acting to produce an organic-rich material (or solum) with distinctive horizons (layers) at shallow depths; it may be formed from weathering of harder rock or older soils; it may be formed on the site in which it currently occurs (*in situ*); it may be of transported material; it may be deposited as a weak geological formation—but, whatever its origin, a soil mass is essentially a particulate material with properties controlled by particle size, shape, grading, contacts, chemical bonds between particles and aggregates, void spaces, and the fluid in the voids. It is usually a continuous body of material with few extensive joints or fissures, although it may contain structural units and fabric in the solum.

These definitions of soil incorporate unconsolidated, young sedimentary bodies such as loess, alluvium, glacial deposits and non-welded pyroclastics as well as regolith and pedogenetic soils. Weak (soft) rock includes Tertiary mudrocks; sandstones, weak chalk, and some ignimbrites which have undergone only moderate depth of burial, (usually less than 2 km), induration, and compaction. The reason for classifying weak rock as belonging to a separate and intermediate category is that it has a range of behaviours which are different from those of jointed rock and different from soil as a result of consolidation and fabric. Weak rocks therefore need to be examined by the methods of both rock and soil mechanics and their behaviour must be recognized as liable to change, over time-spans which are shorter than a human lifetime, where they are exposed or close to the ground surface. For some purposes it is not necessary to treat weak rock as a separate category (Cripps and Taylor 1981; Taylor and Spears 1981).

The *solum* consists of the A and B horizons of soils developed by pedogenetic processes.

Problems of strength measurements

Rock and soil are highly variable materials which may be classified in many ways and for many purposes: a material which is strong enough for one purpose may be too weak for a different purpose. Furthermore strength is not an absolute concept; measurements under compression, tension, and shear give different indications of strength as do measurements made in the field when compared with those made in the laboratory. Even amongst laboratory tests measuring one form of strength, test results may be significantly different for four main reasons.

(1) Laboratory tests require selection of a specimen which is representative of the whole body of material, but representativeness may not be attainable; the size of the specimen, and par-ticularly whether it includes the natural fabric together with voids and weaknesses, has a major effect upon strength values. In general, the larger the specimen the more likely it is to be representative of the whole body, but the more difficult it is to collect. (2) Methods of specimen collection may affect specimen strength, especially of weak soils, by either compressing the specimen, or allowing it to expand to a volume which is not that of the natural state. (3) Storage of a specimen may affect its water-content and volume, and permit chemical and biological activity within it. (4) Testing of the material may be done at various levels of confining stress, rate of stress application, uniformity of stress across the specimen, water-content, and degree of control on the loss or gain of water by the specimen.

It may be said, with some justification, that strength values can be as much a function of the measurement process as a property of the material. In order to obtain consistent and reliable information, therefore, it is essential that standard procedures should be used wherever possible. Many countries publish codes of practice for measurement of soil properties, for example, British Standards Institution, American Society for Testing and Materials. Sections of the codes are brought up-to-date by amendments which are published from time to time. For rock testing, the International Society for Rock Mechanics has produced a manual describing standard testing procedures (Brown 1981) and additions and amendments are published in *International Journal of Rock Mechanics and Mining Sciences and Geomechanics Abstracts*. Because of the existence of these works of reference the discussion of testing in this book will be confined to the principles underlying concepts of strength, the testing methods, and their application to hillslope materials. More detailed studies of testing methods for soils are available in Vickers (1983) and Head (1980–6; 1989). Comprehensive treatments of theory are available in Derski *et al.* (1989), and Bell (1987).

Shear strength parameters

Earth scientists are usually concerned with the shear strength of soil and soft rock, rather than its resistance to failure in tension and compression. The shearing resistance depends upon many factors. A complete equation might take the form:

$$\tau_f = f(e, \phi, C, \sigma', c', H, T, \varepsilon, \dot{\varepsilon}, S \ldots),$$

in which: τ_f is the shearing resistance
      e is the void ratio
      ϕ is the frictional property of the material
      C is the composition
      σ' is the effective normal load holding materials in contact
      c' is the effective cohesion of the material
      H is the stress history
      T is the temperature
      ε is the strain
      $\dot{\varepsilon}$ is the strain rate
      S is the structure of the material.

Many of these components are not independent and many cannot be evaluated quantitatively. Consequently, the two components which can be evaluated—cohesion and friction—are measured in conditions in which water-contents, loads, rates of loading, and confining pressures can be controlled.

Cohesive forces are derived largely from electrostatic bonds, cementing materials and water. Frictional effects are derived from the resistance of particles to sliding past each other; resistance to grain crushing; resistance to rearrangement of grains; and resistance to volume change.

Cohesion

Cohesion is caused by bonding, and not by compressive forces holding particles together. All of the bonding processes, except metallic bonding, discussed in Chapter 2 may contribute to cohesive strength. Chemical bonds are of most importance in rocks and weak rocks; in soils they are of particular importance in cementing nodules and indurated deposits, as in duricrusts. Electrostatic forces are restricted in their operation to clays; the apparent cohesion of capillary forces is of most importance in the granular materials of silts and sands. The types of bonds and their strength are given in Table 5.1.

Chemical bonds are intermolecular forces set up inside mineral particles and between crystals of crystalline rocks and weak rocks with rigid cementing materials. These bonds provide a strength approaching that of mineral grains. All originate from the electrical interaction of atoms and exist as ionic, covalent, and hydrogen bonds. Precipitates such as calcium carbonate, silica, alumina, iron

TABLE 5.1. *Strength of bonds in soils*

Type of bond	Strength of the soil system (kN/m^2)
Chemical, intermolecular ionic, covalent, and hydrogen	10^4–10^5
van der Waals, interaction of polar molecules	<10
Ionic electrostatic, interaction between clay-charged surfaces and cations	<1000
Electrostatic (Coulomb), forces of attraction and repulsion of charged surfaces of particles	1–10
Magnetic forces of ferromagnesian minerals	0.1–1
Capillary, apparent cohesion from surface tension in water films	<400

Source: Vyalov (1986).

oxides, and organic compounds which may fill soil and rock voids are some of the common cementing materials.

The contribution of cementing material to cohesive strength has been studied by highway engineers who are concerned to increase the strength of road foundations. The addition of lime (calcium hydroxide or calcium oxide) is a common practice for such purposes. It has been shown by Lees *et al.* (1982) that the addition of 4 per cent lime to a soil, with Ca-montmorillonite as the dominant clay mineral, increases cohesion by $240 \, kN/m^2$ and by $53 \, kN/m^2$ for a kaolinitic soil. Studies by Dibble (pers. comm.) have demonstrated increases of $10 \, kN/m^2$ in allophanic soils. Major strength gains are achieved in seven days following the addition of lime. The addition of Ca^{2+} to Na-montmorillonite soils is widely recognized as being a useful way of reducing swelling pressure by at least 10 per cent, reducing adsorbed water-content and increasing aggregation. Lime treatment is used for stabilizing soils even though it does not increase the strength of all clays. The smectite clays have a strength order related to the dominant species of adsorbed ion, Fe-clay > Na-clay > Ca-clay (Brown 1977); Na-smectites are therefore not strengthened by lime.

That the gains in cohesion mentioned above are

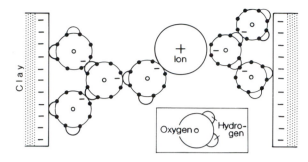

FIG. 5.1. Cohesion between clay particles induced by oriented water molecules and a positively charged ion.

Electrostatic forces between the charged surface of a clay particle and ions in the water around it operate over short ranges of less than 1 nm. The attractions and repulsions between the charged surfaces of clay particles are much weaker than those forces operating between clay surfaces and ions. This emphasizes the importance of ions as links between particles (Fig. 5.1). As the concentration of ions in adsorbed water increases so does the strength of bonding. Ionic attractive forces are therefore at a maximum in dry soils.

Electrostatic bonding has been reported as contributing about 80 per cent of the shear strength of some montmorillonites, 40–50 per cent of that of illites, and <20 per cent of that of kaolinite. For a given water-content at less than saturation the order of strength is montmorillonite > illite > kaolinite (Fig. 5.2) (Brown 1977).

due to cementing materials, rather than cation exchange or saturation of clay surfaces by Ca^{2+}, is indicated by accompanying increases in friction angle of 10° for smectites and kaolinite and 2° for allophane. The nature of the reactions which cause cementing materials to form are not clear. It has been proposed that calcium silicate hydroxide bonds develop between particles and also that tetra-calcium aluminates may be formed (see Brown 1977).

Van der Waals forces are relatively weak but can operate over much greater ranges (1 to several hundred nm) than chemical bonds (0.05 to 0.35 nm).

In pure water the strength of clay soils would be much weaker than it is in the natural state where all water contains some ions. The composition of drainage-water and pore fluids is therefore of great interest to scientists involved in changing soil strength, swelling properties, structure and fertility (see examples in Sridharan and Venkatappa Rao 1979; and Moore 1991).

Magnetic forces are produced by such soil minerals as hematite, goethite, and other metallic

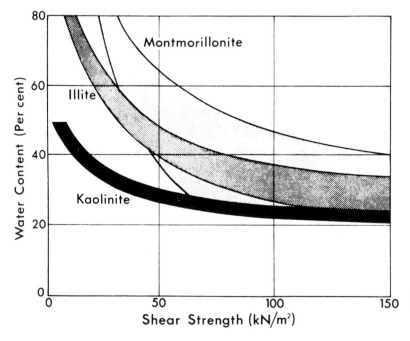

FIG. 5.2. The relationship between shear strength and water-content for selected silicate clay minerals.

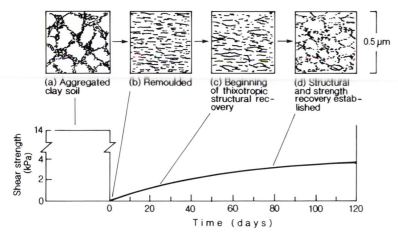

(a) Aggregated clay soil

(b) Remoulded

(c) Beginning of thixotropic structural recovery

(d) Structural and strength recovery established

0.5 μm

FIG. 5.3. Thixotropic recovery of undrained shear strength of a clay soil after remoulding, together with diagrammatic representations of soil fabrics. (Based on Pusch 1979.)

oxides. They occur on particles as surface films 0.05 to 0.5 μm in thickness. The magnitude of the force is low and is probably effective as contributing to aggregation only at the stage of sedimentation. It may be of local importance in some wind-deposited dusts and loess.

Examples of cohesive effects

(1) The significance of cohesion is demonstrated in some marine clays which have a highly structured fabric of clay aggregates. If they are severely sheared (i.e. remoulded) their strength may be reduced to 1/100th or even 1/1000th of its original value (Pusch 1979); such clays are said to be 'quick'. In the remoulded state the aggregates are broken and the particles are dispersed in water. If the suspension is left to stand at constant temperature, volume, and water content, the structure will gradually be recovered and some of the shear strength regained over several months. Recovery has been reported as continuing for over four years (Brown 1977). The energy for thixotropic recovery is entirely electrostatic (Fig. 5.3). It involves the approach of particles with adsorbed water films and the establishment of equilibrium positions in which inter-particle distances are determined by the properties of particle surfaces, and concentrations and species of ions in the water.

(2) The loss of strength under shear has a parallel under compressive load where a thick overburden (>2 km) of marine sediment may squeeze marine clays into a state of greater density and parallel alignment with loss of strong edge–face and edge–edge bonds in aggregates, and replacement by repulsive face–face contacts.

Extreme consolidation also causes squeezing out of some or all of the adsorbed water. Shear strength tests of heavily consolidated marine clays indicate that they commonly have little or no cohesive strength and are therefore frictional materials.

After uplift and erosional removal of the overburden, clay fabrics are not recovered, but stress-release microcracks form. Exposure of the mudrock permits opening of the cracks, swelling of the rock, rotation of the clay blocks, which behave as non-cohesive individual units, and the formation of clay bridges between these units. The clay bridges then provide some cohesive strength to the mass. Continued weathering breaks down the blocks and creates a soil in which silt grains are separated in a matrix of clay aggregates (Fig. 5.4).

(3) Rapidly developed thixotropic behaviour has been recognized in clay pastes subjected to shearing and vibration. Strength is lost immediately upon initial disturbance and regained almost immediately (Osipov *et al.* 1984). The significance of this observation is that disturbed natural soils with high water-contents may be rapidly transformed into weak soils which will fail by landsliding or flowing, but on coming to rest the material will rapidly recover strength.

The addition of water to a remoulded clay-rich soil reduces strength because the strength of the electrostatic cohesive forces is inversely proportional to the square of the distance of separation. Clays in a slurry or mudflow lose some of their strength because of disruption of aggregate structures, and some is lost because of the greater separation distances between particles as water is added; the suction effect of apparent cohesion is also lost.

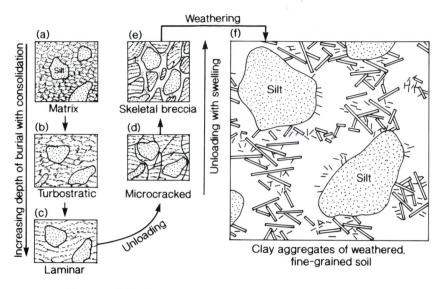

FIG. 5.4. Changes of fabric of a marine clay under an overburden and subsequent unloading. The fracture patterns and clay aggregates formed after unloading and swelling are based on the work of A. G. Beattie (1990); fabrics developed under loading are based, in part, on Huppert (1988).

Apparent cohesion

Apparent cohesion, as distinct from the cohesion of cementation and bonding, is produced by capillary stresses occurring as surface tension in water films between particles, and as interlocking of particles at a microscopic level as a result of surface roughness. Such apparent cohesion is affected by the size of rock and soil particles, their shape and mineralogy, the amount of water present, and the state of packing of particles. In Fig. 5.5 are shown soil particles with water between them. In unsaturated soils attractive forces result from the capillary stress of adsorbed water and the attractive force is inversely proportional to the radius of curvature of the water surface between the soil particles (r). The smaller the radius the greater is the capillary stress and the greater is the apparent cohesion. In a saturated soil, surface tension is completely eliminated and there is no strength available from apparent cohesion. Thus loose packing, large grain sizes, high moisture contents, and low wettability of particles all contribute to low cohesion. (See also Fig. 11.1.)

Apparent cohesion is also produced by mechanical interlocking of surfaces of particles (Fig. 5.6). In shear, microscopic protrusions will have to fail before a macroscopic shear surface can develop.

Friction

The basic control on the strength of soil and of most rocks is the frictional resistance between mineral particles in contact. Frictional strength is thus directly proportional to the normal stress holding grain surfaces in contact and it is influenced by the number of point contacts in a volume of rock or soil (there are probably about five million point contacts in $1\,cm^3$ of a fine sand), by the arrangement, size, shape, and resistance to crushing of grains, as well as by the voids and by dilatancy.

Where the packing of particles is open and particles are of uniform size the points of contact are relatively few and strength is low. Closer packing increases contact between grains, as does variability in grain sizes and more angular grain shapes (Fig. 3.17). Disturbance by shearing forces may require that closely packed grains have to ride up over other grains before shear failure can occur. This necessitates an initial decrease in volume before a new loosely packed state can develop (Fig. 5.7). The less the normal stress the greater is such dilatancy. In shear, therefore, frictional resistance in a soil may be initially high before a lower resistance to continued shear can develop. This sequence is usually described as high initial peak

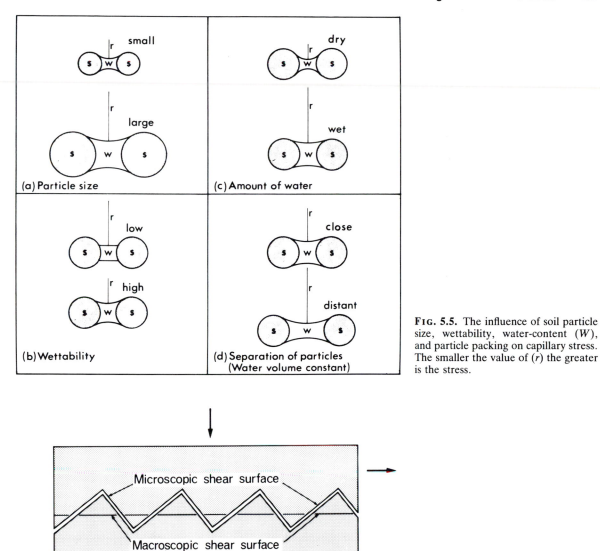

(a) Particle size

(b) Wettability

(c) Amount of water

(d) Separation of particles
(Water volume constant)

FIG. 5.5. The influence of soil particle size, wettability, water-content (W), and particle packing on capillary stress. The smaller the value of (r) the greater is the stress.

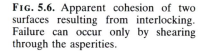

FIG. 5.6. Apparent cohesion of two surfaces resulting from interlocking. Failure can occur only by shearing through the asperities.

strength followed by a lower residual strength of the remoulded material (note that 'strength' is equal to 'resistance' at the moment of failure, and the two terms can sometimes be used interchangeably).

Even polished surfaces have a roughness which induces frictional resistance. When two surfaces are brought into contact they will be supported initially on the summits of the highest asperities. As the load increases beyond a certain value the asperities may fail by brittle fracture, first decreasing resistance to shear and then increasing the area of contact and so the frictional resistance, or they may deform plastically. For quartz the load to produce plastic deformation, or creep, is about $10\,\text{GN/m}^2$ and for diamond it is about $100\,\text{GN/m}^2$.

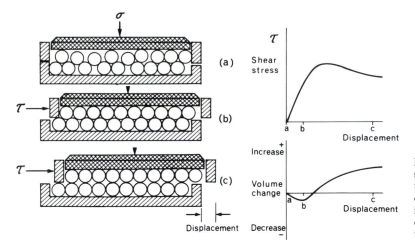

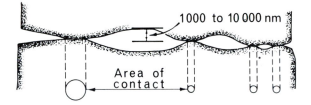

Fig. 5.7. During a shear box test a sample may initially consolidate and then exhibit dilatancy as grains rise over other grains during shearing. This effect is evident from the graphical plot of displacement against volume change in the sample during the shear test.

Creep in the materials at a surface of contact increases the area of contact with time, and thus increases resistance to sliding.

At low loads materials will deform elastically, but as loads rise so resistance increases. The behaviour in shear is also influenced by the relative hardness of the materials in contact, for a hard surface will have asperities which can plough into the surface of a softer material, cutting grooves into it, and thus increasing the area of contact (Scholz and Engelder 1976).

The roughness of a surface is the expression of a succession of asperities and depressions (Fig. 5.8). Highly polished metal surfaces may have a depth of depressions ranging from 1000 to 10 000 nm. Quartz surfaces of mirror smoothness have depressions 50 000 to 500 000 nm deep. The cleavage surfaces of mica may have depression depths of 100 to 10 000 nm. In soils and rocks, surfaces are seldom of mirror smoothness so frictional resistance to sliding is correspondingly high. The mineralogy of particles clearly influences surface roughness as does the history of weathering and transport (Horn and Deere 1962).

Friction angle

Where a block of rock lies upon a horizontal surface the weight of the block (N) generates an equal and opposite reaction (R). N and R together form a compressive stress normal to the plane of contact and the block is immobile (Fig. 5.9).

If a small horizontal stress H (insufficient to move the block) is applied to the rock the reaction R will no longer be normal to the plane of con-

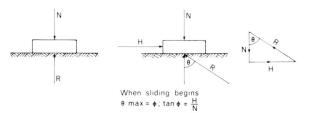

Fig. 5.8. Frictional contact, even along smooth surfaces, occurs at asperities which are of limited area.

Fig. 5.9. Derivation of the coefficient of sliding friction.

tact. It adjusts in magnitude and direction to equal the resultant N and H. The triangle of forces represents, in magnitude and direction, the relationships between N, H, R, and the angle θ.

The normal stress $N = R\cos\theta$, and the stress H cuts across the plane of contact so that $H = R\sin\theta$. If the shear stress H is increased until the block is just about to slide R will increase, and so will the angle θ. At the moment when sliding begins the frictional contact, holding the block and the surface stable along the plane of contact, will be broken and θ will have attained its maximum possible value, on that surface.

That maximum value is ϕ, the angle of plane static friction or angle of shearing resistance, and $\tan\phi = H/N$ and is the *coefficient of plane static friction*. The resultant stress acting along a plane is resolved into a stress, σ, acting normal to the plane of contact and a shear stress, τ, acting along the plane. If the plane is a plane of fracture, in a soil or rock material, movement along that plane in response to the shearing stress will be dependent upon the angle of internal friction of the material and

$$\tan\phi = \tau/\sigma.$$

The friction angle for pure quartz is 26 to 30° and for many clay particles it is close to 13°, but such values have little direct significance for determining the friction angles of soils and rocks because of the great heterogeneity of these materials. The volume of voids and particle sizes have a greater control on the friction angle of soils than does mineralogy (Fig. 5.10).

Water acts as an antilubricant when it is applied to the surfaces of minerals with massive crystal structures—such as quartz and calcite—but this effect is reduced as mineral surfaces become rougher and it is insignificant within soils. Water and clay particles separate large grains and behave like lubricants on the surfaces of platy minerals and reduce friction angles, but they also add apparent cohesion to a soil.

The friction angle, then, contains contributions from several sources; of these interparticle resistance to sliding probably accounts for at least half of the peak strength and much of the residual strength of most soils. The value of the friction angle decreases with increasing plasticity and water-content (see Chapter 7).

The Coulomb equation

The importance of cohesion and friction in determining the strength of materials was recognized and expressed, as early as 1776, by the French engineer Coulomb: the total resistance to shear or strength at failure (τ_f) is given by the normal stress (σ_n) multiplied by the coefficient of friction plus the cohesion (c) of the material or

$$\tau_f = c + \sigma_n.\tan\phi$$

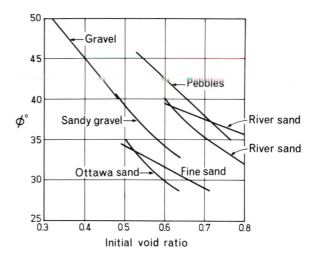

FIG. 5.10. The effect of void ratio on the angle of internal friction (ϕ) for various noncohesive materials. (After Lambe and Whitman 1979.)

for materials without cohesion the Coulomb equation is

$$\tau_f = \sigma_n.\tan\phi.$$

Water and shear strength

In a perfectly dry soil or rock there is no apparent cohesion caused by surface tension and the soil fabric is supported entirely on the point contacts of the constituent particles (Fig. 5.11). The voids in the soil are filled with air and the pressures in the voids are atmospheric; pore-water pressures are then zero. In a moist soil the particles have an apparent cohesion caused by surface tension and are effectively under a suction so pore-water pressures are said to be negative. In a fully saturated soil apparent cohesion is lost because there are no surface tension forces and part of the normal stress of the overburden is transferred from the soil fabric to the soil water. The transfer of load from soil to water is equivalent to a buoyancy or upthrust effect in which pore-water pressures are positive. The effective stress then equals the differences between the total ground stress transmitted through the interparticle contacts and the stress supported by the pore-water (Terzaghi 1936; Skempton 1960).

The value of the pore-water pressure is propor-

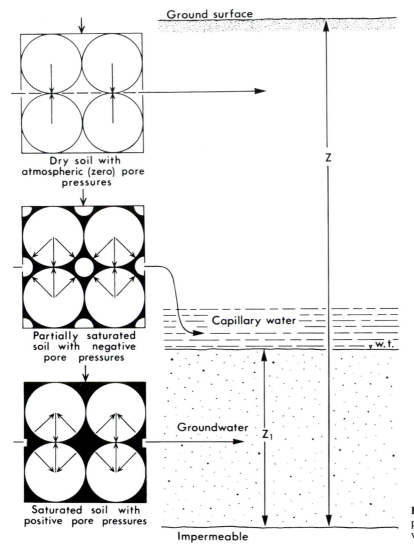

FIG. 5.11. Pore-water pressures depend upon the water-content of soil voids.

tional to the height of the free water column in the soil—known as the hydrostatic head when the water is at rest (i.e. static). This head is measured by inserting tubes, called piezometers, into the soil (Fig. 5.12), and for conditions in which lateral flow or seepage can occur is known as the piezometric head or pressure head.

A rise in pore-water pressure decreases soil strength because although confined water can withstand a compressive stress it cannot withstand a shear stress; further the buoyancy effect counteracts the normal stress in proportion to the pore-water pressure. Thus for saturated soils the Coulomb equation has to be written as:

$$\tau_f = c' + (\sigma_n - u) \cdot \tan\phi',$$

where τ_f is the shear strength at any point in the soil;

c' is the effective cohesion, as reduced by loss of surface tension;

σ_n is the normal stress imposed by the weight of solids and water above the point;

u is the pore-water pressure derived from the unit weight of water and the piezometric head $(\gamma_w \cdot z_1)$;

ϕ' is the angle of friction with respect to effective stresses.

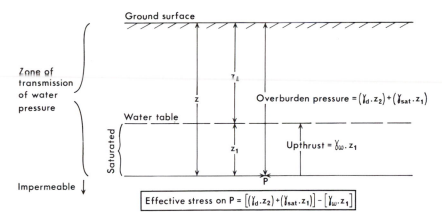

Fig. 5.12. Stresses in a soil resulting from the weight of the dry soil (γ_d) and of saturated soil (γ_{sat}) above point P, and from the buoyancy effect of pore water. If, for example, $\gamma_d = 12\,kN/m^3$; $\gamma_{sat} = 17\,kN/m^3$, $\gamma_w = 9.8\,kN/m^3$, $z = 10\,m$, then: if the soil is perfectly dry the total stress on $P = 120\,kN/m^2$, if it is fully saturated the effective stress is $72\,kN/m^2$, and if $z_1 = 6\,m$ then the effective stress on $P = 91.2\,kN/m^2$. The effect of pore water thus increases with depth as in freestanding water. Capillary water is held suspended above the water-table and thus does not alter the values of the pore pressures below the water-table. The effective stress, however, is increased by the weight of water held within the capillary zone.

Alternatively the equation may be written as:

$$\tau_f = c' + \sigma_n' . \tan\phi',$$

where σ_n' is the effective normal stress ($\sigma_n' = \sigma_n - u$). The symbols c and ϕ thus refer to total stresses and c' and ϕ' to effective stresses. Effective stresses are thus modified by pore-water pressure and conditions of loading or testing and are not fundamental properties of the material.

The above equation is usually related to soil but it can also apply to porous and permeable rock. Rock with low porosity, low permeability, and therefore low water content, will be less affected by saturation. Nearly all rocks which are not deeply buried, however, have numerous clefts in the form of joints, bedding planes, and other fissures. High water pressures in clefts can reduce effective stresses substantially so cleft water pressures are of great importance to rock strength.

Measurement of soil and weak rock strength

Laboratory tests

Most tests of rock and soil samples are carried out in laboratories where carefully controlled conditions permit repeated and comparable tests of many samples. For studies of both soil and rock it is usually important to collect cores or carefully cut specimens whose orientation in the field, with respect to shearing forces, is known.

Soils are not elastic materials, that is, they do not obey Hooke's law, consequently stresses are not easily related to strains by simple mathematical equations which can be used for prediction of behaviour. The behaviour of soils under stress is largely governed by the rate at which stresses are applied and whether the stress is increasing or decreasing. Other factors which influence strength include the size of specimens, their water-content, and the magnitude of the stress applied. In all tests, stress levels, water contents, orientation of the specimen, and rates of stress application should be as close as possible to the conditions occurring in the field. For example, if the test is for determination of strength of soil which may be involved in a landslide the specimen should be taken from the slide plane with that plane passing through the middle of the specimen; the moisture content should be appropriate to that assumed for the landslide, and the chosen normal stress (σ_n) should be close to that provided by the weight of the overburden (thus if the soil above the failure plane is 3.7 m thick (z) and has a unit weight (γ) of $17\,kN/m^3$ then the normal stresses used in testing will be, $\sigma_n = \gamma z$, and close to: $3.7 \times 17 \approx 63\,kN/m^2$ if the ground surface is horizontal. If the ground is sloping at $\beta°$ then the normal stress may be derived from $\sigma_n = \gamma z \cos^2\beta$ (kN/m²).

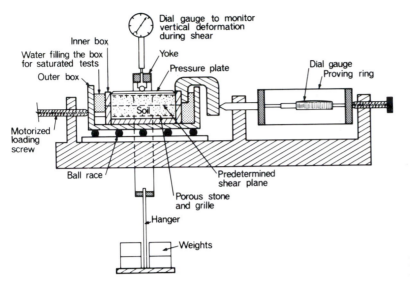

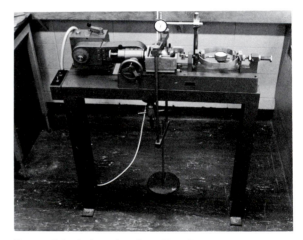

FIG. 5.13. Illustration of the principle of a shear box.

Soil shear box tests

A direct shear box is used to shear a specimen by imposing a shear strain along a plane of failure passing through the centre of the specimen. Values for c and ϕ, or c' and ϕ', are determined from a graph of the test results. It must be recognized that c and ϕ are merely parameters defining the gradient of the graphical line (ϕ) and the intercept (c) on the shear strength axis (Fig. 5.14). Neither c nor ϕ so determined is a physical property of a material. Values of c and ϕ vary according to the range of factors discussed above under the heading 'problems of strength measurements'.

A shear box for soils consists of a square box, without a top or a bottom, split horizontally at the level of the centre of the soil sample. The sample is held between metal grilles and porous stones (Fig. 5.13). A horizontal (or shearing) force is applied to the lower part of the box at a constant rate until the sample fails. The shear load at failure is divided by the cross-sectional area of the sample to give the shearing stress. For determining the shearing resistance under a normal stress a vertical load is applied to the sample by means of dead weight (Plate 5.1).

So that the shearing stress may be applied at a constant rate of strain the lower half of the shear box is mounted on rollers and is pushed forward at a uniform rate by a gear-operated piston. The upper half of the box bears against a steel proving ring, the deformation of which is shown on the dial

PLATE 5.1. A shear box for soil testing. The normal stress is applied by a load hanging below the box.

gauge and indicates the shearing stress. Change in volume of the soil sample during consolidation and shearing is measured by a dial gauge coupled to the normal load. Water escapes or enters the sample through the porous stones and the perforated metal grilles.

The horizontal stress is applied slowly, often at rates of about 0.6 mm per minute in a quick test, but at rates as low as 0.005 mm/minute if it is desired to prevent the buildup of pore-water pressures in the sample during testing. Failure

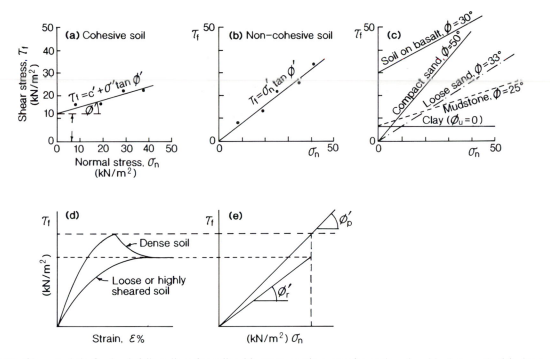

FIG. 5.14. Characteristic Coulomb failure lines for soils with stress–strain curves for peak and residual strengths (d). (*Note*: data are always plotted on arithmetic coordinates with the value being zero at the origin and intervals being of the same value along each axis, as in (a).)

of the sample is indicated by a sudden drop, or levelling off, of the proving-ring dial reading which records the reaction to a shear stress, that is, when the sample shears its resistance to shear stress may drop rapidly.

Tests are made for different values of normal stress and the results are plotted graphically with the shear stress on the vertical scale and the normal stress on the horizontal scale. The value of the effective angle of internal friction (ϕ') is determined from the slope of the line through the plotted points, and the value of effective cohesion (c') is the displacement of the line above the zero point (Fig. 5.14a, b).

For a non-cohesive soil the plotted line passes through the origin of the graph and the Coulomb equation thus becomes:

$$\tau_f = \sigma'_n \cdot \tan\phi'.$$

During a shear test many frictional materials exhibit dilatancy. Densely packed sands, silts, and over-consolidated clays exhibit peak resistance to shear (ϕ'_p) and then decline to a residual or ultimate strength (ϕ'_r). Loosely packed sand and

weathered clays by contrast can deform without the grains riding over one another and although peak strength is achieved there is no decrease in strength with further shearing (Fig. 5.14d, e). The behaviour of clays in shear box tests is also related to the form of the clay crystals with plate-like crystals being readily reoriented parallel to the shear plane (Plate 5.2).

The change in volume of a sample during testing and under different normal loads creates a necessity for communication of the state of packing for which a friction angle is being quoted. The state at which the volume of a soil ceases to change and the shear strength becomes constant is called the constant-volume condition or critical-state condition; they are indicated by the symbols: either ϕ'_{cv}, or ϕ'_{cs}. For sands under shear, a volume will be reached which is constant and equal for samples which were originally in loose or dense states. For clays which are not tabular in form, such as allophane and halloysite, the ϕ'_{cv} values are the same because they behave as fine granular soils. For the platy clay minerals such as the smectites and kaolinites, a constant-volume state, with

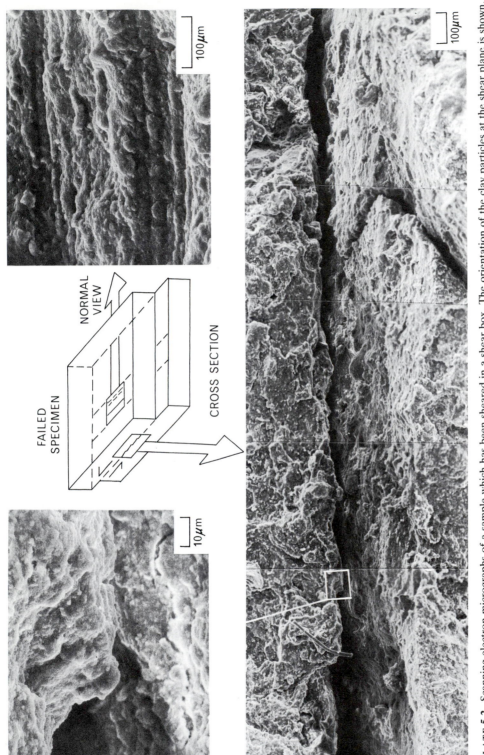

PLATE 5.2. Scanning-electron micrographs of a sample which has been sheared in a shear box. The orientation of the clay particles at the shear plane is shown. (Micrograph by N. W. Rogers.)

random or edge–face contacts dominating, is transient. Once the clay particles reorientate and form parallel alignments along a continuous shear surface the ϕ'_{cv} state is lost and progressive change to a ϕ'_r state occurs. Table 5.2 indicates the range of values for major clay minerals, or soils in which a single clay mineral represents more than 40 per cent of the soil volume.

The shear box test is simple in principle but is open to a number of objections:

(1) The size of shear box normally used for clays and sands is 60 mm square and the sample is 20 mm thick; a larger size, 300 mm square, may be used for gravelly soils. Many soils, however, contain large clasts, plant roots, and structural features which provide major controls on soil shear strength in field situations. For engineering purposes organic matter, soil structures, and large clasts are usually irrelevant as the surface soil is removed before construction takes place. For geomorphic purposes, however, the characteristics of superficial soil materials are of fundamental importance. Tests of small samples in laboratory shear boxes may thus fail to represent natural soil strength which can only be determined in the field using a large shear box and undisturbed soil samples (see below).

(2) Another objection is that the stress distribution across a shear box test sample is complex, and the value of the shearing resistance obtained by dividing the shearing force by the area is only approximate.

(3) For many studies the peak strength of soil is required. This is the case where a 'first-time' landslide is being studied. But where soils on slopes have been repeatedly subjected to landsliding, or shearing of soil, the soil has reached a strength close to that of the residual value. Residual strengths are often difficult to determine accurately, even when the shear box is reversed so that the upper part of the sheared sample travels back over the lower half. Reversal of the direction of shear does not, of course, simulate natural conditions so an attempt has been made to overcome this problem by devising a ring shear apparatus.

(4) Older models of direct shear boxes are not equipped to measure pore-water pressures in specimens during tests, but the use of transducers and electronic controls linked to computers has made it possible to design shear boxes which can control and record pore-water pressures. Many shear boxes without pore-water pressure controls have also been converted to electronic controls on stress application and recording, with the result that proving rings and displacement gauges have been replaced by load cells and transducers.

In a ring shear test an annular ring-shaped sample (Fig. 5.15) is subjected to a constant normal stress. The sample is confined laterally and horizontally by upper and lower confining rings. The lower half of the sample is carried on a rotating table driven by a worm gear. The upper half of the sample reacts via a torque arm against a pair of fixed proving rings that measure the shear force (Plate 5.3).

The two main advantages of the ring shear test are that (1) there is no change in the cross-section of the shear plane as the test proceeds, and (2) the sample can be sheared through an uninterrupted displacement of any magnitude (Bishop *et al.* 1971).

Triaxial compression tests

The triaxial apparatus is possibly the most widely used and most versatile means of measuring the shear strength of soils. A cylindrical specimen of soil is extruded from the sampling tube in which it was collected and placed in the apparatus (Plate 5.4, Fig. 5.16). The specimen is enclosed in a rubber membrane to retain its own moisture and to isolate it from the water which will surround the specimen during the test. The flat ends of the cylindrical specimen may be covered with porous plates to allow water to drain from the specimen during a test and to permit the pore-water pressure to be measured. The perspex cell is flooded with water. A measured pressure head is applied to the water, and the specimen is then enclosed by a cell pressure which may simulate the confining pressure of natural conditions. The cell pressure is the minor principal stress ($\sigma_2 = \sigma_3$). A vertical load is applied to the specimen at a constant rate of strain until the sample fails. The vertical load, together with the cell pressure, which also acts down on the top of the specimen gives the major principal stress (σ_1). The stress applied through the loading ram is ($\sigma_1 - \sigma_3$) and called the deviator stress or principal stress difference.

The major and minor principal stresses at the moment of failure of the specimen are recorded either by a proving ring and dial gauges or electronically.

TABLE 5.2. *Typical soil and rock properties*
(a)

Type and material	Unit weight (Saturated/dry) kN/m³	Friction angle (1) degrees	Cohesion kPa
COHESIONLESS			
Sand			
Loose sand, uniform grain size	19/14	28–34	
Dense sand, uniform grain size	21/17	32–40	
Loose sand, mixed grain size	20/16	34–40	
Dense sand, mixed grain size	21/18	38–46	
Gravel			
Gravel, uniform grain size	22/20	34–37	
Sand and gravel, mixed grain size	19/17	48–45	
Compacted broken rock			
Basalt	22/17	40–50	
Chalk	13/10	30–40	
Granite	20/17	45–50	
Limestone	19/16	35–40	
Sandstone	17/13	35–45	
Shale	20/16	30–35	
COHESIVE			
Clay			
Soft bentonite	13/6	7–13	10–20
Very soft organic clay	14/6	12–16	10–30
Soft, slightly organic clay	16/10	22–27	20–50
Soft glacial clay	17/12	27–32	30–70
Stiff glacial clay	20/17	30–32	70–150
Glacial till, mixed grain size	23/20	32–35	150–250
Rock			
Hard igneous rocks:	(2)		
granite, basalt, porphyry	25 to 30	35–45	35 000–55 000
Metamorphic rocks:			
quartzite, gneiss, slate	25 to 28	30–40	20 000–40 000
Hard sedimentary rocks:			
limestone, dolomite, sandstone	23 to 28	35–45	10 000–30 000
Soft sedimentary rock:			
sandstone, coal, chalk, shale	17 to 23	25–35	1 000–20 000

(b) *Frictional strength of selected clay minerals*

Clay minerals	Effective friction angle for soil at constant volume $(\phi'_{cv})°$	Effective residual friction angle $(\phi'_r)°$
Smectites	15–20	5–11
Kaolinites	22–30	12–18
Allophane	30–40	30–40
Halloysite	25–35	25–35

Notes: 1. Higher friction angles in cohesionless materials occur at low confining or normal stresses.
2. For intact rock, the unit weight of the material does not vary significantly between saturated and dry states with the exception of some materials such as porous sandstones.

Sources: Data from Hoek and Bray (1977); Wesley (1977); Lupini *et al.* (1981); Boyce (1985).

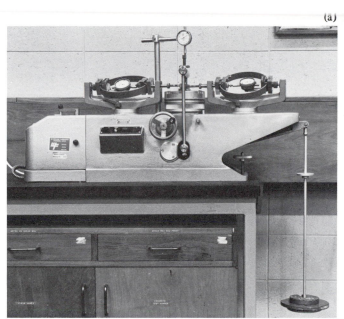

(a)

(b)

PLATE 5.3. (a) A ring shear apparatus; (b) the sample holder.

The principal stresses are plotted on the abscissa and a semicircle is drawn through them. The semicircle is part of a Mohr-circle construction. A minimum of three tests is carried out with different cell pressures for each test. A common tangent to the semicircles is extended to the ordinate, and is known as the Mohr–Coulomb failure line, or Mohr envelope (Fig. 5.17a, b) and should be the same as a line given by the Coulomb equation if triaxial and shear box tests are carried out under similar conditions of drainage. Drawing a common tangent can give rise to inaccuracies, especially if one of the test results gives a lower value than that implied by the common tangent. It is therefore better practice to plot the topmost point of each Mohr circle by erecting a vertical from the centre of the semicircle or plotting $(\sigma_1 + \sigma_3)/2$ as abscissa and $(\sigma_1 - \sigma_3)/2$ as ordinate and calculating the best fit line to the topmost points (Fig. 5.17d, e) this line will then have a slope of $\alpha°$. It can be shown that:

$$\tan\alpha = \sin\phi, \quad \text{or} \quad \phi = \sin^{-1}(\tan\alpha).$$

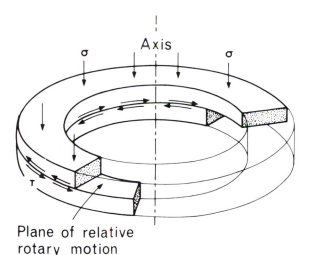

FIG. 5.15. The principle of ring shear. (After Bishop *et al.* 1971.)

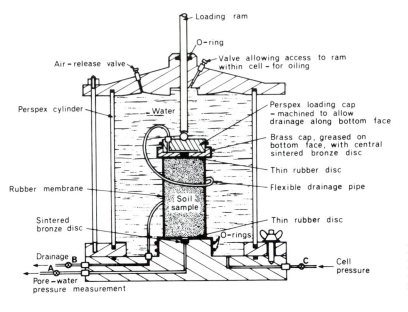

Fig. 5.16. A triaxial cell. The major principal stress (σ_1) is applied as a load and the minor stress (σ_3) through water in the cell.

Plate 5.4. A modern triaxial system (see Menzies 1988) which is computer-controlled (left) and operated through a bank of digital controllers (centre) which regulate and apply liquid pressure, and measure it in the cell (right).

The triaxial test may be carried out under different conditions of drainage of the specimen. The choice of test should be made after careful consideration of the field conditions which are to be simulated and analysed in the laboratory.

An *undrained test* does not allow the specimen to drain during the test. The test may therefore be carried out quickly as there is no need to provide time for drainage. The test is sometimes referred to as a 'quick test'. If the specimen is a saturated cohesive clay, water in it will force particles apart and prevent frictional contact. A saturated

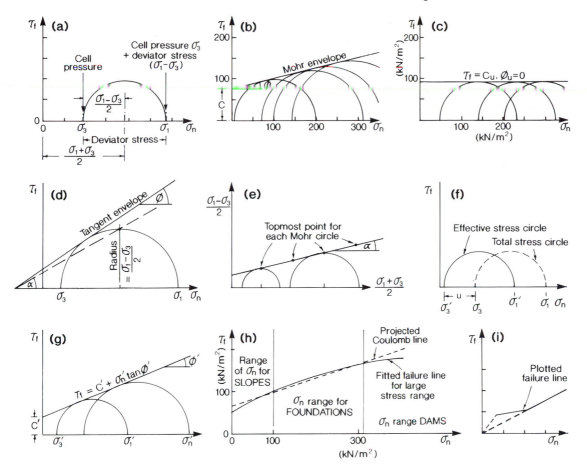

FIG. 5.17. Mohr-circle constructions.

undrained test should therefore give $\phi_u = 0$ (Fig. 5.17c). This is a result of the testing procedure, not a property of the soil. The undrained test is a total stress analysis and is used where short-term values for shear strength are required. In engineering practice it is used as the standard for bearing capacity for foundations because after initial loading the soil will consolidate and gain in shear strength. In studies of hillslope materials the test could be used to simulate conditions in a soil on to which a rockfall had suddenly descended. Such situations have been recognized as causing the loaded soil to fail and flow.

A *drained test* allows drainage of the sample during the test. The specimen therefore has porous plates on its top and bottom surfaces, the drainage valve is open, and the load is applied slowly to allow drainage to occur. This test is therefore sometimes referred to as a 'slow test'. The drained test is an effective stress analysis, as all pore pressures are allowed to dissipate. It is used where long-term values of shear strength are required. Drained tests are therefore relevant to many studies of hillslope stability where soils are granular and free draining or where slope failure is slow and progressive.

A *consolidated undrained test* allows drainage while the sample consolidates under cell pressure and is then sheared rapidly under conditions of no drainage. The pore-water pressure change is measured during the test and an effective stress analysis is used. The results of a consolidated undrained test with pore-pressure measurement are similar to those of a drained test, but the test is more rapid. The main value of the test is that it permits study of changes in pore-pressure and

PLATE 5.5. A cylindrical soil sample at its original size and after shear and barrel failures in triaxial tests.

drainage. It is particularly appropriate in the study of rapid landslides.

Effective stress analyses

It has been seen already that pore-water pressure in a soil produces a buoyancy effect which resists some part of the normal stress; it therefore reduces that stress and on a Mohr-circle plot σ_3' and σ_1' will plot closer to the origin than on a total stress plot. The distance of this displacement will be a measure of the pore-water pressure (u) (Fig. 5.17f, g).

In all types of conventional triaxial test the vertical ram load is increased until there is a clear and marked fall-off in resistance. This is usually indicated by continual axial strain with no increase in deviator stress. Failure will be evident in the specimen either by a shear displacement or plastic failure characterized by a barrel-like lateral expansion with no distinct fall-off in resistance; for plastic failures the strength is taken to be that required to produce an axial strain of 20 per cent. The value of the deviator stress used for drawing a Mohr circle is thus either the maximum deviator stress recorded, or the deviator stress at 20 per cent axial strain (Plate 5.5).

Appropriate parameter

An essential part of all strength testing is choice of the appropriate test, equally important is recording and communication of the results. The tests discussed above lead to the following expressions for strength:

c_u, ϕ_u from undrained tests;
c_{cu}, ϕ_{cu} from consolidated undrained tests

without pore-water pressure measurement;

c', ϕ' effective stresses from consolidated undrained tests with pore-water pressure measurements, or from drained tests;

c_d, ϕ_d from drained tests where volume change occurs during shear due to pore-water expulsion; this situation should be established as simulating field conditions if it is to be used;

c_r, ϕ_r from large-strain tests, i.e. residual strength tests in a ring shear apparatus, or reversed direct shear box.

Variation of c and ϕ

It has been assumed so far in this discussion that, although c and ϕ values are products of the testing process, they are invariable for all conditions of normal stress and density of the soil; neither of these assumptions is valid.

High levels of normal stress force soil particles closer together, increasing soil density and strength. The result is that Mohr–Coulomb failure lines, drawn for a large range of stress levels, are curved. The failure lines in the low stress region therefore indicate higher values for ϕ and lower values for c if failure lines are drawn as straight lines extending back to the shear failure axis (Fig. 5.17h). It is evident, then, that for all tests the normal stresses simulated must bracket those of the field condition which is being studied. Most hillslope investigations are concerned with relatively shallow overburdens and the stress levels are correspondingly low and the failure lines are straight for that range of stresses.

During shear tests it is commonly observed that for low values of normal stress soils such as loess, over-consolidated clays, and clay shales exhibit a characteristic curvature of the failure line (Fig. 5.17i). The initial failure, as stress is applied, is assumed to be purely frictional ($c' = 0$), but at slightly higher stress levels grains and aggregates are forced together and some apparent cohesion is derived from capillary water, and friction values (ϕ') may be reduced by pore-water pressure. At still higher stress levels the materials will behave as normally consolidated frictional materials in which capillary forces are negligible. The above comments should reinforce the warning that values derived from strength testing must be regarded as being greatly influenced by the testing process and not as fundamental properties of the soil.

Field measurements of soil strength

Where it is impossible, or difficult, to obtain soil samples for laboratory tests, or large numbers of readings are required, an indication of shear strength can be obtained by *in situ* testing using a shear vane.

A *shear vane* has four thin rectangular blades set at right angles to each other. Each blade is generally twice as deep as it is wide. The vane is pushed into the soil and rotated slowly (usually at about one revolution/two minutes). The size of the blades used depends upon the stiffness of the soil and the size of the particles within it. Large vanes are used for weak soils (Plate 5.6).

Most hand-held shear vanes have a dial gauge which records the resistance to a turning motion, directly in N/m^2.

Shear vanes may be used successfully only in fine-grained soils without roots, clasts, or strongly developed structures. Soil strength is closely controlled by soil moisture, so soil samples should be taken for determination of moisture content, or all measurements made at a nearly uniform moisture content such as field capacity (the state at which the soil has freely drained under gravity, but retained water in the micropores). As shear vanes indicate total stress they cannot be used to derive independently the values of c' and ϕ'.

Penetrometers have sometimes been used

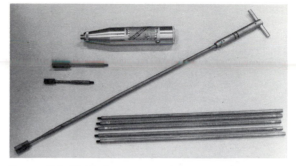

PLATE 5.6. Field shear vanes with extension rods. The torque is measured on the ring in the handle and different size vanes permit measurements of strength in soft to stiff soils. The bottle-shaped instrument is a Schmidt hammer.

to derive an indication of soil strength. A penetrometer indicates the force needed to push or drive a cone or cylinder of known dimensions into the soil. It is a useful indicator of bulk density and for engineering purposes, of bearing capacity, but it does not always measure shear strength reliably and it is at least as susceptible as the shear vane to roots, clasts, soil structure, and moisture variations.

A *field shear box* may be used to determine the strength of very large *in situ* soil samples, in which it is wished to preserve the natural fissures which result from soil structure, and for studying the contribution of plant roots to soil strength. In field shear tests a steel box, either open at the base and sides, or at the base only, is placed on a monolith of soil which is left undisturbed after the surrounding soil has been cut away. A normal load is placed on the box and a shear force is applied by a winch. The shear force is measured by a proving ring or spring balance and the displacement, as the shear force is applied, by a scale. The monolith eventually shears along its base and sides (Plate 5.7). This type of testing is particularly useful in studies of soil erosion by landsliding and for comparing the strength of soils under different kinds of vegetation (Chandler, Parker, and Selby 1981). Field tests in shallow residual soils give values of effective strength because of rapid soil drainage, through macropores.

A similar type of large field shear box can be used on jointed rock with the shear force being applied through hydraulic jacks, but as a single test on a large sample (1 m^3) may cost US $10 000, such testing is rarely undertaken.

PLATE 5.7. A field shear box. The normal stress is applied through lead weights on top of the box and the shearing stress is applied through the winch which is anchored to the ground. The shearing stress is measured by the proving ring and dial gauge.

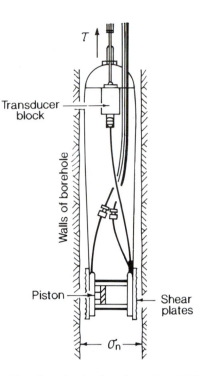

FIG. 5.18. The shear head of an Iowa Bore-Hole Shear Device.

shear plates, effective stresses can be measured. Saturated tests can be run by filling the bore-hole with water. The shear device has the great advantage that it is applied to soil in an undisturbed state at natural pore-pressures. Repeated tests up the bore-hole can be used to identify zones of strength and weakness.

Strength of intact rock

In consideration of the strength of rock a distinction must be made between: (1) the strength of an intact specimen which has no discontinuities within it; and (2) the strength of a rock mass which has natural discontinuities, zones of weathering, and other features which have major effects on the strength of the mass.

The standard laboratory test for classification of rock intact strength is of the unconfined compressive strength (σ_c). It is a measure of the resistance to crushing; it is the test to which most field tests of strength are correlated and as part of the test the moduli of elasticity, E and v, can be

An *Iowa Bore-Hole Shear Device* is designed to be inserted into a bore-hole where the strength of soils forming the sides of the bore-hole can be sheared and the pore-water pressures measured during shearing (Handy 1976). The shear device is an expandable head of two ridged plates which can be forced by gas pressure against the soil (Fig. 5.18). The plates thus apply a normal stress to the bore-hole sides. A shearing stress is applied by pulling the head upwards and so shearing the soil. Measurements are made of the normal and shearing stresses, with σ_n being varied between tests, and a Coulomb failure line plotted. Because pore-water pressure is recorded by a transducer in the

PLATE 5.8. The head of an Iowa Bore-Hole Shear Device.

determined. Separate determinations of these moduli are necessary because they cannot always be predicted reliably from a knowledge of σ_c. However, σ_c is highly correlated with rock density and one can be used to predict the other (Fig. 5.19).

Unconfined compressive strength (UCS) test

The UCS test is carried out in a press in which a cylindrical rock specimen is placed on a platen and an axial load applied to it by a hydraulic ram. The axial length and diametral swelling of the loaded specimen are recorded if Young's modulus is to be determined at the same time as strength is tested. Deformation may be monitored by linear variable differential transformers (LVDTs) or by electrical resistance strain gauges attached to the specimen.

FIG. 5.19. Plots of compressive strength of a wide range of rock types indicate rather weak correlations with elasticity and tensile strength, but a good correlation with rock density (a–d, after Imazu 1986). (e) A logarithmic plot indicates that for a single type of rock a closer relationship may exist; this plot is for Cenozoic marine mudstones from Japan and UK, which give a correlation coefficient of 0.96. (After Yoshinaka and Yamabe 1981.)

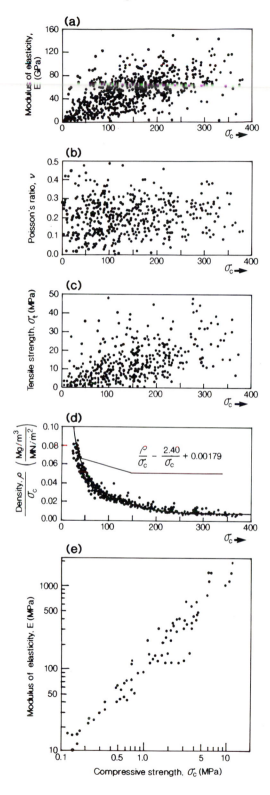

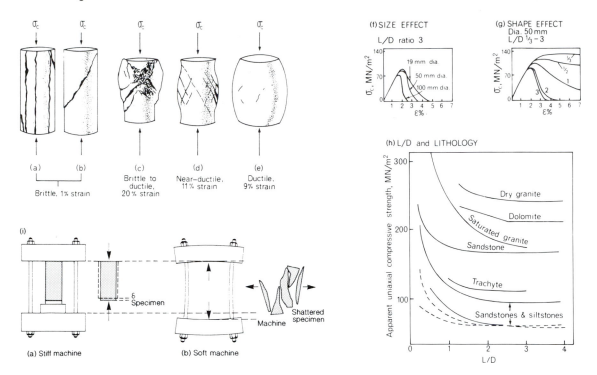

FIG. 5.20. (a–e) Forms of failure of cylindrical specimens under compression. (f–h) The effects of size, shape, moisture content, and lithology on compressive strength; L is length and D is diameter of the specimen. (i) Cartoon of the results of uniaxially compressing a rock specimen in a soft and in a stiff machine. (b–d, based on Spencer 1969; f–h, after Hudson *et al.* 1971; i, based on Bock and Wallace 1978.)

The results of the test are greatly affected by the stiffness of the testing machine, the nature of the platens, the size of the specimen, the smoothness of the specimen faces and whether the specimen ends are parallel to each other. Consequently there are standard requirements which must be observed if an acceptable test result is to be achieved (see Hawkes and Mellor 1970; Brown 1981).

The specimen should have a minimum diameter of not less than 54 mm (NX core size) and a length/diameter ratio of 2.5–3.0. Specimen ends should be cut parallel to each other, normal to the longitudinal axis, and ground smooth. Consequences of not observing these conventions are illustrated in Fig. 5.20.

The platens of the UCS machine should meet standard specifications and the machine's stiffness should be greater than that of the specimen, otherwise the machine will deform during the test and store elastic strain energy. At specimen failure, this energy will be released explosively; the platens of the machine will move rapidly towards one another, the sample will be shattered and all information from a post-failure stress–strain curve will be lost; furthermore, explosive failures are rare in nature and are not simulations of natural conditions. Under controlled conditions the standard deviation for 10 test results should fall in the range of 3.5 to 10 per cent if the rock is homogeneous.

Rock deformation in compression

Two lines of evidence—dilation and microseismicity in a specimen—indicate that during rock deformation leading to breakdown there are successive events which are related to the axial stress and strain (Fig. 5.21). Closing of pores and cracks is followed by elastic deformation, thus indicating that E should be measured at 50 per cent of the peak strength. Thereafter deformation involves propagation of Griffith cracks which link up so that the peak resistance (σ_c) coincides with the onset of continuous crack formation and breakdown with fault dislocation. The noise of crack

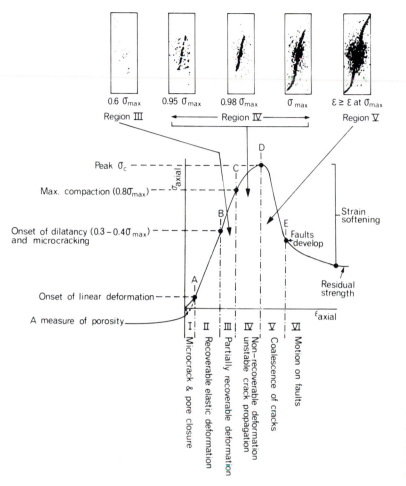

0.6 σ_{max} 0.95 σ_{max} 0.98 σ_{max} σ_{max} $\varepsilon \geq \varepsilon$ at σ_{max}

Region III   ← Region IV →   Region V

Peak σ_c

σ_{axial}

Max. compaction (0.8σ_{max})

Onset of dilatancy (0.3 – 0.4σ_{max}) and microcracking

Strain softening

Faults develop

Onset of linear deformation

Residual strength

A measure of porosity

ε_{axial}

I   II   III   IV   V   VI

I — Microcrack & pore closure
II — Recoverable elastic deformation
III — Partially recoverable deformation
IV — Non-recoverable deformation unstable crack propagation
V — Coalescence of cracks
VI — Motion on faults

FIG. 5.21. (b) Models of failure mechanisms in cylindrical specimens under axial loads in compression (after Hallbauer *et al.* 1973) related to an idealized axial stress–strain curve with descriptions based on Price (1975).

propagation is the clearest indicator of the failure processes and is comparable with that made before and during earthquakes and during failures of mine walls and tunnels.

The specimen failure described above is evidence of brittle fracture. Some rocks, such as evaporites and shales, behave in a ductile manner at low stresses, other rocks, such as marble, become increasingly ductile at high confining stresses. With increasing ductility, brittle behaviour diminishes and plastic deformation increases (Fig. 5.20).

A valuable feature of the UCS test is that it can be used to identify the presence of planes of weakness in a specimen and their effect upon rock strength. Rocks such as shales, slate, schist, and gneiss have obvious fabrics or planes of preferred mineral alignment. Tests using core specimens aligned at various angles to the planes of the fabric are used to study the effects of such fabric on rock strength, fracture propagation, and potential planes of failure (Fig. 5.22). Orientation of the fabric exactly along (0°), or normal to (90°), the principal stress provides the greatest strength. Strength is least at 25–45° to the principal stress direction.

Tensile strength test

The standard tensile strength (σ_t) indicator for rock is the Brazil test in which a disk of rock, cut from a core, is placed between two jaws. The load is applied to the jaws by a UCS machine and the tensile strength is given by:

$$\sigma_t = 0.636\, P/Dt \text{ (MPa)},$$

where P is the load at failure (Newtons), D is

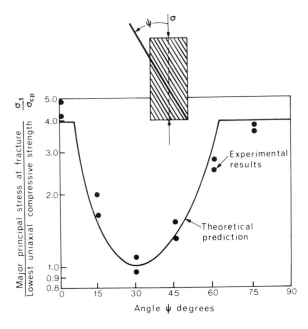

FIG. 5.22. The relationship between orientation of cleavage and the compressive strength of slate. (After Roberts 1977, from an original figure by E. Hoek.)

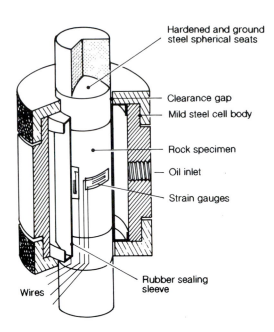

FIG. 5.23. Cutaway view of a Hoek triaxial cell (after Hoek and Franklin 1968). Note the strain gauges, attached to the specimen, for measuring axial and diametral deformation.

the diameter (mm), t is the thickness of the test specimen measured at its centre (mm) (Brown 1981).

Triaxial tests

Triaxial tests of rock strength are performed in a cell designed on the same general principles as a cell used for testing soil, except that the stresses used are of greater magnitude (Fig. 5.23). Confining pressure ($\sigma_3 = \sigma_2$) is applied through an oil-filled jacket which fits around a cylindrical sample cut to the same specifications as those used in the UCS test. In many laboratories the triaxial test is carried out in a UCS machine with the specimen enclosed in a jacket called a 'Hoek cell', which is filled with hydraulic fluid (Plate 5.9).

The strength of the specimen is calculated by dividing the maximum axial load (σ_1), applied to the specimen during each test, by the original cross-sectional area. The confining pressure at failure is plotted on the abscissa and the axial stress on the ordinate. A strength envelope is obtained by fitting a mean curve to the plotted points of at least three tests. Where the range of stresses is large it may be appropriate to fit a straight line to only the most relevant part of the curve, or to fit

PLATE 5.9. A Hoek cell inside a uniaxial compression machine.

several lines to separate parts of the curve, as envelopes for rock material are generally concave downwards. It must be stressed, however, that laboratory testing should use a range of applied stresses which simulate those relevant to the study problem. In studies of hillslopes the stress range is generally small and fitting a straight line to data points is appropriate. A large range of stress levels is relevant to tunnel and mining situations (see Brady and Brown 1985).

Shear strength of intact rock

Most standard soil shear boxes are designed to operate at lower stresses than those which will shear intact rock specimens. Shear strength tests on intact rock may be carried out in a jig which fits into a uniaxial machine. This is a satisfactory method provided that strain measurements are not required during the test. Shear strength measurements of jointed rock are discussed in Chapter 6.

Shear testing dies for a uniaxial machine

Each die is made of two halves which act jointly as a specimen holder. They may be made to accept rock cubes or cylinders. Cylinders may be aligned (1) to shear across the long axis to produce two half-length cylinders, or (2) along a diametrical plane along the long axis. The first method is preferred to ensure shearing in the required plane.

To derive values of c and ϕ it is necessary to carry out tests over a range of inclinations of the shearing plane with respect to the compressing force. Sets of wedges may be employed to achieve this or several sets of dies are required. The dies have roller bearings on the ram face or on both the ram face and base plate. The recommended inclinations of the shear plane fall in the range of 30° to 50°. Greater or lesser angles cause distortions (Fig. 5.24).

As cylindrical specimens are easier to cut than cubes, to an accuracy of 0.1 mm diameter, they are used most commonly. If the samples are irregular they should be rejected and minor roughness should be accommodated by use of a gasket in the sample holder. Cored specimens should be of 54 mm diameter or larger and have a length equal to the diameter.

Determination of shear strength

The compressing force P applied, through the die, to the potential shear surface can be resolved into a normal component N and a shearing (tangential)

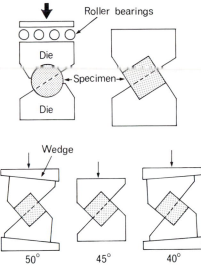

Fig. 5.24. The arrangement of dies and wedges for shear testing of rock cubes or cylinders in a uniaxial compression machine.

component T (Fig. 5.25a). The normal (σ_n) and shearing stresses (τ) acting across and along the shear plane respectively and over the area A of that plane are:

$$\sigma_n = \frac{N}{A} = \frac{P \sin\alpha}{A} = q \sin\alpha$$

$$\tau = \frac{T}{A} = \frac{P \cos\alpha}{A} = q \cos\alpha,$$

where: α is the angle of inclination of the shearing plane with the direction of the applied compressing force,

$q = P/A$ (in units of N/m²).

The method of plotting the Mohr diagram representing the state of stress in the shear area is shown in Fig. 5.25. The points can be found either by plotting the values of σ_n and τ on the coordinate axes, or by q along a line passing through 0 at the angle α to the ordinate axis.

Plotting the Mohr diagram points for various die-bevel (shearing-plane inclination) angles (Fig. 5.25c) permits the construction of the Mohr-circle envelope curve. The curvature of the envelope is sometimes disregarded, and the envelope is considered a straight line. In such a case two points are sufficient for two values of α (Fig. 5.25). By draw-

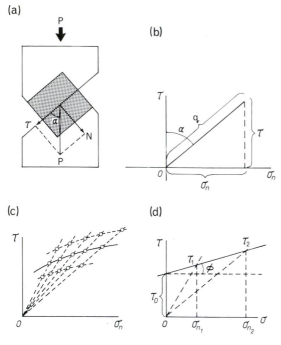

(a)

(b)

(c)

(d)

FIG. 5.25. Computation of values for internal friction angle and cohesion from data from shear tests using bevelled dies. (Based on Protod'yakonov and Koifman 1969.)

ing a straight line through the two points and extending it to intersect the ordinate axis the so-called cohesion intercept τ_o and the internal friction angle ϕ can be found:

$$\tau_o = \frac{q_1 q_2 \cdot \sin(\alpha_2 - \alpha_1)}{q_2 \cdot \sin\alpha_2 - q_1 \cdot \sin\alpha_1}$$

$$\tan\phi = \frac{q_2 \cdot \cos\alpha_2 - q_1 \cdot \cos\alpha_1}{q_2 \cdot \sin\alpha_2 - q_1 \cdot \sin\alpha_1}.$$

Details of alternative methods of shear testing are given in Vutukuri *et al.* (vol. i, 1974).

Water and rock strength

In strong, dense rocks the effect of water is most noticeable as 'cleft' water pressure in partings. This effect is consequently discussed under rock mass strength and slope stability in later chapters. Pore-water pressure does not significantly reduce the strength of strong rocks for two reasons: (1) the pores within the rock mass are not fully connected and hydrostatic pressure is not, therefore, transmitted through them; consequently there is little or no buoyancy effect; and (2) rock cohesive strength

is high and resists any buoyancy effect where pores are connected.

In bodies of rock with moderate and low strength, pore-water pressure has major effects upon strength, as it also has on strong rocks which are highly fissured. Provided that fluid pressure is fully communicated through the pore space, the effective normal stresses that limit crack opening and extension, and frictional resistance, are reduced by the magnitude of the fluid pressure. Pore-water pressure is recognized (see Kirby 1984, and other papers in the same volume) as having a major influence in promoting brittle fracture, hydraulic fracturing, large-scale faulting, and earthquakes within the Earth's crust.

Chemical effects of water reduce rock strength, over time, by solution. In laboratory tests, cracks in mica grow more rapidly in moist air than under a vacuum or in dry air, and quartz, calcite, marble, and gypsum are greatly weakened by the presence of water. In silicate minerals, replacement of strong silica oxygen bonds with much weaker hydrogen bonds in the presence of water is a major weakening factor (Atkinson 1984). During shearing processes, water may reduce the hardness at points of particles and so reduce frictional resistance and promote deformation.

At saturation, under both triaxial and UCS testing, the reduction in strength from the dry state may be in the range of 10–25 per cent for very hard rocks, and for shales and moderately strong sandstones as much as 40 per cent. It is therefore essential that the moisture contents should be reported for all tests. Broch (1974) has reported that saturated cores may lose as much as 25 per cent of their water in as little as ten minutes and 50 per cent in an hour. Loss of water from a core can be prevented by use of plastic adhering film. Water-content control should be part of all testing procedures.

Reductions in strength may be reported as a softening factor (S_f):

$$S_f = \frac{\text{Ultimate oven-dry strength}}{\text{Ultimate saturated strength}} = \frac{\sigma_{c,\,dry}}{\sigma_{c,\,sat}}.$$

Hard dense rocks of low permeability may have low softening factors of 1.01–1.2. Shales, weak sandstones, and some ignimbrites may have softening factors in the range of 2–5 (Dobereiner and De Freitas 1986; Moon, in Selby *et al.* 1988). Over-consolidated mudrocks may suffer extreme

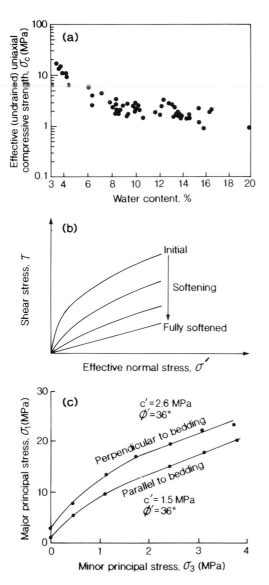

FIG. 5.26. (a) Softening of mudrocks in relation to water-content for Melbourne Mudstone (after Johnston and Chiu 1984). (b) Characteristic Coulomb failure lines for an over-consolidated mudstone at various stages of softening (after Yoshida *et al.* 1990). (c) A non-linear failure envelope for a saturated sandstone, tested parallel and perpendicular to the bedding (after Dobereiner and De Freitas 1986).

softening during stress release, fissure opening, the drawing of water into fissures, and the onset of weathering, with softening factors in the range of 10–20 (Johnston and Chiu 1984). Fig. 5.26 indicates the changes of strength with water content for the Melbourne mudstone and characteristic

Mohr–Coulomb failure lines for shear tests on such rocks and for weak sandstones; the influence of bedding is apparent for the sandstones.

Field measurements of intact rock strength

The *Schmidt hammer test* was devised in 1948 by E. Schmidt for carrying out *in situ* non-destructive tests on concrete. The test hammer measures the distance of rebound of a known mass impacting on a rock surface. Because elastic recovery of the rock surface depends upon the hardness of the surface, and hardness is related to mechanical strength, the distance of rebound is a relative measure of surface hardness or strength (see Plate 5.6).

Three types of Schmidt hammer have value in geomorphic studies. The 'P' type is a pendulum hammer intended for testing materials of low hardness with compressive strengths of less than 70 kPa.

The 'L' type hammer has a head which is driven by a spring against a rock surface. The amount of rebound of the hammer-head is recorded and used as an index of rock hardness. This hammer has been adopted as the standard for rock mechanics (Brown 1981).

The 'N' type hammer is commonly used for geomorphic purposes. It is similar to the 'L' type but has a higher impact energy and can provide data on a range of rocks from weak to very strong. It is theoretically capable of testing rocks with compressive strengths in the range of 20 to about 250 MPa, but it is rather unreliable in weak rocks with strengths below about 30 MPa.

The 'N' type hammer has a scale from 10 to 100. Rocks like chalk have a rebound number (R) of 10 to 20 and at the other end of the scale basalts, gabbros, and quartzites have values of 55 or higher (see Table 5.3).

The advantages of the hammer are that it is light, weighing only 2.3 kg, very easy to carry, easy to use, and relatively cheap. Large numbers of tests may be made in a short time and data collected from a variety of rock surfaces. Weathering rinds, case hardening, individual large clasts, and rock matrices may be readily tested.

Among the disadvantages of the Schmidt hammer are that it is extremely sensitive to discontinuities in a rock, even hair-line fractures may lower readings by 10 points, hence fissile, closely

TABLE 5.3. *Approximate strength classification of cohesive soil and of rock*

Description	Uniaxial compressive strength, MPa	Point load strength $I_{s(50)}$, MPa	Schmidt hammer, N-type, 'R'	Characteristic rocks
Very soft soil—easily moulded with fingers, shows distinct heel marks.	<0.04			
Soft soil—moulds with strong pressure from fingers, shows faint heel marks.	0.04–0.08			
Firm soil—very difficult to mould with fingers, indented with finger nail, difficult to cut with hand spade.	0.08–0.15			
Stiff soil—cannot be moulded with fingers, cannot be cut with hand spade, requires hand picking for excavation.	0.15–0.60			
Very stiff soil—very tough, difficult to move with hand pick, pneumatic spade required for excavation.	0.6–1.0	0.02–0.04		
Very weak rock—crumbles under sharp blows with geological pick point, can be cut with pocket knife.	1–25	0.04–1.0	10–35	Weathered and weakly compacted sedimentary rocks—chalk, rock salt
Weak rock—shallow cuts or scraping with pocket knife with difficulty, pick point indents deeply with firm blow.	25–50	1.0–1.5	35–40	Weakly cemented sedimentary rocks—coal, siltstone, also schist
Moderately strong rock—knife cannot be used to scrape or peel surface, shallow indentations under firm blow from pick point.	50–100	1.5–4.0	40–50	Competent sedimentary rocks—sandstone, shale, slate
Strong rock—hand-held sample breaks with one firm blow from hammer end of geological pick.	100–200	4.0–10.0	50–60	Competent igneous and metamorphic rocks—marble, granite, gneiss
Very strong rock—requires many blows from geological pick to break intact sample	>200	>10	>60	Dense fine-grained igneous and metamorphic rocks—quartzite, dolerite, gabbro, basalt

Sources: Deere and Miller (1966); Piteau (1971); Robertson (1971); Broch and Franklin (1972); and Hoek and Bray (1977).

foliated, and laminated rocks cannot be assessed with this instrument unless samples are clamped in a vice. It is also very sensitive to water content, especially of weak rocks. To eliminate as much variability as possible, test impact sites should be more than 60 mm from an edge or joint and surfaces should be flat and free from flakes and dirt. The surface texture of surfaces may influence test results (Williams and Robinson 1983). Smooth planar surfaces give higher readings than rough or irregular surfaces. It is recommended that the hammer should be used on the same spot for five impacts and the highest reading, which is usually the last, be used for that site. The number of impacts on each sample area of about $2\,m^2$ should be 20 to 50 with the larger number being undertaken if the rock is variable. The most reliable results are obtained if the lower 20 per cent of impact readings are ignored and measurements are continued until the deviation from the mean value

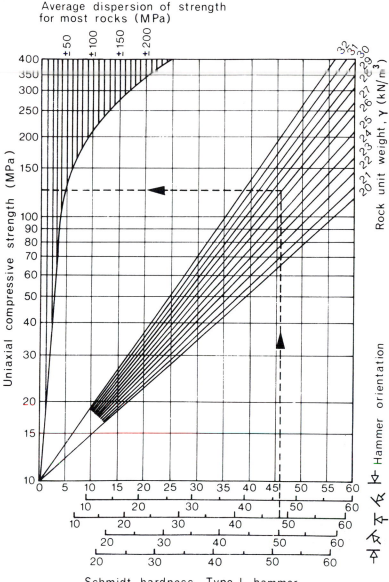

FIG. 5.27. Relationship between Schmidt hammer reading and the uniaxial compressive strength of rock for the 'L' type hammer. (Modified from Hoek and Bray 1977, after an original figure by Deere and Miller 1966.)

of the remainder does not exceed ±3 points. The R value should be corrected for the inclination of the rock face by using either the manufacturer's chart or the table given by Day and Goudie (1977).

Each Schmidt hammer has a slightly different rebound so the instrument number should be recorded and all instruments regularly calibrated against a test anvil.

On each test hammer there is a printed calibra-

tion curve giving a plot of unconfined compressive strength against the rebound number (R). This is unfortunate as the correlation is not universal, being affected by both the unit weight of the rock and its moisture content. A more reliable correlation chart which relates R, hammer orientation, rock unit weight, dispersion of strength, and unconfined compressive strength, has been provided by Deere and Miller (1966) (Fig. 5.27). For many purposes, however, it is unnecessary to convert R values to strength values as the R values alone can be used as indices (Table 5.3).

In the absence of any of the three instrumental methods of determining intact rock and soil strength an approximate indication may be obtained by a subjective field assessment. In Table 5.3 criteria for such an assessment are provided together with related values derived from the recommended strength tests. These values should not be taken as exact equivalents as they are dependent upon such properties as rock unit weight, moisture content, and anisotropies. The terms 'strong' and 'weak' are used here to apply to intact strength alone and not to mass strength.

The *point load test* was developed in Russia to provide a rapid strength test of irregularly shaped rock specimens (Protod'yakonov 1960). The International Society for Rock Mechanics (1985) has subsequently adopted the Protod'yakonov test as a standard technique. In the test, specimens approximately 50 mm in diameter are fractured between two cone-shaped platens (Plate 5.10). The maximum stress (I_s), occurring at the centre of the specimen, may be related to the applied load (P) and the distance between platens (D) by an expression with the form: $I_s = k \cdot P/D^2$. The constant k is found to assume values which are dependent upon the geometry of the specimen. Correction charts for specimens of various sizes are provided by ISRM (1985). Strength indices are usually expressed as for a rock cylinder of 50 mm diameter ($I_{s(50)}$).

Point load strength tests have been carried out on a variety of rock types ranging from the strongest rocks to weathered materials which have the characteristics of strong soils. The irregularity of sample sizes causes scatter in strength data, but this can be compensated for by testing a large number of samples (15–20) and by trimming irregularities off samples with a hammer. If size correction curves are used the error is unlikely to exceed 15 per cent and in most cases should be much smaller. Anisotropies in materials can be studied by orienting samples so that they are fractured across or along planes of weakness and, in many cases, mean values may be used in a classification of rock mass strength.

Studies by Broch and Franklin (1972) indicate that there is similar scatter in the data from point load tests and in data from unconfined compressive tests and that both tests are very dependent upon

PLATE 5.10. A point-load strength tester.

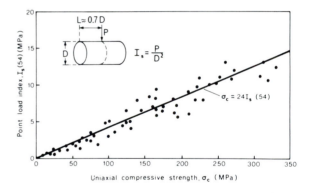

FIG. 5.28. The relationship between point load index and uniaxial compressive strength for 54 mm diameter cores (Data from D'Andrea *et al.* 1965; Broch and Franklin 1972; Bieniawski 1975.)

rock moisture content. Sandstones often undergo a strength reduction of 20–30 per cent as water content increases from a dry to a saturated state and even granite may lose more than 13 per cent of its strength. Field studies should normally be made at 'natural' moisture content even though this involves a different strength rating in an arid and a humid climate.

The point load strength test is easily carried out with equipment which can be carried in a vehicle or, if necessary, on a back pack frame. Its results

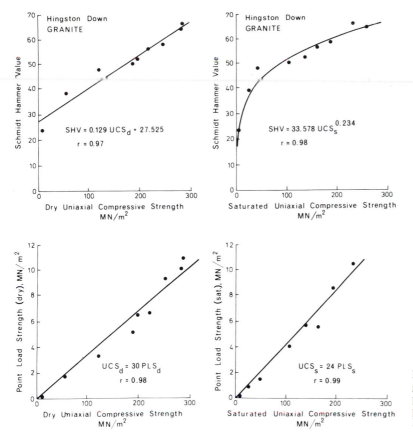

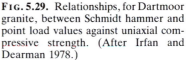

FIG. 5.29. Relationships, for Dartmoor granite, between Schmidt hammer and point load values against uniaxial compressive strength. (After Irfan and Dearman 1978.)

TABLE 5.4. *Typical rock properties*

	Strength (MN/m²)			Bulk density (Mg/m³)	Porosity (%)	Velocity of compression waves, V_p (m/s)
	Compressive	Tensile	Shear			
Granite	100–250	7–25	14–40	2.6–2.9	0.5–1.5	3000–5000
Basalt	150–300	10–30	20–60	2.8–2.9	0.1–1.0	4500–6500
Gneiss	50–200	5–20	—	2.8–3.0	0.5–1.5	—
Slate	100–200	7–20	15–30	2.6–2.7	0.1–0.5	3500–5500
Marble	100–250	7–20	—	2.6–2.7	0.5–2.0	3500–6000
Shale	100–200	7–20	15–30	2.0–2.4	10–30	1400–3000
Sandstone	20–170	4–25	8–40	2.0–2.6	5–25	1400–4000

Note: Density of water = 1 Mg/m³

Source: Data from Attewell and Farmer (1976).

are highly correlated with unconfined compressive strength tests (Figs. 5.28 and 5.29): published correlation coefficients range from 0.88 to 0.95 (D'Andrea *et al.* 1965; Irfan and Dearman 1978).

A *cone indentor*, designed to test the strength of very small specimens of weak rocks, has been developed by the UK Mining Research and Development Centre of the National Coal Board (1979).

It has been found to be applicable to highly fractured and fissile rocks which are very difficult to test by other methods.

Dynamic measures of elasticity

The elastic properties of rock may be measured by transmitting ultrasonic waves or seismic waves through rock bodies (Carroll 1969; Obert and Duvall 1967).

There are two basic types of elastic waves: body waves which travel through the interior of the rock body, and surface waves which can only travel along the surface of the rock. Body waves can be subdivided into two modes: compression or primary (P) waves and shear or secondary (S) waves. P-waves induce longitudinal oscillatory particle motions similar in many ways to simple harmonic vibrations and, when they impinge on a free boundary in any direction other than head-on, one of the resultant effects of the displacement is the induction of S-waves in which the particles move in a transverse direction without compressing the material. P-waves, of course, travel in any direction in a material which resists compression, but since S-waves depend upon the ability of the transmitting material to resist changes in shape, they can only exist in solids.

The velocity equations of these waves are as follows (Vutukuri *et al.* 1974).

$$V_p = \left[\frac{E(1-v)}{\rho(1+v)(1-2v)} \right]^{1/2}$$

$$V_s = \left[\frac{G}{\rho} \right]^{1/2} = \left[\frac{E}{2\rho(1+v)} \right]^{1/2},$$

where: V_p = velocity of compression waves, m/s;
$\quad\;\; V_s$ = velocity of shear waves, m/s;
$\quad\;\; E$ = dynamic modulus of elasticity, Pa;
$\quad\;\; G$ = dynamic modulus of rigidity, Pa;
$\quad\;\; v$ = Poisson's ratio;
$\quad\;\; \rho$ = density, Mg/m^3.

By the ultrasonic pulse method:

$$E = \frac{\rho V_s^2 (3V_p^2 - 4V_s^2)}{V_p^2 - V_s^2}$$

$$v = \frac{V_p^2 - 2V_s^2}{2(V_p^2 - V_s^2)}.$$

Body wave velocities are usually greater in igneous and crystalline rocks than in sedimentary

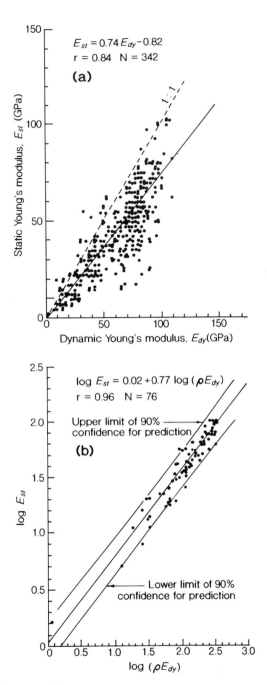

FIG. 5.30. (a) Relationship between static and seismically determined dynamic Young's moduli of rocks. A 1:1 correspondence is indicated along the dashed line. (b) A better estimate of the static Young's modulus is obtained from the logarithmic relationship with the product of rock density and dynamic modulus. (Data from many sources plotted by Eissa and Kazi 1988.)

rocks. In sediments it has been observed that the wave velocity tends to increase with increase in the depth of cover and increase in lithification. It is also common for stratified rocks to display considerable anisotropy in seismic velocity. The wave velocity in a direction parallel to the stratification may be 10–15 per cent greater than that normal to the stratification planes. Typical longitudinal wave velocities for rocks are listed in Table 5.4.

The sonic velocities in a rock mass may be expected to vary very considerably from those determined by laboratory test, because they depend on the ambient stress level and other environmental factors, such as porosity and pore fluid content. The rock-mass properties should therefore be investigated *in situ*, and this may present problems in that it is sometimes difficult to recognize the arrival of the shear wave unless sophisticated and expensive equipment is available for this purpose.

At atmospheric pressure and low confining stresses the static elastic modulus for dry rocks is generally less than the dynamic. The reason for the difference is ascribed to microcracks which affect the strain of the static sample in a UCS test more than they affect the propagation of ultrasonic waves during a dynamic measurement (King 1983). Where cracks are few, the static and dynamic moduli converge. For saturated rocks the static modulus is decreased, together with strength, but the dynamic modulus is increased because the velocities of compressional and shear waves are higher in the saturated rock than in dry rock (Toksoz *et al.* 1976).

Equipment for determining dynamic elastic moduli in the field and laboratory is available commercially and has been tested (e.g. Allison 1988; McDowell and Millett 1984). The use of such equipment to estimate strength of rocks requires careful evaluation because of the uncertainty of correlations between elasticity and strength (Fig. 5.19) and between dynamic and static values of E (Fig. 5.30a). The value of static modulus of elasticity can best be predicted from the relationship:

$$\log_{10} E_{st} = 0.02 + 0.77 \log_{10}(\rho E_{dy}),$$

in which the values of static modulus (E_{st}) and dynamic modulus (E_{dy}) are expressed in GPa and density (ρ) is expressed in g/cm^3 (Eissa and Kazi 1988) (Fig. 5.30b).

6 Properties of Rock Masses

Joints and other partings

Hillslopes formed on rock are virtually never developed under conditions in which the intact strength of the rock is the dominant control on the resistance of the rock to failure; rather, it is the resistance along the natural partings which influences the development of the hillslopes. The normal stresses acting across partings are produced by the weight of the overburden. Within a cliff, with a vertical face, a horizontal plane through the base of the rock body is therefore subjected to a stress described by:

$$\sigma_n = \gamma H,$$

where: γ is the unit weight of the rock (kN/m^3);
H is the height of the cliff (metres).

An unjointed rock body will be stable against failure under the compressive overburden load as long as the load does not exceed the strength of the rock (σ_c). In a condition for failure, $\sigma_c = \sigma_n$. The critical height (H_c) for failure is then:

$$H_c = \sigma_c/\gamma.$$

A strong intact igneous rock, with a uniaxial compressive strength of 250 MPa and a unit weight of 28 kN/m^3, will have a critical height, for stability against failure in compression, of nearly 9 km. That no cliffs of such height exist on Earth is due to the presence of partings in all rock masses. Without such fractures even weak rock could support vertical cliffs of several hundred metres (for $\sigma_c = 5$ MPa and $\gamma = 20$ kN/m^3, $H_c = 250$ m).

Partings cause concentrations of stress, control the movement of groundwater in the rock body, and permit weathering to penetrate and weaken the body. At an extreme they may be so closely spaced that the rock is effectively a granular material. Understanding the strength and other properties of rock bodies is consequently not possible without an understanding of partings.

The terms 'fracture', 'parting', and 'discontinuity' are synonyms for planes of separation in rocks, they have no specific or genetic definition. The terms 'joint', 'fault', 'cleavage', 'fissure', and 'foliation' have specific definitions.

Joints are breaks of geological origin in the continuity of a body of rock, occurring either singly or in a set or system, with no shearing or lateral displacement parallel to the surface of the joint. Faults are planes of displacement within rocks or soils; they may be surfaces of macroscopic shear or zones of displacement formed during plastic yielding. Cleavage is the property or tendency of a rock to split along secondary aligned fractures which are planar, persistent and closely spaced; cleavage is produced by deformation or pressure metamorphism and is visible in rocks such as slates in which alignment of mineral grains has occurred. A fissure is a fracture along which there is a distinct opening which may become filled with secondary deposits. Foliation is a set of short, planar or wavy, closely spaced laminae, usually showing mineral alignment, as in schistose and gneissic rocks.

Partings may be classified in several ways, of which two are most relevant to geotechnical investigations: (1) according to the stresses or processes which are responsible for creating the parting (Aydan and Kawamoto 1990); and (2) according to the characteristics of the feature (e.g. joints, faults, etc.) (Hencher 1987).

(1a) *Partings caused by tensile stresses* include those from: cooling of magma from fluid to solid phase; dessication shrinkage of sediments; freezing of solids at very low temperatures, (a) without intervention of water, (b) by formation of ice in cracks, and (c) by extension of pre-existing cracks;

doming and stretching of rock bodies during folding; formation of *en echelon* cracks propagating away from faults; stress relaxation after unloading.

(1b) *Partings caused by shear stresses* include those from: viscous frictional drag along walls of intrusions; fault displacements, often with off-setting of rock units; displacement of rock in the limbs of folds; conjugate joints resulting from regional stresses, or in apices of folds.

(1c) *Partings formed during sedimentation* may result from: planes developed where there has been a change in particle size; a hiatus or change in rates of deposition.

(1d) *Partings developed during metamorphism* include: planes of schistosity developed by re-crystallization of aligned or platy minerals; cleavage in slates and shales resulting from alignment of crystals and particles under compression; and folding and faulting associated with intrusions (Fig. 6.1f).

(2a) *Lithological boundaries* often mark major changes in rock type and hence rock properties such as strength, permeability, and jointing and they may act as boundaries to water movement.

(2b) *Tectonic joints* are the result of regional stresses. The joints form systems or sets which have constant orientations over considerable distances and are independent of lithology.

Attempts have been made to apply Mohr–Coulomb forms of analysis to tectonic joints with the expectation that compression would result in the convergence of opposite wedges and formation of conjugate pairs of shear joints intersecting at an angle of 45–60° along the principal stress direction (Fig. 6.1a, b). Any tensile fracture in a compressive stress field would be parallel to the principal stress direction. In a tensile stress field conjugate pairs may form but tensile failure would be normal to the principal stress and parallel to the minor stress direction (Fig. 6.1c).

A clear example of conjugate joints and tensile joints has been reported by de Sitter (1956) from eastern Algeria in an area of Cretaceous and Tertiary sedimentary rocks forming open anticlines and synclines (Fig. 6.1d, e). The regional stress is tectonic and compressive but local gravitational stresses have controlled the development of joints along the crests of anticlines and in rocks forming the floors of perched synclines.

Patterns of regional joints have been recognized as having considerable influence on orientation of river channels, fluvial and glacial valleys and on the form of mountain peaks (Gerber and Scheidegger 1973, 1975; Gerber 1980; Scheidegger 1963, 1980; Scheidegger and Ai 1986).

Relationships between joints and regional stresses are, however, not always clear for three major reasons: (1) in any region rock bodies are likely to be anisotropic with the result that rock features will influence the formation of joints in a variety of ways; (2) stress directions change over time with the result that it is common to find several joint systems in the landscape and interpreting them is then a very complex problem; and (3) existing joints and faults are likely to be exploited by successive generations of stresses so that, for example, a joint system or fault originating from extension may accommodate strain resulting from a later compressional regime, with a principal stress direction which is oblique to that of the original tensile stress. The most recent stress pattern may therefore give rise to deformation along inherited joint systems.

(2c) *Faults* may have displacements of any scale from millimetres to hundreds of kilometres. Studies undertaken in mines (Robertson 1982) indicate that shearing processes form gouges and breccias of broken and ground rock, from the fault walls, which have a thickness (t) which is proportional to the displacement (d) with $d \approx 100t$ so that a displacement of 100 m will create a gouge 1 m thick (Fig. 6.1g). This relationship holds for all rocks, except limestones which do not produce breccia but behave as ductile materials, and for all normal and reverse faults that have steep dips and displacements of less than 5 km. It does not apply to strike-slip and low-angle overthrust faults which have extreme variation in their displacements and breccia thicknesses.

The importance of faults is that they create zones of weakness, zones of altered permeability, frequently create lithological boundaries where displacements are large, become zones of preferential weathering and may become zones of large groundwater or hydrothermal water flow.

(2d) *Unloading joints* have been discussed in Chapter 4. They may occur in massive rocks with few other joints but they may also occur in rocks with other types of joint, usually indicating that the unloading, or sheeting, joint developed first.

(2e) *Bedding-planes* are formed in sedimentary rocks, parallel to the original surface of deposition and are therefore originally nearly horizontal. They are often flat and persistent over consider-

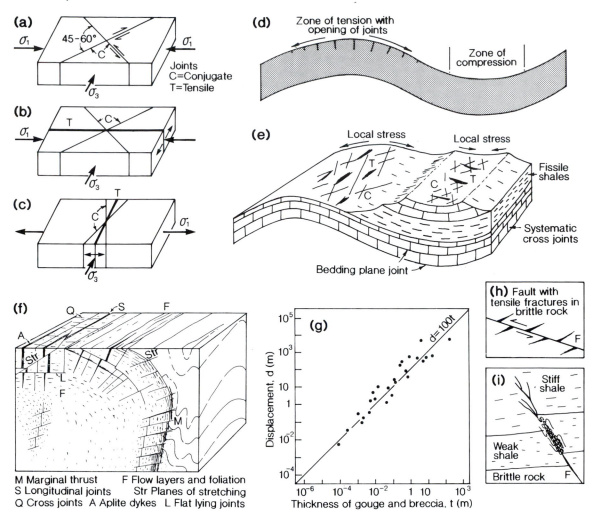

(f)

M Marginal thrust
S Longitudinal joints
Q Cross joints A Aplite dykes L Flat lying joints

F Flow layers and foliation
Str Planes of stretching

FIG. 6.1. (a–c) The formation of conjugate pairs of joints and tensile joints under compressive and tensile stresses. (d, e) Local joint patterns formed in gently folded sedimentary rocks (based on de Sitter 1956). (f) Joints developed along the margins and in the roof of a granitic intrusion (based on Cloos 1936, and Balk 1937). (g) Displacement plotted against thickness of gouge and breccia of normal and reverse faults with displacements of <5 km (after Robertson 1982). (h, i) Features of faults in rocks of different strength.

able distances. Bedding-planes are tightly closed, with considerable cohesion, until stress relief permits the formation and opening of joints which then become planes of weakness.

Stress release and other processes commonly contribute to the formation of cross-joints normal to the bedding-planes. In clay- and silt-rich sedimentary rocks which have been deeply buried, shaley cleavage and fissuring results from swelling and alternating wetting and drying.

(2f) *Cooling joints* form in solidifying magma.

These joints are networks of interconnected tension fractures that divide the rock into prisms or columns with diameters ranging from a few millimetres to a hundred metres or more. Thermally induced joints first develop at the outer boundaries of magma bodies, which cool first, and grow inward as magma cools and solidifies (De Graff and Aydin 1987). Growth is incremental, as is indicated clearly where diametral banding is visible across a column.

(2g) *Metamorphic cleavage* has high cohesive

strength across the planes before separation but, after opening, cleavage planes become zones of weakness where minerals are aligned or surfaces are coated with platy minerals such as mica, talc, and chlorite. The close spacing of cleavages also creates weakness and potential zones of weathering. Metamorphic cleavage is caused by regional stresses and is often recognizable over considerable distances.

Shear strength of partings in rock

Analyses of rock-slope stability usually involve estimates of the shear strength of pre-existing partings on which movement is likely to occur. With few exceptions potential failure planes are partings which are continuous, or nearly so, and hence have a resistance to sliding which is dependent upon the frictional contact between the opposite walls of the critical parting. Partings have a shear strength which is substantially less than that of intact rock, but the total strength available to resist shear is extremely variable as three major factors are involved: (1) the basic frictional strength of the rock material; (2) the roughness of the walls of the parting; and (3) the thickness and strength of any infill material in the partings (Figs. 6.2, 6.3, Table 6.1). Rough, closed joints are stronger than smooth joints, and as the thickness of the fill increases towards the amplitude of the asperities along the joint wall, the resistance to shear becomes progressively that which is due to the infill alone. Infills of clay-rich soil are commonly causes of landslides and this is particularly so where the clays are smectites with high water-adsorption capacity and very low frictional strength. Infills can be studied by using conventional shear boxes, but rough joint surfaces are more difficult to sample and test (see Hencher and Richards 1989, for discussions of laboratory testing).

Shear strength of joint wall surfaces

The frictional resistance along partings is composed of two components: (1) the angle of frictional sliding resistance of the rock as measured in a shear test along a macroscopically smooth-cut surface (ϕ_μ) which does not dilate under shear testing, and (2) the average angle i of the asperities along the parting with a plane through the base of the asperities (Fig. 6.4). The asperities may be categorized as large, or of first-order roughness (or waviness), and as minor, or of second-order roughness (or unevenness). The angle i is a measure of the dilation, of the parting, which must occur for sliding without shearing through the asperities. For a continuous parting the shear strength at failure is thus represented by:

$$\tau_f = \sigma_n . \tan(\phi_\mu + i) = \sigma_n . \tan\phi_j, \quad (6.1)$$

where σ_n is the normal stress across the parting, and ϕ_j is the joint friction angle.

The fundamental work on joint shear strength was done by Patton (1966) who carried out investigations at 101 sites on sandstone and 146 on carbonate rocks for slopes close to failure. Strength tests showed that for sandstone ϕ_μ values range from 22° to 31° and for limestones 25° to 39°. Values of i for sandstone had a median of 3° and for limestone of 5°. In all cases i was derived from measurements of the first-order roughness only. Later studies by Barton (1973) have shown that the values of i are related to the normal stress across

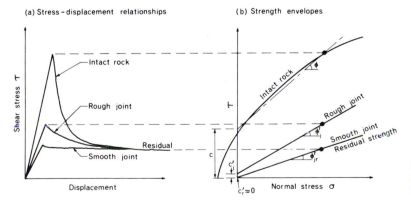

(a) Stress–displacement relationships

(b) Strength envelopes

Shear stress τ

Intact rock

Rough joint

Residual

Smooth joint

Displacement

Intact rock

Rough joint

Smooth joint

Residual strength

Normal stress σ

Fig. 6.2. Cohesion and friction angles for joints and intact rock. (After Lo and Lee 1975.)

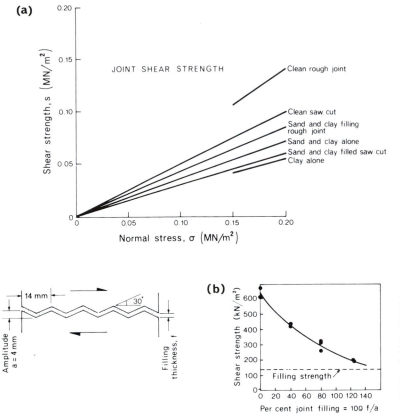

(a)

JOINT SHEAR STRENGTH

Clean rough joint

Clean saw cut

Sand and clay filling rough joint

Sand and clay alone

Sand and clay filled saw cut

Clay alone

Shear strength, s (MN/m²)

Normal stress, σ (MN/m²)

(b)

Shear strength (kN/m²)

Filling strength

Per cent joint filling = 100 f/a

14 mm

30°

Amplitude a = 4 mm

Filling thickness, f

FIG. 6.3. (a) Direct shear test results for joint strengths with varying roughness and fill of the joint (data from Kutter and Rautenberg 1979). (b) Influence of joint filling thickness on the shear strength of an idealized sawtooth joint; as the filling increases in thickness, strength along the joint reduces towards the strength of the fill alone (after Goodman 1970).

TABLE 6.1. *Joint friction angles*

Joint features	$\phi_j°$
With clay infill	Infill controls
Flat rock surfaces	26–38
Slightly rough bedding joints	66–73
Rock joints in strong rock	72–77
Very rough joints	77–80

Notes: 1. All rock surfaces are unweathered,
2. Normal stresses are in the range 150–700 kPa.

Sources: Data from tests by Goodman (1970); Paulding (1970); Rengers (1971).

the joints in the slopes Barton studied. At low normal stresses the second-order roughness is effective and raises the values of *i* above those found by Patton. In Table 6.1, values of $(\phi_\mu + i)$ are given for a number of strength tests. Assuming that the rock friction angle (ϕ_μ) is approximately

30°, the effective roughness angle *i* is shown to vary between 30° and 50° for low normal stresses.

Shear box tests, at a range of normal stresses, have confirmed that at very low normal stresses the asperities remain intact but, as the normal stresses increase in magnitude, the second-order asperities are crushed or sheared off and the first-order asperities take over as the controlling factors on joint friction angles. As the normal stresses increase still further the first-order asperities also will be sheared and the condition may eventually be reached in which the effective roughness angle is reduced to zero. An ultimate condition may occur in which the joint is filled with crushed debris or it may become polished (or slickensided) by progressive failure so that a residual strength condition is reached in which the joint residual angle of friction (ϕ_{jr}) equals ϕ_r of the intact rock.

The work of Patton (1966a), together with that of Einstein *et al*. (1970), indicates that a realistic shear test of a natural joint will produce a failure

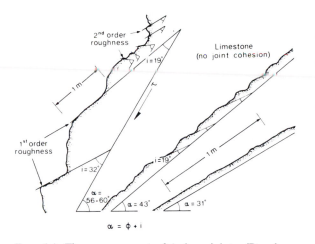

FIG. 6.4. The measurement of i along joints. (Based on Patton 1966.)

After interlocked asperities have suffered a certain measure of dilation the strength of the asperities may be exceeded and they will shear through. This stage can be represented by:

$$\tau_f = K + \sigma_n \tan\phi_r, \qquad (6.2)$$

where: ϕ_r is the angle of residual shearing resistance, and

K is the constant, equal to the ordinate of the intersection with the shear stress axis of the straight line which can be used to approximate the $\tau-\sigma_n$ curve at relatively high normal stresses (the value of K is sometimes spoken of as an apparent cohesion in reference to a Coulomb failure criterion).

The account given above has a number of assumptions hidden in it: (1) that the asperities are originally entirely interlocking; (2) that the normal stress is evenly distributed along the joint surfaces; (3) that the joint walls are unweathered and of equal compressive strength; (4) that there is no gouge or infill in the joint; and (5) that water has not affected the strength of the joint wall materials. In natural rock, however, joints have two very important characteristics: (1) the walls of open joints are not in contact along the whole wall surface but are supported upon the apices of asperities which may occupy only 1/10th to

envelope with the general form indicated in Fig. 6.5. At low stresses the failure line has a steep angle representing the angle of sliding resistance along a rough surface, which will approximate the angle of residual shearing resistance of the material plus the angle that the asperity surface makes with a plane through the bases of the asperities. At low normal stresses the asperities slide over one another and shearing resistance can be represented by equation (6.1) above.

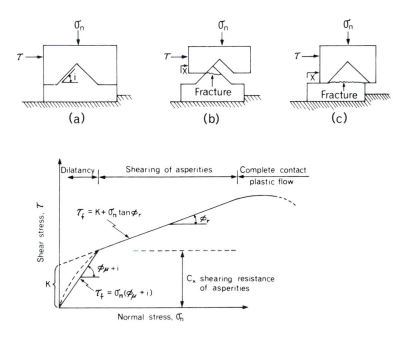

FIG.6.5. A direct shear test of a single asperity: (a) the first-order roughness; (b) dilation of the joint with failure of the tip of the asperity; and (c) fracture through the base of the asperity. The lower diagram presents a failure line for the three conditions shown above with development of residual strength and, finally, plastic deformation.

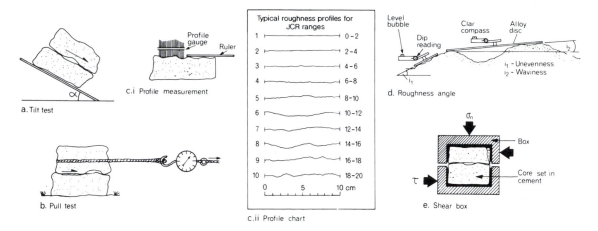

Fig 6.6. Methods of measuring joint shear strength and joint roughness. (c.ii, after Barton and Choubey 1977; d, after Brown 1981.)

1/1000th of the area of the joint wall (Jaeger 1971), consequently the normal stress is concentrated on the apices and can cause them to fail at average normal stresses which are relatively low; (2) joints are zones of preferential weathering and water circulation, consequently the joint wall compressive strength may be anything from 100 to 10 per cent of the intact rock compressive strength. Methods, for measuring the joint shear strength, are needed which will take these factors into account.

Measurement of joint shear strength

There are no wholly satisfactory ways of measuring the frictional resistance of joints in rock. Ideally a large-field shear box should be used which would simulate the stresses which are likely to induce failure in the field, and which would test the available strength along the whole of the joint, or a representative section of it. Large *in situ* tests have been carried out as part of the investigation for major foundation engineering projects, but the high cost of such tests—commonly several thousand dollars each—makes them impossible for most slope stability investigations. The commoner substitutes are described below.

A *tilt test*, Fig. 6.6, requires that two blocks, which represent the upper and lower mated walls of the parting which is likely to become a failure plane, be tilted to an angle at which sliding of the upper block over the lower may just begin. This angle is the angle of plane sliding friction for that parting. The tilt test has a number of major dis-

advantages. (1) It must sample a potential slide plane in the direction of sliding and over the length of surface contact. Selecting and obtaining adequate samples may be impossible on a steep slope as the blocks must be large enough to be representative of roughness along the failure plane—a Herculean requirement. (2) There is increasing evidence that the value derived for (ϕ_j) is strongly dependent upon the length of the sample chosen for the test. This is due to the mobilization of larger, but less steeply inclined asperities as the specimen size is increased. There is a reduction in both the dilation effect and asperity failure effect, as the sample becomes longer, with a corresponding decrease in the value of i so that, for a very long joint, the total frictional resistance is likely to be that of the basic angle of friction of the rock, ϕ_μ, plus two or three degrees (Fig. 6.7). (3) Single large asperities on the walls of one block may interlock with a mated depression in the opposite surface, especially where the slide plane has a 'stepped' profile. The range of angles achieved in tilt tests of several samples from one potential failure plane may thus vary from about 30–75° depending upon the presence of such steep asperities and their frequency. (4) A tilt test does not measure the effect of the characteristic normal load acting across the parting. (5) Measuring a tilt angle to the nearest 5° is often the best that can be achieved in difficult field conditions involving large blocks (Plate 6.1a, b, c).

A *pull test*, Fig. 6.6b, involves pulling a loose block of rock, with a joint surface forming its base,

W is the mass of the block; and
α is the angle of inclination of the joint.

As with the tilt test the experiment should be repeated for a number of block and joint surfaces, and a mean value derived for ϕ_j. The limitations of pull tests are similar to those of tilt tests.

A *profile gauge* can be used to derive roughness directly (Fig. 6.6c.i). As indicated in Fig. 6.6d, joint roughness may be divided into two components: a waviness which results from the large asperities, and an unevenness which results from the small asperities. The profile gauge (also called a feeler gauge) is most reliable when it is used to measure waviness as the needles do not fill all depressions in the rock and so ignore minor roughness. The gauge can record the roughness along many lines drawn on the surface of the joint and each record can be drawn from the gauge on to a field-data sheet. The drawn profile may be used to derive directly measurements of i using the relationship of Turk and Dearman (1985):

$$i = \text{arc} \cos\left(\frac{L_1}{L_2}\right), \qquad (6.4)$$

where: L_1 is the straight line distance between any two points on the joint face chosen to represent that joint, L_2 is the distance between the same two points as measured with the profile gauge.

Alternatively, a profile derived from the gauge may be compared with the published *comparison chart* (Barton and Choubey 1977) Fig. 6.6c.ii, and a joint roughness coefficient (*JRC*) estimated.

Where the focus of interest is the large-scale waviness of the joint surface, an estimate of i may be made by the use of special plates as shown in Fig. 6.6d. The plates are circular discs of aluminium alloy, with diameters of 5, 10, 20, and 40 cm. These are fitted in turn to an inclinometer. The Clar compass, manufactured by Breithaupt, is recommended as it incorporates a levelling bubble and a lid which is used to record dip directly. The waviness angles are obtained by using the largest disc held against the joint surface in at least twenty-five positions, and recording dip angles and directions for each position. Up to one hundred positions may be needed to record small asperity angles with the smallest disc (Brown 1981).

A *laboratory shear box* is the only readily available device for measuring joint frictional strength under known stresses. Most boxes are so

PLATE 6.1. (a) Measuring the angle of inclination after a tilt test. (b) A pull test. (c) A profile gauge on a bedding plane joint with waviness.

along an exposed joint plane. The force required to just move the block is recorded by a spring balance or similar instrument. The spring balance may also be suspended from a pole and used to measure the mass of the block. The frictional resistance along the joint is determined from:

$$W \cdot \cos\alpha \cdot \tan\phi_j = F + W \cdot \sin\alpha$$
$$\tan\phi_j = (F + W \cdot \sin\alpha)/W \cdot \cos\alpha, \qquad (6.3)$$

where: $\tan\phi_j = (\phi_\mu + i)$;
F is the spring balance reading;

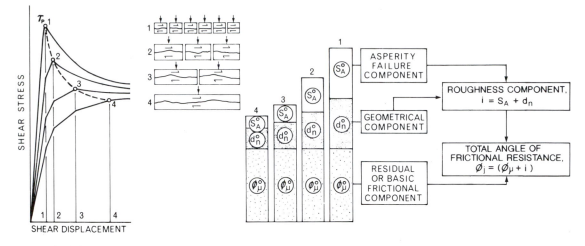

FIG 6.7. The influence of scale on components of shear strength along a joint. At all scales there is a basic friction angle component; at large scales the steep asperities shear off and the inclination of the controlling roughness decreases. Shearing through progressively larger asperities is possible as the normal stress and shear displacement increase. (After Bandis *et al.* 1981.)

small that repeated testing is required if the whole joint surface is to be represented in the tests. In tests on models and specimens of rock it has been found that large asperities have waviness angles which are relatively low (<20°), and an unevenness due to small asperities with angles which may be as high as 60°, but more commonly fall in the range 40–50°. Which of these components of roughness is significant depends upon the strength of the asperities and the length of the joint being studied (see Fig. 6.7).

It has been stated already that the peak shearing resistance (τ_p) is related to a measure of joint roughness and the resistance of the joint wall asperities to crushing or shearing by an applied stress, as well as to ϕ_μ of the intact rock. These components have been incorporated into an empirical formula by Barton and Choubey (1977):

$$\tau_p = \sigma_n . \tan[JRC . \log_{10}(JCS/\sigma_n) + \phi_\mu], \quad (6.5)$$

where: *JRC* is a joint roughness coefficient, and
JCS is the joint wall compressive strength.

For an estimation of the value of *JRC* two methods are available:

1. The roughness of the natural joint surface may be evaluated by pulling a natural joint block from a rock face and comparing the roughness of the joint with the chart (Fig. 6.6c.ii) of characteristic joint surface profiles provided in Barton

and Choubey (1977) and in Brown (1981). This chart indicates the typical range of *JRC* values associated with the classes of roughness profiles. Users of the chart are recommended to ensure that the chart is at true scale before comparisons are made. It has been pointed out by Tse and Cruden (1979) that rather small errors in estimating *JRC* could result in serious errors in the evaluation of peak shear strength, and Barton and Choubey have recommended that tilt tests be used.

2. In a tilt test:

$$JRC = \frac{\alpha° - \phi_{jr}°}{\log_{10}(JCS/\sigma_{no})}, \quad (6.6)$$

where α is the tilt of the joint at failure, ϕ_{jr} is the residual friction angle of the joint, and σ_{no} is the normal stress induced by self weight of the sliding block.

The tilt angle (α) is measured with an inclinometer. The residual friction angle of the joint may be derived from the relationship:

$$\phi_{jr} = (\phi_\mu - 20°) + 20(r/R), \quad (6.7)$$

where ϕ_μ is the basic friction angle of the intact wall rock determined either by tilt tests on dry, unweathered, sawn surfaces of the particular rock, or estimated from the values given in Table 6.2; *R* is the uniaxial compressive strength indicated by Schmidt hammer tests on smooth or sawn, dry,

TABLE 6.2. *Basic friction angles of flat un-weathered rock surfaces*

Type of rock	Dry	Wet	Friction angle (°)
Sandstone	•		26–35
		•	25–34
Shale		•	27
Siltstone	•		31–33
		•	27–31
Conglomerate	•		35
Chalk		•	30
Limestone	•		31–37
		•	27–35
Basalt	•		35–38
		•	31–36
Fine-grained granite	•		31–35
		•	29–31
Coarse-grained granite	•		31–35
		•	31–33
Porphyry	•		31
		•	31
Dolerite	•		36
		•	32
Gneiss	•		26–29
		•	23–26
Slate	•		25–30
		•	21

Source: Barton and Choubey (1977).

unweathered, rock surfaces; and r is the uniaxial compressive strength indicated by Schmidt hammer tests on wet, weathered, joint surfaces. A correlation chart for expressing L-type Schmidt hammer readings as compressive strength (MPa) is given as Fig. 5.27. A problem still to be overcome is the unreliability of the Schmidt hammer on weak, weathered material (Shehata and Eissa 1985).

A typical rock with $\phi_\mu = 30°$ and limited weathering ($r/R = 30/40$) has a theoretical minimum ϕ_{jr} value of 25°. If joint weathering has been more severe ($r/R = 20/40$), ϕ_{jr} is 20°. This method gives ϕ_{jr} values within 1° of the value derived by large scale laboratory tests.

The normal stress induced by self weight of the sliding block is given by:

$$\sigma_{no} = \gamma . h . \cos^2\alpha, \qquad (6.8)$$

where: γ is the unit weight of the rock (kN/m³), and h is the thickness of the upper block

(metres). For example, if the following values have been measured:

$\alpha = 51°$ (tilt angle)
$h = 500$ mm (block thickness), hence
$\sigma_{no} = 0.005$ MPa
$\gamma = 25$ kN/m³
$JCS = 50$ MPa (estimated using a Schmidt hammer on a dry joint)
$\phi_{jr} = 23°$

from equation 6.6:

$$JRC = \frac{51° - 23°}{\log_{10}\left[\dfrac{50}{0.005}\right]} = 7.0.$$

It may be noted that an underestimated ϕ_{jr} value results in an overestimated *JRC* value, and vice versa. This automatic compensation of errors is one reason for the provision of accurate estimates of the peak shear strength by a method which is otherwise rather imprecise. The tilt test is performed on dry joints to prevent problems induced by water films on the joint surfaces. Consequently the appropriate *JCS* value is that measured on the dry joint surfaces used in the tilt test. Three tilt tests are performed on each joint and the mean value is used for estimating *JRC*. Because of the very low stresses across the joint there is no visible damage and the test can be repeated on the same samples. In the determination of shear strength we are concerned with the minimum strength for stability so the *JCS* value used to evaluate equation 6.5 is that determined by a Schmidt hammer test on a saturated joint.

The method of determining joint roughness of Barton and Choubey (1977) has been substantially validated by shear testing of rough joints in a large ring shear box (Xu and De Freitas 1990a,b).

It is evident from the preceding discussion that roughness—not mineralogy—largely controls the available strength along a joint. Exceptions to this statement occur in the presence of silicate sheet minerals such as mica, talc, serpentine, and silicate clays, which may all reduce the basic friction angle to about 12° compared with the normal 25–40° for rock surfaces.

Effective stresses along partings

Shearing resistance is also influenced by the presence of water under pressure, and an effective stress condition occurs unless high normal stresses largely close the joint so that the actual area of inter-particle contact is a significant proportion of the total area of a discontinuity.

For a normal joint with an overburden γz (where γ is the unit weight of the rock and z is the thickness of the overburden)

$$\sigma_n = \gamma z \quad \text{and} \quad u = \gamma_w . z \qquad (6.9)$$
$$\text{then} \quad \tau_p = \gamma z . \tan\phi_j \text{ for dry conditions}, \quad (6.10)$$
$$\text{and} \quad \tau_p = (\gamma - \gamma_w)z . \tan\phi_j \text{ for water-filled joints.} \qquad (6.11)$$

If $\gamma = 27 \, \text{kN/m}^3$ and $\gamma_w = 10 \, \text{kN/m}^3$ then hydrostatic pore pressures reduce shear strength along the joint by about 35 per cent. If σ_n is reduced by removal of the overburden without free drainage permitting a reduction in joint-water pressures (also called cleft-water pressures), then the effect is worse still, with a critically high stressed region developing at the base of a slope. Dilation of joints as a result of the buoyancy effect of the cleft-water pressures may lead to development of a landslide at the base of a slope. These matters are dealt with more fully in Chapter 15.

Rock-mass strength

It has been stated above that the strength of rock masses is controlled by a number of factors, of which intact strength and shear resistance along joints have been discussed already in some detail. The problem of determining total mass strength was first overcome by engineers engaged in tunnelling, mining, quarrying, and with designing foundations for structures in rock, by incorporating into one classification those parameters which contribute to rock-mass strength and weakness.

Classification parameters

Numerous classifications of rock-mass strength have been proposed for engineering purposes (Müller 1958; Pacher 1958; Deere and Miller 1966; Piteau 1971, 1973; Robertson 1971; Wickham *et al.* 1972; Bieniawski 1973, 1989). Most of the commonly used classifications include several of the following parameters:

(1) strength of intact rock;
(2) state of weathering of the rock;
(3) the spacing of joints, bedding planes, faults, foliations, or other partings within the rock mass;

(4) orientation of the partings with respect to a cut slope;
(5) width (or aperture) of the joints, bedding planes, or other partings;
(6) lateral or vertical continuity of the partings;
(7) gouge or infilling material in the partings;
(8) movement of water within or out of the rock mass.

These eight parameters have been incorporated into one classification and assessed for field studies in geomorphology (Selby 1980). A further three parameters are used in some engineering classifications:

(9) residual stresses within the rock;
(10) the angle of internal friction along partings;
(11) waviness and unevenness along partings.

Because residual stresses are usually greatly reduced or eliminated by the development of open joints near the ground surface they may be disregarded in a geomorphic study. Neither the friction angle nor the roughness of rock surfaces along joints are easily measured in the field, especially where exposure of joint surfaces is limited. Measurement of joint shear strength is therefore likely to be undertaken in studies where it is specifically required, as in investigations of slope stability, but for general investigations an assessment of joint orientation is more appropriate.

Most classifications for mining purposes indicate that it is possible to divide the values of each parameter into five classes and to apply a numerical rating to each. This practice has been followed in the geomorphic classification.

Intact strength

Rock intact strength is measured in the field using either a Schmidt hammer or point load strength tester. Care is taken to ensure that the strength of unweathered rock is assessed.

Weathering

The state of weathering of a rock mass has a major influence upon its strength, permeability, and ease with which its material may be deformed (Figs. 6.8 and 6.9). Permeability and porosity may increase and strength decrease by more than two orders of magnitude during the conversion of unaltered rock to residual soil. It is, however, frequently difficult to quantify this change because weathering proceeds preferentially along rock fissures and zones

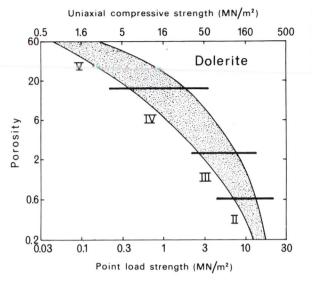

FIG 6.8. The relationship between porosity and strength in a weathering profile. The horizons are indicated by Roman numerals. (Data from Dearman 1974.)

of weathering may be extremely irregular. For field classification several schemes have been devised for the description of granitic rock and these can be generalized to be applicable to a wide range of rock types. Many schemes recognize six grades of weathering ranging from fresh unweathered rock to a residual soil. The three main criteria for classification of weathering grades are: degree

of rock discoloration; ratio of rock to soil; and presence or absence of original rock texture. Descriptions of these grades are given in Fig. 6.10 and Table 6.3. It should be noted that grade VI—the surface soil—lies outside any consideration of rock strength. Major papers on this subject include those by Dearman (1974, 1976); Baynes *et al.* (1978); Moye (1955); and Ruxton and Berry (1957).

Spacing of partings within a rock mass

For the sake of simplicity all partings will now be referred to as joints as their mechanical properties, not their origin, control their influence on rock strength.

All intact rocks have cohesive and frictional strength. As joints in rocks open cohesion is lost and frictional strength is reduced, although the strength along a joint in which rock grains are interlocking may still be high. Consequently the effective strength of a rock mass containing only microfissures may be nearly as great as that of an unfissured rock.

It has been shown by Hoskins *et al.* (1967) that in low stress situations, such as exist close to the ground surface, the strength along a joint may be described by a Mohr envelope in a straight line form, and that the frictional strength is independent of the area of contact and directly dependent upon the normal load. They also discovered that cohesion values are appreciably

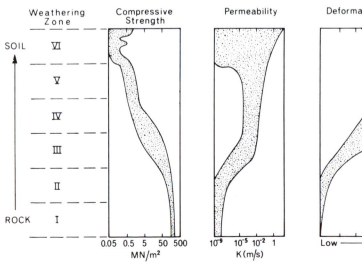

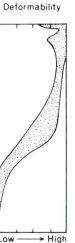

FIG 6.9. An idealized weathering profile with some characteristic properties. (Based upon Dearman 1974.)

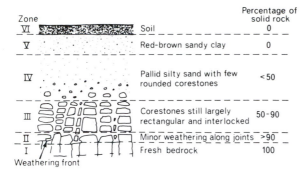

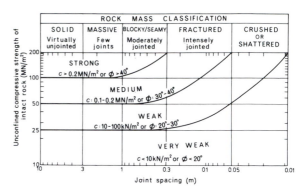

FIG 6.10. Features of a full weathering profile developed on granitic rocks. (Modified from Ruxton and Berry 1957.)

FIG 6.11. Strength classification of jointed rock masses. (Developed by Bieniawski 1973, from an original classification by Müller 1963.)

large for clean closed joints and often lie between 350 and 1400 kN/m² (Fig. 6.2). This phenomenon may be explained either by cohesive interlocking of rock grains on either side of a 'hair-line' joint or, as Patton (1966) suggested, by showing that the Mohr envelope is in fact curved at very low normal stresses and cohesion is consequently negligible. Since cohesion is destroyed even after very small movements, and there is no way of measuring it in the field, it is always assumed to be zero. The possibility that it does exist, however, suggests that very tightly closed hair-line joints should not

normally be included in an assessment of joint spacing (Plate 6.2).

The more closely spaced are joints the weaker is the rock mass, and the greater is the opportunity for water pressures and weathering processes to weaken it further. Any joint spacing classification thus has a simple logic behind it. The classification of Deere (1968) has been widely used by engineering geologists and is adopted here as it provides a convenient five-class scale (Fig. 6.11).

TABLE 6.3. *A scale of mass weathering grades*

Grade	Class	Description
VI	Residual Soil	A pedological soil containing characteristic horizons and no sign of original rock fabric.
V	Completely weathered	Rock is discoloured and changed to a soil but some original rock fabric and texture is largely preserved. Some corestones or corestone ghosts may be present.
IV	Highly weathered	Rock is discoloured throughout; discontinuities may be open and have discoloured surfaces and the fabric of the rock near to the discontinuities may be altered so that up to one half of the rock mass is decomposed and disintegrated to a stage in which it can be excavated with a geological hammer. Corestones may be present but not generally interlocked.
III	Moderately weathered	Rock is discoloured throughout most of its mass, but less than half of the rock mass is decomposed and disintegrated. Alteration has penetrated along discontinuities which may be zones of weakly cemented alteration products or soil. Core stones are fitting.
II	Slightly weathered	Rock may be slightly discoloured, particularly adjacent to discontinuities which may be open and will have slightly discoloured surfaces; intact rock is not noticeably weaker than fresh rock.
I	Unweathered fresh rock	Parent rock showing no discoloration, loss of strength, or any other weathering effects.

Source: Modified from Dearman (1974, 1976).

PLATE 6.2. Joints with varying degrees of openness, continuity, spacing and, therefore, significance for rock-mass strength.

Müller (1963) and Bieniawski (1973) have demonstrated how the spacing of joints modifies rock-mass strength in comparison with intact rock material. Thus a rock with high intact strength (100 to 200 MPa) but intensely jointed, with joints 50 mm or less apart, has a weak rock mass.

On many sedimentary rocks weathering processes produce a close fissuring of the outcrop surface. This pattern may obscure the more widely spaced true joint pattern. Superficial fissuring does not reduce mass strength, even though it may aid the production of talus material, and it is frequently essential to remove the shattered rock debris before measurements can be made of true joint spacing, orientation, and width (Plate 6.3).

Orientation of joints

The significance of joint orientations may be gauged from Terzaghi's (1962) comment that the stability of a slope formed on chemically intact rock, with uniaxial compressive strength greater than about 30 MPa, depends primarily on the orientation of the major continuous joints with respect to the hillslope. The theoretical critical hillslope angle for stability is, in the absence of cohesion across the joint, controlled by the friction angle along the joint (ϕ_j°) in the direction of potential shear. Where the major continuous joints have a random orientation the critical hillslope angle for stability on igneous rocks is commonly about 70° because of the high friction angle of interlocking joint blocks. In sedimentary rocks the major joints are usually along bedding planes and hence essentially planar. The friction angle in planar joints is commonly close to 30° (Table 6.1) and if bedding planes dip towards a valley at an angle smaller than ϕ_j° the theoretical critical hillslope angle (for stability) is 90°. For bedding planes dipping towards a valley at angles greater than ϕ_j° the critical hillslope angle is equal to the angle of dip of the bedding planes (these relationships are explored in more detail in Chapter 15).

PLATE 6.3. An outcrop of sandstone showing the clean face and widely spaced joints exposed by a recent block fall, and the superficial spalling and blocky disintegration which gives the appearance of more closely spaced fissures. The strength of the rock mass is related to the spacings of the main joints, not to the weathering features.

The theoretical condition seldom applies exactly to a field situation because the water-content of a joint varies and produces a range of values of ϕ_j as well as upthrust, and hence decreased stability, from cleft water pressures. Joint roughness also varies; joint blocks may be cemented together or interlocked and hence stability increased. It has been shown experimentally by Barton (1973) that rough joints have a friction angle which is greater than that of a smooth joint in intact rock in proportion to the amplitude of the asperities along the joint.

It is rare for outcrops to provide sufficient exposure for a full evaluation of joint roughness, so this parameter is included in the rock-mass classification to only a very limited extent.

It must be recognized that high water pressures, very smooth joints—such as cleavages in slates—and cementation may produce considerable variation from theory, but some guide to orientation significance may be obtained from Table 6.4. In this system horizontal joints are taken as an intermediate case with greater strength being attributed to joints dipping into the slope, and low strength ratings being adopted for increased dips out of the slope. Vertical jointing is classed with horizontal joints for rocks with high compressive strength, but it is regarded as unfavourable in low compressive strength rocks which may fail by buckling. Random orientation of joints is regarded as favourable in hard rocks with rough joints produced by tensile failure, but fair or unfavourable in shattered rock in which shear failure may occur along multiple intersecting smooth joints, foliation, or cleavage. Table 6.4 is an approximate guide only and anyone using the classification should be primarily guided by local field evidence, particularly of slope failures. It should also be noted that in rocks with strong planar joints dipping steeply into a slope, failure is likely to

TABLE 6.4. *Strength classification for joint orientations*

	Mode of joint formation	
	Tensile (rough)	Shear (smooth)
Very unfavourable	Joints dip out of the slope: planar joints 30–80°; random joints >70°.	Joints dip out of the slope: planar joints >20°; random joints >30°.
Unfavourable	Joints dip out of the slope: planar joints 10–30°; random joints 10–70°.	Joints dip out of the slope: planar joints 10–20°; random joints 10–30°.
Fair	Horizontal to 10° dip out of the slope. Nearly vertical (80–90°) in hard rocks with planar joints.	Horizontal to 10° dip out of the slope.
Favourable	Joints dip from horizontal to 30° into the slope: cross joints not always interlocked.	
Very Favourable	Joints dip at more than 30° into the slope: cross joints are weakly developed and interlocking.	

PLATE 6.4. Joints in dolerite. The main joints dip into the cliff but the control on cliff stability is exercised by cross joints nearly parallel to the cliff face.

occur in cross joints rather than along the continuous joints, and the cross-joint angle should then be considered as the control (Plate 6.4).

The accuracy with which dip measurements, on limited outcrops, can be made is indicated by studies of the accuracy of joint surveys in tunnels and open-pit mines (Piteau 1973). The estimated average maximum error is ±5°. For planar joints this may be improved to ±1° where dips are greater than 70°, and ±3° where inclinations are between 30 and 70°.

Width, or aperture, of joints

Separation is important to mass strength because it largely controls the frictional strength along the joint as well as the flow of water and the rate of weathering of the wall rock. Widely open joints have no inherent strength and their resistance to shear is only that of the joint infill. Where the infill is of clay then the stability of a rock mass with joints dipping out of the hillslopes may be little more than that of the clay.

Tightly closed joints may have high frictional and some cohesive strength, while those with widths between the two extremes have strengths dependent upon the joint roughness and contact between the wall rocks.

In engineering classifications joint width is usually judged according to that likely to be encountered in the high stress conditions of newly opened tunnels or mines. In geomorphic situations joints exposed at the groundsurface may be several centimetres open. Furthermore such joints can be further opened by weathering and erosion.

TABLE 6.5. *Geomorphic rock-mass strength classification and ratings*

Parameter	(1)	(2)	(3)	(4)	(5)
	Very Strong	Strong	Moderate	Weak	Very Weak
Intact rock strength (N-type Schmidt Hammer 'R')	100–60	60–50	50–40	40–35	35–10
	r:20	r:18	r:14	r:10	r:5
Weathering	Unweathered	Slightly weathered	Moderately weathered	Highly weathered	Completely weathered
	r:10	r:9	r:7	r:5	r:3
Spacing of joints	>3 m	3–1 m	1–0.3 m	300–50 mm	<50 mm
	r:30	r:28	r:21	r:15	r:8
Joint orientations	Very favourable. Steep dips into slope, cross joints interlock	Favourable. Moderate dips into slope	Fair. Horizontal dips, or nearly vertical (hard rocks only)	Unfavourable. Moderate dips out of slope	Very unfavourable. Steep dips out of slope
	r:20	r:18	r:14	r:9	r:5
Width of joints	<0.1 mm	0.1–1 mm	1–5 mm	5–20 mm	>20 mm
	r:7	r:6	r:5	r:4	r:2
Continuity of joints	None continuous	Few continuous	Continuous, no infill	Continuous, thin infill	Continuous, thick infill
	r:7	r:6	r:5	r:4	r:1
Outflow of groundwater	None	Trace	Slight <25 l/min/10 m^2	Moderate 25–125 l/min/10 m^2	Great >125 l/min/10 m^2
	r:6	r:5	r:4	r:3	r:1
TOTAL RATING	100–91	90–71	70–51	50–26	<26

Source: Selby (1980).

Consequently the width which has to be considered is that of the joint at a depth of 10 or more centimetres into the outcrop. Another problem is that 'hair-line' fractures may have little effect in reducing strength and may have to be ignored (Plate 6.2). Judgement has to be exercised in assessing width and a conservative view is advocated in interpreting the classes of Table 6.5.

Continuity and infill of joints

The significance of continuity of joints is illustrated by the observation of Orr (1974) that where relative continuities of joints are less than 60 per cent then shear along their surfaces is very improbable. Terzaghi (1962) analysed the significance of the cohesive strength of rock in cliff slope stability. He suggested that the effective cohesion within a jointed rock mass is given by:

$$c_e = c\,\frac{(A - A_j)}{A},$$

where: c_e is the effective cohesion of the rock mass;

c is the intrinsic cohesion of intact joint blocks;

A is the total area of the shear plane;

A_j is the total joint area within the shear plane.

Thus the larger the bridges of intact rock between joints the greater is the effective cohesion (i.e. the smaller is A_j). Continuous joints thus reduce cohesive strength, provide zones of shear, and permit the circulation of water.

Because of the difficulty of measuring fill in inaccessible joints which are seldom exposed in three dimensions an entirely qualitative classification is used (Table 6.5).

Groundwater

The flow of water out of a rock face is probably of greater significance in reducing the stability of rock exposures in mines and tunnels than in landforms. However, seepage forces develop along joints and the incidence of rockfalls from cliffs in mountains is frequently correlated with high water pressures and rainfall maxima.

Subsurface water promotes instability in at least four ways:

(1) pore water in joint filling materials promotes weathering and solution and thus reduces the cohesive and frictional strength of the infill;
(2) cleft-water pressures along joint surfaces produce uplift forces along potential failure planes, thereby reducing frictional resistance and effective normal stresses;
(3) water in the pores of intact rock promotes weathering and solution and decreases rock compressive strength, especially where confining stresses have been reduced;
(4) mudstones, shales, and argillites may disintegrate by slaking and be converted to slurries.

The extreme effects of water upon shale slopes, in the De Beers diamond mine near Kimberley, were noted by Jennings (1971) who reported that they will stand at angles greater than 45° when drained, but undrained slopes in the same material will stand at only 18–29°. A similar situation occurs in cohesionless sand which will support a hillslope of about 30° when it is dry, but of only half that when the water table has risen to the slope surface.

A rock-mass strength classification should, ideally, use pore- or cleft-water pressures within the rock mass at the least favourable season of the year. This is seldom, if ever, possible in a geomorphic study and instead it is recommended that an estimate be made of the volume of water flowing out of each $10\,m^2$ of rock face in the wettest season. It is recognized that in many situations it may be impossible to do better than express the flow as 'none, trace, slight, moderate, or great'.

Unified classification and rating of parameters

In a total classification of rock-mass strength all of the parameters have to be combined in a graded scale. For this purpose five classes are used: this number of classes can be readily recognized in the field and is commonly used in engineering classi-fications. Class 1 denotes a very strong rock mass and class 5 a weak mass.

The parameters which are assessed in the field provide semi-quantitative data for the classi-fication. The parameters do not necessarily con-form to the same class. Thus at one exposure the intact rock strength may be class 2; the joints may be widely spaced (class 1), discontinuous (class 1), but the dip of the strata may be unfavourable (class 4); the same rock mass may be only slightly weathered (class 2), but have a moderate outflow of ground water (class 4).

Not all parameters are of equal importance in producing rock strength and it is thus necessary to assign a numerical weighting to each. The final rock-mass strength class will then be the sum of the weighted values determined for the individual parameters at each site. Higher numbers reflect greater mass strength.

Intact rock strength is given a 20 per cent rating; separation of joints 30 per cent; joint orientations 20 per cent; joint width, continuity, and water flow collectively 20 per cent. Weathering is accorded only 10 per cent because much of the effect of weathering is subsumed in the other parameters: thus loss of strength and opening of joints at a rock face are largely weathering phenomena, and water movement and formation of joint infill both promote and are an effect of weathering.

The classification, with the ratings (r) for each parameter and each class is presented in Table 6.5. Field experience indicates that most of the ratings can be applied without much difficulty. The greatest problem arises where the spacing of joints along a face is variable and more than one class of spacing is present. This problem may be resolved either by dividing the exposure into several units and classifying each one separately, or by using an intermediate rating value between the set values of each class. The latter procedure is most relevant to the joint spacing parameter where the differences in rating between classes 2, 3, 4, and 5 are six or seven points.

In this classification the lowest possible score is 25, leaving open the possibility of adding soil into the lower range of a comprehensive strength scale. Very rarely will very weak rock masses be exposed in the field for they will usually be vegetated in humid climates and in arid climates the absence of a high water outflow will prevent application of the lowest possible rating.

The classification as set out above has been

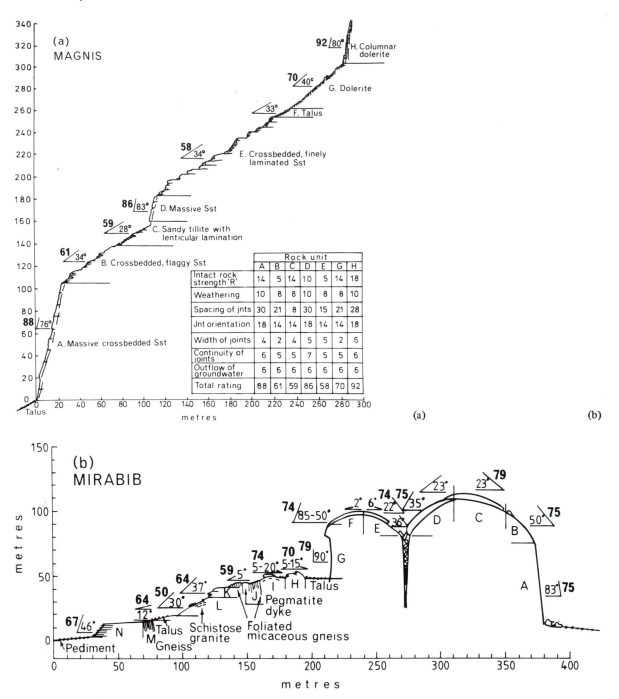

The table within figure (a):

	Rock unit						
	A	B	C	D	E	G	H
Intact rock strength 'R'	14	5	14	10	5	14	18
Weathering	10	8	8	10	8	8	10
Spacing of jnts	30	21	8	30	15	21	28
Jnt orientation	18	14	14	18	14	14	18
Width of joints	4	2	4	5	5	2	6
Continuity of joints	6	5	5	7	5	5	6
Outflow of groundwater	6	6	6	6	6	6	6
Total rating	88	61	59	86	58	70	92

FIG **6.12.** (a) Scaled profile of a rock slope in Magnis Valley, Transantarctic Mountains, Antarctica, showing the strength ratings for each rock unit and the slope angle. This slope is shown in Plate 6.5; (b,c) profiles of two granite bornhardts in the Namib desert showing slope angles and mass strength ratings.

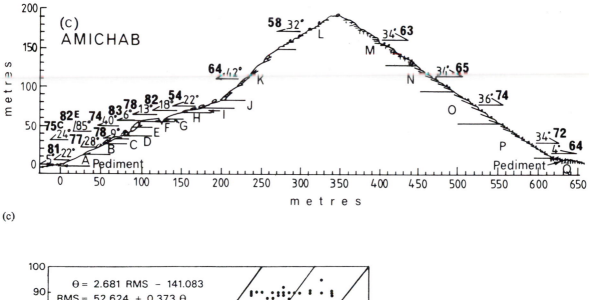

(c)

FIG 6.13. The relationship between slope inclination and rock-mass strength with 95 per cent confidence limits defining the strength-equilibrium envelope. All data are either from publications or unpublished field-work of M. J. Selby and B. P. Moon for slopes which were recognized as being in strength equilibrium. Data plot and statistical analysis by Abrahams and Parsons (1987).

independently tested and validated by Moon (1982), who also suggested subdivisions of the intact strength and joint spacing ratings (Moon 1984a). A subsequent study (Allison and Goudie 1990: 142) has cast doubt on the usefulness and applicability of these subdivisions.

The classification is intended primarily to aid the study of rock hillslopes. If rock slope profiles are examined and the Rock Mass Strength rating and the mean hillslope angle are recorded for every rock unit of uniform characteristics, or every hillslope unit of uniform mean inclination, then the profile may be recorded as in Fig. 6.12 and the inclination (θ) plotted against RMS for all slope units (Fig. 6.13).

Strength-equilibrium slopes

On many rock slopes there is an approximate condition of limiting equilibrium between the inclination of the hillslope and the resistance to failure of the rock mass as measured by its RMS rating. Slopes in this condition have been designated strength-equilibrium slopes (Selby 1982a).

PLATE 6.5. Rock slopes for which the relationship between rock-mass strength and slope angle has been determined, Transantarctic Mountains.

The RMS classification has been applied to over 300 rock units in three continents, in a wide range of climates and to sixteen lithologies. Strength-equilibrium slopes have been found to be widespread. Data from them has been plotted by Abrahams and Parsons (1987) and it can be seen from Fig. 6.13 that the data points lie in an envelope defined by the 95 per cent confidence limits about the regression line. It must be emphasized that all of the plotted data points come from slope units which were recognized to be in strength equilibrium by the field evidence. Many rock slopes are not in equilibrium because they are controlled by such factors as: structure; buttressing; undercutting by streams, waves, or glaciers; relict and no longer effective processes; solution processes; regolith covers; and talus mantles. Such matters are considered in Chapter 16.

Field procedures

The only items of equipment needed for the application of the classification are: a tape measure, an inclinometer, and a strength tester—in most cases a Schmidt hammer is the most convenient device.

The classification can be applied to any rock mass with sufficient exposure for measurements to be made of the joints. The outcrop must be divided into a number of regions, each having similar characteristics; for example, the same joint spacing, dip, and lithology. The boundaries of a region will usually correspond with a prominent feature such as a distinct break of slope, a fault, dyke, or change in lithology.

Measurements of joint spacing are made in both a vertical and horizontal direction and the area of outcrop measured is usually about $10\,\text{m}^2$, although this has to be adjusted to suit field conditions. By using Table 6.5 the ratings for each parameter may be decided in the field but it is recommended that a detailed description and field sketch should also be made of each rock unit so that the operator can check on the consistency of his, or her, measurements.

The classification may be used to further an understanding of changes in slope profiles, the location of hills and depressions, the location of stacks, tors, inselbergs, roches moutonnées, or other features. Where rock exposure is adequate the rock mass classification rating can be incorporated into a geomorphic map. Examples of applications are given in Fig. 16.11.

Strength criterion for rock masses

Hillslopes rarely fail as a mass; more commonly failure occurs along joints which are critical for stability. Consequently the strength of rock masses as defined in triaxial stress conditions is seldom relevant to geomorphic study, although it is relevant in mining, dam-foundation engineering, and large rock-slope excavations. It could also be used in some natural slope failure studies. The problem of testing huge representative samples from rock masses has been overcome by the use of Bieniawski's Rock Mass Rating as a way of developing an appropriate Mohr–Coulomb failure envelope for a rock mass (Hoek and Brown 1980; Hoek 1990). Researchers needing to use this criterion should refer to these papers.

7 Properties of Soils

The physical properties of soils are strongly influenced by their mineralogy, texture, and fabric. The properties are not constant over time, with water-content and void space capable of changing very quickly and other properties changing more slowly. Nor are properties constant in space with major variations in structure, fabric, and mineralogy being identifiable over distances of a few metres (Beckett and Webster 1971; Culling 1986).

A knowledge of soil physical properties is an important foundation for classification of soils and a major component of any capacity to predict the behaviour of soils in response to applied stresses and variations in water content.

Many physical properties are expressed in the results of index tests. An index is a quantity which expresses or indicates a physical property in terms of a standard (Martin 1986). For practical purposes simple index tests may be expressed semi-quantitatively as a number of ordered groups or classes, or quantitatively on a continuous numbered scale. Semi-quantitative indices are commonly used as aids to classification, as in weathering grades of altered rock. Quantitative index numbers may be used for classification and to develop an understanding of behaviour by the use of correlation statistics and other comparative methods.

The existence of many standard index tests requires that, for effective communication and application, the user should follow standard procedures for determining the index numbers and also that the nature of the test and its implications should be thoroughly understood. It must also be recognized that although there are now available many advanced analytical facilities, our ability to predict field behaviour is still inadequate (Mitchell 1986).

Phase relationships

Soil has three phases: solids, water, and air (Fig. 7.1). The total volume (V) of the soil material is expressed as the sum of the volume of these components:

$$V = V_s + V_w + V_a,$$

and the total mass (M) is the sum of the mass of solids and mass of water:

$$M = M_s + M_w.$$

Unit weight of a soil and its natural moisture content are the simplest and most commonly used indicators of the state of the material and of its characteristic strength and other properties. A sample of undisturbed (i.e. natural structure, void space, and water-content) soil is collected in a tube of known weight and internal volume. The *bulk density* (ρ) of the soil is then M/V, expressed in units of kg/m^3 or Mg/m^3.

The unit weight (γ) of the soil is its mass multiplied by the acceleration due to gravity (g) to give the weight (W) for the volume of that soil:

$$\gamma = \frac{M \cdot g}{V} = \frac{W}{V} \quad \text{or} \quad \gamma = \rho \cdot g,$$

where: $g = 9.81 \, m/s^2$ and soil weight is expressed in Newtons. Unit weight of soils is conveniently expressed in units of kN/m^3 (for many practical purposes a value of $g = 10 \, m/s^2$ may be adopted).

Moisture content is obtained by drying a core sample of soil at a standard temperature for a standard time; usually at $105 \, °C$ for 24 hours, but some soils, such as those containing allophane, gypsum, or peat, undergo fundamental changes in their properties if dried at such temperatures and for them longer drying at lower temperatures may be required.

TABLE 7.1. *Maximum and minimum densities for granular soils*

Description	Void ratio		Porosity (%)		Dry unit weight (kN/m³)	
	e_{max}	e_{min}	n_{max}	n_{min}	$\gamma_{d,min}$	$\gamma_{d,max}$
Uniform spheres	0.92	0.35	47.6	26.0	—	—
Clean uniform sand	1.0	0.40	50	29	13.0	18.5
Uniform inorganic silt	1.1	0.40	52	29	12.6	18.5
Silty sand	0.90	0.30	47	23	13.7	20.0
Fine to coarse sand	0.95	0.20	49	17	13.4	21.7
Silty sand and gravel	0.85	0.14	46	12	14.0	22.9

Source: B. K. Hough (1957), *Basic Soils Engineering* (New York: Ronald Press Company).

$$\text{Moisture content} = m = \frac{\text{mass of water}}{\text{mass of solid}}$$

$$= \frac{M_w}{M_s} \, (\%)$$

$$\text{Degree of saturation} = S = \frac{\text{volume of water}}{\text{volume of voids}}$$

$$= \frac{V_w}{V_a + V_w} = \frac{V_w}{V_v} \, (\%).$$

It should be noted that m can be greater than 100%; at full saturation $S = 100\%$ and if the soil is dry $S = 0\%$:

$$\text{Voids ratio} = e = \frac{\text{volume of voids}}{\text{volume of solid}} = \frac{V_a + V_w}{V_s}$$

$$= \frac{V_v}{V_s} \, (\%)$$

$$\text{Porosity} = n = \frac{\text{volume of voids}}{\text{total volume of soil}}$$

$$= \frac{V_a + V_w}{V_a + V_w + V_s} = \frac{V_v}{V} \, (\%)$$

and

$$n = \frac{e}{1 + e}.$$

Specific gravity of the solid matter (G_s) is the ratio of the mass of the solid with the mass of water occupying the same volume as the solid matter in the soil:

$$G_s = \frac{M_s}{V_s \cdot \rho_w},$$

where ρ_w is the density of water at $20\,°C = 1000 \, \text{kg/m}^3 = 1 \, \text{Mg/m}^3$.

The packing of particles has a major effect on the values of these indices for all granular soils, with the extreme values being for the loosest possible stable packing in cubic arrangement and the densest state resulting from the closest possible packing. Maximum and minimum values of index values for granular solids are given in Table 7.1. For cohesive soils index values are potentially more variable than for granular soils, because under high confining pressures clays can deform plastically and close voids. Sodium montmorillonite is an extreme example: at low confining pressure it can have a water-content (by weight) of 900 per cent and void ratio of 25 per cent; at high confining pressure the water-content may be as low as 7 per cent and the void ratio <0.2 per cent.

Behaviour and water-content

Fine-grained soils composed of clay-size and silt particles and colloidal organic matter can exist in solid, semi-solid, plastic, and liquid states, depending upon their water-content. The water-content of remoulded soils at which there is a transition from one phase to the next has become one of the most widely used properties in soil mechanics. The transition water-contents for any soil are the shrinkage, plastic, and liquid limits (Fig. 7.1). There is a general relationship between the capacity of clays and silts to attract and hold water at their surfaces and the limits calculated for a soil, but there is not a direct quantitative

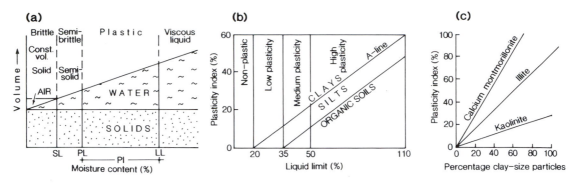

FIG 7.1. (a) The three-phase system of soils and the influence of water-content on the Atterberg limits indices. (b) A simplified standard classification chart indicating the zones in which soils of various plastic properties will plot. (c) The influence of the amount and species of clay in a soil on its plasticity index. (After Skempton 1953*b*.)

relationship between the limits and the volume of adsorbed and double layer water. A strong statistical relationship between the liquid limit and surface area of soils has been established by Muhanthan (1991) and is evident in the data presented in Table 7.2. The index limits are best used as a means of classifying soil.

Atterberg limits

The idea of a standard index of soil behaviour was first proposed by Atterberg (1911), and it is now widely known as the 'Atterberg limits'. Water-content is readily measured but behaviour is less easily defined. Methods for determining plastic and liquid limits were designed by Casagrande (1948) and have been in wide use. The Casagrande methods, however, are dependent upon the skill of the operator (Sherwood and Ryley 1970) and a drop-cone penetrometer has since been adopted as a standard instrument in many laboratories (Plate 7.1).

A good correlation between Casagrande methods and drop-cone penetrometer results has been achieved by many workers over the range of plastic behaviour, but closer to liquid behaviour the correlation is weaker (Wasti and Bezirci 1986). It is therefore necessary to state which method has been used for determination of Atterberg limits and to exercise caution in comparing data derived from different testing methods. The apparatus and methods are described in standard texts such as Vickers (1983).

The drop-cone penetrometer has a standard mass of 80 g and a 30° cone head which is lowered until it just touches the surface of the soil sample

which has been remoulded into a paste. The cone is then released to sink into the surface of the soil for 5 seconds. The depth of penetration of the cone into the soil is recorded. The test is repeated at various moisture contents to give the relationship between cone penetration and moisture content and a graph of the relationship is drawn to a linear scale. The liquid limit is taken as the moisture content at which the standard cone penetrates 20 mm into the soil.

Determination of the plastic limit is still contentious. Towner (1973) suggested a penetration of 2 mm; Allbrook (1980) has suggested 2.8 mm; Campbell (1976), Wroth and Wood (1978), and Davidson (1983) have proposed that the minimum turning point of the U-shape curve, derived from the graph of penetration against moisture content, should be used (Fig. 7.2); differences between these methods are usually <5 per cent.

The range of water-content between the liquid (LL) and plastic limits (PL) is known as the plasticity index (PI):

$$PI = LL - PL.$$

All standard Atterberg limits tests are carried out on remoulded soils in which all fabric has been deliberately destroyed, consequently a liquidity index (LI) is used to compare the soil's plasticity with its natural field-moisture content (NMC):

$$LI = \frac{NMC - PL}{PI} \, (\%).$$

If LI = 100% the natural soil is at its liquid limit, and if LI = 0% the soil is at its PL.

The use of a drop-cone penetrometer to deter-

PLATE 7.1. Equipment for determining Atterberg limits. At top left is a Casagrande cup for determining liquid limits by counting the number of times the cup needs to be dropped on its base plate in order to close the groove cut in the soil with the standard tool. Top right are threads of soil rolled on a glass plate; at the plastic limit a thread 3 mm in diameter can be formed. In the centre is a drop-cone penetrometer.

mine liquid and plastic limits indicates that these index tests are also measures of shear strength at those water-contents. It has been suggested by several authors that the index values can be expressed as undrained strengths; Whyte (1982) proposed the values:

$$\text{at liquid limit, } S_u = 1.6 \, \text{kN/m}^2$$

$$\text{at plastic limit, } S_u = 110 \, \text{kN/m}^2.$$

Other workers have derived slightly different values in the ranges of:

$$\text{LL} = 1\text{--}3 \, \text{kN/m}^2 \quad \text{and} \quad \text{PL} = 110\text{--}130 \, \text{kN/m}^2.$$

The shrinkage limit defines arbitrarily the change of state from a semi-solid to a solid state (Fig. 7.1) and is the moisture content below which the soil remains at a constant volume on drying

out. Shrinkage limit is determined by filling a brass mould, with semicircular cross-section of standard 12.5 mm radius and with a length of 140 mm, with a paste of soil near its liquid limit. The linear shrinkage (LS) after drying under controlled conditions is given by:

$$\text{LS} = 1 - \left(\frac{\text{length after drying}}{\text{initial length}} \right) \times 100\%,$$

and the test is continued until no further shrinkage occurs.

$$\text{PI} \approx 2.13 \, \text{LS}.$$

Activity is another useful index. It is the ratio of the plasticity index to the percentage of particles of clay-size. Activity is therefore a measure of the plasticity of the clay-size particles; it is controlled

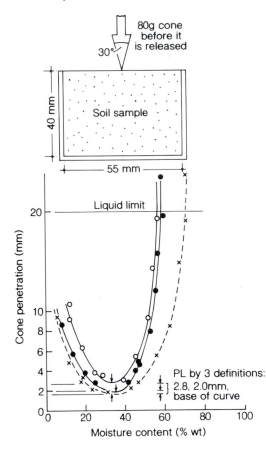

Fig 7.2. (a) The cone of a drop-cone penetrometer placed for release into a soil sample. (b) Three definitions of plastic limit show a range of about 5 per cent moisture content between three curves, each representing a different soil type.

by the dominant clay mineral species in the soil and is therefore a useful indicator of the presence of those species (Skempton 1953a):

$$\text{Activity} = \frac{\text{PI}}{\% \text{ of soil} < 0.002 \text{ mm (by dry mass)}}.$$

Skempton defined 'inactive' clays as having values of <0.75 and non-clay minerals and kaolinitic clay minerals fall in this range. The 'normal' clays have values in the range of 0.75–1.25 and this range includes some illites. 'Active' clays have values >1.25 and include allophanic soils and smectites. The relationship between activity and clay species is not precise owing to the influence of ions in the pore fluids and organic soil materials, but it is a quick indicator of chemical

reactiveness and is correlated with cation exchange capacity (CEC), Table 7.2.

Interpretation of Atterberg limits data must be based on the clear recognition of several conditions. (1) The soils are always remoulded and are dried and sieved to remove coarse-grained particles, consequently the data relate to soil without fabric, natural permeability, or fissures. (2) Because of their stable voids and fabric, it is possible for natural soils to behave as solids holding water in quantities well above the liquid limit. Disturbance of such soils by earthquakes and other applied stresses may cause destruction of fabric and catastrophic landslides with natural remoulding and flow behaviour. (3) The species and concentrations of adsorbed ions on clay minerals can change as a result of leaching, weathering, and variations in soilwater-content. Index values can therefore change over time.

In general, high values of LL and PL indicate the presence of smectite clays; low values the presence of kaolinitic clays (Table 7.2), but PL is influenced also by the amount of clay-size material in the sample and PI is controlled by that amount for each clay mineral species (Fig. 7.1c). Within each clay species, adsorbed ions produce index values in the order Fe < Mg < H < Ca < K < Na. Atterberg limits are, therefore, not constant properties of soils, they are indicators only and are best used for comparisons of soils in local areas and guides as to which bodies of soil require further investigation of their properties.

Viscosity

Viscosity is the property of a material that characterizes its flow behaviour. Flow is the continuous deformation occurring under the influence of gravity or an applied stress. Viscosity of a soil is a measure of the cohesive and frictional resistance as influenced by the water-content. It has high values at low moisture content and low values at high moisture content (Fig. 7.3a). The shear stress required to produce and maintain flow is related to the rate of shear and is therefore not a unique quantity. For Newtonian fluids such as water, solutions, emulsions, dilute suspensions, and gases the shear stress is directly proportional to the rate of shear, but structured soils are non-Newtonian and are usually Bingham plastics for which the applied shear stress is proportional to rate of shear only when an initial yield stress is exceeded (Dinsdale and Moore 1962).

TABLE 7.2. *Properties of clay minerals*

Clay-mineral species	Particle thickness (nm)	Liquid limit (%)	Plastic limit (%)	Activity	ϕ'_r (°)	Specific surface area (m²/g)	CEC (me per 100 g clay)
Montmorillonite	2	100–1300	50–100	Ca: 1.5 Na: 6–13	4–10	750–800	80–150
Allophane	Spherical	120–250	100–140	>3	2–40	500–700	25–70
Illite	20	60–120	35–60	0.5–0.9	10	80–130	10–40
Halloysite (hydrated)	Tubular, wall = 3	50–70	47–60	0.1–0.4	25–40	40	40–50
Halloysite (dehydrated)	Tubular, wall = 3	35–55	30–45	0.5	25–40	40	5–20
Chlorite	—	44–47	36–40	0.3–0.5	—	5–50	10–40
Kaolinite	30–100	30–110	25–40	0.3–0.5	12–18	10–20	3–15

Note: Atterberg limits values are influenced by type and concentration of adsorbed cations; liquid limits are highly correlated to specific surface area.

Sources: Grim (1968); Lambe and Whitman (1979); Wesley (1977); Lupini *et al.* (1981); Boyce (1985).

Soils with natural structures behave commonly as Bingham plastics as long as their structure is retained, but remoulding during flow may break down their structure and water may be added to the flow by rain or streams. A transition may then occur to dilatant plastic or Newtonian behaviour. Stiff clays may be remoulded in landslides and the slide may be transformed downslope into a flow.

There is no standard method for measuring viscosity, but rotational viscometers are used widely. These machines have a cone or cylinder which is rotated at various controlled rates within a cylindrical container of material, this material is sheared by the drag set up within it and its resistance to shear is recorded.

Behaviour and loss of structure

Loss of soil structure, particularly where soil moisture contents are high, can lead to soil behaving as a suspension of single grains with consequent loss of strength and radical changes in behaviour. Many forms of behavioural change are recognized, but sensitive, collapsing, dispersive, and expansive behaviours are those most commonly associated with soil erosion and various forms of landsliding.

Sensitive soils

Sensitive soils are recognized as a distinct class because they may remain stable bodies of material for thousands of years then suddenly lose strength, sometimes catastrophically, with little or no increase in moisture content and only low values of applied stress.

Sensitivity (S_t) is defined as:

$$S_t = \frac{\text{Undisturbed, undrained strength}}{\text{Remoulded, undrained strength}}.$$
(at the same moisture content)

It is usually measured with a shear vane, but in extreme cases the structured soil turns to a liquid on remoulding and then a drop-cone penetrometer is used for measurements. Marine clays in Canada and Norway have been studied in some detail: those with high to moderate salinity have S_t values in the range 2–29 and leached marine clays have values in the range 10–680 (Torrance 1983). The extreme recorded value appears to be 1500 (Penner 1963).

Several classifications using descriptive terms related to S_t values are in use (e.g. Skempton and Northey 1952, (Table 7.3); Rosenqvist 1953). In all classifications soils with S_t values of unity are insensitive and for those described as quick clay S_t

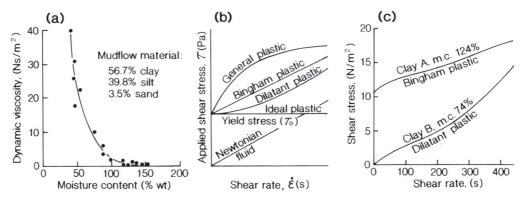

Fig 7.3. (a) Viscosity of soils is greatly influenced by water-content. (b) Curves indicating the major types of plastic behaviour together with the relationship for a Newtonian fluid. (c) Measured shear-stress–shear-rate curves for two soils showing that one is behaving as a Bingham plastic and the other as a dilatant plastic material. (After Bentley 1979.)

Table 7.3. *Sensitivity scale of Skempton and Northey (1952)*

Sensitivity	Designation
1	Insensitive clays
1–2	Clays of low sensitivity
2–4	Clays of medium sensitivity
4–8	Sensitive clays
8–16	Extra-sensitive clays
>16	Quick clays

is >16. All quick clays have remoulded strengths of <0.5 kPa.

Soils with high sensitivity have been described from a number of countries but the greatest intensity of investigation has occurred in Canada and Norway where sensitive soils have been involved in many large landslides (see Plate 13.4). These soils have particles predominantly of clay-size, but not all such particles are clay minerals; a large proportion (>50 per cent) are ground quartz and feldspar from 'rock flour' transported by glaciers and sedimented through sea water, with chlorite and illite being the most important clay minerals. In Japan sensitive marine clays are of non-glacial origin and are predominantly iron-smectites (Torrance 1984). The important characteristic of such sedimented clays is their open, flocculated structure with cementation bonds developed at contact points in sea water with high concentrations of divalent cations which become adsorbed to the clay particles.

After deposition the clay beds were raised above sea-level by isostatic recovery on deglaciation or by tectonism. Saline pore water has been replaced by fresh water at some sites, but not all. It is widely recognized as being significant that large land-slides occur only in clays in which the pore water salinities have been reduced below approximately 2 parts per thousand (Torrance 1988). Low salinity is believed to result in leaching and loss of divalent cations and their replacement with organic and inorganic anions, which act as dispersants, and with monovalent cations. The resulting weak bonding leads to collapse of the structure on disturbance and flow failures are common because of the high natural moisture content which often exceeds the liquid limit (Fig. 7.4c).

Sensitive behaviour is, however, not an inevitable consequence of uplift of marine clays. Weathering, sometimes called ageing, may result in strength gains by processes described by Lessard and Mitchell (1985) and Mitchell (1986). Lessard worked on silty clays at La Baie, Quebec, which are 88 per cent non-clay primary minerals and 12 per cent clay minerals, mostly illite. The chloride concentration in the pore water is some 1800 times less than that in sea water, indicating intense pore-fluid replacement. High sodium and bicarbonate concentrations and high pH are characteristic of these clays and of reducing environments. Black bands in the clays indicate the presence of iron sulphides.

Samples exposed to the air show increases in pore-water salinity, concentration of divalent cations, and decreases in pH during periods of

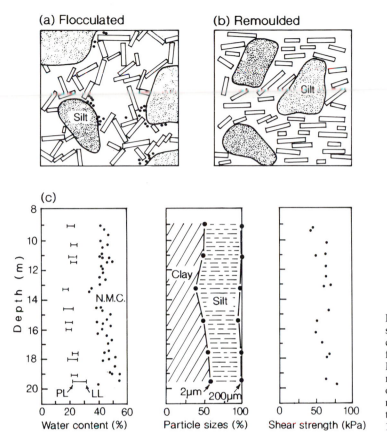

(a) Flocculated

(b) Remoulded

(c)

FIG 7.4. (a) Cartoon of a flocculated soil showing the arrangement of silt particles and clay minerals. Divalent ionic bonds have formed at the surfaces of some particles. (b) Loss of structure and bonding in the soil after remoulding. (c) Geotechnical characteristics of the quick clay from La Baie. Note that the natural moisture content (NMC) far exceeds the liquid limit. (After Lessard in Mitchell 1986.)

about a year. Collectively, these changes are responsible for increases of remoulded strength and decreases in liquidity index, as double layers around clay minerals are depressed, with decreases in interparticle repulsion.

The causes of the changes are thought to be initial formation of iron sulphides in reducing environments as clays and silts were deposited in arms of the glacial-age seas occupying what is now the St Lawrence Lowlands. Iron and sulphates from rock minerals and organic matter were reduced to the iron sulphides FeS and FeS_2 (pyrite). Simultaneously, formation of carbon dioxide by bacterial oxidation of organic matter caused an increase in pH and reductions in pore-water Ca^{2+} and Mg^{2+} as they were precipitated as Mg-calcite. After uplift fresh-water leaching decreased the salt content, at the same time oxygen oxidized organic matter to form carbonic acid which dissolved some of the calcite, so increasing the Ca^{2+}, Mg^{2+}, and bicarbonate in the pore water. This effect may also

have been enhanced by oxidation of the sulphur from the sulphides with the formation of sulphuric acid which also attacked the calcite. The total effect on exposure is therefore an increase in divalent cations, which causes increases in remoulded strength and liquid limit, with decreases in sensitivity and liquidity.

This type of strength gain is not thixotropic but from chemical weathering in the presence of oxygen. It suggests that artificial exposure to oxygen and chemical treatment may change quick clays *in situ* into clays of lower sensitivity, with less dangerous properties (Rosenqvist 1984).

Volcanic-ash soils from Japan (Mochinaga 1941) and New Zealand (Jacquet 1990) have been reported as having high sensitivities in the range of $S_t = 5-55$. The causes of this behaviour have been variously attributed to: thixotropy in which collapsed structure and strength are regained over time; presence of iron oxides which develop cementation bonds in the virgin soil, but with these bonds being irretrievably lost on remoulding; losses of bonded water, as recognized by decreases in water tension after remoulding; and to disruption of strongly bound aggregates in the remoulding process.

It has been demonstrated by Jacquet (1990) that New Zealand volcanic-ash soils which have various proportions of halloysite, allophane, and imogolite are, at most, only weakly thixotropic as the strength recovery after resting for one year is only 2–9 per cent and this is again lost after further remoulding. Water re-adsorption probably explains the 2–9 per cent recovery. The high sensitivity values for allophanic soils are attributable to the structure which, seen under the scanning electron microscope, is of aggregates and particles linked by fibres of imogolite. In some cases the clay fibres form networks covering particles and aggregates. In the remoulded state, the networks have disappeared and a featureless gel appears to cover the body of material. Halloysites have tubular forms and in the undisturbed state these form bundles; after remoulding the particles are separated. After four months of resting some slight degree of aggregation with edge–face bonds seems to occur but the original structure is not recovered.

Sensitivity of many volcanic-ash soils, therefore, is attributable to loss of structure by remoulding clay minerals which owe their natural strength to formation of aggregates through ionic and electrostatic bonds, but which cannot recover their structure after remoulding. Some volcanic-ash soils may lose strength by other processes but these require further investigation.

Liquefaction

The sudden loss of all strength of saturated non-cohesive sediment is called liquefaction. It occurs in saturated silts and sands which have high void space as a result of loose packing and uniform grain sizes. It is rarely observed in gravels, which usually allow pore-water pressures to dissipate almost instantaneously, nor in plastic clays which inhibit the production of high pore pressures by shearing. The most common trigger mechanism is earthquake shock waves, the passage of which causes shear stresses in alternating horizontal directions as the waves pass through the soil. These cyclic shear stresses cause progressive rises in pore-water pressures until failure occurs.

In conditions prevailing prior to an earthquake a body of soil in level ground is subjected to a confining stress due to the weight of the overburden. The weight of the overburden is transmitted to the bedrock through the point contacts between individual particles. In Fig. 7.5a this is indicated by the total stress gauge which is recording the load due to the overburden, plus the sand grains, plus the water filling the voids. The effective stress gauge, however, is immersed in the water and is recording the immersed weight of the sand, plus the weight of the overburden. After enough earthquake shock waves have passed through the soil to break the point contacts between the sand grains, the grains settle into a denser state of packing. This reduction in volume cannot occur immediately because the water cannot escape and the load of the overburden is transferred to the water. This produces an instantaneous rise in the pore-water pressure to a level which is equal to the initial confining stress. As no intergranular stresses are acting on the sand grains, the grains are suspended in the water, effective stresses are then zero, this is the state called liquefaction. The suspension has no shear strength.

As sand grains start to settle, the overburden is liable to crack and water with suspended sand may be ejected to form sand boils, up to 3 m high, or sand ridges along surface fissures. The expulsion of

FIG. 7.5. a–c. Changes from a state of (greatly exaggerated) loose packing, to a suspension during liquefaction, to a state of denser packing. The total stress is indicated by the lower gauge and is constant because there is no gain or loss of material. The upper gauge is recording the effective stress due to the load transmitted from grain to grain of the saturated sediment. In (b) the effective stress is zero (modified from a figure by Ishihara 1985). (d) Formation of a sand boil. (e) Idealized stresses on an element of soil: (i) before, and (ii and iii) during transmission of a shock wave. (f) Schematic record of a pulsating load on loose sand, and its effects (based on Seed 1968). (g) The relationship between cyclic stress ratio tending to cause liquefaction and the Standard Penetration Resistance of the soil (modified from Obermeier 1989). (h) Curves for earthquakes of various magnitudes (after Seed *et al.* 1983).

Note: The Standard Penetration Test (SPT) is a field test in which a sampling tube is driven into the ground by dropping a 65 kg mass from a height of 760 mm. The penetration resistance is reported as the number of blows of the mass required to drive the tube 300 mm.

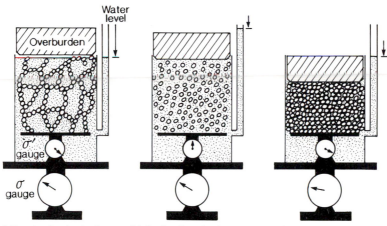

(a) Sand soil prior to lique-faction carries overburden stress

(b) During liquefaction water carries all over-burden stress; sand is in suspension

(c) Surface settles after expulsion of water; densified sand carries overburden stress

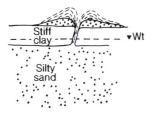

(d) Sand boil created by ejection of water and sand

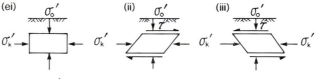

σ_o' Initial effective overburden stress
σ_k' Effective confining stress
τ Earthquake induced horizontal shear stress

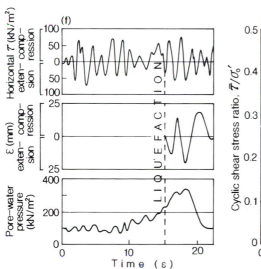

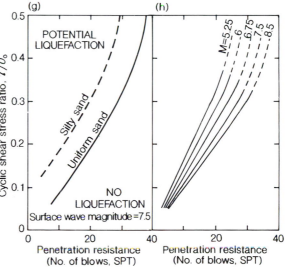

pore water causes the ground to settle above the densified body of sand.

The onset of liquefaction is dependent upon the magnitude and number of cycles of the applied shear stress (Fig. 7.5e, f). The duration of the state of liquefaction depends upon the duration of the cyclic shear stresses and on the drainage conditions of the deposit. The thicker the deposit, the finer the sand or silt grains, and the thicker the over-burden mass, the longer will be the drainage time and the duration of excess pore-water pressures and, therefore, the longer is the state of lique-faction. High magnitude, long duration earth-quakes, in areas where thick bodies of saturated sand have thin coverbeds, are likely to produce the conditions for large-scale liquefaction (Seed 1968).

The highest susceptibilities to liquefaction occur in very young fluvial and eolian sands and silts, loess, and uncompacted artificial hydraulic fill. Dry sand above the water-table will not liquefy; where the water-table is more than 5–7 m deep, lique-faction is unlikely because pore water is likely to be dissipated into the pores of the surface soil and be recorded as a water-table rise. Liquefied soils are not likely to be erupted through cover beds thicker than about 7 m because thicker beds produce a greater effective overburden pressure than the effective pressure in the liquefied sand (Hu and Wan 1987).

Much attention has been paid to liquefaction because flowslides caused by earthquake-generated liquefaction have had major and catastrophic effects on buildings on many alluvial plains, deltas, estuaries, and loess uplands. Lenses or beds of low-density soil within hills have also contributed to hillslope failures; many examples are discussed by Seed (1968). A major objective has been to identify areas which are liable to fail by lique-faction. Studies by Obermeier (1989); and Elton and Hadj-Hamou (1990) provide examples of the methodology.

The effect of an earthquake on the soil may be characterized by the ratio of the average earthquake-induced shear stress (τ) to the effective confining pressure due to the overburden (σ_o') (Seed and Idriss 1971). This ratio, the cyclic stress ratio (CSR), is obtained from:

$$CSR = \frac{\tau}{\sigma_o'} = 0.65 \frac{\sigma_o}{\sigma_o'} a_{max} \cdot r_d,$$

where: σ_o is the total overburden pressure;
       a_{max} is the maximum ground acceleration;

r_d is a factor which recognizes the soil flexibility, varying from 1.0 at the surface to 0.85 at a depth of 10 m and given by Iwasaki et al. (1981) as:
$$r_d = 1 - 0.015d$$
where: d is the depth (in metres)

Liquefaction is likely if the cyclic stress ratio exceeds a critical value which is derived by plotting CSR against the standard penetration resistance of the soil (Fig. 7.5g: see the note below the caption). This figure is based upon a magnitude 7.5 earth-quake. Versions of the figure can be produced for earthquakes of other magnitudes (Fig. 7.5h).

The size of the earthquake is critical. The smallest known to produce liquefaction had a magnitude close to 5 and such an event is likely to produce peak horizontal acceleration of $0.1g$ (where g is the acceleration due to gravity) and cause strong ground shaking up to a kilometre away for less than ten seconds. For an event of magnitude 7.5, the peak acceleration of $0.5g$ may cause ground shaking over 100 km away for a minute or more. Prolonged shaking is not only necessary for creating liquefaction (Fig. 7.5f) but also causes liquefaction to greater depths, perhaps exceeding 20 m, and therefore larger and more prolonged surface effects.

Prediction of liquefaction is, however, not a simple matter for lenses or seams of low density sand can liquefy more readily than large bodies. In lenses the stresses are concentrated at the edges of the lenses and are thus magnified. Finding the lenses and predicting their behaviour are problems requiring further research.

Protective measures against liquefaction include lowering the water-table, compacting soil by vibration, and placing structures on deep piles.

Collapsible soils

Collapsible soils are those which suffer an appre-ciable loss of volume upon wetting, load applica-tion, or a combination of both. These soils have a predominance of silt or sand grains with minor amounts of colloidal clay, calcium carbonate, aluminium and iron oxides, gypsum, or salts which act as bonding materials at the points of contact of the soil grains (Fig. 7.6a). The silts and sands may be residual and *in situ* from the weathering of silicate rocks such as granites, but more commonly are transported materials deposited by wind, sheet wash, mudflows, or floods. The deposit is left in a dry loose state in predominantly semiarid and arid

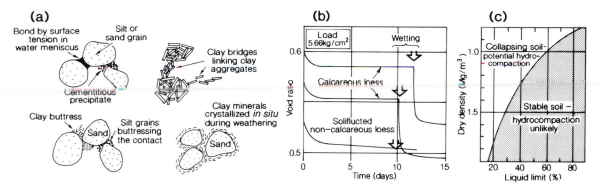

FIG. 7.6. (a) Bonding materials which are subject to loss of strength or rupture on wetting or loading (based on Barden *et al.* 1973). (b) Consolidation tests on two loess samples from Prague, showing structural collapse on wetting for the two calcareous loesses (after Feda 1966). (c) Soils which are subject to collapse by hydrocompaction as indicated by their dry density and liquid limit (after Gibbs and Bara 1962).

climates where water does not penetrate deeply into the soil and does not transport fine-grained material into soil voids. Precipitation of colloids and salts near the soil surface forms crusts which also limit water penetration.

In dry climates, deposition by wind or water is intermittent, as is soil wetting. Consequently, no soil layer has to support an added overburden of more than a few centimetres while in a wet and weakened state. Compaction under the influence of water (hydrocompaction) is therefore negligible and low-density soils can accumulate to considerable thicknesses.

Perhaps the most widespread soil which is liable to collapse is loess. It is well recognized that loess varies in grain size with distance from its source (Lin and Liang 1982; Lutenegger 1982). Away from the source area where deflation is occurring, loess deposits decline in thickness and show a corresponding increase in fineness as the sand-content decreases and the clay-content increases. This creates a cohesionless thick deposit near the source and a cohesive thin deposit further away. In China, the north-western deserts are the source, in North America and Europe the glacial-age beds of such rivers as the Mississippi and Danube are the sources. The zone in which collapse is most probable lies between the extremes of the source and distal regions if that zone has, or has had, a dry climate, and soil porosities are greater than 40 per cent. Old and deeply buried loess is likely to have suffered collapse in the past as a result of overburden stresses and of wetting; modern collapse is

therefore largely confined to upper, young loess beds less than about 10 m thick.

Collapse occurs as a result of the weakening or loss of bonds between the silt and sand grains. This is most likely to occur if water can penetrate the deposit, soften the clays, create thick water lenses between particles, with larger radii of curvature of menisci, or take calcite and salts into solution. The effect is most obvious when a laboratory test is carried out to determine the compaction under load in the dry state and then by addition of water. For this test a sample with natural undisturbed structure is placed under an applied load. It will initially consolidate a little but, upon moistening, the collapse will occur suddenly. In Fig. 7.6b the effect upon calcareous loess can be compared with that on loess which has no calcium carbonate forming bonds between grains.

In field conditions, collapse has been most widely reported as being a result of human activity when drainage water from irrigation, effluent, or runoff from roads has occurred. Under natural conditions, extreme climatic conditions or gradual rises of the water-table may promote collapse. Earthquakes with prolonged cyclic stresses are another possible cause of failure, particularly in the hill country of China where loess soils may collapse and move as dry flows with internal friction reduced by the entrapment of the air from the original void space.

Identification of collapsible soils in the field involves recognition of old collapse features from air photos, records, and field inspection. At a site, the pinhole-size pores are visible in soils studied with aid of a hand-lens. The 'Jennings sausage test' (Brink 1979) may be carried out by carving two small cylindrical samples of undisturbed soil as nearly as possible to the same volume, wetting and kneading one of the samples and remoulding it into a cylindrical shape of the original diameter. An obvious decrease in volume, when compared with the undisturbed twin sample, will confirm a collapsible grain structure. Backfilling a pit with wetted remoulded material from that pit will reveal collapsible soil if the wetted soil does not fill the pit.

Laboratory tests involve the consolidation test (Fig. 7.6b), in which a collapse potential of more than 10 per cent is indicative of severe subsidence. Low dry densities ($0.9-1.5\,Mg/m^3$) associated with low liquid limits (<45 per cent) (Fig. 7.6c), and clay-contents in the range 5–20 per cent, are other indicators (Gibbs and Bara 1962). On saturation, the moisture content will then exceed the liquid limit and the soil will behave as a very low strength suspension.

Remedial measures involve inducing hydro-compaction by flooding in conjuction with vibration and, if it can be done safely, rolling. Infiltration wells and loads of gravel embankments have been used on building sites where collapsible soils occur at depths greater than about 5 m. Alternatively, it may be cheaper to exclude water from construction sites in arid areas (Clemence and Finbarr 1981; Waltham 1989).

Dispersible soils

Dispersion is a process which separates clay particles by breaking the bonds between them. In both laboratory practice and in field conditions this is usually the result of providing the monovalent cation sodium (Na^+) to replace divalent cations such as Ca^{2+} and Mg^{2+} on the exchange complex of a clay. This results in a more extensive diffuse double layer around each of the clay particles since the extent of the double layer varies inversely with the valence (charge) of the cation. The clay particles are prevented by this extended double layer from coming close enough to cohere and hence they separate as individuals.

Dispersion occurs naturally when soils containing clays with a high exchangeable sodium percentage (ESP) are immersed in pure water. ESP is exchangeable sodium expressed as a percentage of CEC. Soils which contain a clay fraction of 30–50 per cent Na-montmorillonite are particularly prone to dispersion because of the thick double layers of the clay particles; in comparison kaolinite is much less prone to dispersion because of its lower CEC and thinner double layers. Increasing salinity of water increases availability of Ca^{2+} and Mg^{2+}, or other divalent cations, so dispersion is <u>reduced</u>. The sodium adsorption ratio ($Na/\sqrt{0.5(Ca + Mg)}$) for the soilwater is then one of the useful indicators of the likelihood of dispersion (Sherard *et al.* 1976*a*).

Several tests of dispersiveness are available. (1) A small crumb of soil is immersed in distilled water and the tendency for the clay particles to go into colloidal suspension is observed; water which takes on a milky appearance indicates high dispersion. (2) A pinhole test is carried out on a small core of soil, in its natural state, held between two coarse filters in a cylinder. The core has a hole punched in it and distilled water is passed through the hole. Dispersion is indicated by cloudiness of the water (Sherard *et al.* 1976*b*). (3) Other tests include chemical analyses of soil water; and comparison between samples dispersed by (a) a sodium dispersion agent and (b) by mechanical agitation. These methods have all been compared and have been shown to give results which are in agreement (Tuncer *et al.* 1989).

The importance of dispersion is particularly great in areas with arid and semiarid climates in which exchangeable sodium can accumulate. Dispersion is a major cause of the formation of natural pipes in soils. Pipes, once started, can enlarge by hydraulic action of water passing through them and collapse of the roof of a pipe can be a major cause of gully initiation. Piping within earth dams and embankments has caused a number of collapses with high costs of human life and economic damage.

Remedial treatment involves using fine filters in dams and embankments to prevent loss of the dispersed soil and more generally by replacing sodium with calcium and magnesium ions. This is done most effectively by the use of gypsum ($CaSO_4.2H_2O$) at regular intervals (Shanmuganathan and Oades 1983).

Volume change in clay soils

Perhaps the most widespread problem for soil engineers is that caused by clays which change

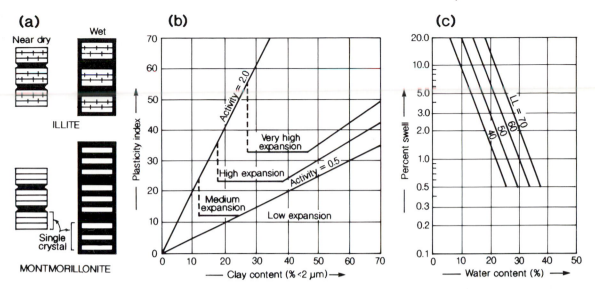

FIG. 7.7. (a) The contrast in swelling and shrinking potential between different clay-mineral species depending upon whether water adsorption occurs only on crystal outer surfaces or also within single crystals. (b) Estimation diagram for the degree of expansiveness of a clay soil. (c) Prediction of percentage swell for natural clays in relation to their liquid limit and water content. (After Popescu 1986.)

their volume as a result of gains and losses of double-layer water. The clays which cause these problems are variously called stiff active clays, expansive or swelling clays, and shrinking clays. The emphasis on expansion is most common in semiarid climates in which the common state is low soil moisture content and problems are created in wet periods when soils swell and engineering structures are subject to differential soil heave. The emphasis on shrinkage is common in predominantly humid and temperate climates when unusually dry summers cause soils to shrink and crack with resulting settlement and fracture of structures. Soil movements would be most easily assessed if soils in a locality had uniform characteristics, but this is seldom the case. Soil depths, clay mineral species, amounts of clay, and moisture content all vary over short distances, as may slope angle, drainage conditions, and vegetation cover. Vegetation is particularly important, for some species of plants withdraw from soils far more moisture than others; many species of eucalypts, for example, have such high moisture demand that they can cause differential ground settlement in clay soils, great enough to damage buildings on shallow foundations.

On hillslopes, volume change of clay soils pro-
motes soil creep and the opening of tension cracks during dry periods. Deep cracks may, in subsequent storms or wet periods, become filled with water adding a driving force to soil movement downslope, as well as developing high water pressures in the soil and so promoting landslides.

Because of the relationships between clay mineral species, clay amounts, Atterberg limits, soil volume change, and soil moisture content, it is possible to estimate the volume change potential in the soil (Fig. 7.7). It is evident from these figures that montmorillonite, and especially Na-montmorillonite, produces the greatest potential for volume change. The charts in Fig. 7.7 are based on correlation and are indicators rather than quantitative predictors of swelling and shrinking. For semi-quantitative estimates of volume change both simple and more complex methods are available. A quick, simple method is calculation of the Duncan free-swelling coefficient (ε_D) (Olivier 1979), by saturating a previously oven-dried core sample for two hours:

$$\varepsilon_D = \frac{\text{Change in length after swelling}}{\text{Initial length of specimen}}.$$

The bar shrinkage mould can be used to estimate shrinkage as water is lost by evaporation and the mould and its contents are weighed and measured until no further water loss occurs.

The relationship between water-content and swelling and shrinking is, however, far more complex than the simple methods described above imply. Clay soils containing montmorillonite show an almost reversible swelling and shrinking on wetting and drying, but sedimented clays containing kaolinite and illite show an initial large volume decrease on drying with only limited swelling on rewetting. With further wetting and drying the kaolinite and illite show net decreases in volume after each wetting and drying cycle until an equilibrium is reached where swelling and shrinking occur between constant limits. Such behaviour is related to the microfabric of the clays and the distribution of the water within the micro-fabric (Yong and Warkentin 1966). Attempts to determine swelling and shrinkage properties of a natural soil may therefore require many samples to be tested over long periods in specially designed cells in which small changes of soil volume and water suction or pressure are recorded. It may take many weeks to fully hydrate an oven-dry soil.

Control of soil volume change is best achieved by measures which keep soil moisture as constant as possible. In humid climates, this is usually done by keeping soil moist by reducing exposure to air and by controlling the vegetation. In dry climates, it is usually better to exclude water from the soil.

Compression and consolidation

For soil engineers compression and consolidation of soils are topics of major concern because accurate predictions are needed of the total amount of settlement of loaded foundations of structures, and of the rate of settlement. For researchers studying hillslopes the interest is virtually confined to study of soils which will collapse on wetting and to studies of hillslopes formed on marine clays and mudrocks.

The distinction between compression and consolidation is readily understood by consideration of the passage of a heavy roller over a moist clay road. After one passage the soil is compressed with all cracks and obvious voids closed; if no further traffic passes over the road the compression will gradually be lost as the soil swells. If, by contrast, the roller is repeatedly driven over the moist clay, water will be driven out of the soil, voids will be closed, clay particles will be deformed and forced into dense states of packing; the result will be a permanent loss of volume and the soil is said to be consolidated.

The progressive conversion of marine muds to mudrocks beneath an overburden of 3000 m, or more, occurs over long periods of perhaps 2–5 My. The conversion is a direct result of the overburden pressure (p_o) which at any point is given by:

$$p_o = \sigma_v - u,$$

where σ_v is the total vertical pressure exerted by all of the particles and water above the point considered and u is the pore-water pressure at that point (Skempton 1970b). The material at any point in the accumulating column may be said to be normally consolidated by gravitational compaction.

If the mudrock is raised by tectonism and eroded into a hill and valley landscape, some of the overburden will be lost but the mudrock will not swell and increase its void space to a volume adjusted to the new and smaller overburden pressure. With respect to that overburden pressure it will be over-consolidated. The degree to which a mudrock or marine clay can swell, and draw in water and soften, is of primary interest if landslide risk or landslide mechanisms are being studied and it may be necessary, therefore, to determine the degree of over-consolidation. This is expressed as an over-consolidation ratio (OCR) defined as:

$$\mathrm{OCR} = \frac{\text{maximum previous effective pressure}}{\text{existing effective pressure}}$$

For normally consolidated soils the ratio is unity, and for over-consolidated soils is greater than unity.

The consolidation apparatus, known as an oedometer is illustrated in Fig. 7.8. The soil sample is encased in a steel cutting ring and placed between porous discs which are saturated with air-free water. A vertical load is applied and the resulting compression recorded by means of a dial gauge or displacement transducer at intervals of time until the sample has reached full consolidation. As the sample is compressed water escapes from it through the porous discs and its amount is measured. Depending upon what is being studied, the test may be interrupted by removal of the load to determine what swelling occurs and then the load may be reapplied. The changing thickness of the sample is obtained to calculate the void ratio of the soil for each stage

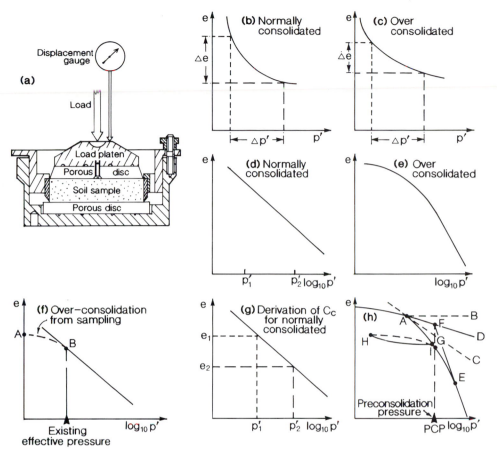

FIG. 7.8. (a) Major components of one style of oedometer. (b, c) Consolidation curves for normally and over-consolidated soils on arithmetic scales. Note the smaller change in void ratio, for the same change in confining pressure, for the over-consolidated soil. (d, e) Consolidation curves on semi-logarithmic plots. (f) One method of determining the existing effective pressure. (g) Derivation of the compression index for a normally consolidated soil. (h) Two methods of determining the preconsolidation pressure. Point F obtained from extension of the straight line below E to F, to intersect AD. Point I obtained from an unloading curve following the full concave-upwards line from G to H, which on resumption of loading rejoins the original compression curve at I.

of the consolidation. Because the clay sample contains water, the applied pressure is an effective pressure p' which can be plotted against the void ratio, e, to give an $e-p'$ curve.

A normally consolidated soil subjected to an effective pressure increase compresses more than does an over-consolidated soil subjected to an identical pressure increase (Fig. 7.8b, c). The data may also be plotted as e against $\log_{10} p'$. This gives an instant indication of whether a soil is normally consolidated from display of a linear curve, or over-consolidated, from display of a convex upward curve (Fig. 7.8d, e).

For a normally consolidated soil the existing

effective pressure is indicated by an initial curve AB (Fig. 7.8f) which is derived from overconsolidation of the sample during collection. The point B, at which the linear relationship begins, indicates the existing field effective pressure.

For a normally consolidated soil a compression index (C_c) is derived from the gradient of the semi-logarithmic plot (Fig. 7.8g) or:

$$C_c = \frac{e_1 - e_2}{\log_{10}(p_1') - \log_{10}(p_2')} = \frac{\Delta e}{\log_{10}(p_1'/p_2')}.$$

For an over-consolidated clay the curved semi-logarithmic plot will not yield a constant value for

PLATE 7.2. A consolidation cell in which soil drainage and pore-water pressures can be measured. Pressure is applied hydraulically. Below is the sample holder, and the pressure diaphragm can be seen inside the top of the cell. This is a modern computer-controlled system which operates through the same digital controllers shown in Plate 5.4.

C_c. From the e-$\log_{10} p'$ curve it is possible to derive an approximate value for the preconsolidation pressure which existed before removal of the overburden. In Fig. 7.8h, first estimate the point of greatest curvature A, then draw a horizontal line through A (the line AB) and the tangent to the curve at A (AC). Bisect the angle BAC to give the line AD, and locate the straight part of the compression curve (commences at E in the figure). Project the straight part of the curve upwards to cut AD in F. The point F then gives the value of the preconsolidation pressure. An alternative method is to remove the load on the sample and allow it to swell. Only partial recovery of the void space will occur and the unloading curve will follow a path GH. If the compression load is again applied the load curve will follow the path HI with I representing the rejoining point with the original compression curve. Both F and I should lie upon the line indicating the preconsolidation pressure or overburden pressure to which the sample was subjected in its natural state before tectonic uplift and erosion. The value of $\log_{10} p'$ of H is the existing effective pressure, and the OCR is then the effective pressure value at I divided by the value at H.

Soils with high values of OCR are dense and can only draw in water, in wet field conditions, slowly. The response time of the soil to changing groundwater is therefore lagged. As soil strength is greatly influenced by moisture content and pore-water pressure, the degree of soil over-consolidation has a major influence on soil behaviour under changing groundwater conditions.

8 Weathering Processes

Weathering is the process of alteration and breakdown of rock and soil materials at and near the Earth's surface by physical, chemical, and biotic processes. Igneous and metamorphic rocks, as well as deeply buried and lithified sedimentary rocks, are formed under a regime of high temperature and/or pressure. At the groundsurface the environment is dominated by temperatures, pressures, and moisture availability more characteristic of the atmosphere and hydrosphere: thus rocks are altered by weathering to new materials which are in equilibrium with surface conditions. Weathering has three very important results: it is the process which renders resistant rock and partly weathered rock into a state of lower strength and greater permeability in which the processes of erosion can be effective; it is the first step in the process of soil formation; and during weathering the release or accumulation of iron, calcium, aluminium, and silicon takes place—where concentrated after initial solution these form indurated oxide shells on rocks, or layers in the soil which may become hard and resistant to erosion.

The alteration of rock by weathering occurs in place, that is *in situ*, and it does not directly involve removal processes. It may be characterized by the physical breakdown of rock material into progressively smaller fragments without marked changes in the nature of the mineral constituents. This disintegration process leads to the formation of a residual material comprising mineral and rock fragments virtually unchanged from the original rock (Plate 8.1). By contrast chemical alteration may induce thorough decomposition of most or all of the original minerals in a rock, resulting in the formation of material composed entirely of new mineral species, particularly of clay minerals (Plate 8.2). Biological weathering induced by biophysical and biochemical agencies is largely confined to the upper few metres of the Earth's crust in which plant roots are active.

It must be appreciated that physical, chemical, and biological processes usually operate together. Also erosion takes place from the surface of the ground, and within the soil by solution, almost continuously so that, although we speak of weathering as a process of decomposition *in situ*, transport of the residuum of weathering may be simultaneous and assists in the continuation of weathering.

Factors affecting weathering

Few generalizations can be made about the rate of weathering of minerals because of the numerous factors which can influence the process. However, climate and the physical and chemical composition of the parent rock are of outstanding significance.

Climatic influences

Climatic conditions determine the temperature and moisture regime in which weathering takes place. Under conditions of low rainfall, mechanical weathering is dominant and, therefore, comminution of particles occurs, with little alteration of their composition. With an increase in precipitation, more minerals are dissolved and chemical reactions increase so that chemical decomposition of minerals and synthesis of clays becomes more important. In humid temperate climates clay minerals are both formed and altered.

The strength of rock is significantly reduced by saturation in water (Fig. 5.20h) and there is increasing evidence that moisture within rock voids can exert tensile stresses which are of sufficient magnitude to disrupt the strongest rock (Winkler 1977). Furthermore, moisture can enhance the

PLATE 8.1. A residual soil, derived from glacial moraine, in which there has been little or no chemical weathering and hence no clay-mineral formation, Antarctica.

expansion of some rocks, especially limestones; Hudec and Sitar (1975) found a large isothermal expansion upon wetting, which, on average, was equivalent to the dry thermal expansion produced by a temperature increase of 78 °C; Hockmann and Kessler (1950) found a similar effect in granites. The availability of moisture, even in the driest deserts, is greater than is commonly perceived, with dew, fog droplets in coastal deserts, and capillary water in shaded sites being sufficient to permit hydration effects and salt weathering to be a nearly continuous process. In Antarctica, liquid water may be produced from melting of snow in contact with dark rock when the air temperature is −10 °C; such water is then available for frost weathering.

Increases in temperature alter the stability of minerals. Quartz, for example, is highly resistant to weathering in temperate climates, but is more readily dissolved at the higher temperatures of the tropics. In such climates the solubility of iron and aluminium hydrous oxides is reduced. Iron and aluminium oxides, therefore tend to accumulate in tropical soils: the red colour is caused by iron (III) oxides. Temperature also increases the rate of any weathering reaction: a rise of temperature of 10 °C usually doubles or trebles the reaction rate.

Another effect of climate is to control the vegetation and its production of litter. In humid tropical climates the production of organic matter is high—3300 to 13 500 kg/ha per year from tropical forests—compared with temperate forests that produce 900 to 3100 kg/ha per year. This means that the supply of organic compounds to take part in chemical weathering is high in the tropical forests and low in the temperate ones. The appearance of the soils of these forests suggests the reverse because dark humus can accumulate in the cool forests but in the tropical ones organic matter is broken down very rapidly and much

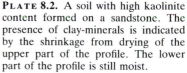

PLATE 8.2. A soil with high kaolinite content formed on a sandstone. The presence of clay-minerals is indicated by the shrinkage from drying of the upper part of the profile. The lower part of the profile is still moist.

of the humus has a pale colour which makes it difficult to identify. The turnover of tropical forest humus is about 1 per cent per day compared with 0.1 to 0.3 per cent in temperate forests.

The significance of climate has prompted the idea of weathering regions which approximately correspond to the distribution of major zonal soil groups (Fig. 8.1). This type of generalization is a useful model but it has to be qualified. Variations in soil type and weathering rates and depths depend not only upon differences in the kind of processes prevailing in climatic zones, but also upon the intensity of those processes. Broad schemes also have to be modified because they

apply only to tectonically stable areas with adequate drainage. Uplands and depressions give rise to distinctive erosional and depositional processes which may mask zonal weathering processes. A further qualification is that zonal processes have been modified by climatic changes in large areas of the Earth so that, although modern processes are occurring under modern climates, there may be relict weathering products in many areas derived from earlier climatic regimes.

In spite of these reservations we can detect areas of very thin weathering profiles in polar and desert zones where the absence of water and plants produces low weathering rates; intermediate rates

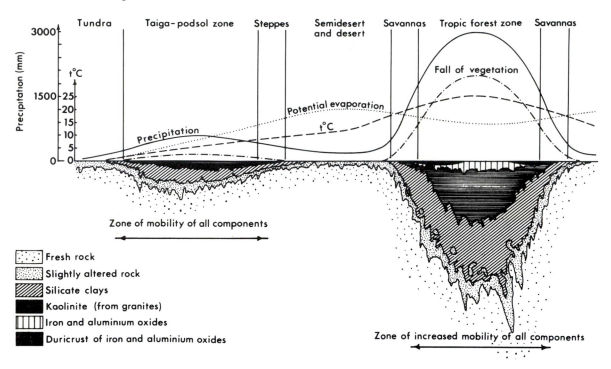

FIG. 8.1. The formation of weathering mantles in areas of tectonic stability and low relief. This scheme demonstrates a relationship between climatic factors, vegetation cover, depth of weathering, and dominant profile horizons. It does not consider relict effects. (After Strakhov 1967.)

occur in the temperate latitudes; and high rates in the humid tropics where weathering profiles are commonly deep and the formation of insoluble secondary clay minerals is at a maximum.

The physical characteristics of rocks

Particle size, hardness, permeability, and degree of cementation, as well as mineralogy all influence weathering processes.

Particle size is important because chemical weathering is mainly the result of surface reactions between solutions and mineral grains. The rate of weathering is therefore largely dependent upon the surface area of the grains. Thus it is relatively slow in sands but fast in silts.

Hardness, mineralogy, and cementation affect the rate at which weathering can reduce the rock to smaller particles. A siliceous sandstone, because of its quartz constituents, hardness, and cement, will be more resistant than a calcareous sandstone. Permeability, however, is probably a more important characteristic for it will control the rate at which water can seep into the rock and the area of

TABLE 8.1. *The porosity and permeability of selected rocks and sediments*

Rock	Porosity (%)	Permeability m/day
Unconsolidated		
Clay	45–60	10^{-6}–10^{-4}
Silt	20–50	10^{-3}–10
Sand	30–40	10–10^4
Gravel	25–40	10^2–10^6
Indurated		
Shale	5–15	10^{-7}–10
Sandstone	5–20	10^{-2}–10^2
Limestone	1–10	10^{-2}–10
Conglomerate	5–25	10^{-4}–1
Granite	10^{-5}–10	10^{-7}–10^{-3}
Basalt	10^{-4}–50	10^{-5}–10^{-2}
Slate	10^{-4}–1	10^{-9}–10^{-6}
Schist	10^{-4}–1	10^{-9}–10^{-5}
Gneiss	10^{-5}–1	10^{-9}–10^{-6}

Source: Gregory and Walling (1973).

STAGES OF WEATHERING OF ROCK MATERIAL

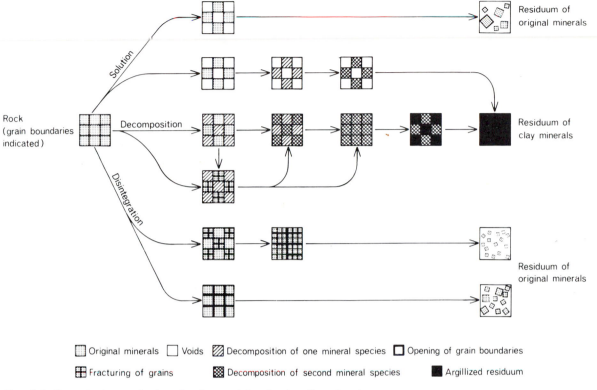

Fig. 8.2. Stages in the weathering of rock material under the effect of various processes.

the surface on which it can act. Rocks display a very large range of permeabilities, as is shown by the approximate values in Table 8.1.

The chemical properties of rocks and soils are of fundamental importance in determining the rate of processes and the products of weathering—these influences are discussed below in the section on chemical weathering. Other major factors which influence weathering are plants which produce CO_2 from their roots, and organic compounds from decay of plant tissue. Site factors—especially the rate of soil drainage—also have an influence on weathering.

Processes of weathering

The formation of joints in rock is perhaps the most important single weathering process, even though it is seldom classed as such. It is by the formation of joints that stresses, produced by cooling and by the pressure of tectonism and an overburden, are released. The presence of stress in rocks is illustrated dramatically, and sometimes disastrously, in deep mines when shells of rock burst from the walls, floors, and roofs of galleries. Rock-bursts are a response to the opening of the gallery which permits the rock to expand and so release its internal stress.

Once joints have developed in a rock, physical and chemical processes operate together in nearly all environments and at nearly all stages, although physical weathering may be dominant in early stages and chemical weathering dominant once a regolith has formed. The product of weathering has a composition which is determined by the dominant process. Physical disintegration produces a residuum of fractured and comminuted particles of the original rock minerals. Solution processes remove the soluble minerals and leave a

STAGES OF WEATHERING OF A ROCK MASS

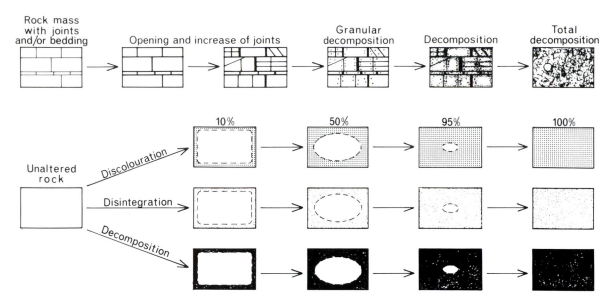

FIG. 8.3. The visible effects of weathering at successive stages.

residue of primary minerals which may be further altered by the chemical processes (Fig. 8.2). Chemical decomposition ultimately results in the production of clay minerals and complete alteration of original rock minerals. Stages in the weathering sequence are illustrated in Fig. 8.3.

Physical weathering processes

Physical weathering processes are most evident where rock is exposed at the ground surface, and are thus particularly obvious in hot or cold deserts, and on cliff faces. The result of physical weathering is a chemically unaltered fragment of rock which may range in size from a fractured single rock crystal to a massive joint block. Stress release is almost certainly the only weathering process which can produce massive joint blocks, but small rock spalls, fragments, and separated crystals may be produced by a variety of processes, and because they leave the rock chemically unaltered it is often very difficult to determine which processes are responsible for a particular product. The most commonly recognized physical processes are those resulting from internal rock stress, insolation, frost action, salt crystal growth, and wetting and drying.

Internal stresses exist in all rocks which have

been subjected to high temperatures and pressures during their formation. In igneous rocks, such as granite, stresses are set up during cooling. Feldspar crystals form first from the melt; then quartz crystallizes in a form known as high quartz, and with further cooling this converts to low quartz. The difference between these two types of quartz is dependent upon the arrangement of the atoms in the crystal. Solid silica is built from a tetrahedral unit consisting of one silicon atom surrounded by four oxygen atoms. This unit is connected to other similar units through the oxygen atoms, which allows for some variety in the possible structures. Low quartz differs from high quartz in one major respect: the bond angle at the oxygen is $180°$ in high quartz, but about $150°$ in low quartz (Fig. 8.4a). The transition from high to low quartz in a cooling melt occurs at about $573°C$ and involves a change in crystal shape. As the quartz crystals are constrained by the surrounding crystals tensile stresses develop in the rocks (Krinsley and Smalley 1972). Once the confining stress of the surrounding rock is removed by erosion, the stresses within the rock may be relieved by fracturing.

Few measurements have been made of stresses

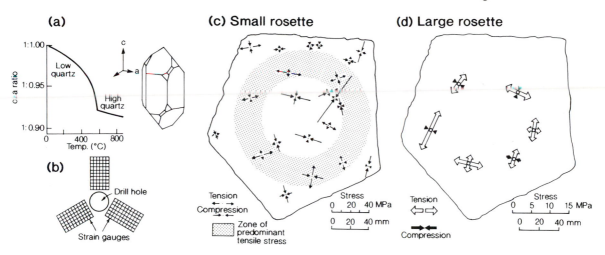

FIG. 8.4. (a) During crystallization from a melt, high quartz changes to low quartz. These two forms of quartz have different ratios of the length of their *c* to *a* axes, different bond angles, and hence different volumes (after Krinsley and Smalley 1972). (b) The pattern of strain gauges around a drill hole. (c) The distribution of tensile and compressive stresses across a transverse section of a pentagonal joint block of basalt, where small strain-gauge rosettes are used. (d) Where large rosettes are used the strain magnitudes are lower and the general pattern of tensile stresses, within an annular zone, is clearer, but the stress directions are different from the pattern in (c). (b–d after Bock 1979.)

within joint blocks, but those by Bock (1979) show that, for a rock to be in a stable state of static equilibrium, regions of compressive stress must be compensated by other regions in tension. In the example shown in Fig. 8.4b, c, d, a joint block of Miocene basalt of pentagonal cross-section has been prepared for measurement of internal stresses. The block was formed at or near the ground surface in a lava flow and is not visibly weathered. It is assumed, therefore, that any stresses within it result from original crystallization and not from overburden or tectonic loading.

The block has been cut across its transverse axis and rosettes of strain gauges mounted around a site in which a drill hole was then cut. Expansion of the rock around the drill hole was recorded and resolved into stresses along two axes (σ_1 and σ_2). Measurements made with small strain gauges (1.5 mm long, around a 1.5 mm diameter hole) produced high values of -12.6 MPa in tension and $+15.2$ MPa in compression, with one extreme value of -30.6 MPa. The tensile stresses were found in an annular zone which separated inner and outer zones of compression. The somewhat irregular stress pattern and rather high magnitude of individual stresses suggest that overall the stresses are balanced and the rock is not subject to stress-relief fracturing, as is confirmed by its field

appearance, but that there is a possibility of small crack development in the tensile zone if the confining compressive zone becomes weathered. When larger strain gauge rosettes were used (6 mm long individual gauges) a different pattern was found (Fig. 8.4d), suggesting a ring of tensile stresses acting in a tangential direction and very small compressive stresses acting radially. The stress magnitudes for the large-scale tests were about half those of the small-scale tests. From these measurements it may be concluded that: (1) stresses large enough to cause crack extension exist even in apparently unweathered joint blocks; (2) stress directions are varied within a block; (3) joint blocks may have balanced stress states but that fracturing a block, as in a rockfall, may create an unbalanced state and give rise to crack propagation. Swolfs *et al.* (1974), for example, separated a core of quartz-diorite by drilling and found that, after only nine months, it suffered granular disintegration while the original rock mass suffered no disintegration, presumably because the residual stresses in the mass were still in equilibrium.

A loss of stress equilibrium has been recognized as the cause of the separation of both large and small plates of rock from an unjointed rock mass as well as individual crystals and grains (Plates 8.3, 8.4, 8.5). Concentric plate separation is

PLATE 8.3. Arching of a sheet of arkosic sandstone on the flank of Ayers Rock, central Australia. The sheet is buttressed at each end.

PLATE 8.4. Granular disintegration of a coarse-grained granite; note the grus of individual grains and rock crystals, Namib desert.

particularly important in producing rounded boulders both within chemical weathering profiles and on the ground surface. Separation and exfoliation of the outer plates may involve arching as the plate swells (see below). Plates may be of many shapes and degrees of curvature and commonly cut through individual crystals, indicating that the stresses operate in the rock body not just at grain boundaries (Plates 8.6, 8.7).

Sheeting has already been described as a response to unloading by erosional removal of overburden or laterally confining rock masses (see Figs. 4.10 to 4.14). Large-scale sheeting is particularly common in granitic rocks and massive sandstones in which cross joints have not developed or are widely spaced. Sheeting joints develop nearly parallel to the ground surface, and observations in quarries indicate that the thickness of individual concentric sheets increases with depth into the rock mass to depths of 30 m on domes and 100 m in flat terrain. Sheeting joints pass through dykes and xenoliths and are therefore younger than those features, but they terminate at pre-existing joints and boundaries with plastic rocks.

Hypotheses of the origin of sheeting joints include: dilation of the rock mass as overburden is removed (Gilbert 1904); high compressive stress parallel to the ground surface (Dale 1923); and

PLATE 8.5. Exfoliation of plates of fine-grained granite, Namib desert.

PLATE 8.6. Exfoliation of thick plates from a granodiorite, Antarctica.

PLATE 8.7. Spheroidal weathering of basalt showing the successive rock layers.

heating by solar radiation or by fire. It is evident that severe heating and cooling have been observed to split rock and that in quarries fire is sometimes used to cause localized expansion and splitting of rock slabs (Holzhausen 1989), but these are exceptional situations which certainly do not apply on valley sides in Antarctica (Selby 1977b) and Greenland (Oen 1965) where sheeting in granites occurs parallel to valley walls and even follows the slope form around convex and concave curves, indicating that sheeting develops in conformity with the valley form and does not control it.

Quarry workers have long recognized that granite will split readily along planes parallel to sheet jointing. Holzhausen (1989) has shown that microfractures are aligned in such planes and Folk and Patton (1982) have described granular dis-aggregation as an extreme result of microsheeting. A mechanism for separation of sheets has been demonstrated by Folk and Patton (1982). It is a relatively common observation in quarries in massive rock that thin sheets of rock can suddenly spring upwards, sometimes with enough force to throw people off their feet. The result is an A-tent form, or a ridge with a cleft along its apex (Plate 8.8). Folk calculated the amount of lateral expansion which is required to produce arching of a thin beam from the relationship:

$$x^2 = (\ell_e)^2 - (\ell_o)^2.$$

where x is the height of the arch, ℓ_e is the length of one half of the arch roof, and ℓ_o is half the length of the floor of the arch (Fig. 8.5b, c). Thus, for a beam with an original length of 100 m undergoing a 1 per cent expansion, the arch will be 7 m high (Table 8.2). For such a process to operate, the bending beam must be constrained by lateral buttresses, otherwise there would be a simple lateral expansion of the slab without bending. Lateral expansion of 0.1 per cent has been recorded in quarried blocks from Stone Mountain, Georgia, and expansions of 15–20 mm in 30 m long slabs are not uncommon (Dale 1923). Assuming a Young's modulus of 50 GPa, these strains

PLATE 8.8. Arching of quartzite to form a small anticline, Kalahari.

TABLE 8.2. *Amount of arching resulting from a given amount of expansion*

If a 100 m bar expands: %	length	It must arch up (m):	Rise (%)
1	1 m	7	7
0.1	10 cm	2.2	2.2
0.01	1 cm	0.7	0.7
0.001	1 mm	0.22	0.22
0.0001	0.1 mm	0.07	0.07
0.00001	0.01 mm	0.022	0.022

Source: Folk and Patton (1982).

correspond to *in situ* compressive stresses of 17–35 MPa. Such estimates of stress have been corroborated by direct measurements. These stresses are commonly twice the uniaxial tensile strength of granite (Holzhausen 1989). It is important to recognize that the compressive stresses are produced by biaxial loading in which the two principal stresses, σ_1 and σ_2, are parallel to the ground surface and at least one of these is highly compressive. A third stress, σ_3, is perpendicular to the ground surface and has an insignificantly small magnitude. Consequently the sheet can arch with only a low tensile strength to be exceeded in the plane of fracture. Sheeting does not occur at depth where the rock is in a state of high triaxial compression. Complete three-dimensional unloading without buttressing does not give rise to sheeting or spalling if the internal stresses are in equilibrium.

Steps may occur in sheets (Fig. 8.5d, Plate 8.9) if crack propagation is deflected, either by an anisotropy in the rock or by a change in stress distribution if a unit of a higher slab slides or falls off a rock face. The crack is diverted towards a free face and extends towards the ground surface until it is parallel to, and perhaps no more than a few centimetres below, the upper surface of the sheet, at which point fracture propagation ceases.

The phenomena of rock-bursts and sudden arching indicate that fracture propagation may be very rapid, but evidence from quarries suggests that it may also be a gradual process. Similarly, rock bodies along a single valley wall may respond to removal of confining rock quite differently, with some areas showing rapid development of sheeting joints and other areas showing no such joint 10 000 years after deglaciation or other effective processes of removal.

Induced fracture can occur when a large rock rests on a smaller rock in which the induced unconfined stress exceeds the tensile strength (Fig. 8.5e). The resulting failure may be from a vertical fracture, of the kind induced in a Brazil test, if the lower rock is relatively thin and the stress is concentrated at opposing point contacts. A boulder resting on a very narrow point may suffer induced fracture in which lens-shaped plates may split off (Ollier 1978). The zones of tensile stress in each case are indicated in Fig. 8.5e, f.

Thermal expansion from set fires to induce fracturing in quarries, and the observation (Ollier and Ash 1983) of spalling and cracking of boulders

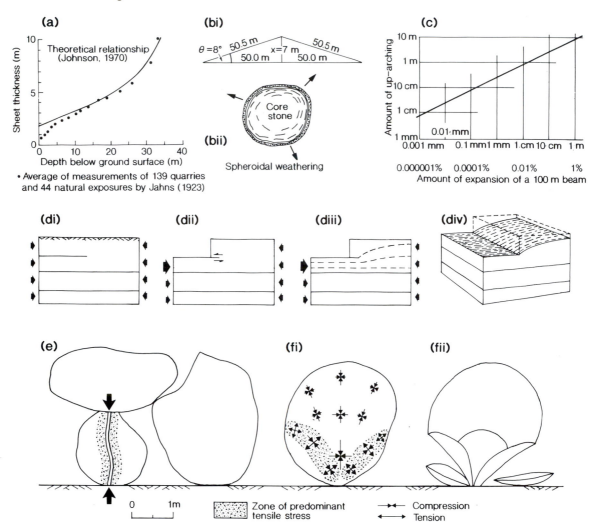

Fig. 8.5. (a) The increase of thickness of sheets in granite with depth (after Johnson 1970). (b, i) The geometry of a buttressed arch formed by rock expansion. (b, ii) Concentric exfoliation produced by buttressed expansion. (c) Amount of arching as a function of the amount of expansion of a 100 m beam (b, c after Folk and Patton 1982). (d, i) Development of a step in a sheet. The arrows indicate surface-parallel compression. (d, ii) A portion of the upper sheet is removed, redistributing the stresses and causing fracture propagation in an upward direction (d, iii). (d, iv) Shows the newly stepped sheet after failure (d, after Holzhausen 1989). (e) Tensile failure in a boulder under load. (f) Tensile stress patterns and splitting of plates. (e and f, ii are based on Ollier 1978.)

during forest fires in Australia, has rekindled interest in direct heating as a mechanism of weathering, particularly in deserts.

Surface temperatures on dark rocks in deserts can reach 79 °C (Peel 1974) but more commonly maxima are in the range 50 to 60 °C (Kerr *et al.* 1984). The penetration of a heat wave may exceed one metre and temperature gradients below the

rock surface can reach 1 °C/cm (Roth 1965). In a savanna zone, thermal amplitude over a year, at a depth of 20 mm in granite, has been recorded as 37.3 °C (Mietton 1988). In these conditions, linear expansion of granite may be 0.025 per cent and for some rocks it may be higher. Table 8.2 suggests that where buttressing occurs considerable arching may develop from solar insolation alone.

PLATE 8.9. Upward-stepping of a sheeting joint. Removal of the large sheet will leave a rollover, Namib desert.

PLATE 8.10. Possible insolation shattering of a quartzite boulder, central Sahara. The scale is 30 cm long.

A more general process of rock breaking was investigated experimentally by Blackwelder (1933) and Griggs (1936) whose data appeared to demonstrate that thermal expansion and contraction was, at best, unlikely to weather rocks. They repeatedly heated and cooled rocks over temperature ranges greater than occur naturally. Their specimens were, however, unconfined and the effects of water and salt in rock cracks were not investigated in detail (see below for comment on fatigue failure and stress corrosion).

In spite of the early experimental work the

existence of clearly split boulders in desert areas (Plate 8.10) has promoted a continuing belief in the effectiveness of thermal expansion and contraction. Aires-Barros *et al.* (1975) repeated the early experiments and noted some microfracturing of rocks under dry heating, but substantially more fracturing and spalling when the temperature changes occurred in the presence of water. It seems possible that the volumetric expansion of water trapped in rock capillaries may disrupt rocks.

Two main factors, apart from heat supply, influence the heating of a rock: the colour of that rock and therefore its albedo; and the thermal conductivity, or rate of heat transfer through the material. External factors such as shade, wind, and air temperature have major effects within and between sites. In a study involving specimens of basalt, sandstone, granite, and chalk, McGreevy (1985) showed that there are striking differences between surface temperatures for the four rock types after one hour, with the range being about 9 °C (Fig. 8.6). Dark basalt had the highest temperature and white chalk the lowest. The difference between surface temperatures and those at a depth of 50 mm was least for chalk, and increased in the order: chalk, granite, sandstone, basalt. It is evident that a dark colour and low ability to transfer heat to depth both operate to create high surface temperatures at the surface of basalt.

Low heat transfer creates a high temperature gradient between surfaces and interiors, and between pale and dark crystals, with consequent differences between the expansion and contraction of different parts of a boulder. Buttressing and arching may then, in theory, operate, particularly around the outer surfaces of a boulder. This situation is exactly that which Griggs (1936) was investigating and found to be ineffective, it may, however, be effective where on a rock face, a block or a clast in a conglomerate protrudes and is heated far more than the rock mass, and cracks. Such a process has been suggested as occurring over the face of the domes of the Olgas of central Australia (Ollier and Tuddenham 1962). Blocky disintegration, pebble cracking, and similar forms which are evident in deserts may occur only where water and/or salt is present in the rock (Schattner 1961; Whitaker 1974). In forest fires temperatures in the range 300–800 °C may occur at rock surfaces. Thermal expansion, and particularly differential expansion, may then be extreme and

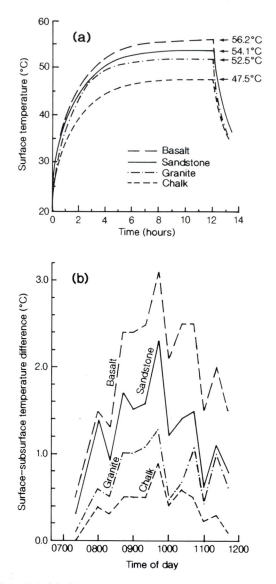

FIG. 8.6. (a) Change in rock-surface temperature, under closely controlled conditions, for four rock types. (b) Differences in temperature between the rock surface and at a depth of 50 mm, with all four specimens receiving the same insolation. (After McGreevy 1985.)

cause large-scale cracking and splitting of surfaces and boulders. The evidence for insolation weathering as a general process on exposed rock faces is, however, equivocal and certainly not well-established by experiment.

Crystal growth, whether those crystals are of ice, salts, or newly formed minerals, can exert

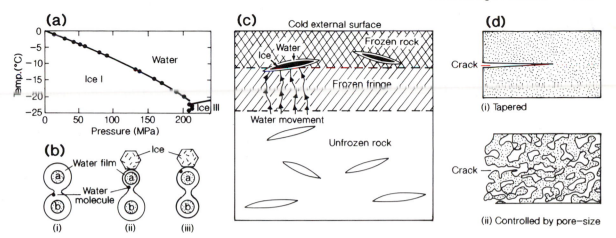

FIG. 8.7. (a) The increase of pressure in confined pure water at lowered freezing temperatures (data from Bridgman 1911). (b, i) Water films in equilibrium around two soil particles; (ii) ice attracts molecules from out of the film around the nearest particle until (iii), a new quasi-equilibrium is established with thinner films (after Selby 1985). (c) Idealization of freezing of cracked rock. Discus-shaped cracks, in the frozen zone, develop lenses of ice separated from the rock by very thin films of water. Water migrates to the freezing zone, drawn by the ice in the cracks (after Walder and Hallet 1985). (d) Tapered and rectangular cracks (after Tharp 1987).

pressures, on the walls of cracks in rock, which are of sufficient magnitude to cause rock fracturing.

Frost weathering is a result of the phase change of liquid water to crystalline ice; this phase change causes an increase in specific volume of 9 per cent at 0 °C and 13.5 per cent at −22 °C, and molecules become arranged into a rigid hexagonal crystalline network. It was once thought that this simple ice formation was itself capable of forcing open rock fissures and that, as temperature falls, the pressure that ice exerts on the rock would be proportional to the increasing pressure in the ice itself (Fig. 8.7a). This simplistic view has been recognized, for some time, as being inadequate to explain the variety of responses of different types of rock to frost action at a range of environmental conditions. Work by Hallet (1983), Walder and Hallet (1985, 1986), and Tharp (1987) has changed fundamentally the understanding of the formation and effect of ice on physical weathering of rock. This modern work has concentrated on the fracture mechanics of ice growth in cracks and the propagation of cracks under stresses induced by the ice; it is an extension, therefore, of the processes outlined in Fig. 4.8.

The work of Hallet, Walder, and Tharp is based upon mathematically modelling rock properties (including elastic moduli, crack shape, and length) environmental conditions (including temperature,

temperature gradient, rate of temperature change, and water pressure), and the stress required to extend cracks. Hallet and Walder studied the situation of discus-shaped cracks within a rock, and Tharp the open cracks formed at the exterior of a rock body. Their theoretical work has yet to be confirmed by experimental studies in field situations, but has already helped explain many field observations (e.g. Hall 1987).

The wedging effect of ice growth in closed cracks is not attributed primarily to the volumetric expansion accompanying the phase change of water to ice, but to the induced pressure which is thought to develop in cracks as water migrates towards freezing centres and permits the growth of ice lenses in the crack, even at temperatures well below the freezing-point. The basis for crack growth is thus perceived to be the same condition which permits the growth of ice lenses and frost heave in soils. In soils, water migrates because ice crystals draw water from the water films on the surfaces of soil particles. Water films in contact then develop a new equilibrium by 'sharing' the remaining water molecules between them; the effect is to draw water towards the freezing front (Fig. 8.7b). Rock surfaces act similarly, and because the water at the particle, or rock surface, is strongly bonded it will not freeze at temperatures above −22 °C, below this temperature there is a change to a dense form of ice (Ice III). The direct result of water movement along rock surfaces is that ice lens growth can continue as temperature falls, as long as water can be drawn into the crack, but the growth will be slower as water becomes increasingly viscous at lower temperatures and as water films become thinner.

In an idealized model crack (Fig. 8.7c), a thin ice lens is separated from the rock by a film of water which is only a few micrometres thick. In soils, water migration and the growth of segregated ice lenses create heaving pressures which may exceed 20 MPa. It is assumed that the same pressure levels may be reached in rock; such pressures exceed the tensile strength of most rocks. For crack propagation a critical stress intensity (K_{ic}) is required and this is a function of the crack geometry (with r being the maximum crack radius) and the internal ice pressure (P_i), then at propagation:

$$K_{ic} = \left(\frac{4r}{\pi}\right)^{1/2} P_i.$$

The pressurized crack opens into an oblate ellipsoid (discus-shape) with the maximum opening related to the maximum crack width (w), the Poisson's ratio (v), and shear modulus of the rock (G) thus:

$$\frac{w}{r} = \left(\frac{4}{\pi}\right)\left(\frac{1-v}{G}\right)P_i.$$

To sustain crack propagation water must continue to flow into the crack, otherwise the stress intensity (K_i) and ice pressure (P_i) will decline. Experimental studies indicate that cracks propagate distances similar to rock-grain sizes (0.1 to 10 mm) in about 1–100 days. These rates are highest for the temperature range $-4°$ to $-15\,°C$ and very low for the range $-15°$ to $-20\,°C$. Oscillation of temperature about the freezing-point has no effect on crack growth. Rates of crack growth are strongly influenced by the permeability of the rock, which is very low for dense crystalline rocks (Table 8.1), so prolonged freezing periods favour cracking. As with all other forms of stress-induced fracture, the linking of cracks gives rise to larger planes of failure, which are evident in Plate 8.11.

The study by Tharp of cracks on the outer surfaces of rocks recognized that cracks have a variety of forms (Fig. 8.7d) related to the grain-size, density, and composition of the rock. Crack

shape is expressed as an aspect ratio (maximum aperture/crack length); for a spherical crack this ratio has a value of unity, but for cracks in most rocks aspect ratios fall in the range 1×10^{-3} to 7×10^{-4}.

In open cracks two conditions for failure are recognized: (1) when water in a crack is confined, and has no connection with the atmosphere because of plugging by ice, the ice pressure is due to its volumetric expansion on freezing and is an order of magnitude higher than in (2) the unconfined condition, in which free water remains at atmospheric pressure and adsorptive force conditions prevail. Capillarity generally does not produce stresses of sufficient magnitude to propagate cracks.

In microcracks and short tapered joints in crystalline rock, the dilation necessary to raise the stress intensity at a crack tip to the critical level (K_{ic}) is commonly greater than the dilation produced by the expansion of the thin layer of water in the crack. Crack propagation will then not occur until additional water is pulled into the crack and will not occur if crack length is less than several centimetres. The larger and more rectangular cracks common in sedimentary rocks favour crack propagation which can then occur in cracks as short as 1 mm. In general, for surface cracks the effects of rate of freezing and temperature change are subordinate to the effects of shape and length of the crack.

These findings confirm field observations of the high resistance of crystalline, impermeable, and dense rocks in dry conditions (Plate 8.12), and the susceptibility of porous, weak, foliated, argillaceous, and fissured rock in damp conditions (Mugridge and Young 1983; Fahey and Dagesse 1984; K. Hall 1986). There are, however, still many unanswered questions about frost weathering: the effects of rock anisotropy, of freeze penetration, and water movement; the rate of temperature fall in zones of different degrees of water saturation; the period of time needed for a particular freezing temperature to be maintained for freezing effects to occur; the temperature range over which water can migrate; the significance of ions in the water for the strength and other properties of ice; and, for open surface cracks, the importance of dust and rock particles in increasing ice strength—all of these, and other aspects of freezing processes require further investigation.

The work of Hallet, Walder, and Tharp casts

PLATE 8.11. Frost-shattered dolerite. The joint blocks in this out-crop had a blade-like shape and fractured readily, Antarctica.

PLATE 8.12. Weathering of dolerite, Antarctica. The coarse-grained variety is disintegrating but the fine-grained rock is nearly immune to the prevailing weathering processes.

PLATE 8.13. Felsenmeer of quartzite on a high plateau (2000 m), Antarctica.

doubt on the validity of many older ideas about the importance of the duration of freezing, rate of temperature change, oscillation across the freezing-point, hydrofracturing due to capillary water being forced towards the tips of cracks, and shock waves being induced by sudden phase changes and by cavitational collapse of air bubbles within freezing water. Only further work by mathematical modelling, laboratory simulations, and field measurements will properly test these ideas.

An aspect of frost weathering which has received little attention, outside USSR, is the comminution of single rock and soil grains. In his review of work carried out in the Soviet Union, Konishchev (1982) has emphasized that, contrary to trends in temperate or warm climates, quartz is characterized by a high degree of instability in cryogenic (frost weathering) conditions and is the source of much of the silt fraction (0.05 to 0.01 mm) in soils and especially in loess. Heavy minerals and feldspars

also break down to silt-size particles, but much less readily than quartz, because they have greater protection from films of unfrozen water adsorbed on their surfaces. The Soviet work emphasizes cold-desert weathering as a source of loess, rather than the rock 'flour' from glacial abrasion which has been regarded in western Europe and North America as being the main source (see Selby 1976, for a review). There is also increasing evidence of an abundance of silt in tropical soils, derived from splitting of quartz grains, which suggests that some silt in cool climatic zones may be from eroded deep weathering profiles (Nahon and Trompette 1982).

The results of frost weathering are accumulations of angular fragments with forms which may be rod and plate-like if rocks have a notable preferential aligned fabric, or more equant if there is no control by fabrics. Shattering is largely confined to rock outcrops and a regolith greatly reduces or eliminates its effectiveness. In cold deserts or alpine areas with no regoliths the entire ground surface may shatter to produce a surface of broken rock fragments called a *felsenmeer* (a German word meaning 'rock sea') (Plate 8.13).

Salt weathering results from the crystallization of salts in rock pores. Growing crystals can exert high pressures on the confining rock and cause exfoliation and rock fracturing. The salts involved in this type of weathering may be derived from the sea and blown inland in spray or carried by snow and rain, or they may be derived from chemical weathering of rock. In desert areas particularly, drainage waters may evaporate to leave a salt-rich sediment which will be deflated by wind and the salt redeposited on rock surfaces.

Salts may be carried in solution into pores and fissures in rocks by percolating water. Once in the rock the salts may contribute to rock failure by any of three processes which can create stresses resulting from: (1) the growth of crystals from solution; (2) thermal expansion; or (3) hydration.

(1) The growth of salt crystals from saturated solutions has now been studied in a number of experiments which have been reviewed by Evans (1970); Selby (1971a); Cooke (1981); and by McGreevy and Smith (1982). It has been shown that in a fissure large salt crystals will always grow at the expense of smaller ones and that they will continue to grow until they completely fill the pores in a rock. A crystal will grow in a large pore until the pressure builds up to such an extent that mechanical fracture occurs, or the confining pressure is sufficient to make the crystal grow down a capillary pore. Whether or not fracture occurs will depend upon the tensile strength of the rock. For rocks of equal inter-pore strength those with large pores separated from each other by microporous regions will be most liable to destruction by crystal growth.

(2) Thermal expansion of crystals which have already formed in rock fissures is a result of a rise in temperature causing the crystal volume to increase. Cooke and Smalley (1968) measured the thermal expansion of five salts ($NaNO_3$, $NaCl$, KCl, $BaSO_4$, $CaCO_3$) and found that in all cases except one ($CaCO_3$) the expansion was considerably greater than that of granite (Fig. 8.8a). For a common salt such as sodium chloride a rise of $54\,°C$ would give a volumetric change of 1 per cent, which is considerably greater than that in the surrounding rock. The stresses caused in the rock are concentrated at the inner extremity of the crack

in which the salts occur and fissures may be progressively opened by this process even when there is a considerable confining pressure.

The temperature range of $54\,°C$ may seem extreme but rock surfaces experience considerably higher and lower temperatures than the surrounding air and may well be subjected to such a range during a 24-hour period in extreme environments. The low thermal conductivity of rock probably limits this process to the outer few centimetres of any outcrop.

(3) Hydration of crystals occurs when some salts take up moisture within their lattices. The resulting expansion may be considerable and produce very large pressures. The hydration pressures of a number of salts have been calculated by Winkler and Wilhelm (1970). The hydration of thenardite (Na_2SO_4) to mirabilite ($Na_2SO_4 \cdot 10H_2O$) following a change of relative humidity from 70–100 per cent at a temperature of $20\,°C$ can exert a pressure of nearly $50\,MPa$, and hydration of bassanite ($CaSO_4 \cdot \frac{1}{2}H_2O$) to gypsum ($CaSO_4 \cdot 2H_2O$) with a change of relative humidity from 30 to 100 per cent at a temperature of $0\,°C$ can exert a pressure of $200\,MPa$ (Fig. 8.8b, c). Such temperature and humidity ranges can occur in many arid regions and at a local scale on coastal cliffs and rock platforms which are frequently wetted by spray and then rapidly dried in the sun.

Experimental studies indicate that crystal growth is the most effective of the three salt weathering processes with both hydration and thermal expansion being of minor importance (Goudie 1974; Sperling and Cooke 1985; Fahey 1986a).

Granular disintegration is readily recognized as being a result of salt crystal growth (Plate 8.14) but it has also been shown that flaking is likely to occur where moisture absorption and loss is directed through one exposed surface (Smith and McGreevy 1983). Such flaking may be enhanced by rock fabric, but it can also occur in equigranular rocks, and commonly occurs parallel to the existing rock surface.

Experimental studies have not always simulated natural conditions but there is increasing recognition that useful experimental work must concentrate on conditions of natural rock moisture content, availability of salts, and conditions of shade and aspect.

In hot and cold deserts, $CaCO_3$, $CaSO_4$, and $NaCl$ are relatively common. Desert sulphates are commonly related isotopically to sea-water

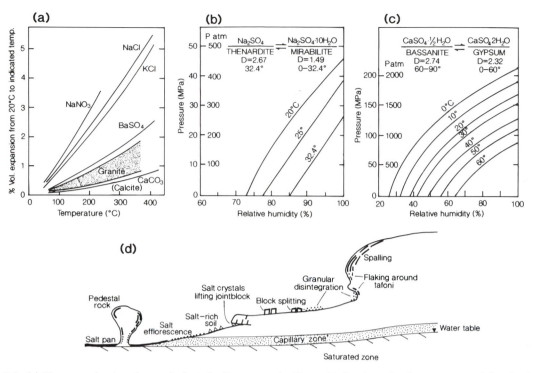

FIG. 8.8. (a) The expansion in volume of selected salts compared with granite for given rises in temperature (after Cooke and Smalley 1968). (b) Pressures exerted in the hydration of thenardite to mirabilite, and (c) of bassanite to gypsum (b, c after Winkler and Wilhelm 1970). (d) Features recognized as being the results of salt weathering in deserts.

sulphate, suggesting that studies which have placed much emphasis on $MgSO_4$ and Na_2SO_4 may be concentrating on salts which are relatively uncommon in most inland deserts (Keys and Williams 1981; McGreevy and Smith 1982). NaCl is undoubtedly an important salt in coastal zones and may be carried far inland in rain, but much of its effect may be due to enhanced solution of quartz in the presence of chloride ions, rather than to crystal growth (A. R. M. Young 1987).

Salt crystal growth has been described as being responsible for a number of small landforms; granular disintegration, flaking, spalling, pebble splitting, cavernous hollows and tafoni, and undercutting of outcrops with formation of pedestal rocks and tors. In virtually all cases it is difficult or impossible to be sure whether salt action has operated alone or in association with other processes such as chemical alteration, solution of rock, frost action, wetting and drying, and stress fracturing. What is clear, however, is: (1) the abundance of soluble salts in semiarid and arid environments, even at high altitude and far from

the sea; (2) the mobility of salts in solution in runoff, groundwater, and in capillary water, with the result that salts may crystallize within rocks as well as on their surfaces; (3) the rapidity with which rocks disintegrate in salt-rich environments; and (4) the effectiveness of salt crystallization in breaking down rocks at the scales of large joint blocks to small pebbles, and single crystals and grains (Whalley *et al.* 1982; Goudie and Watson 1984). A further conclusion, (5) is that porous and fissured rocks are more susceptible to salt weathering than fine-grained and impermeable rocks (see Plate 8.12). At very large scales it may be effective in rock disintegration which creates fine-grained material which is then available for removal by wind so that large deflation hollows and depressions can be excavated down to the water-table.

Alternate wetting and drying has been suggested as a process responsible for the disintegration, or slaking, of fine-grained rocks (Ollier 1969). The nature of the processes involved is not fully understood but it may involve the effects of both water

PLATE 8.14. Granular disintegration of a marble boulder, Antarctica.

and air (Taylor and Spears 1970; Franklin and Chandra 1972). Water molecules which are adsorbed on to the negatively charged surface of a crystal through their positively charged hydrogen ions may eventually force rock particles apart, and then permit a collapse to a lesser volume in a subsequent drying of the rock; repeated adsorption and loss of moisture creates 'fatigue' effects, or a lowered resistance to fracture propagation and, eventually, disintegration may occur. Such effects are most notable in argillaceous rocks (Mugridge and Young 1983; Fahey and Dagesse 1984) but they are also evident in quartz (Moss *et al.* 1981) and in schist (Fahey 1983).

Air breakage may occur when, during dry periods, evaporation from the surfaces of rock fragments promotes high suctions within the rock pores. At extreme desiccation the bulk of the voids will be filled with air which, on rapid immersion in water, becomes compressed by capillary pressures developed as the rock becomes saturated. High

disruptive internal air pressures may cause cracks to be opened and extended from their tips. Such processes are particularly effective in breaking down aggregates on bare soil surfaces.

The expansion and contraction of clays, with wetting and drying, is not normally included in a list of physical weathering processes, but, as clay minerals of the montmorillonite group can experience an increase in volume of up to fifteen times on wetting, their expansion and contraction in clay-rich rocks such as mudstone can be a major weathering process, and may promote rapid losses of strength and landsliding (Plate 8.15).

Comparative studies of physical weathering processes indicate that, where conditions for the existence of saturated solutions are suitable, salt weathering is a most effective cause of rock breakdown and more effective than frost-induced shattering. Frost action and salt weathering in conjunction have been observed in some studies (Goudie 1974) to produce a higher rate of break-

PLATE 8.15. Weathering of a mudstone showing the effect of swelling and shrinking with wetting and drying.

down than salt action alone but other studies contradict this conclusion (McGreevy 1982). It is probable that the different experimental results reflect different availabilities of salt, with a high availability being necessary for rapid rock breakdown, and varying water-content of the rock; saturation is likely to cause maximum weathering rates (Jerwood *et al.* 1990).

Salt and frost weathering are far more effective as weathering agents than insolation and wetting and drying, and for all processes major controls are rock characteristics such as water adsorption capacity, bulk density, and rock fabric, with rocks that have linear alignments, such as in schists, producing elongated rod or plate-like fragments, and rocks without fabric producing equant fragments.

Fatigue failure and stress corrosion

The effect of repeatedly applied cyclic stresses, at levels far below the instantaneously determined strength of a material, is widely recognized as a major contributor to failure in metals. For iron and steel, a plot of stress against the number of

reversals of stress—an S–N curve—(Fig. 8.9) shows that there is a fatigue limit, that is, a level of stress below which the metal will not fail by fatigue, but above that limit there is a definable number of reversals of stress that the material can withstand before failure. The fatigue limit is therefore a valuable property of steel. Nonferrous metals do not generally have a definite fatigue limit, but instead the S–N curve trails off gradually.

The fatigue process begins on the surface of a metal, usually at a pre-existing stress concentration such as the tip of a crack; the crack then propagates slowly from the surface down into the metal. Once this crack reaches a critical length it propagates suddenly and explosively across the metal body. The significance of fatigue failure in such machines as aircraft is self-evident. In metals, the number of stress cycles required to cause failure is large for low-stress levels and small for high-stress levels.

Studies of rocks show that the number of cycles needed to cause failure increases with confining pressure. Fatigue failure is therefore most likely in

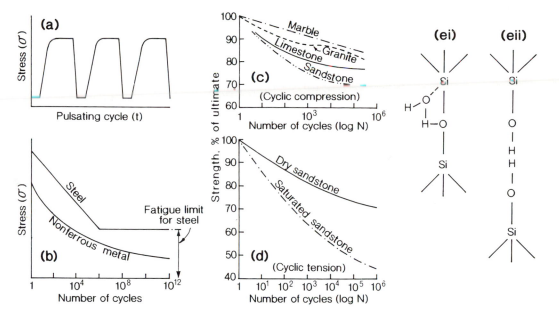

FIG. 8.9. (a) A pulsating cycle of constant stress level and frequency. (b) Typical fatigue (S–N) curves for steel and non-ferrous metals. (c) S–N curves for selected dry rocks in compression (based on Haimson 1974). (d) S–N curves for dry and saturated sandstone in tension (based on Brighenti 1979). (e) Model for stress corrosion by water at a crack tip, showing strained Si–O–Si bond (based on Michalske and Freiman 1983).

rocks exposed at the groundsurface. In rocks such as sandstone, marble, and limestone the fatigue limit occurs at 80–60 per cent of the ultimate strength, with no further decline in fatigue limit strengths after a number of cycles which range from 10^3 to 10^6. Some granites do not appear to develop a fatigue limit (Burdine 1963; Hardy and Chugh 1971; Haimson 1974). Work by Brighenti (1979) has shown also that fatigue failure in saturated specimens occurs at much lower cyclic stress levels, that is the fatigue limits are lower, than for dry specimens.

The reason for lower strengths in saturated specimens has been investigated in studies of stress corrosion, which is defined as a cracking process caused by the simultaneous action of corrodent and stress. There appear to be few data on corrodents in rocks, but the action of water is widely recognized and is evident in its effect in reducing strength. Michalske and Freiman (1983) suggest that this effect is caused by water enhancing the rate of crack growth by the replacement of Si-O bonds of tetrahedral silicate crystal units by O-H bonds. The bond energy of the O-H bond (464 kJ/mol) exceeds that of the Si-O bond

(368 kJ/mol) so the replacement would seem to be energetically favoured.

Stress cycling is recognized in many experiments carried out on salt, frost, and insolation weathering, but the importance of stress level, number of cycles, and presence of corrodents, other than water, has seldom been studied by geomorphologists and even rock-moisture content is not always adequately monitored. It is evident, however, that fatigue failure is a major factor in physical weathering. Furthermore, stress corrosion is probably a major, but largely unassessed component. Descriptions of physical weathering as being due to one form of weathering, such as frost action, are therefore likely to be inadequate.

Chemical weathering processes

The disintegration of rocks, such as granite, in the early stages of weathering is partly a physical and partly a chemical process. Studies using a scanning-electron microscope (Baynes and Dearman 1978) reveal that weathering is usually initiated along primary cracks, pores, and open cleavages. Such microfissures permit solutions to penetrate into the rock and commence chemical weathering. Quartz

PLATE 8.16. A scanning-electron micrograph of a feldspar crystal weathered in a soil profile. Removal of ions has left the crystal surface deeply pitted. (Micrography by Dr M. J. Wilson, Macaulay Institute for Soil Research, Aberdeen.)

grains suffer microfracturing and pit etching. Microfracturing may occur when solution of neighbouring feldspar crystals causes an increase in porosity and permits expansion of quartz with the release of locked-in residual stresses produced during the early cooling history of the rock from a magma. Feldspars also suffer pit etching (Plate 8.16) and solution along structurally determined planes, while biotite undergoes decomposition to clay minerals and expansion of the crystal lattice. Clay-mineral formation results from the removal of ions in solution; if only small amounts of cations are removed cation-rich clay-mineral species, such as montmorillonite and illite, can form but intense flushing and continued weathering leaves only kaolinite or gibbsite.

Chemical weathering is seldom a simple process, because: (1) rock and soil minerals are seldom 'pure' but contain minor components of varying chemical reactivity; (2) even simple reactions occur in stages, with the result that compositions change during reactions; (3) removal of weathering products is essential for continuity of reactions; (4) heat and water are required for all chemical reactions and availability of each is variable; and (5) minerals vary in their solubility and reactiveness so that the most soluble are lost first and bulk compositions of soils progressively change towards dominance by the most stable minerals.

Solution is usually the first stage of chemical weathering. The effectiveness of water as a solvent for ionic compounds is a result of distribution of electrical charges of its own molecules. A water molecule can always rotate and confront an ion with a complementary charge (Fig. 8.10a). The amount of solution depends upon: (1) the volume of water passing the surface of a particle; (2) the solubility of the solid being dissolved; and (3) the availability of hydrogen ions (expressed by pH, the negative logarithm of the hydrogen ion concentration). The pH of alkaline soils is in the range 8–10, sea water is 8–9, rain-water and river water 6–7, acid soils 4–6, and acid hot springs about 2. The solubility of iron is about 100 000 times greater at pH6 than at pH8.5. Silica is slightly soluble at all pH values, whereas alumina (Al_2O_3) is only readily soluble below pH4 and above pH9 (Fig. 8.10b). Alumina therefore tends to accumulate in the

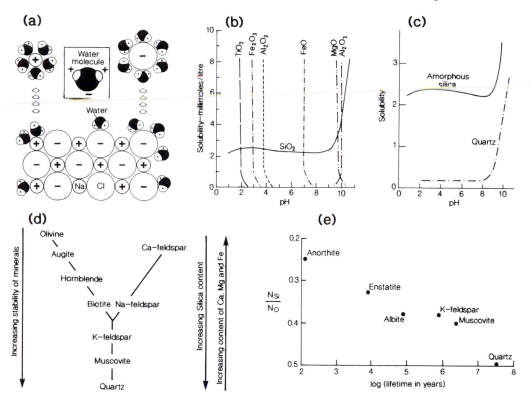

FIG. 8.10. (a) A water molecule can rotate to confront an ion with a complementary charge (after Cotterill 1985). (b) Solubility in water of selected rock and soil minerals as a function of pH (after Siever 1959). (c) Solubility of quartz and amorphous silica as a function of pH (data from Siever 1959). (d) Stability of minerals to dissolution and chemical weathering (based on Goldich 1938). (e) 'Lifetime' of minerals determined from laboratory simulations of weathering, in relation to silica content (e.g. $N_{si}/N_o = \frac{1}{2} = 0.5$ for quartz) (data from Lasaga 1984), Anorthite, $CaAl_2Si_2O_8$; Enstatite, $(Mg, Fe)SiO_3$; Albite, $NaAlSi_3O_8$; K-feldspar, $KAlSi_3O_8$; Muscovite, $KAl_2(OH)_2Si_3AlO_{10}$; Quartz, SiO_2.

clayey residuum during soil weathering and silica is slowly leached. Amorphous silica is about ten times more soluble than quartz. This difference is important because amorphous silica is frequently a cement in sandstones. Solution of the cement produces separate grains of quartz sand. At pH values above 9 the solubility of silica increases, but such pH values are rare in soils and confined to very alkaline desert conditions.

The solution processes which result in the formation of an alumina-rich residuum with varying amounts of silica are complex, because they involve the initial release of ions into solution and then the reactions of these ions with other ions or minerals to form new mineral combinations. The addition of organic acids further complicates the processes so that even a simple order of solubility

of ions is not applicable in all situations. The most commonly quoted order of solubility (Polynov 1937) is:

$$Ca > Na > Mg > K > Si > Al > Fe,$$

but the only invariable rule is that Ca, Mg, Na, and K are all more soluble than Al, Si, and Fe.

Similarly, and because rocks have variable compositions, there can be no fixed order of rock resistance to chemical weathering but general trends are indicated in Fig. 8.10d. Minerals vary in their response to the attack of acid water: some are nearly insoluble, some become gels, and some partly dissolve leaving fragments of the framework for the formation of clay minerals. During prolonged attack the frameworks of silica tetrahedra and alumina octahedra are disintegrated. Laboratory simulations of weathering indicate that the rates of mineral dissolution and thus the lifetimes of crystals of a given size (Lasaga 1984) concur with the relative thermodynamic stability order (cf. Figs. 8.10d, e).

The solution of limestones and relatively 'pure' sandstones are the most important of chemical weathering processes which produce distinctive landforms, but in neither case is 'pure' water involved. Dissolved carbon dioxide plays a major part in solution of limestones (see under 'carbonation' below) and silica is soluble primarily at high pH values and particularly where chloride ions are present.

Hydration is the addition of water to a mineral and its adsorption into the crystal lattice. The adsorption may make the mineral lattice more porous and subject to further weathering. Iron oxide, for example, may adsorb water and turn into hydrated iron oxide or iron hydroxide:

$$2Fe_2O_3 + 3H_2O \rightleftharpoons 2Fe_2O_3 . 3H_2O.$$
$$\text{hematite} \qquad\qquad \text{limonite}$$

The hydration reaction is frequently reversible, and because it involves a considerable volume change it is important in physical weathering.

Hydrolysis is a chemical reaction between a mineral and water, that is between the H^+ or OH^- ions of water and the ions of the mineral. In hydrolysis, therefore, water is a reactant and not merely a solvent.

The concentration of hydrogen ions (measured as pH) is of fundamental importance in all weathering processes because: (1) it determines the solubility of silica, and metal hydrous oxides; (2) the H^+ ions combine with OH^- ions thus removing them from crystal surfaces and permitting further hydrolysis; and (3) H^+ ions replace other cations in the mineral crystals. The major sources of H^+ ions in soils are acid clays (a clay with a high proportion of H^+ in its cation-exchange sites) and living plants. The plant roots exchange H^+ ions for nutrient ions. Soils may also become acidic by the photosynthetic fixation of carbon dioxide and its subsequent respiration from plant roots and the bacterial degradation of plant debris. The carbon dioxide in the soil atmosphere dissolves and dissociates in the soilwater, yielding bicarbonate ion:

$$H_2O + CO_2 \rightarrow H^+_{(aq)} + HCO_3^-_{(aq)}.$$

The formation of clay minerals can be written as a hydrolysis reaction that also requires hydrogen ions. For example, the weathering of feldspars may be represented as:

$$2NaAlSi_3O_8 + 2H^+ + 9H_2O \rightarrow Al_2Si_2O_5(OH)_4$$
$$\text{albite} \qquad\qquad\qquad\qquad \text{kaolinite}$$
$$+ 2Na^+_{(aq)} + 4H_4SiO_4_{(aq)}.$$

Soilwaters, in fact, are rarely pure water. The dissolved carbon dioxide may cause the decomposition of minerals as suggested by Curtis (1976):

$$2NaAlSi_3O_8 + 11H_2O + 2CO_2 \rightarrow 4H_4SiO_4_{(aq)}$$
$$+ Al_2Si_2O_5(OH)_4 + 2Na^+_{(aq)} + 2HCO_3^-_{(aq)}.$$

Alternatively, rather than this carbonation reaction, it may be considered that the photosynthetic fixation of carbon dioxide provides the hydrogen ions to maintain the hydrolysis reaction.

The concentration of other species in the water may affect the type of clay mineral formed. For example, the feldspar orthoclase could weather to kaolinite:

$$2KAlSi_3O_8 + 2H^+ + 9H_2O \rightarrow Al_2Si_2O_5(OH)_4$$
$$+ 2K^+ + 4H_4SiO_4,$$

or to the mineral illite.

$$6KAlSi_3O_8 + 4H^+ + 24H_2O \rightarrow$$
$$K_2Al_4(Si_6Al_2O_{20})(OH)_4 + 4K^+ + 12H_4SiO_4.$$
$$\text{illite}$$

From Fig. 8.11a, it is clear that to form illite rather than kaolinite requires higher concentrations of K^+ or higher pH of soil waters for comparable dissolved silica, or higher silica for comparable pH and concentration of K^+.

Similarly the dissolution of limestone and dolomite can be represented as either hydrolysis reactions, e.g.

$$CaCO_3 + H^+ \rightarrow Ca^{2+} + HCO_3^-,$$

or as a carbonation

$$CaCO_3 + H_2O + CO_2 \rightarrow Ca^{2+} + 2HCO_3^-.$$

Sulphate ion may also enter soilwater by similar dissolution of gypsum and anhydrite, but more commonly enters the weathering profile as a result of the oxidation of metal sulphides, especially pyrite. This oxidation can be represented by the reaction:

$$2FeS_2 + 4H_2O + {}^{15}/_2O_2 \rightarrow Fe_2O_3 + 4SO_4^{2-} + 8H^+$$
$$\text{pyrite} \qquad\qquad\qquad\qquad \text{sulphuric acid.}$$

Oxidation involves the release of electrons. The oxidation of pyrite to sulphate ion can be represented by a half reaction.

$$2FeS_2 + 19H_2O \rightarrow Fe_2O_3 + 4SO_4^{2-} + 38H^+ + 30e^-.$$

This release of electrons can be countered by the uptake of electrons by oxygen, which is itself reduced to water:

$$O_2 + 4H^+ + 4e^- \rightarrow 2H_2O,$$

or to balance the number of electrons:

$$^{15}/_2O_2 + 30H^+ + 30e^- \rightarrow 15H_2O.$$

In the natural environment, oxygen is the most common oxidizing agent, and one of the most readily oxidized elements is iron. In its native state, iron has an oxidation state of 0. In many ferromagnesian minerals iron has an oxidation state of 2 (ferrous), but in the red and brown oxides present in soils an oxidation state of 3. The progressive oxidation can be shown as half reactions:

$$Fe^{(0)} \rightarrow Fe^{2+}_{(aq)} + 2e^-$$
$$3H_2O + 2Fe^{2+}_{(aq)} \rightarrow Fe^{(111)}_2O_3 + 6H^+ + 2e^-.$$

Combining these two half reactions yields:

$$2Fe^{(0)} + 3H_2O \rightarrow Fe_2O_3 + 6H^+ + 6e^-.$$

A combination of the half reaction with that of the reduction of oxygen is a representation of the rusting of iron. The overall reaction involves the flow of electrons, so in principle it could be used as the basis of a galvanic cell which would have an electromotive force or electrode potential characteristic of the reaction. The potentials of the half reactions are considered with reference to an assigned zero potential for the half reaction:

$$H^+_{(aq)} + e^- \rightarrow \tfrac{1}{2}H_{2(g)} \quad E^0 = 0.$$

Lists of reduction potentials (E^0) are given in the conventional form with the oxidized species on the left-hand side, e.g.

$$Fe_2O_3 + 6H^+ + 2e^- \rightarrow 2Fe^{2+} + 3H_2O \quad E^0 = +0.66 \text{ volts}$$
$$O_2 + 4H^+ + 4e^- \rightarrow 2H_2O \quad E^0 = +1.23 \text{ volts}.$$

These values can be used to calculate the standard redox potential for the overall reaction:

$$Fe_2O_3 + 4H^+ \rightarrow 2Fe^{2+} + 2H_2O + \tfrac{1}{2}O_2$$
$$E^0 = -0.57 \text{ volts}.$$

The redox potential at some non-standard state (Eh) varies with the concentration of the reacting substances and if H^+ or OH^- is involved the Eh varies with the pH of the solution. The stability fields of iron oxides and of dissolved ionic species Fe^{2+} and Fe^{3+} at various values of Eh and pH are shown in Fig. 8.11b. Positive values of Eh are indicative of an oxidizing environment, negative values a reducing environment (Fig. 8.11c).

Where ferrous iron links silica tetrahedra in a silicate structure, the dissolution may be enhanced by oxidation of the Fe^{2+}; the resulting ferric oxide accumulating within the soil or on an exposed rock outcrop.

In bogs and other waterlogged environments, the negative Eh and low pH indicate reducing and acid conditions. Ferrous ion predominates over ferric, and thus iron compounds are grey and green in colour, contrasting with the red and brown colour of iron compounds formed under oxidizing and aerobic conditions. In the low Eh, low pH conditions much of the reduction is carried out by anaerobic bacteria, which require organic compounds, particularly chelating agents, to act as electron donors. An example of this is the reduction of ferric ion in the presence of gallic acid (a tannin derivative). The relevant half-reaction for gallic acid is given in Fig. 8.11d and is combined with the half reaction:

$$Fe(OH)_3 + e^- + 3H^+ \rightarrow Fe^{2+}_{(aq)} + 3H_2O.$$

Seasonal fluctuations in the water-table may result in alternate oxidation and reduction of iron in response to changing Eh. The soils become mottled when reduced iron (ferrous) is oxidized to the red ferric oxide. Processes of chemical weathering are discussed in more detail by Ollier (1984) and Yatsu (1988).

Secondary mineral formation is the normal consequence of mineral weathering. It involves both the modification of crystal structures by cation-exchange and cation substitution to produce new minerals, and also the precipitation of new minerals from solution. The precipitation of calcium carbonate in soils of dry regions is an example of the latter process.

The formation of clay minerals from primary silicate minerals is probably the most important single weathering process of humid environments. The chalky whiteness found on granite gravestones is kaolinite which develops a thickness of a few micrometres in 100 years (see Siever 1988 and

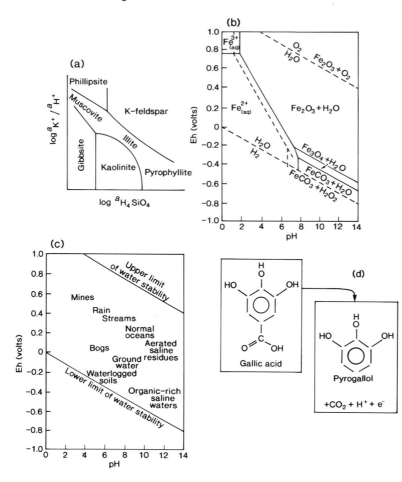

Fig. 8.11. (a) Stability relationships in the system $K_2O\text{-}Al_2O_3\text{-}SiO_2\text{-}H_2O$ assuming that illite may be represented as a solid solution between muscovite and pyrophyllite (from Drever 1982). (b) Eh-pH diagram showing the stability fields of iron oxides (hematite, Fe_2O_3; magnetite, Fe_3O_4), siderite ($FeCO_3$), and dissolved ionic species $Fe^{2+}_{(aq)}$, $Fe^{3+}_{(aq)}$ for a total ion activity of 10^{-6} M at 25 °C and 1 atm total pressure and the partial pressure of CO_2 as 0.01 atm. Broken line is to show effect of changing ion activity to 10^{-4} M. Stability fields are shown only within the limits of water stability (from Henderson 1982). (c) Approximate Eh-pH ranges of some natural environments. (d) Half reaction for the oxidation of gallic acid to pyrogallol.

Fig. 8.12). Clay mineral composition of a soil is an indicator of its weathering state and dominant environmental conditions (see Chapter 3 and Table 8.3).

Biotic weathering

Biotic weathering is a combination of chemical and physical weathering effects of which the following are the most important: (1) breakdown of particles by the action of roots and burrowing animals (Plate 8.17); (2) transfer of material by animals; (3) simple chemical effects as when solution of rock is increased by the CO_2 released into the soil during respiration; and (4) complex chemical effects such as chelation, and the formation of complexes of organic-mineral substances.

As mentioned earlier soil bacteria have an important role in the weathering of minerals in many soils. Chemotrophic bacteria which can oxidize

TABLE 8.3. *Clay minerals produced from primary minerals in relation to climate*

Climatic regime	Biotite	Plagioclase	K-feldspar
Arid	Smectite	Kaolinite	
Temperate	Vermiculite	Smectite / Kaolinite	
Humid-tropical	Kaolinite	Kaolinite	Kaolinite
Very humid-tropical	Kaolinite	Gibbsite	Kaolinite / Gibbsite

Source: Wild (ed.) (1988: 175).

chemical species involving sulphur, iron, and manganese are particularly active in weathering in waterlogged soils. For example the oxidation of sulphide:

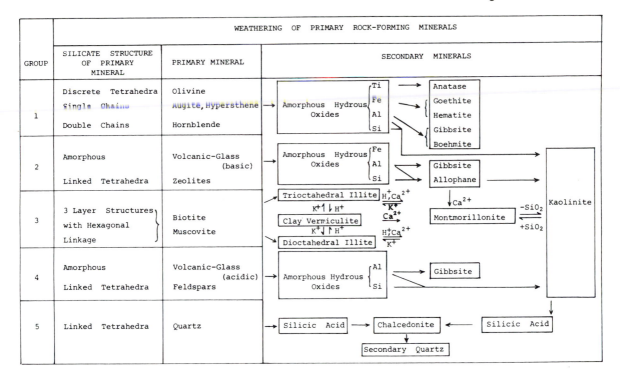

FIG. 8.12. The weathering sequence of primary rock-forming minerals. (After Fieldes and Swindale, 1954.)

$$S^{2-} + 4H_2O \rightarrow SO_4^{2-} + 8H^+ + 8e^-,$$

may be driven by the bacterium *Thiobacillus ferroxidans* acting on organic molecules derived from plants. This can be shown, in principle, for a carboxylic acid being reduced to an aldehyde by bacteria:

$$RCOOH + 2H^+ + 2e^- \rightarrow RCOH + H_2O,$$
carboxylic acid   aldehyde

or even, under some conditions, to methane, as in:

$$CH_3COOH + 8H^+ + 8e^- \rightarrow 2CH_{4(g)} + 2H_2O.$$
acetic acid   methane

The organic acids and other compounds are derived from the vegetation. Many are chelates, chemical species that are able to surround or co-ordinate to metal cations. By preventing access to the metal cation of OH^- and other anions with which it might precipitate, as well as effectively increasing its radius and decreasing charge and thereby reducing its ability to participate in adsorption or ion exchange processes, chelation increases the solubility of cations. If the movement

of soil solution is downwards, chelation assists in the leaching of ions, particularly aluminium and iron. This washing down, within a soil profile, of chelates (i.e. cheluviation) is a process of major importance in podzolization, especially under coniferous trees and podcarps which produce poly-phenolic compounds which are effective chelating agents.

The role of lichens and mosses, which can grow on and within rock surfaces is not well-understood. Some species have organs which penetrate cracks, but it is unclear whether these organs develop a mechanical force capable of prising grains from rock surfaces (Winkler 1966). Endolithic lichens formed of unicellular green algae and filamentous fungi have been found in surface layers of trans-lucent rocks such as sandstone, marble, and granite in both Antarctica and the Canadian Arctic (Friedmann 1982; Eichler 1981). They live in the natural pore space and in cracks. The carboxylic acids that are released by such organisms may form chelates to cations, particularly iron and aluminium (Fig. 8.13). This may explain the enhanced, experimentally determined, rate of

PLATE 8.17. Tree roots have grown in an open joint and prised loose the plate-like joint block so that it fell.

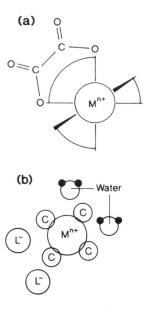

FIG. 8.13. (a) The oxalate ion $C_2O_4^{2-}$ is a bidendate ligand. The negatively charged oxygens can form an ionic bond to the metal cation from both ends of the molecule. Three such ions may be grouped octahedrally about the cation. (b) Schematic representation of how a metal cation (M^{n+}), that has formed a complex with the surrounding chelates (C), is difficult for both the water molecules to approach, and, more importantly for anions (L^-) to precipitate (ML_n).

silicate mineral dissolution in the presence of organic acids (Huang and Kiang 1972). By this process silica may be mobilized and in Antarctic and Arctic regions concentrate iron oxides at the rock surface. The result of these chemical changes is formation of thin flakes of rock which readily break down to a single-grain grus. As the lichens live on the rock faces which receive greatest insolation, with consequent availability of melt-water, the weathering forms show preferential orientation with aspect. In extreme cases they play a major part, with salt weathering, in the formation of tafoni.

Endolithic lichens are also found in hot deserts (Shachak *et al.* 1987) where they, and the surface rock in which they live, may be munched by snails. It has been estimated that in the Negev Desert, snails feeding on lichens in limestone etch grooves 0.4 mm deep leaving powdered rock and faeces rich in calcium, which creates about 1 tonne of new soil per hectare per year; this is an amount of material which is similar to the dust deposition from all sources.

Human interference with the environment, and especially by releasing industrial pollutants into the atmosphere, has had a marked effect upon rates of weathering. The release of sulphur compounds, which become sulphuric acid, and CO_2 into the atmosphere has greatly increased the acidity of rainfall close to and downwind of industrial areas of Europe and North America. As a result soils have become more acid and rocks, especially limestones, are undergoing more rapid weathering. In cities, also, building stones are being attacked by acids, and by salt weathering resulting from the reaction of calcite in the rock with sulphuric acid to produce gypsum ($CaSO_4.2H_2O$) (Winkler 1970).

Hydrothermal alteration

Hydrothermal alteration of rocks by migrating geothermal fluids is not a weathering process, but the end-products of weathering and hydrothermal processes may be similar. Granite, for example, may be altered to kaolinite by both processes. It is now possible to use the isotopes of hydrogen and oxygen to determine which process is responsible for a deposit. In the field, the best indicator is that the proportion of clay in the material formed by weathering increases towards the ground surface, and that from hydrothermal alteration increases with depth. Alteration of igneous rocks can result in lenses and crack infills of clay minerals, including swelling clays, and these may influence slope stability where they occur within rock bodies near the ground surface.

Sequence of weathering processes

Under a humid temperate climate the weathering of an igneous rock might be summarized as:

(1) mechanical fracturing of rock by stress release and opening of fissures between grains by physical processes;

(2) pitting, etching, and cracking of mineral grains and crystals;

(3) at the same time feldspars, micas, and ferromagnesian minerals will suffer carbonation or hydrolysis to produce clay minerals, while releasing insoluble quartz to form an inert mineral skeleton of the soil; iron is oxidized and hydrated;

(4) porosity is increased, bulk density is reduced, and calcium, magnesium, sodium, and potassium ions are removed in solution;

(5) the secondary hydrated silicate clay minerals will be further weathered and iron and aluminium oxides will increase in proportion in the residual soil mass.

Sedimentary rocks break down initially into particles which have a size determined by the original particle size and by the nature of the cement which binds the grains. Sandstones always produce a preponderance of sand, but individual sand grains are fragmented into silt grains as solution of silica occurs along cracks in the sand grains (Pye 1983). Shales and mudrocks may initially break down into aggregates of various sizes but continued weathering breaks down the aggregates and clay, silt and sand-size particles result. Soluble rocks like limestone lose their soluble components to leave an insoluble residue. Weathering always tends towards stability and insolubility.

Indices of weathering

Terms such as 'slightly weathered' or 'moderately weathered' are useful for descriptive purposes but need to be quantitatively tested against objective measures of chemical composition, or mechanical strength and behaviour, if they are to be accepted for technical purposes.

Chemical indices of the alteration of rocks in humid climatic environments may be expressed as a weathering index, W_I, ratio of unweathered minerals to weathered minerals in a volume of material, or as the ratio of the chemically more mobile to the chemically less mobile species, in the general form:

$$W_I = \frac{\text{Proportion of chemical reactants}}{\text{Proportion of residual products}},$$

or more specifically in a form such as:

$$W_I = \frac{\text{feldspar} + \text{mica} + \text{calcite}}{\text{clay minerals} + \text{quartz}}.$$

Such relationships can be quantified cheaply and easily by undertaking X-ray diffraction (XRD) and X-ray fluorescence (XRF) analyses of samples of fresh rock and of the altered material.

The indices proposed by Parker (1970) and Miura (1973) recognize that carbonation and hydrolysis result in removal of the mobile elements in rock and lead to physical disaggregation of the minerals. The most mobile elements are Na, Ca, Mg, and K, which are likely to be exchanged for H, with or without loss of silica which may be depleted or incorporated into newly created silicate clay minerals.

The Parker weathering index, W_p, has the form:

$$W_p = \frac{a_{Na}}{0.35} + \frac{a_K}{0.25} + \frac{a_{Mg}}{0.90} + \frac{a_{Ca}}{0.70},$$

where a_E is the atomic proportion of the element, E, defined as atomic percentage divided by atomic weight, and the denominators are the relative oxygen to element bond strengths.

The Miura weathering index, W_m, has the form:

$$W_m = \frac{\text{MnO} + \text{FeO} + \text{CaO} + \text{MgO} + \text{Na}_2\text{O} + \text{K}_2\text{O}}{\text{Fe}_2\text{O}_3 + \text{Al}_2\text{O}_3 + \text{H}_2\text{O}}.$$

For both indices superscripts may be used to indicate original unweathered material (W_p^o and W_m^o), and weathered material (W_p^w and W_m^w). For a weathered rock the relationship to the original

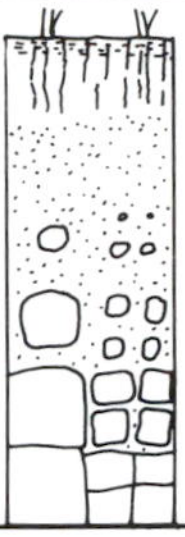

Description of weathering		Weathering grade	Miura index	Shore hardness	Point load (kPa)	Slake durability (7 cycles %)	Seismic velocity (m/s)
Solum		VI	–	–	–	–	–
Completely		V	0.06–0.1	< 5	–	–	< 550
Highly		IV	0.21–0.33	12	–	57.5	550–1650
Moderately		III	0.38–0.4	41–51	1.5–2.5	95.2	1650–3800
Slightly		II	0.6–0.8	47	7	99.0	3800–5000
Fresh rock		I	> 0.8	86–88	10	99.5	5000–5500

Fig. 8.14. Composite data indicating relationships between weathering grades and indices for sandstones. (Data from Hodder and Hetherington 1991; and Brink 1983.)

material can be expressed as the dimensionless indices:

$$W_p = \frac{W_p^w}{W_p^o} \quad \text{and} \quad W_m = \frac{W_m^w}{W_m^o}.$$

Hodder (1984) has assessed the Parker and Miura indices and produced a susceptibility index which can be used to compare different rock types for their relative resistance to chemical weathering. The Hodder susceptibility index, W_h, is:

$$W_h^o = (W_m^o - W_p^o)/W_m^o . W_p^o.$$

Correlation of this chemical weathering susceptibility index with physical durability tests of abrasion resistance and mineral hardness have indicated close relationships between chemical and physical indices (Hodder 1984; Hodder and Hetherington 1991).

Physical indices of weathering have been expressed in several ways: capacity of material to absorb water, the softening effect of water, the swelling and slaking which result from water absorption; changes in strength or hardness; velocity of ultrasound (seismic) waves in material; and by observation and recording of weathering products.

The effects of water are a result of several weathering processes. Microcracking of minerals and removal of mobile ions in solution increase porosity and rates of water absorbtion; the relative amounts of water absorbed in one hour can be compared for materials of different weathering grades by comparing the weight of a sample at the oven-dry state with its weight after a period of immersion (Hamrol 1961). Fractures and voids, air entrapment within blocks, and clay minerals all enhance softening, swelling, and weakening,

as water is absorbed into the voids in weathered materials. The slake durability test (Franklin and Chandra 1972) and geodurability classification (Olivier 1979) are indices of these effects.

Changes in strength in various grades of weathered material have been assessed using Schmidt hammers; a Shore scleroscope (a laboratory instrument, with a drop-hammer with a diamond tip 0.1 to 0.4 mm in diameter, which can measure the hardness of individual crystals); a Los Angeles abrasion mill; and a point-load strength tester (Hodder 1984; Hodder and Hetherington 1991). A rather close relationship has been found between weathering grade, strength, and Miura weathering index (Fig. 8.14) for sandstones. The Miura index and Shore hardness (*Sh*) are most closely correlated and have the relationship:

$$Sh = 118.2 \, (\log W_m) + 91.4.$$

Some scatter in the data is expected because of the survival of some less weathered material in a generally more highly weathered zone, and of more weathered material in a generally less weathered zone.

The velocity of ultrasound elastic waves is influenced by porosity, fractures, and water-content of materials. Ilier (1966) has proposed a weathering index, *K*, based on these factors:

$$K = (V_o - V_w)/V_o,$$

where V_o is the ultrasound velocity in unweathered rock and V_w is the ultrasound velocity in weathered material.

The description of weathering forms and weathering profiles according to a standard 6-point scale (Irfan and Dearman 1978), as set out in Figs. 6.10 and 8.14, has added considerably to the value

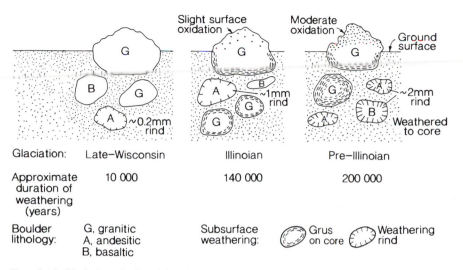

Increasing relief and oxidation of surfaces of boulders due to weathering ⟶

Glaciation:	Late–Wisconsin	Illinoian	Pre–Illinoian
Approximate duration of weathering (years)	10 000	140 000	200 000

Boulder lithology: G, granitic A, andesitic B, basaltic

Subsurface weathering: ◯ Grus on core ◯ Weathering rind

FIG. 8.15. Variation of subaerial and subsurface weathering, as a function of time, on tills in the Sierra Nevada, California. (After Birkeland 1984.)

of field observations and precision of communication about weathering.

Rates of weathering

Available data on rates of weathering have been reviewed by Ollier (1984), Colman and Dethier (1986), and Kukal (1990). The data are derived from a wide range of observations and measurements using a great variety of methods and indicators. There is no widely accepted method of reporting except that for the production of the solutes transported by rivers, which are then treated as indicators of chemical weathering. Solution calculations are most readily applied to areas of limestone bedrock and karst relief, and to areas of quartzose sandstones.

Some of the best indicators of the initial weathering of rocks are derived from study of building-stones and tombstones for which the date of quarrying and cutting is known. Geikie (1880) made a survey of Edinburgh churchyards and found that good quality sandstone weathered very little in 200 years, a slate of 90 years had clear engraving, but marble of that age had partly crumbled to sand. Recent similar studies are described in Colman (1981). Dated buildings indicate that sandstones may show strong weathering in 100 to 900 years. Granite in Egypt (Barton 1916) shows virtually no change in 4000 years, flaking and granular disaggretion in 5400 years, but below soil-level the alteration is much greater, perhaps 5 to 10 mm in 5000 years.

Chemical weathering of lava flows of known age is indicated by the changing composition of the rock. Weathering of Krakatoa Island lava in 45 years (Mohr and van Baren 1954) is shown by decreased silica content of nearly 6 per cent, Al_2O_3 content increased by 1 per cent, and total iron content by about 3 per cent. In the West Indies lavas have weathered to depths of 0.4 to 0.6 metres in 1000 years.

Rates of weathering and soil formation on hill-slopes are particularly difficult to assess because of the extreme variability induced by drainage conditions. On flat surfaces, such as those of ground moraines, terraces, and floodplains, more reliable comparisons may be made. Birkeland (1984) has compared weathering forms and rinds on exposed boulders and those within the soil for three tills in the Sierra Nevada (Fig. 8.15). Over a period of 200 ky, or more, weathering of granitic clasts within the soil exceeds or equals that on exposed faces. Where salt weathering is prevalent, subsurface weathering exceeds surface weathering. Similar studies in New Zealand (Chinn 1981) have shown that rind thickening is greater on coarse-grained, but less on fine-grained rocks and those of high metamorphic grade.

It is evident from a number of studies (e.g. Chinn 1981; Colman 1981) that long-term rates of weathering of mineral grains or clasts decline, whether the initial reaction rate is controlled by processes at the site of reaction or by diffusion of reactants to and away from the site. Unless the weathering products are removed, they may inhibit access to new sites and the diffusion of reactants that are in solution. In principle, this has the same effect as if the grain size of the dissolving minerals were to be increased. Another control, which is usually underestimated in weathering, is the fabric of the rock. Rock weathering rates are largely controlled by permeability, not by the chemical reactivity of minerals. Rock bodies which are massive and free of fissures will therefore weather far more slowly than those which are highly fractured, even if the former have minerals which are more reactive.

The effect of climate on weathering rates has been assessed by using rock discs from the same source. It has been shown (Day *et al.* 1980) that, in a two-year period, weathering in the humid tropics of Malaysia occurred 3.5 times faster than in the cool humid temperate environment of Wales.

The few examples, listed above, are indications only of recognized rates. Natural variability in climate, rock, and soil characteristics limits the capacity to generalize, but so does the inadequate state of reporting, which is often very subjective and without standardized criteria.

Rates of soil formation

Soils, in the sense of a body of material forming a solum, seldom develop from bare rock. They usually form in pre-weathered grus, saprolite, or sediments such as alluvium, colluvium, eolian, lake, estuarine, and glacial deposits. Rates of soil formation are difficult to assess because soils form over very long periods, for example, 10^3 to 10^5 years for development of a clay-enriched, Bt horizon (Birkeland 1984; 1990). Such periods are also long enough for many environmental changes to occur, with the result that it is seldom possible to fully assess the effects of the controlling factors of soil formation, or to assess the importance of addition of eolian dust to profiles, even though such additions are known to be very important (e.g. Yaalon and Ganor 1973; Gile *et al.* 1981).

Furthermore, dating the initiation of profile development is also difficult. Estimates of profile ages from their own properties has been assessed as being subject to errors of ±50 per cent (Birkeland 1990).

Methods used to assess profile development include: thickness of the solum, amount of clay in a B horizon, formation of clay skins, soil colour, ratio of Fe and Al to Si, and, in semiarid zones, accumulation of $CaCO_3$. A comparative study of soils in hot desert and humid tropical environments indicates the application of such methods (Cooke and Warren 1973). After 5000 years, in the humid tropics mechanical disintegration of bedrock has occurred; in desert gravels calcium carbonate concretions have begun to develop with disturbance of the original structure. In 20 000 years, humid tropical weathering has caused major chemical alteration with enrichment in Fe and Al accompanied by loss of SiO_2; in the desert, some clay formation has occurred with clay coatings formed on clastic grains.

There has been some success in quantifying the rate of formation of soils from dated volcanic ash, particularly in establishing the rate of formation of the clay mineral allophane from dacitic tephra (Ruxton 1988), andesitic tephra (Neall 1977), and rhyolitic tephra (Hodder *et al.* 1990). There is an exponential relationship between the amount of allophane formed and the time for weathering, usually the interval between successive tephras. This corresponds to first-order reaction kinetics, for which the usual expression is:

$$C/C_o = \exp(-k_1 t),$$

where C and C_o are the concentration at time t and 0 respectively, and k_1 is the first-order rate constant. Typical values for the rate constant for clay formation are about 10^{-12}/s.

There are obvious differences in the soils discussed above but the data are not strictly comparable. A proposal for use of a 'soil development index' using combinations of soil properties may permit more reliable statements in future (Harden and Taylor 1983; Harden 1990). Along with improving methods of dating and more information on environmental change, an index and standard methodology could greatly improve knowledge of rates of soil formation.

9 Landforms from Weathering, Soils, and Duricrusts

Landforms from weathering processes

The direct production of a distinctive landform created by weathering is most readily seen in desert areas or on steep cliffs where little regolith can accumulate. Because continued weathering depends upon the removal of debris most weathering forms are relatively small. If attention is confined to these small forms, however, a false impression will be obtained of weathering significance and the variable resistance to it of different rocks. Much of the relief of the land which is not directly attributable to tectonic forces is due to differential weathering and erosion acting most effectively in weak rocks and along lines of structural weakness. Many examples of these processes are offered elsewhere in this book.

Some landscapes are the result almost entirely of solution processes; karst formed in limestone is well known and increasingly there is recognition that karst landscapes can develop, under certain circumstances, in highly siliceous rocks. Weathering, particularly to great depth, is also responsible for development of many distinctive landforms which may be revealed by stripping of the weathered mantles. In certain climatic zones, weathering crusts also create distinctive landforms.

Case-hardening, weathering rinds, and rock varnish
Case-hardening is the development of a shell or outer casing which is more resistant than the immediate interior of the rock. This greater resistance may be due to an increase of strength relative to the original strength of the rock or to less weakening of the outer rock surface than of the rock core. Core softening has been recognized in Antarctic granite but nearby sandstone has gained in absolute hardness by crystallization of secondary silica cement in the near-surface rock (Conca and Cubba 1986). Both Antarctic and Nevada sandstones have developed greater absolute hardness from cemented crusts of calcium carbonate with wind-deposited kaolinite (Conca and Rossman 1982) and the Bishop Tuff ignimbrite, California, has developed much harder outer casing as a result of devitrification of its glassy matrix along joint planes (Conca and Cubba 1986).

Case-hardening can be a contributor to formation of 'visors' and raised lips on outer surfaces above and below tafoni, but so few detailed examinations of rock strength have been carried out of these features that the significance of case-hardening is unclear.

Formation of polygonal cracks on rock surfaces is common in arid, semiarid, and temperate zones but rare in the humid tropics and periglacial zones. The origins of cracking are unclear, but one hypothesis (Robinson and Williams 1989; Williams and Robinson 1989) is that cracking results from the shrinkage of silica gel, during formation of siliceous crusts, as a result of changing temperature and moisture conditions.

Weathering rinds develop as the early stages of weathering of rock outcrops and boulders. They are easily recognized on many split pebbles and cobbles, for example, as an outer thin rind of different colour from the original rock and interior of the cobble. Rind colours may be white indicating formation of hydroxides or clay minerals, or reddish indicating iron oxides. Weathering and formation of rinds indicates reduction in molecular percentages of calcium, magnesium, sodium, and potassium, lesser depletion of SiO_2, relative concentration of iron and aluminium oxides, oxidation of iron, and incorporation of water (Colman 1982) (see Fig. 8.15).

Rock varnish is different from a weathering

PLATE 9.1. Pan formed in quartzite, central Sahara. Note the pitting of the rim. The scale is 30 cm long.

rind in that its material is derived wholly, or predominantly, from sources external to the varnished rock. Rock varnish usually has a thickness in the range 5 to 200 μm but has been found as thick as 1.1 mm. It is composed, typically, of over 50 per cent clay minerals, about 20–30 per cent manganese and iron oxides, and the remaining about 30 per cent is of other minor and trace elements (see Dorn and Oberlander 1982). The abundance of manganese and iron in the outer few micrometres generally controls the colour of the varnish. Manganese, in similar or greater concentrations than iron, gives a black colour; a predominance of iron gives dusky-red to orange colours. Manganese concentrations increase relative to iron as effective moisture increases, but highly alkaline environments inhibit the concentration of manganese and its oxidation by bacteria (Dorn and Dragovitch 1990). Rock varnish is extremely variable in thickness and composition, even within small sites, because it takes several thousands of years to form, is eroded in the presence of most plants and is removed by scaling and spalling from rock surfaces. The presence of varnish, therefore, indicates stable rock surfaces.

Rock varnish may form in many environments, but is most evident in arid to semiarid environments lacking plants, such as lichens and mosses, on rock surfaces.

The clays of varnishes commonly produce a lamellar structure in the varnish, but perhaps of greater significance are the manganese-oxidizing micro-organisms. Because these micro-organisms

are formed partly of carbon they can be dated by AMS ^{14}C (AMS-accelerator mass spectrometry); this has already provided clear evidence that, on desert hillslopes in Australia, there are sites which have been stable for more than 38 000 years alongside others which have been actively eroded in the last few thousand years. Such techniques may answer many questions about rates and forms of slope evolution and rates of weathering. The rate of scaling on one site on Ayers Rock, central Australia, is calculated at 1 cm/10 ky. AMS ^{14}C dating can also be supplemented by using the relative weathering index—$(K^+ + Ca^{2+})/Ti^{4+}$—of varnishes, as this cation-ratio decreases with time (Dorn 1989).

Pits, pans, caverns, and rills

Many outcrops have a microrelief attributable to a great variety of miniature depressions which have formed either along lines of noticeable weakness, such as joints, or on sites which appear to be very uniform. Many of the minor features occur on a variety of rock types although the readily soluble rocks, such as limestones, are particularly prone to various forms of etching. The silicate-rich rocks such as granite, granodiorite, schist, quartzite, sandstone, and gneiss also bear many minor weathering forms, as does basalt.

Weathering pits of flat surfaces (German, *opferkessel*; Portuguese, *oriçanga*; Polish, *kokiolki*; Australian, *gnamma*) range in size from a few centimetres in diameter to several metres (up to 7 m have been reported) and from a few centimetres in depth to a metre or more. Pits may be hemispherical or flask-shaped hollows in which the walls are overhanging, or they may be flat pan-like forms; they may be circular or irregular in plan (Hedges 1969) (Plate 9.1). As the slope of rock surfaces increases the shape and variety of weathering pits changes.

The walls of deep pits are often smooth and appear to retreat by spalling rock flakes; their smoothness is not like that produced by the swirling of debris-laden water, and the rock debris on floors of pits is not rounded, so these features are evidently not produced by the pot-holing action of streams or air. The walls of some shallow pans are irregular and, in granitic rocks especially, the rock is laminated.

Spillways lead away from some pans and pits and indicate that water is involved in the formation of many of these features. In a few pans on very soluble rocks small features that look like splash marks surround the pits and suggest that solution of the rock by water is the main process but, as will be seen below, this is only one of several possible processes contributing to the formation of these forms.

Honeycomb or *alveolar weathering* is characterized by numerous pits a few millimetres or centimetres in diameter and in depth. They may enlarge to produce a fretwork or honeycomb structure. Alveolar weathering occurs most frequently on granular rocks such as tuffs and sandstones and presumably involves selective granular disintegration of rock faces (Plates 9.2 and 9.3).

Weathering processes do not act uniformly through a rock body, but are commonly more intensive along joints and other fissures. Fissures may thus become zones of greater weakening through decomposition and solution of minerals, or zones of strengthening by deposition of cements, such as iron oxides. In either case differential processes lead to greater disintegration and losses in the weakened material resulting in rims of strengthened material creating a 'box-work', or bosses of stronger material standing out from elongated depressions along weakened fissures (Plate 9.4).

Tafoni (sing. *tafone*) are hollows which may be several metres in depth, width, and height, although they are generally smaller. They are cut into steeply sloping rock faces and may have overhanging entrances, called visors. The floors of tafoni are often flat and may be littered with a

PLATE 9.2. Small tafoni in granite, Antarctica. Scale is given by the gloves and ice-axe.

debris of rock fragments. The shapes of caverns are sometimes influenced by structural features of the rock, but this is not always the case.

The *origins* of the various kinds of cavernous weathering are varied—they occur on a great variety of rocks; in all climates; on outcrops with a great variety of forms, and with both large and small exposed surfaces around them. Attempts to classify them according to shape, rock type, or climatic environment (e.g. Wilhelmy 1964) have, therefore, not been successful: exceptions seem to be as common as the cases which fit the classification schemes.

Most pits are assumed to be initiated at some depression or weakness in a rock surface and to enlarge gradually, perhaps by consuming other pits. The rates at which they change are varied and range from deepening at a rate of 1 cm/year on some coastal cliffs in weak rock to an imper-

PLATE 9.3. Sculpturing below a granite overhang showing exfoliation, central Namib desert. The cave is 2 m high.

PLATE 9.4. Honeycomb weathering penetrating the centres of joint blocks and 'boxwork' relief along joints as a result of deposition of iron oxides along those joints.

ceptibly slow rate. In Antarctica progressive development of tafoni, with increasing size of features and abundance on boulder surfaces, has been discerned in rocks on the surface of glacial moraines which can be ranked in order of age (Calkin and Cailleux 1962). The processes of tafoni enlargement include granular disintegration and exfoliation of rock flakes. The removal of this debris, which is necessary for the continuation of weathering, may be by wind or under gravity, and for pans it may be aided by the overflow of water.

A variety of processes have been identified as causing cavernous forms. In some tafoni in the central Sahara it is obvious that granitic rocks are decaying by the decomposition of feldspars and mica to clay minerals, to leave a grus which is composed largely of quartz sand and fine gravel. Elsewhere, for example in Antarctica and the Namib desert, there is no trace of chemical decomposition in the rock spalls and the debris on the floor of each tafone is composed of unaltered quartz and feldspar crystal fragments (Selby 1977b). In some cold climates frost action may contribute to rock weathering, for snow and ice will melt in contact with a rock surface warmed by insolation (perhaps to +20 °C) when the air temperature is at zero or below. Freezing will occur as soon as the rock is again in shadow.

Weathering pits on schist rocks in Otago, New Zealand, have been attributed to frost weathering and hydration effects (Fahey 1986b). A study of tafoni in Spain (Sancho and Benito 1990) concludes that initiation is by salt weathering and wetting–drying acting on the most easily weathered minerals on a rock face. Cavernous weathering in coastal areas of Australia has been attributed to solution etching of quartz in the presence of NaCl which increases the rate of dissolution (A. R. M. Young 1987).

The most general statement about cavernous weathering and tafoni formation is probably that of Conca and Rossman (1985) and Conca and Astor (1987). They have studied the porosity and permeability of igneous rocks in California and Antarctica. They recognize that the interiors of rock bodies have a higher permeability than the exterior, either because of core softening by weathering or because of formation of impermeable surface crusts. The higher internal permeability permits greater flow of capillary water in the interior and therefore greater dissolution and other weathering processes, which may be chemical or physical.

Organic action by algae, lichens, and fungi appears to be particularly important in the development of some cavernous forms. By releasing CO_2 these plants may encourage carbonation and particularly the decay of feldspars. Some lichens may also be involved in the extraction of iron, and perhaps silica, from rock minerals (Franzle 1971) and are responsible for superficial rock staining to black and brown colours. This process weakens rock surfaces and does not cause case-hardening. As the lichens shrink and swell with alternate dehydration and hydration they can also tear away rock particles and so enlarge hollows.

The high pH of water in many weathering pits suggests that alkaline waters, in which the solubility of silica is greatly increased, are responsible for much of the enlargement; soluble salts may contribute to weathering by salt weathering. It has been pointed out by Twidale and Bourne (1975b) that some forms of cavernous weathering are initiated and develop within the regolith in the zone of ground water where chemical processes are almost certainly responsible for their production. Partly undercut or flared slopes may also originate in this way. Many forms of cavernous weathering, however, develop in areas where there is no trace of a deep regolith and no evidence that one ever existed on the present landsurface—as in Antarctica and the Namib.

It is evident that many processes may contribute to cavernous weathering, although some are not active in certain environments, and even in one area the significance of such processes as hydration, alkaline water, and organisms will vary with slope, shade, and other factors. About the only valid generalization on the causes of tafoni is that water is nearly always involved in the weathering process (Dragovich 1969).

In many environments cavernous weathering forms are present but no longer developing. This may be because they were formed under climatic conditions which no longer exist, as in parts of the northern Sahara where tafoni formed at the foot of some cliffs when water-tables were higher than they are now (B. J. Smith 1978), or tafoni may become inactive when they are so deepened that the diurnal fluctuations of moisture or surface heating, which were effective in shallow recesses, cannot reach the shaded and protected surfaces at the back of deepened hollows.

Rills, rillen, grooves, or gutters, as they are variously known (Plate 9.5), can also form in many different types of rocks and in many environments. Rills may form a regularly spaced pattern with individual rills being 5–30 cm deep and 20–100 cm wide, where they form the drainage network for evenly sloping rock surfaces, or they may be single features. Some are connected with lines of pits which, in rocks such as granites, show no structural influence on their alignment, but on rocks with structural control may be accordant with that structure (Plate 9.6). The example of Ayers Rock indicates that large rills may be regarded as microvalleys. Features up to 10 m wide and 1–2 m deep have been described on granite domes in Zimbabwe (Whitlow and Shakesby 1988) where they are regarded as predominantly fossil features developed over long periods of removal of overburden and exposure of the bedrock.

Solution by alkaline waters, perhaps enhanced by algal slimes and lichens is regarded as a significant contributor to rill development. Deposition of silica coatings at margins of rills and breaks of slope contributes to formation of rims to channels and pools.

PLATE 9.5. Rills on limestone, northern New Zealand.

PLATE 9.6. Linked pits developed along bedding units of arkose, Ayers Rock, central Australia. The bedding has been tilted into a near-vertical plane.

PLATE 9.7. Landforms of the Bungle Bungle, on the eastern margin of the Kimberley Plateau region, north Western Australia.

Convergence of forms

It has been stressed in this section, and will be stressed again with reference to other landforms, such as tors and bornhardts, that similar landforms can be produced by a variety of processes in a variety of environments. This convergence of landforms of various origins towards a similar geometry demonstrates that it is seldom possible to study the shape alone of the landform and so deduce its origin. It also implies, where detailed evidence is lacking, that it may be impossible to determine how a particular feature has evolved.

Landforms from dissolution

Karst landforms developed on limestones are the best known and understood of those landforms resulting predominantly from solution processes. They are described comprehensively in a number of major texts (see Sweeting 1972; Jennings 1985; Trudgill 1985; Ford and Williams 1989). Karst landforms also develop on highly siliceous rocks and, because these are poorly known and little understood, brief comment is made here.

Two major areas of karst on pure quartz sandstones and quartzites have been described: in the Guiana Highlands of Venezuela ten large mesa mountains, of which Mt Roraima (2810 m) is the highest, have caps (300 m thick) of quartzitic rocks and in them are developed deep gorges, and depressions surrounded by castellated joint-controlled ridges and isolated blocks. Some caves are also present (White *et al.* 1967; Chalcraft and Pye 1984; Galan and Lagarde 1988; Briceño and Schubert 1990). In Australia, an area known as the Bungle Bungle (Plate 9.7) consists of complex groups of towers, sharp ridges, and escarpments with a vertical relief up to 300 m and with tube and cave systems draining into flat-floored valleys (R. W. Young 1986, 1987, 1988). Some similar features have been described in the Sahara (Mainguet 1972).

The Roraima massif is in an area of humid tropical climate with a regional annual rainfall of 2800 mm but the mountain tops may receive as much as 7800 m. A major feature of the breakdown of the quartzite is direct solution of both quartz grains and siliceous cement with alteration of feldspars and micas to kaolinite. The concentration of silica in the drainage waters is so low that the landscape, clearly, has developed over a very long period. Whether or not solution is aided by the presence of organic molecules is unclear, although it has been postulated that this is the case (Pouyllau and Seurin 1985), and work by Bennett and Siegel (1987) indicates that silica can be complexed and mobilized by some organic acids in waters which have close to neutral pH.

It has been demonstrated, by Young (1986, 1987, 1988), that the Bungle Bungle sandstones are composed of quartz grains cemented by overgrowths of silica which have not eliminated all pore

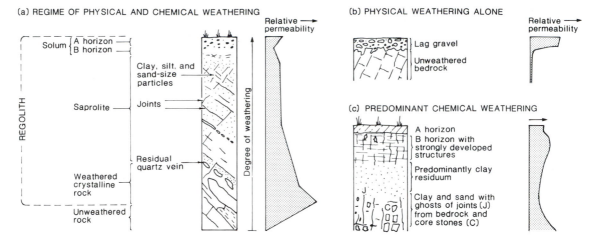

FIG. 9.1. Some major features of weathering profiles.

space. Circulating water has consequently etched the grains and dissolved much of the overgrowth silica leaving a body of rock formed of interlocking grains with few cementing bonds. As a result, the rock has moderate compressive strength (35–55 MPa) as applied loads are transmitted by point-to-point contacts between grains, but low tensile or shear strengths because of weak cementation.

Climate in the Bungle Bungle area is semiarid with mean annual rainfall of about 600 mm and a long dry season. Whether or not solution has taken place under relatively dry conditions, with alkaline pH and presence of chlorides enhancing the process, is unclear. Paleoclimatic data from Australia indicate that much of the continent had tropical rainforest cover until the middle Miocene (Kemp 1981) and the long-term tectonic stability and low relief of the region indicate that landform development may be both ancient and prolonged.

Weathering profiles

Some of the major features of weathering profiles are shown in Fig. 9.1. Very deep and complete profiles can be divided into three zones. (1) At the groundsurface is the solum formed of distinct soil horizons developed through combined weathering and plant-induced processes acting to transform, add, and transfer organic and mineral matter. (2) Saprolite is the product of chemical weathering of mineral matter below the solum. It is generally,

but not exclusively, formed in an oxidizing environment. The solum and saprolite zones together form the regolith. (3) Below the water-table, alteration can occur in a reducing environment by hydrolysis and penetrate to considerable depths. Many deep weathering profiles on stable continental surfaces have developed over very long periods, some in Australia are as old as the Mesozoic and began forming before the breakup of Gondwanaland. Very deep profiles may be formed in an oxidizing environment if the water-table falls progressively—usually as a result of tectonic uplift—or below the water-table.

The solum

Processes of solum formation, called pedogenesis, are discussed in many textbooks (e.g. FitzPatrick 1983; Birkeland 1984). Their significance for studies of hillslopes lies in the close relationship between solum profile properties, hillslope hydrology, transport of material downslope, and geomorphic processes of erosion, transport, and deposition. These topics will be discussed further in later chapters. Some key points only will be made here.

Ideas of soil development, which were dominant through the middle part of the twentieth century were first expressed in a semiquantitative way by Jenny (1941), who viewed the four environmental factors of climate (cl), organisms (o), relief (r), parent material (p), plus time (t) and a few unspecified subsidiary factors ($. . .$) as independent variables for formation of soil (S), hence:

$$S = f(cl, o, r, p, t, \ldots).$$

The time factor was seen as being implicitly linked to a development model leading to a stable and mature state in which soil horizon development was pronounced and relatively complete. It was, of course, recognized that erosion and other forms of disturbance may occur but these were regarded primarily as local variations within a broadly progressive system.

Subsequent work, and recognition of the extreme variability of climate over short, medium, and long terms, has forced a change in thinking towards recognition that most soils are polygenetic and have formed by the operation of a large number of processes acting with varying intensity to create soil profiles which may thicken or become thinner, may increase in their distinctiveness of horizons, or may suffer mixing and other processes causing loss of horizons. Such ideas have been expressed most clearly by Johnson and Watson-Stegner (1987) and Johnson *et al.* (1990) in a model expressed as:

$$S = f\left(D, P, \frac{dD}{dt}, \frac{dP}{dt}\right),$$

where: S is the state or degree of soil profile evolution;

D is the set of dynamic factors;

P is the set of passive factors;

dD/dt and dP/dt denote change through time.

The dynamic factors include: energy flux at a site, due to insolation and air masses, and resulting oxidation; mass fluxes of water, gases, and solids into the soil; frequency of wetting and drying events; activity of plants and animals; and processes which stir, move, and transfer materials within the profile.

The passive factors include: parent material from which soil is developed with its characteristics of texture, structure, weathering state, and chemical composition; soil chemical environment; permanently low water-tables; stability of geomorphic surfaces; and the characteristics of existing soil horizons at any time.

Soil development is, consequently, not seen as an inevitable progression towards a mature or zonal state, but one of constant change in which one set of factors may be dominant for a while, yet be replaced later by another set. In general, because many soils have considerable depth and

distinct horizons, it is reasonable to recognize progressive development of profiles with greater horizon differentiation, physicochemical stability with soil deepening, and downward migration of the weathering front. In progressive development, soils become increasingly leached and pass through successive recognizable stages of development, but at varying rates during this development.

At any stage there may be a shift towards regressive pedogenesis such as: (1) rejuvenation of profiles as bioturbation or new plant species provide inputs of organic matter rich in the alkali metals (Na^+, K^+) and the alkaline earths (Mg^{2+}, Ca^{2+}); (2) profiles become shallower as erosion occurs at the surface; or (3) profiles are added to at the surface by accretions of dust, tephra, flood deposits or colluvium, at rates which permit incorporation of additional sediment into pedogenetic processes. Such concepts stress the importance of recognizing the polygenetic nature of nearly all soils, and especially those formed on hillslopes where water and soil movement are at a maximum.

In Figs. 9.2 and 9.3 soil processes are shown as being predominantly due to downwards movement of water, containing organic acids, leaching soluble ions and enhancing clay formation and accumulation in B horizons. In the podzol, iron and humus may also accumulate in B horizons. These processes create horizon differentiation and also major changes in permeability, usually with the clay-rich B horizons having lower permeability than those above them. Changes of permeability and shear strength within profiles have major influences on vertical and lateral movement of water, location and propagation of shear planes, location of natural pipes, and form horizons of varying resistance to erosion by flowing water. Soil structures (fabrics) are largely due to clay formation and deposition, with swelling and shrinking, and selective root activity between structural units. Volume change is related to climatic events and may be seasonal. Soil permeabilities may also vary with soil climate and so influence geomorphic processes. Such phenomena are characteristic of humid climates. In drier climates, materials are not leached and tend to accumulate within profiles, sometimes creating impermeable subsoils. In tundra zones vertical movement of water is restricted by ice lenses or permafrost.

On any hillslope, soil profile characteristics vary across and down the slope both as a cause of, and a consequence of, varying concentrations of water,

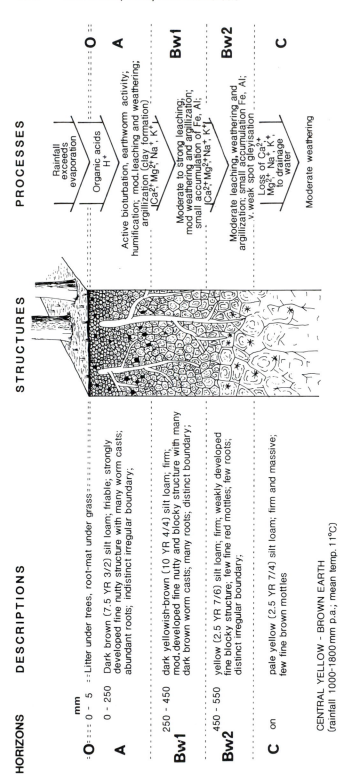

HORIZONS

DESCRIPTIONS

mm

O 0 - 5 Litter under trees, root-mat under grass

A 0 - 250 Dark brown (7.5 YR 3/2) silt loam; friable; strongly developed fine nutty structure with many worm casts; abundant roots; indistinct irregular boundary;

Bw1 250 - 450 dark yellowish-brown (10 YR 4/4) silt loam; firm; mod.developed fine nutty and blocky structure with many dark brown worm casts; many roots; distinct boundary;

Bw2 450 - 550 yellow (2.5 YR 7/6) silt loam; firm; weakly developed fine blocky structure; few fine red mottles; few roots; distinct irregular boundary;

C on pale yellow (2.5 YR 7/4) silt loam; firm and massive; few fine brown mottles

CENTRAL YELLOW - BROWN EARTH
(rainfall 1000-1800 mm p.a.; mean temp. 11°C)

STRUCTURES

PROCESSES

Rainfall exceeds evaporation

Organic acids H⁺

O

A Active bioturbation, earthworm activity; humification; mod.leaching and weathering; argillization (clay formation) [Ca²⁺, Mg²⁺, Na⁺, K⁺]

Bw1 Moderate to strong leaching; mod weathering and argillization; small accumulation of Fe, Al; [Ca²⁺, Mg²⁺, Na⁺, K⁺]

Bw2 Moderate leaching, weathering and argillization; small accumulation Fe, Al; v. weak spot gleyisation

Loss of Ca²⁺ Mg²⁺, Na⁺, K⁺ to drainage water

C Moderate weathering

Fig. 9.2. Characteristics and processes of a soil developing on sedimentary rocks in a humid temperate climate in North Island, New Zealand.

HORIZONS   **DESCRIPTIONS**   **STRUCTURES**   **PROCESSES**

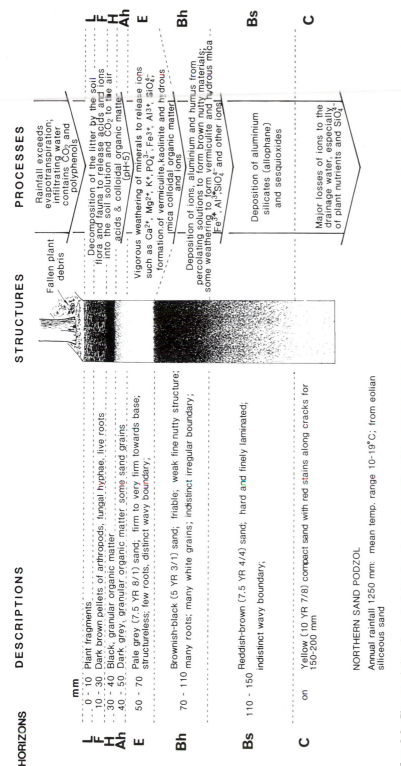

PROCESSES:

- Rainfall exceeds evapotranspiration; infiltrating water contains CO_2 and polyphenols
- Decomposition of the litter by the soil flora and fauna to release acids and ions into the soil solution and CO_2 to the air — acids & colloidal organic matter (pH<5)
- Vigorous weathering of minerals to release ions such as Ca^{2+}, Mg^{2+}, K^+, PO_4^{3-}, Fe^{3+}, Al^{3+}, SiO_4^{4-}, formation of vermiculite, kaolinite and hydrous mica colloidal organic matter and ions
- Deposition of ions, aluminium and humus from percolating solutions to form brown nutty materials; some weathering to form vermiculite and hydrous mica [Fe^{3+}, Al^{3+}, SiO_4^{4-} and other ions]
- Deposition of aluminium silicates (allophane) and sesquioxides
- Major losses of ions to the drainage water, especially of plant nutrients and SiO_4^{4-}

STRUCTURES:

Fallen plant debris

DESCRIPTIONS:

mm

L   0 - 10   Plant fragments
F   10 - 30   Dark brown pellets of arthropods, fungal hyphae, live roots
H   30 - 40   Black, granular organic matter
Ah 40 - 50   Dark grey, granular organic matter some sand grains

E   50 - 70   Pale grey (7.5 YR 8/1) sand; firm to very firm towards base; structureless; few roots, distinct wavy boundary;

Bh 70 - 110 Brownish-black (5 YR 3/1) sand; friable; weak fine nutty structure; many roots; many white grains; indistinct irregular boundary;

Bs 110 - 150 Reddish-brown (7.5 YR 4/4) sand; hard and finely laminated; indistinct wavy boundary;

C   on   Yellow (10 YR 7/8) compact sand with red stains along cracks for 150-200 mm

NORTHERN SAND PODZOL

Annual rainfall 1250 mm: mean temp. range 10-19°C; from eolian siliceous sand

Fig. 9.3. Characteristics of a podzol developed on dune sands under evergreen broadleaf forest in a humid warm-temperate climate, North Island, New Zealand.

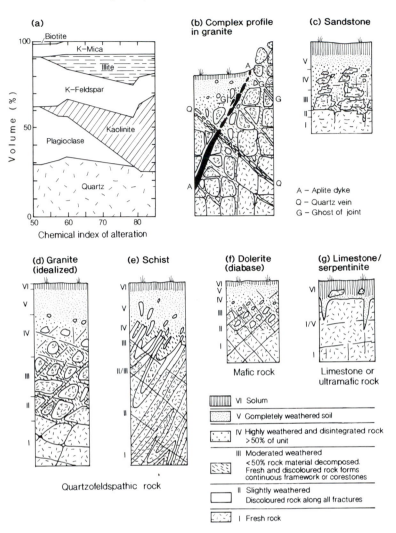

(a)

Volume (%)

Biotite
K–Mica
Illite
K–Feldspar
Kaolinite
Plagioclase
Quartz

Chemical index of alteration

(b) Complex profile in granite

A – Aplite dyke
Q – Quartz vein
G – Ghost of joint

(c) Sandstone

(d) Granite (idealized)

Quartzofeldspathic rock

(e) Schist

(f) Dolerite (diabase)

Mafic rock

(g) Limestone/ serpentinite

Limestone or ultramafic rock

VI Solum
V Completely weathered soil
IV Highly weathered and disintegrated rock >50% of unit
III Moderated weathered <50% rock material decomposed. Fresh and discoloured rock forms continuous framework or corestones
II Slightly weathered Discoloured rock along all fractures
I Fresh rock

FIG. 9.4. (a) Percentage change in volume of minerals over time, in a weathering profile from granite, as a function of weathering intensity indicated by a chemical index of alteration (after Nesbitt and Young 1989). (b–g) Features of weathering profiles developed on selected bedrock lithologies.

preferential sites of erosion, and areas of deposition (see Chapter 10).

Saprolith

Saprolite is the material forming the saprolith. Saprolite is formed *in situ* by chemical weathering which creates secondary clay minerals and sesquioxides (Al_2O_3, Fe_2O_3) from primary minerals such as quartz, feldspars, phyllosilicates (micas), amphiboles, pyroxenes, and volcanic glass. Such minerals, and particularly the most abundant minerals—quartz and feldspars—are primarily found in igneous and metamorphic rocks. Sandstones consist primarily of quartz grains and silica overgrowths binding the grains; mudrocks are primarily composed of clay minerals with lesser proportions of silt-size quartz and feldspars; and limestones are predominantly carbonates. These rocks weather to produce different types and thicknesses of regolith depending largely upon the chemical reactivities and solubility of their major mineral components in an oxidizing environment.

Quartzofeldspathic rocks, such as granites, schists, and metasedimentaries, retain a relatively inert skeletal framework of quartz and potassium feldspars even after most of their other minerals have been converted to clays or their component chemical species removed by drainage water. Consequently they retain 'ghosts' of the rock fabric, such as quartz veins and joints, and the same volume as the original rock; that is, they weather isovolumetrically.

Bulk compositional changes to quartzofelds-pathic rocks are relatively simple (Fig. 9.4a) and are not noticeably modified by climate, although climate does modify the rate at which those changes occur (Nesbitt and Young 1989). Plagioclase, a feldspar of low potassium content, is the most abundant mineral of the exposed, unweathered continental crust and among the more rapidly weathered silicates; it weathers to kaolinite. The potassium feldspar, orthoclase, weathers more slowly and some remains in the profile after all plagioclase has disappeared. Potassium feldspar commonly forms illite and kaolinite. Biotite changes to vermiculite or, less commonly, to kaolinite, and muscovite may remain unaltered during the early stages of weathering. The ultimate products of chemical weathering of granitic rocks are kaolinite and/or gibbsite.

The upper parts of saprolites are converted to a quartz grus with varying proportions of clay minerals, while the lower parts of the saprolite contain a greater proportion of the original primary minerals, thus indicating that saprolites increase in depth by penetrating into the bedrock. The use of chemical weathering indices is then useful in comparing the relative ages of separate saprolite profiles of the same rock type or in comparing weathering regimes.

Mafic rocks, such as dolerite (diabase in US terminology), contain no chemically resistant minerals, such as quartz and muscovite, their plagioclase and augite alter to structurally weak clays and iron hydroxides. Consequently saprolites on such rocks are thin, do not preserve ghosts of joints, and consist of an oxide-rich clay 'tonguing' downwards along joint planes into a zone of spheroidally weathered corestones.

Ultramafic rocks, such as serpentinite, likewise contain no chemically resistant minerals and because they have few joints and foliation they form few corestones and little or no saprolite (Pavich *et al.* 1989).

Sedimentary rocks such as pure quartz sandstones, and also metasedimentary quartzites, produce relatively thin weathering profiles consisting largely of a sandy grus created by dissolution of the amorphous silica binding the quartz grains of sandstone, and some surviving unaltered rock fragments.

Mudrocks, such as marine mudstones and shales,

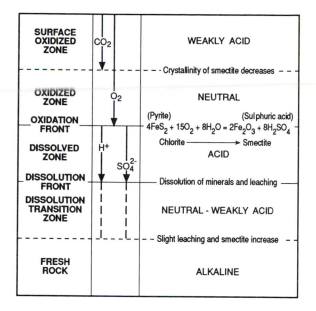

FIG. 9.5. Schematic sketch showing the major weathering processes occurring in a mudstone. (After Chigira 1990.)

have a large proportion of their material formed of clay minerals, but they may be altered further to more stable clay minerals by carbonation and the generation of acids such as sulphuric acid from pyrite, as shown in Fig. 9.5. Mudrocks therefore have zones of oxidized and softened clay with development of fissures, extending downwards into unaltered rock.

Limestones consist of carbonates with various proportions of more resistant mineral grains. Much of the bulk rock is lost by solution and the residuum is most commonly incorporated into the solum with little or no development of a saprolith.

Saprolite formation is clearly most effective in quartzofeldspathic rocks. Such rocks make up a large proportion of the basements of continents and particularly of the cratonic cores of continents. As a result deep saprolites and truncated relics of saprolites survive in many areas.

Rates of saprolite formation are little understood, but Pavich (1989) has found rates of 4–20 m/My for uplands on foliated metamorphic rocks of the Appalachian Piedmont, with an average rate of descent of the weathering front into bedrock of 1 m/100 ky. As chemical weathering rates are enhanced by groundwater circulation, it is reasonable to assume that active saprolite formation is favoured by high relief and high hydraulic gradients within hillslopes. The survival of saprolites will therefore depend upon a balance between erosion rate and rate of chemical alteration, with both being enhanced by uplift and high relative relief. Thick saprolites (10 m or so) can therefore coexist with tectonic uplift of a few metres to tens of metres per million years. This appears to be the condition in the Appalachians (Pavich *et al.* 1989) which Hack (1960) called dynamic equilibrium.

Deep weathering

Deep weathering involves the alteration of rocks and minerals to depths of tens or even hundreds of metres. It may be the result of a progressively falling water-table and hence extension downwards of the zone exposed to oxidation, but weathering can occur below the water-table (Ollier 1988) in reducing conditions.

Below the water-table hydrolysis is the dominant process. Silicate minerals are formed of metallic ions (M) in combination with a silica group (SiO_n). The reaction may be represented by:

M-O-Si-Al + H_2O → M^+
M-Al-silicate + water → metal ion
    + H-O-Si-Al + OH^-
    + *clay*   + *hydroxyl ion,*

or, if the silicate mineral is olivine:

Mg_2SiO_4 + $4H_2O$ → $2Mg^{2+}$
olivine   + water → magnesium ion
    + $4OH^-$ + H_4SiO_4
    + *hydroxyl ion + silicic acid.*

The production of hydroxyl ions results in alkaline conditions which enhance the breakdown of silicates. The cations and hydroxyl ions cannot be leached away, but can be lost by diffusion and movement of groundwater. Nevertheless, the increased concentration of ions may serve to shield other ions from participating in precipitation reactions. The solubility of many salts is enhanced in solutions of higher ionic strength: silicate minerals probably behave similarly. Ferrous iron migrates in the same way, but if it reaches the oxidizing zone it is there converted to ferric hydroxide and is precipitated. By this means a layer of ferricrete, one metre thick, may be formed in 10000 years.

Deep weathering below the water-table takes place in a zone where temperature is more likely to be controlled by geothermal heatflow than atmospheric climate. Consequently it is largely independent of climate and can occur anywhere there is groundwater. In the Great Artesian Basin of the Australian desert zone, groundwater extends to a depth of 3000 m (Habermehl 1980), so deep weathering can occur to that depth. The groundwater is of meteoric origin, as indicated by isotopic analysis, and at the flow rates known for the area could have taken 2 My to travel from its source area. References to deep weathering as 'tropical weathering' or as indications of former humid tropical climates are not always appropriate although some deep profiles may have developed in humid tropical conditions. In Australia some deep profiles did develop in Late Mesozoic and early Cenozoic times when tropical rainforest cover was widespread.

Depths of weathering mantles are influenced by many factors such as: stability of the climate and vegetation; effectiveness of erosion at the ground surface; depth of penetration of groundwater as influenced by water availability, joint and fault density, porosity and permeability of the rock; by the time which has been available for weathering at a site; and the balance between erosion and saprolite formation. Surface relief has both an active and a passive influence on weathering: sites of active erosion are common on valley sides and sharp-crested ridges; sites of accumulation of regolith occur primarily on plateaux, plains, and terraces of low relief. Weathering profiles are, as a result of these various influences, often highly irregular in depth and may lack one or more of the six weathering zones which have been described as characteristic of deep profiles (Fig. 8.14). Four examples of hillslopes with varying depths of weathering profiles and related landforms are shown in Fig. 9.6. Further examples are included in the 'Geological Society Engineering Group Working Party Report' (1990).

Weathering fronts and etching

The base of a weathered mantle is known as the weathering front. This front may be very irregular, especially if it develops in an area of closely-spaced joints in bedrock where weathering processes can penetrate downwards, or in massive rocks it may be a relatively smooth undulating surface. Rock at the front shows various degrees of alteration with rinds and laminae of alteration products around cores of fresh rock.

Separated corestones are commonly formed in granite but they are also found in basalts, andesites, dolerites, and some sandstones. The feature of the

FIG. 9.6. Varying depth of weathering profiles on continental, and especially cratonic, surfaces. Surface relief, weathering front, and stream channels may be unrelated.

parent rock which controls the production of corestones is the variation in permeability produced by jointing and the spacing of joints. Where joints are very close together, or where the passage of solutions is through pores in the rock, there are few zones of preferential water movement; where joints are widely spaced and the massive rock has low porosity and water movement is along joints, weathering becomes concentrated in those joints and works progressively into joint blocks. The unaltered core of the joint block thus becomes the corestone. In Fig. 9.7 the progressive breakdown of the joint block from its exterior towards the interior is illustrated. The formation of clay minerals through the breakdown of feldspars occurs close to the joint so that no trace of rock structure survives, then successively into the block are zones of decreasing alteration.

Spheroidal weathering is a feature of many corestones formed in granites, basalts, andesites, and dolerites. It is rare in sedimentary rocks which tend to have uniform compositions. There is still controversy regarding the origin of such weathering forms. One group of hypotheses (e.g. Jocelyn 1972) suggests that the concentric shells may be a result of residual thermal stresses developed as igneous rocks cooled; the work of Bock (Fig. 8.4b, c, d) indicates that there may be some validity in this idea. A second group of hypotheses suggests that the shells are formed by chemical processes (e.g. Augusthitis and Otteman 1966) which may be aided by the formation of microcracks along which chemical solutions can migrate (Bisdom 1967).

GRANITE

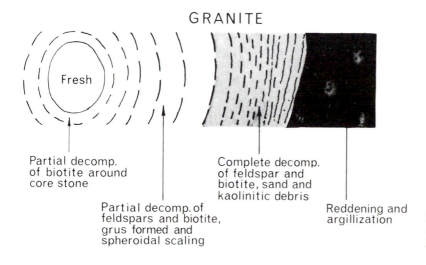

Partial decomp. of biotite around core stone

Partial decomp. of feldspars and biotite, grus formed and spheroidal scaling

Complete decomp. of feldspar and biotite, sand and kaolinitic debris

Reddening and argillization

FIG. 9.7. The progressive development of weathering zones around a granitic corestone.

PLATE 9.8. The progressive rounding of a joint block by thickening of weathering skins at corners and by weathering along fractures at corners. The rock is a basaltic tuff.

The rounding off of corners of an original angular joint block (Plate 9.8) permits the gradual formation of spherical shells around a corestone. A feature of many spheroidally weathered joint blocks is that the rock shells vary in colour, sometimes with whitish and reddish-brown shells alternating. The reddish-brown colours are often found to be the result of deposition of limonitic iron between rock granules, and the whitish rings are depleted of iron but relatively enriched in silica, potassium, and aluminium with the formation of clay minerals. It seems probable that these migrations and sites of deposition are influenced by changes in pH of the migrating solutions so that once separation has started the precipitates create the pH which is suitable for further deposition of ions of their own species. Solution must be a very important part of the total process of change for, according to Blackwelder (1925), the chemical weathering of a granite to new mineral species would be accompanied by a 50 per cent increase in volume if soluble products were not removed by water. The close fit of neighbouring blocks and

survival of aplite and quartz veins indicate clearly that spheroidal weathering is an isovolumetric process. Spheroidal weathering occurs within the regolith and is distinct from exfoliation of rock spalls from exposed surfaces.

Boulder fields may be the result of the stripping away of deeply weathered regolith materials to leave exposed corestones or they may result from exfoliation of boulders at the groundsurface. It is only when a regolith containing corestones and spheroidal weathering forms is found in association with the boulder fields that their origin can be determined with confidence.

Mantle stripping and inherited forms

The recognition that rounded boulders, and other minor and major forms, have developed within weathering mantles and at the weathering front has a long history which goes back to the beginnings of modern geology (see Twidale 1990).

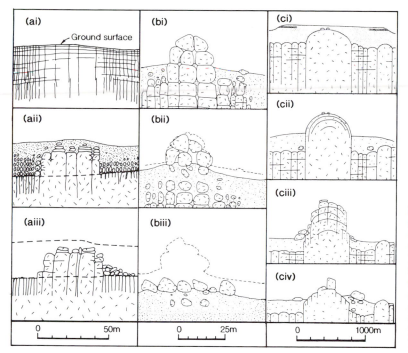

FIG. 9.8. (a) Development of tors as a result of selective weathering in areas of closely spaced jointing followed by saprolite stripping (based on Linton 1955). (b) Separation of tors from their bedrock as a result of chemical weathering, in the soil below them, occurring at a greater rate than on the exposed rock. (c) The exposure and subsequent decay of a bornhardt in Nigeria as joints open. (b, c, after Thomas 1965.)

PLATE 9.9. Granite boulders and a tor exposed at the ground surface in the savanna zone of Cameroon, West Africa.

The major features exposed by stripping are collectively known as etchforms (Wayland 1933). Two of the most prominent etchforms are tors and koppies of rounded boulders and also some domed inselbergs, called bornhardts (Fig. 9.8; Plates 9.9, 9.10, 9.11).

It was recognized by Falconer (1911) that the inselberg landscapes of northern Nigeria have not been developed at the groundsurface but by the penetration of chemically active waters into irregularly jointed rock with development of irregular weathering fronts. Such ideas have been elaborated into a system of geomorphological evolution of the surfaces of continental, and especially cratonic,

PLATE 9.10. A tor of granite boulders revealed by mantle stripping, Devil's Marbles, central Australia. Note that the boulders have retained their position relative to each other.

PLATE 9.11. A large domed inselberg formed in conglomerate, the Olgas, central Australia. Note the large vertical joints which divide the massif into compartments. Sheeting is developing in the uppermost rock unit of the dome.

surfaces by Büdel (1957, 1982) who emphasized the importance of the two surfaces of geomorphic activity: (1) the weathering-front etch processes; and (2) surface erosion of the soil. Stripping of the weathering mantle exposes the etched surfaces over large areas and leaves only pockets of relict regolith as evidence of the existence of the earlier mantles. Weathering mantles on parts of the Gondwana cratons are known to be of great age: in Australia, for example, mantles of Mesozoic and early Cenozoic ages are recognized (Idnurm and Senior 1978). Relicts of deep weathering profiles have been found in the Piedmont and Blue Ridge areas of the Appalachians, with average depths of 18 m and a maximum of over 90 m (Hack 1979); A. Hall (1986) recognized preglacial saprolites in Buchan, Scotland, up to 50 m deep, and in the Gaick area of the Grampian Mountains up to 17 m (Hall 1988); Bouchard (1985) has found saprolites up to 15 m thick at protected sites which have been covered by Quaternary ice-sheets on Canada. In brief then, deep weathering mantles, or relicts of

them, are widespread, especially on the surfaces of cratons (the cores of continental crust formed predominantly of Precambrian rocks), and where the mantles have been stripped they have revealed etchsurfaces which have been modified to varying degrees by late Cenozoic erosion processes.

The presence of such remnants of weathering profiles has not only given support to the ideas of Büdel but become the basis of many geomorphological explanations of continental landscapes (e.g. Ollier 1988; Thomas 1989a,b; Twidale 1990). Periods of weathering mantle development have been seen as times of continental and sea-level stability which may be interrupted by periods of lower base-levels, rejuvenation of rivers, land erosion, changing climate, and therefore mantle stripping (Fairbridge 1988). The difficulty in such concepts is the problem of dating the weathering mantles, the periods of climatic and sea-level change, and of relating the events. Care should also be taken over extending the concepts, for which there is evidence in one region, into areas which apparently retain no evidence of weathering mantles. The western coasts of central and southern Africa provide an example of lack of evidence. In Angola there are deep weathering profiles with koppies and bornhardts standing above areas of deep regoliths and partly exposed etchplains. Further south, in the Namib desert, similar bornhardts and koppies occur but the ground surface is of gravel or exposed bedrock with the total thickness of superficial deposits rarely exceeding 1 m in depth; nowhere has a deep saprolite been found even though many deep pits, and tens of kilometres of shallow exploratory drives, have been formed during mineral exploration (Selby 1977a). The absence of saprolites does not prove that the landsurface has not suffered etching and mantle stripping, but equally the existence only of irregular exposed rock surfaces, some of the kind which have elsewhere been demonstrably associated with deep weathering, do not provide evidence of mantle stripping and the creation of a landscape which is inherited from the forms developed at an ancient weathering front.

Major inherited forms include the regular and irregular elevated groups of joint blocks, rounded boulders, and domes of various degrees of elevation. Discussion of these phenomena fills many books and papers. Further reference on modern work should be made to Ollier (1984), Thomas (1989a,b), and Twidale (1982, 1990).

Duricrusts

An important group of landforms owes its origin to the presence and dissection of indurated crusts, of various mineral compositions, which have formed as a result of weathering processes, by the redistribution of weathering products, or by cementation of detrital materials such as colluvium, taluvium, and alluvium. Crusts frequently form resistant caps to hills and terraces, and underlie extensive plains and plateaux, especially of the more stable continental platforms with tropical, subtropical, or temperate climates. The crusts, classed together as duricrusts, can also occur in other environments but seldom with as great a thickness or forming major features of the landscape (Goudie 1985).

The terminology of duricrusts is very confused and is still undergoing revision. The confusion results from disagreement over the origin of duricrusts and also because many original definitions have been changed as knowledge has extended. Many genetic definitions are now known to be misleading or wrong.

A duricrust is a product of processes acting to cause the accumulation of iron and aluminium oxides, silica, calcium carbonate, or, less commonly, gypsum and halite. The accumulation of these compounds may be either a result of removal of other materials to leave an enrichment of the crust-forming minerals, or it may be an enrichment caused by deposition from water or of windborne minerals which can then accumulate and harden to form a duricrust. This secondary or depositional origin can be of materials derived locally by material moving down a hillslope from a higher crust, or it may be a much more widespread redistribution by streams or wind.

In the twentieth century there has been a growing tendency to use terms which indicate the cementing agent or composition of duricrusts, and to add to the indicator of the cement the ending '-crete' or the word 'crust' to the composition term: thus 'calcrete' or 'calcitic crust' is used for a duricrust composed largely of calcium carbonate (see Table 9.1). This terminology avoids the very large number of local names which are in common use and also, for crusts rich in iron and aluminium oxides, the term 'laterite' which has caused much confusion. The application of such classifications is, however, not always simple because the chemical content of duricrusts can vary over short distances in response to change in rock mineralogy and site factors. Local names for duricrusts are given in Table 9.2.

Profiles

Duricrust profiles are generally of three types—those in which the crust is derived from: (1) overlying soil; (2) underlying weathered rock; and (3) deposits in which the crust-forming material is detrital, that is, it has been transported and deposited or precipitated.

Ferricretes are particularly important in humid tropical and subtropical regions and also occur in some temperate areas such as eastern Australia. Their profiles have been described as having four zones from the surface downwards: (1) a soil, often sandy and containing nodules or other concretions; (2) a crust of reddish or brown, hardened, or slightly hardened, material, with vermiform (or vermicular) structures (i.e. having tube-like cavities 20–30 mm in diameter) which may be filled with

TABLE 9.1. *Classification of duricrusts*

By cementing agent	By dominant content	Essential chemistry	Typical crystalline minerals
Silcrete	Silitic crust	SiO_2	Quartz (90–95%)
	Siallitic crust	SiO_2, Al_2O_3	Quartz (aluminous compounds often amorphous to crypto-crystalline)
	Fersilitic crust	Fe_2O_3, SiO_2	Hematite, quartz
Ferricrete	Fersiallitic crust	Fe_2O_3, $FeO(OH)$, SiO_2, $Al_2O_3 . nH_2O$, + $AlO(OH)$	Hematite, goethite, quartz, gibbsite, +boehmite
	Ferrallitic crust	Fe_2O_3, $FeO(OH)$, $Al_2O_3 . nH_2O$, $AlO(OH)$	Hematite, goethite, gibbsite, boehmite
	Ferritic crust	Fe_2O_3 (up to 80%), $FeO(OH)$,	Hematite, goethite
	Fermangitic crust	Fe_2O_3, MnO_2	Hematite, pyrolusite/psilomelane
Alcrete	Tiallitic crust	TiO_2, $Al_2O_3 . nH_2O$	Rutile/anatase, gibbsite
	Allitic crust	$Al_2O_3 . nH_2O$ (up to 60%), $AlO(OH)$	Gibbsite, boehmite
Calcrete	Calcitic crust	$CaCO_3$	Calcite (60–97%)
	Calcsilitic crust	$CaCO_3$, SiO_2	Calcite (silica often chalcedonic, etc.)
Gypcrete	Gypsitic crust	$CaSO_4 . 2H_2O$	Gypsum
Salcrete	Halitic crust	$NaCl$ (usually impure)	Rock salt

Note: Gibbsite is $Al_2O_3 . 3H_2O$; boehmite, $AlO(OH)$; goethite, $FeO(OH)$; hematite is α-Fe_2O_3.
Source: Based on Dury (1969).

TABLE 9.2. *Local names for duricrusts*

	Iron-rich and aluminium-rich[*]	Silica-rich	Calcium-carbonate-rich
South Africa	Ferricrete	Surface quartzite	Sheet limestone, tufa[†]
Australia	Pisolite	Porcellanite, grey billy	Travertine[‡]
USA	Plinthite		Caliche
France	Cuirasse		Croûte calcaire, carapace calcaire
Other	Canga (Brazil) Murrum (Uganda)		Tosca (Argentina) Kankar (India) Kankur (India)

[*] Al-rich crusts (alcretes) are widely called bauxite crusts
[†] Also widely used for spring deposits
[‡] Widely used outside Australia for $CaCO_3$ deposits from springs.

kaolin; less cemented horizons may be pisolithic (i.e. formed of pea-sized grains of red-brown oxides); (3) a mottled zone of white clayey kaolinitic material with patches of yellowish iron and aluminium oxides; and (4) a pallid zone of bleached kaolinitic material (Plates 9.12, 9.13).

It is now clear, however, that there is seldom, if ever, a genetic relationship between the crust and the underlying mottled and pallid zones. Furthermore, the crust is commonly separated from the underlying zones by a stone-line and either, or both, of the mottled and pallid horizons may be

PLATE 9.12. An exposure of ferricrete crust above a pallid zone, northern Nigeria.

absent. The logical conclusion is that all, or most, ferricretes have formed on a groundsurface where oxides, predominantly iron, have been precipitated from water which has moved laterally, on horizontal surfaces, or downslope where slope-foot ferricretes have developed (Ollier and Galloway 1990).

Ferricrete is commonly found on plateaux and at crests of flat-topped hills where it forms a resistant cap. Such locations are unlikely to have had such relief when iron was transported there by water and precipitated as hydrous oxides, and it seems probable, therefore, that inversion of relief has been common and repeated. Oxides have formed cements around soil grains on lower slopes and formed indurated layers. Erosion later removed softer soils from above and around the ferricrete leaving it as a cap on the edges of mesas and plateaux.

Alcrete profiles have received far less attention than ferricrete profiles from geomorphologists, perhaps because they are weaker, do not form caps, and are less common. Their main importance is as a source of aluminium.

Silcrete profiles consist of an indurated silicified layer which may be up to 3 m thick, with as much as 95 per cent SiO_2, with quartz grains set in a matrix of fine quartz or opaline silica. The silica-rich horizon commonly has columnar structure and is grey, yellow, or brown in colour. The rock has a vitreous appearance and non-vesicular varieties have a conchoidal fracture. When struck with a hammer silcrete may emit a pungent smell. Silica

PLATE 9.13. A soil profile in the rainforest of central Zaïre. The pisoliths are pea-sized concretions of iron oxides.

crusts may overlie weathering profiles, they may be formed of cemented gravels, and they may form a rock layer which lies upon stream or lake deposits.

Silcrete is widespread in western and central Australia and in south-eastern Africa. Isolated silicified boulders from parts of France, Central Otago (New Zealand), and southern England—where they are called sarsen stones—may be remnants of ancient silcretes. There is no conclusive evidence that silcretes are forming anywhere today. The extensive silcretes of central Australia developed during the Tertiary (Exon *et al.* 1970), and thin silcrete horizons formed round the bases of many hills in the Adelaide to Lake Eyre region during the Pleistocene (Wopfner and Twidale 1967).

Calcrete profiles are widely distributed in arid and semi-arid lands. On a world scale calcretes have about 80 per cent calcium carbonate (Goudie 1973), and range in thickness from nodules in a soil profile to massive sheets over 10 m thick. Netterberg (1967) has suggested a sequence of calcrete types which is essentially an evolutionary sequence (Plates 9.14, 9.15).

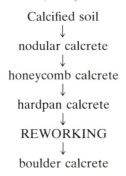

Calcified soil
↓
nodular calcrete
↓
honeycomb calcrete
↓
hardpan calcrete
↓
REWORKING
↓
boulder calcrete

Calcium carbonate accumulation occurs within soils at the present time in semi-arid zones where evapotranspiration is in excess of precipitation for most of the year. Leaching through the soil profile is negligible and the carbonate is deposited at a level in the profile controlled by water removal by plants, by evaporation, or by the CO_2 content of the soil atmosphere (if CO_2 is released from solution $CaCO_3$ is precipitated). The removal of water by plant roots sometimes causes the carbonate to be precipitated around the roots where it forms casts called rhizomorphs (Plate 9.16).

All calcretes require a source for their calcium carbonate and this is usually locally derived from limestone, marble, or calcareous sandstones.

Where the local rocks are deficient in lime, calcretes can form by deposition from flood waters, and such a fluvial origin probably accounts for the thick valley calcretes of the Kalahari and some of the calcrete cemented gravels of the inland Namib desert.

Both calcrete and silcretes may be formed, or at least added to, by volcanic ash or windblown dust which can accumulate in soils, and from which solutes may pass through the organic cycle before being precipitated in the soil (Fig. 9.9).

Gypcrete and salcrete profiles are exceedingly rare because of the high solubility of gypsum and halite. They are preserved only under arid climates and, because they are usually precipitates from the evaporation of sea water or saline desert streams, their profiles usually consist of a white or buff-coloured crust overlying lake or marine muds.

Origins of ferricretes, alcretes, and silcretes

The basic problem of understanding the genesis of ferricretes and alcretes is to determine how the chemical separation of iron, aluminium, and silicon can occur. These three elements are normally relatively insoluble yet they are removed from their parent silicate and ferromagnesian rocks and then selectively concentrated. The commonly offered explanations for selective removal of Si and weakly soluble metals include considerations of: (1) rates of reactions; (2) increased solubility at particular pH-ranges or with increased ionic strength; (3) preferential solubility of one element due to organic complexing; and (4) the availability of oxygen in groundwater for oxidizing processes—a condition usually equated with Eh. While any of these considerations may be relevant to particular situations it is most probable that (2) and (4) provide the most generally valid explanations of solubility.

Selective removal from the soil of alkalis, alkaline earths, and silicon occurs at a certain combination of pH and Eh where the solubility of aluminium and metal oxides, and hydroxides, is less than that of quartz. Selective removal of iron from the soil with retention of aluminium requires unique Eh and pH conditions (Norton 1973). At low pH aluminium is mobilized, but iron requires both low pH and low Eh before it is mobilized. At high Eh and pH iron is immobile and becomes enriched, while at high pH aluminium is mobile. Given the availability of Fe and Al the formation of an aluminium-rich soil (bauxite) or an iron-rich soil

PLATE 9.14. Nodular calcrete, near Kalkrand, Namibia.

PLATE 9.15. Massive hardpan calcrete, near Kalkrand, Namibia.

PLATE 9.16. Calcium carbonate accumulations around fossil grass roots in ancient dune sands, central Namib desert. Such root casts are called rhizomorphs.

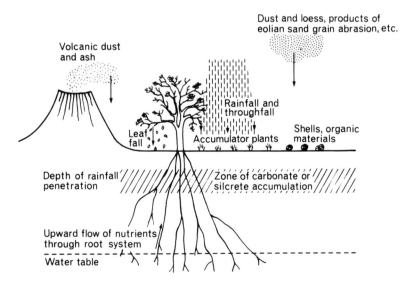

FIG. 9.9. Possible sources for accumulations of soil carbonate and silica. These sources do not include deposition from flood waters. (After Goudie 1973.)

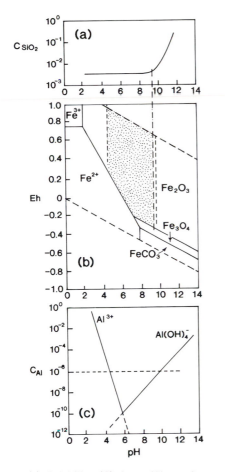

Fig. 9.10. (a) Solubility (C) in mol/litre of amorphous silica as a function of pH. Note marked solubility increase at pH > 9. (b) Eh–pH diagram for iron. Hatchured region satisfies conditions for co-precipitation of aluminium and ferric oxides. (c) Solubility of aluminium as Al^{3+} for low pH, $Al(OH)_4^-$ at high pH. (a, b based on Drever 1982.)

Knowledge of the reactions which produce mobility of various elements in soils having organic materials within them is still limited and the discussion above is, no doubt, greatly oversimplified. Our ignorance is increased by the certainty that many iron- and aluminium-rich soils and crusts are of considerable geological age (see Van de Graaff *et al.* 1977 for a review). These crusts may have developed under vegetation covers unlike any existing today and in climates that were probably warmer, and perhaps with different rainfall regimes from those of the present humid tropics.

Recognition that most, or all, ferricretes are not genetically related to the materials they lie on eliminates many old hypotheses of ferricrete formation. Attention is now focused on processes of ferricrete thickening and development. It has been proposed by McFarlane (1976) that many ferricretes develop in the sequence: (1) original iron and aluminium oxide precipitates form as pisoliths within the soil in the relatively narrow range of fluctuation of the groundwater-table; (2) as the land surface is lowered, by surface erosion and stream incision, the water-table sinks and further pisoliths form below the existing precipitates; (3) when stream downcutting ceases the water-table stabilizes and a massive variety of vesicular or vermiform ferricrete develops; and (4) deepening of a ferricrete profile can continue because iron is mobilized from the surface by the action of vegetation and is reincorporated at the base of the ferricrete horizon, and because additional oxides are available from newly weathered rock below the ferricrete. In this hypothesis pallid zones are neither necessary nor related to ferricrete formation, but they develop beneath ferricretes by leaching, especially when the land surface is incised and groundwaters can move downwards through permeable duricrusts, and then laterally through pallid zone materials towards the streams (Fig. 9.11).

The extensive silcrete deposits of central Australia were formed in mid-Tertiary to early Pleistocene times. It has been suggested by Stephens (1971) that the silica of which they are composed was derived from the selective leaching occurring beneath the ferricretes and alcretes of northern and eastern Queensland. He suggested that the drainage waters carrying the silica in solution, or as a gel, seeped and flowed towards central Australia where the silica was precipitated under a drier climate in areas of low relief, and

will depend upon a critical combination of Eh and pH. A secondary enrichment of Al may be brought about by an increase in pH of groundwater, whereas secondary enrichment of Fe may be caused by an increase in pH or an independent Eh increase. The effect of chelating organic materials is to change Eh and pH conditions towards those favouring increased solubility of iron and aluminium.

Silica as quartz crystals has solubilities ranging from 6 to 14 ppm, but quartz dust and amorphous silica have solubilities of up to 140 ppm (see Figs. 8.10c, 9.10a). This suggests that the form of silica is very important in influencing its solubility.

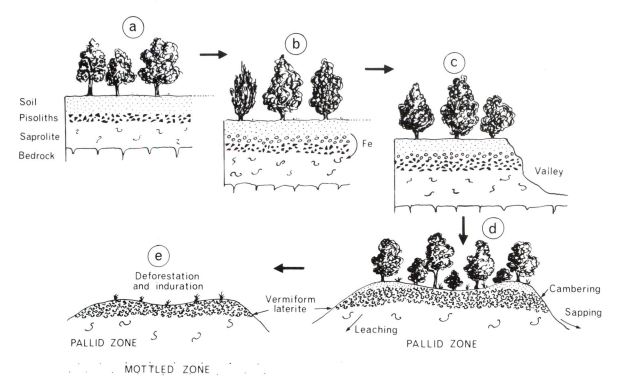

Fig. 9.11. McFarlane's (1976) model of laterite (ferricrete) formation below a downwearing surface a–c, followed by incision and the formation of a pallid zone, and then by deforestation, soil erosion, and induration of exposed ferricrete.

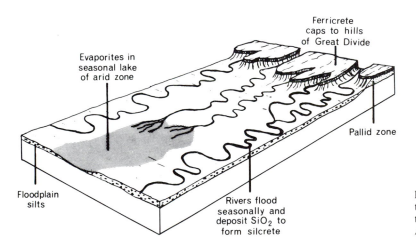

Fig. 9.12. A schematic representation of Stephen's (1971) hypothesis of the formation of silcrete in central Australia.

restricted surface drainage (Fig. 9.12). The silcretes were thus discontinuous, but because of the low relief, none the less extensive. This hypothesis certainly accounts for many silcretes which are clearly detrital in origin and explains the thick lenses of gypsum found in many silcrete profiles, for gypsum is commonly formed in desert salt lakes. The flood-deposit hypothesis is less satisfactory for silcretes above a related kaolinized profile unless it can be established that the silica-

PLATE 9.17. A section through a massive silcrete, with a mottled silicified saprolite below (level of the machine) and granular soil above it, Coober Pedy, South Australia.

rich flood waters invaded old and deep weathering profiles over large areas. If this did occur then it is possible that some silcretes result from silicification of ferricrete and others may be the hardened remnants of truncated weathering profiles. Gravels derived from silcrete now form the surface of extensive desert gibber plains, such as those of the Sturt Desert. The formation of opal in cavities in silcrete by rhythmic precipitation of silica has permitted the development of extensive mining in areas around Coober Pedy from where 90 per cent of the world's opals are derived (Plate 9.17).

The concentration of ferricretes and alcretes in the humid tropics, of silcretes in the subtropics, and calcretes in areas with a precipitation of less than 500 mm per year and temperate or tropical temperature regimes, has produced ideas that duricrusts may be classified on a zonal basis. Watkins (1967) suggested a scheme for part of Australia. This idea of simple zonation has been questioned for a number of reasons: (1) modern mapping shows that over large areas of eastern Australia silcretes and ferricretes occur in the same zone (Young 1985); (2) ferricretes, alcretes, and silcretes may occur in the same landscape and have a closer relationship with the rocks from which their minerals were derived than with climate (Taylor and Ruxton 1987); (3) most duricrusts are of great age and have suffered many climatic changes since their formation. A closer approximation to climatic zonation occurs in West Africa where the dry margin of ferricrete formation is almost coincident with the wet margin of calcrete, at 500 mm of rain per year. The absence of obviously developing modern silcrete makes reliable comment upon the climate in which it evolved virtually impossible.

Rates of formation and hardening

Goudie (1973) suggested that rates of crust enrichment may reach about 1 m in 0.3 to 2.3 million years for ferricretes, and 1 m in 0.3 million years for calcrete. The figure for ferricrete is largely based upon the work of Trendall (1962) who calculated how much iron could be derived from the bedrocks in Uganda and concluded that 10 m of ferricrete could be derived from 200 m of bedrock over a period of about 35 million years. If, however, the oxides are transported considerable distances before precipitation, Trendall's figures are questionable.

It has been suggested by Persons (1970) that, in the Cameroons (West Africa) under a seasonal rainfall averaging 3250 mm per year, 2 m of complete induration of ferricrete, after deforestation and exposure, would take about 100 years. Superficial hardening can occur in a few months or years. The process of hardening is, at least partly, due to dehydration of the iron and aluminium oxides and to the development of crystallinity in goethite and hematite (Maignien 1966). Harder ferricretes are usually found to have higher iron contents and to have been exposed for greater lengths of time. The process of reversal of hardening is very slow: it can usually occur only where there is an accumulation of superficial deposits and the establishment of vegetation on them.

Calcrete hardening is a result of cementation as precipitates form in soil voids, and by recrystallization of the carbonates.

Landforms and duricrusts

Duricrusts form most readily on surfaces with gentle slopes. As a result they provide resistant caps to plateaux and hills and act like cap rocks to buttes, mesas, and, if warped by earth movements, can form the dip slopes of hills. Ferricretes have very high porosity so water may infiltrate through them very rapidly. As the water moves laterally in, or above, a kaolinized zone below, it can cause high seepage pressures and produce natural pipes which may give rise to springs on slopes beneath a dissected plateau surface. The slopes beneath an exposed ferricrete are, therefore, frequently remodelled by rilling and seepage in the weak materials of a weathered zone, and by the undercutting of the resistant cap (Fig. 9.13). Undercut ferricrete and silcrete caps may sag or collapse when caves are eroded in the pallid zone beneath them, and a complex landscape of collapse depressions and natural arches may form.

Ferricretes, silcretes, and calcretes which may be originally formed in hollows can produce a form of inverted relief when the less resistant rocks around them are eroded away and the crust is left capping a residual hill (Plates 9.18, 9.19).

Benched or terrace landscapes may develop

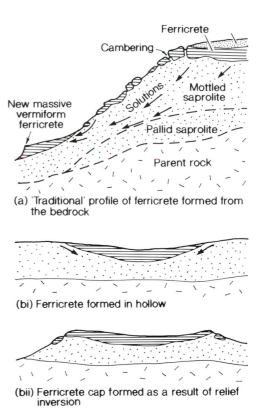

(a) 'Traditional' profile of ferricrete formed from the bedrock

(bi) Ferricrete formed in hollow

(bii) Ferricrete cap formed as a result of relief inversion

FIG. 9.13. Some features of ferricretes and their influence on landforms.

where solutions containing iron move laterally towards a seepage zone, hollow, or terrace edge, and into a zone of wetting and drying in which the iron is precipitated. A new massive ferricrete may be formed there. Slope debris may form cemented ferricrete taluses or, more commonly, recemented footslope, pediment, or detrital ferricretes (Fig. 9.13). Pavements of recemented ferricrete fragments are relatively common on valley floors and lower slopes below ferricrete-capped hills.

Calcretes may produce a very distinctive landscape when re-solution of the lime opens joints and forms pipes through which surface water can drain. In parts of southern Africa areas of discontinuous surface drainage are features of calcrete outcrops, while areas with impermeable rocks nearby have a complete drainage network.

Duricrusts as resources

It has already been pointed out that alcretes are the major sources of aluminium ores. Ferricretes

PLATE 9.18. Calcrete cap to a small mesa, northern Sahara.

PLATE 9.19. Silcrete forming a resistant cap above slopes formed on saprolite, South Australia.

have been used as sources of iron in East and West Africa as well as in India. Ferricretes and calcretes are widely used for road-making and both have been used for building materials: the famous temple of Angkor Wat in Cambodia is made of ferricrete. Most duricrusts, however, have soils of low fertility and hinder agriculture by their resistance to cultivation.

10 Hillslope Stratigraphy and Form

Stratigraphy and slope deposits

Stratigraphy is the study of natural bodies of rock and soil with the intention of using them for interpreting their chronological succession, distribution, origin, and environment of deposition and formation. The stratigraphy of hillslopes is concerned with the study of soils, weathering products, colluvium, talus and fan materials with the aim of understanding the history of hillslope forms and the influence on them of former processes of erosion and deposition. The fundamental principle of stratigraphy is that of superposition: this states that in any succession of layered geological deposits, not severely deformed, the oldest stratum lies at the bottom. Hillslope deposits are most commonly removed from upper hillslope positions by erosion and deposited towards the base of the slope. The lowermost deposit is therefore related to the oldest erosional event of which a record is preserved.

The environment in which erosion, deposition, and subsequent stability of the landscape have occurred has to be deciphered from study of the texture, fabric, composition, and stratification of the hillslope deposits, and from the nature of the solum which has developed in them.

Solum characteristics and environment

Solum features are the best indicators of pedo-genetic and geomorphological processes which have contributed to its development and hence of the environment in which that development occurred.

Some features of solum profiles are indicators of geomorphic processes: thin A horizons on upper slopes and thicker A horizons on lower slopes usually indicate that surface wash has eroded upper slopes and deposited material on lower slope units (Plate 10.1); mottling and gleying of B horizons indicates seasonal or permanent soil saturation and subsurface water flows; vertical veins of clay indicate vertical percolation of water and, where the veins are horizontal, they indicate lateral flow; swelling clays cause cracks to close and open with wet and dry soil conditions, and clay skins on crack walls indicate seasonal or regular variations in soil moisture.

Climatic and weathering environments may be interpreted from other features: organic matter and root traces indicate plant life and animal activity; salts indicate arid, semiarid, or coastal areas of dominant evaporation; carbonate nodules indicate semiarid climate with limited soil drainage; clay movement and deposition in B horizons indicates strong chemical weathering and leaching in a humid environment; iron oxide formation and movement usually indicates strong weathering and may be associated with good drainage and seasonal humidity; structural features (Fig. 10.1) are particularly important because structure development takes considerable time and indicates a degree of landsurface stability, chemical weathering, and vegetation cover.

Soil structures consist of aggregates of various sizes ('peds') which incorporate grains into distinct forms separated from each other by a network of voids and irregular planes ('cutans' in pedological terminology). A common form of cutan is a clay skin formed where clay has been washed down a crack and deposited as a lining. Clay skins are virtually restricted to clay-rich B horizons and usually form in soils which are close to saturation for part of the year. Granular and crumb structures are indications of abundant biological activity and fine crumb structures are particularly common in soils of grasslands. Domed columnar peds form in soils saturated with sodium ions and therefore

PLATE 10.1. Profile through the toe of a slope showing a sharp boundary between the calcareous sandstone bedrock and the soil above. Thickening of the soil profile downslope indicates the effect of erosion on upper slopes and accumulation near the slope base, Hawke's Bay, New Zealand.

indicate desert, salt pan, and salt marsh conditions. Soils formed under waterlogged conditions usually lack structure (Retallack 1988).

Structure is a particularly important feature of the solum because it commonly survives burial beneath younger deposits (Plate 10.2). As structure is a major feature of B horizons, these horizons are the most obvious indications of a paleosol, that is, a buried solum or one which has formed under a vanished environment. A horizons may be partly preserved because of crumb or granular structures, but their organic matter content may disappear and the horizon become compacted.

Soils and topography

Local relief has a major influence on the soil types found within a landscape; microclimate, parent material, geomorphic processes, and soil-forming processes are other factors which may be influential. The influence of relief is expressed through other factors. (1) Aspect influences insolation and therefore available heat; it may also influence precipitation through rain-shadow effects and sites of snow accumulation; heat and water availability influence soil moisture and therefore hydraulic conductivity, soil strength, and rates of weathering and pedogenesis. (2) Slope gradient has a major influence on the distribution of water

Columnar

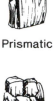
Prismatic

Prism breaking
down to blocky

Blocky

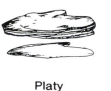
Platy

Crumbs and
granules

FIG. 10.1. The major types of structural units found in soils.

PLATE 10.2. Paleosols developed in tephra. The best developed soils have formed thick B horizons which are rich in clay and now stand out from the cutting face. The accumulation of eolian sediment, and the development of the paleosols, indicates that the site has been one of predominant stability for several hundred thousand years.

on and within the soil, as well as on the rate at which water moves during and after rainfall; gradient also influences the processes and rates of soil movement down the slope. (3) Pedogenetic and geomorphic activities of the past have been responsible for creating much of the parent material in which modern soils are forming, consequently they may influence the water-holding capacity and hydraulic conductivity of modern soils. (4) Vegetation and organic processes are both influenced by, and themselves influence, soil moisture, geomorphic processes of erosion, transport and deposition, as well as pedogenetic processes. (5) Pedogenesis may be regarded as a product of the interaction of various factors, but it also modifies those factors; for example, movement of bases, organic matter and clay down a slope depletes soils high on the slope, reducing soil depth, and creates thicker soils at the base of the slope (Plate 10.1); pedogenetic transfers change the physical properties of soil and so modify future

resistance to geomorphic processes. The processes acting on slope materials are consequently interlinked and all are modified by the action of other factors and the results of their own activity. It is not surprising then, that there have been many attempts at developing schemes which will aid classification, mapping, and explanation of soil and geomorphic processes on hillslopes. Perhaps the most widely used such concept is that of a catena or toposequence.

A *catena* was originally defined as: 'a unit of mapping convenience . . . a grouping of soils which while they fall wide apart in a natural system of classification on account of fundamental and morphological differences, are yet linked in their occurrence by conditions of topography and are repeated in the same relationship to each other wherever the same conditions are met with' (Milne 1935: 197). The phrase, 'same conditions are met with', is both the key for use of the concept and the reason why it is not always applicable; many conditions occur on hillslopes and not all are conducive to development of catenary sequences of soils.

Catenary sequences are most clearly recognized where water movement and storage is a dominant control on pedogenesis. For example, soil profiles may be thin on upper slopes where creep and wash transport material downslope; towards footslopes soil seasonal moisture abundance may give rise to alternating reduction and oxidation processes with development of mottling in the lower parts of the profile; in the toeslope area nearly permanent saturation may form permanent reducing conditions and gleying.

The catena concept has also been found useful in restricted circumstances, such as in studies of fault scarps which undergo periodic displacement, with creation of new parent materials, at the scarp foot, in which pedogenetic processes can operate. The degree of soil development has been valuable in establishing tectonic history (Menges 1990). Over very broad areas such as the Appalachians (Ciolkosz *et al.* 1990), and the New Zealand Southern Alps (Tonkin and Basher 1990) general relationships between soil types, landforms, and regional climate are evident.

Some studies have thrown doubt on the general validity of the catena concept (e.g. Gerrard 1990; Pennock *et al.* 1987). At the scale of individual slope units the soil types commonly form a mosaic in which little pattern is discernible because of the great variability in the distribution and intensity of pedogenetic and geomorphic processes. The variability is particularly evident in areas of steep slopes, high rainfall, and active erosion (e.g. Basher *et al.* 1988).

In spite of some reservations about the value of the catena concept, soil mapping units under titles such as 'series', 'associations', 'sequences', and 'land systems' have been based commonly upon a recognition of a relationship between soil type, topography, and geomorphic processes. The most explicit expression of this relationship is in the 'nine-unit landsurface model' of Dalrymple *et al.* (1968) in which the soil topography relationship is termed a 'landsurface catena' (Conacher and Dalrymple 1977).

Hillslopes, catenas, and paleosols

The *nine-unit landsurface model* treats the hillslope system as a three-dimensional complex extending from the drainage divide to the centre of the channel bed, and from the groundsurface to the uppermost boundary of weathered rock. Each of the nine slope units is defined in terms of form and the dominant processes currently acting on it. In reality it is unusual to find all nine units occurring on one slope profile; they do not necessarily occur in the order shown in Fig. 10.2 and individual units may recur in a single profile.

Concave-convex hillslopes are relatively common in many temperate environments so the order of units may be 1, 2, 3, 6, 8, 9; on steep faces with repeated rock outcrops the order might be, for example, 1, 2, 3, 4, 5, 4, 5, 4, 5, 6, 8, 9. A cliff above a river might have only units 1, 2, 3, and 4. The model thus provides a means of describing, and a means of mapping, slopes to show how they vary along the contours; it also relates processes to slope forms. Such relationships can be represented in drawings of profiles in block diagrams and maps (Fig. 10.3). The main deficiency of the model is its mixture of morphological and process terminology for slope units; unit 3, for example, is called a convex creep slope, but convexity on some slopes in this position may be due to cambering, not to creep.

A landsurface catena is defined as a three-dimensional pedogeomorphic body extending from the centre of the interfluve to the centre of the channel bed or lowest position on the slope, thus permitting its extension to whole landsurfaces, and from the bedrock, or saprolite, to the landsurface.

Not all landscapes have slope units which are as sharply delineated as those in Figs. 10.2 and 10.3. It may be necessary to recognize intergrade zones which share the properties of neighbouring zones, but the intergrades are minor in area compared with the major units.

The nine-unit landsurface model is concerned with modern processes and their influence on hillslopes. Most hillslopes, however, have a long history of development which, if understood, can provide information on rates of change, frequency of past land-forming events, past environments, and the nature of features in the landscape which are relict from past events. The evidence, if it has survived, of past events lies in the soils and deposits left by those events. Interpretation of ancient soils and deposits requires an understanding of their position in an ancient landscape, in other words their place in a paleocatena (Valentine and Dalrymple 1975).

A *paleocatena* may be defined as 'a group of paleosols, on the same buried surface, whose original soil properties differ owing to their differential original landscape position and soilwater

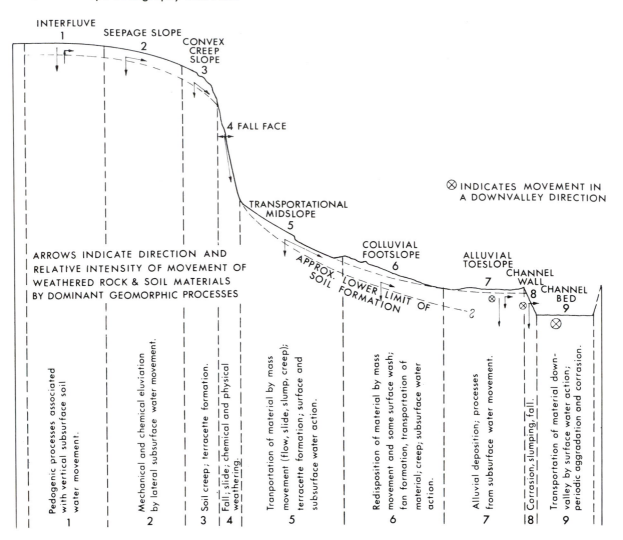

FIG. 10.2. The hypothetical nine-unit landsurface model. (After Dalrymple *et al.* 1968.)

regimes'. Using paleosols to reconstruct buried landsurfaces was first developed as a distinct methodology by Butler (1959) who introduced the concept of K-cycles.

K-cycles refer to bodies of sediment in which soils have developed; they represent 'groundsurfaces which have undergone sequences of unstable environmental conditions, in which erosion has occurred in some parts of the landscape with deposition in other parts, followed by a stable phase in which soils develop in erosion scars and in the deposited sediments'. Repeated cycles of instability and stability are recognized in an accumulation of sediment containing paleosols. The modern groundsurface is everywhere the K_1 groundsurface; older surfaces, in stratigraphic order are designated K_2, K_3 ... K_n. For recognition of K-cycles it is necessary to establish that every groundsurface is represented by an independent sedimentary deposit with a soil profile developed in it. The deposit, with its soil, is studied as a mantle which extends for some distance as a distinct organized body even though such mantles may be, and usually are, discontinuous. It is the nature of the discontinuities and contacts which are significant for interpretation of past events.

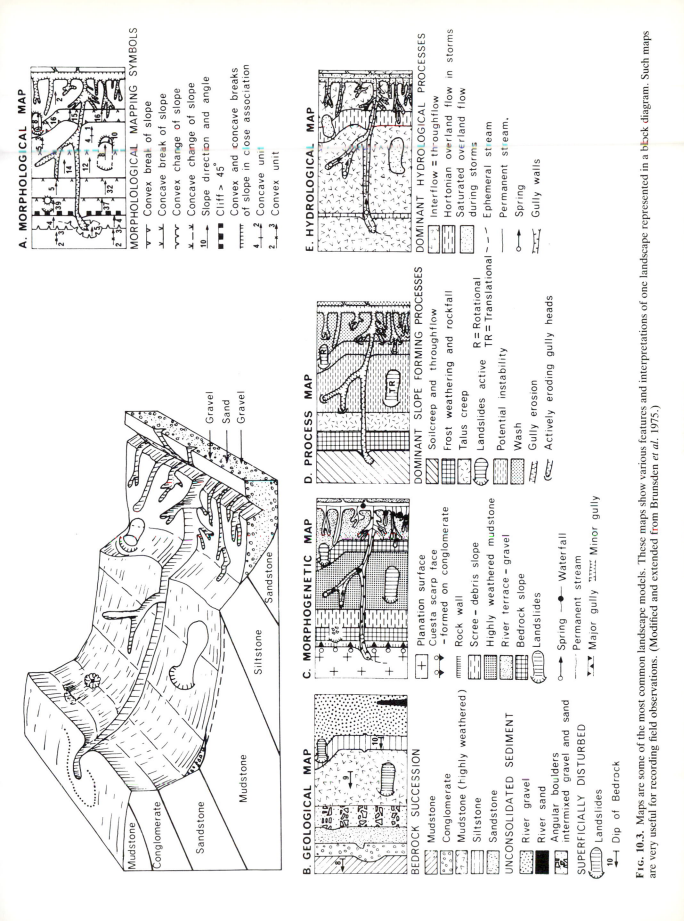

F ɪ ɢ. 10.3. Maps are some of the most common landscape models. These maps show various features and interpretations of one landscape represented in a block diagram. Such maps are very useful for recording field observations. (Modified and extended from Brunsden *et al.* 1975.)

Soil development takes considerable time and occurs primarily on a stable groundsurface, which may be said to be a Ks phase in which such environmental factors as climate and vegetation cover are rather constant. Unstable periods, K_u phases, may be caused by periods of erosion, loess and tephra deposition, and extreme flooding.

K-cycle deposits and soils are most clearly preserved in sites where sediments accumulate at rates which exceed the rates of soil formation. Soil mantles with distinct profiles, including such features as B horizons, are most readily preserved in floodplain, loess, and tephra deposits which may cover large areas. They are less commonly preserved in talus, colluvium, and fan deposits which are usually limited in their distribution and may be the result of local rather than regional geomorphic events.

Examples of K-cycles were first recognized in eastern Australia (Fig. 10.4). In the terminology of the nine-unit model, transportational midslope units are likely to suffer stripping of soil mantles during erosional phases so that either there is no developed soil present, or the soil has a patchy distribution, or there is a nearly complete mantle of the K_1-cycle. Soil mantles on transportational slopes commonly have thinner profiles, weaker horizon development, especially of clay-rich B horizons, and coarser textured soils than occur on flatter slope units.

Footslopes and toeslopes commonly preserve the most easily deciphered record of K-cycles because the soils are successively buried by colluvium and alluvium. Where deposits do not separate soils and groundsurfaces, there may be evidence of gradual accumulation of wash deposits and dust causing progressive thickening of A horizons. Such accumulations may suggest gradual and low energy changes in the environment rather than the severe changes implicit in the K-cycle concept.

An application of soil stratigraphy to interpretation of landscape history is that of Leslie (1973, Fig. 10.5). In the present landscape of Otago Peninsula, New Zealand, Leslie recognized two loess units on toeslopes, colluvium on footslopes, and relict solifluction deposits on transportational slopes. There is sufficient overlap of these deposits to place them in stratigraphic order and the buried soils have preserved evidence of former environments; radiocarbon dates have confirmed the chronology. The morphology of buried soils indicates that climate during an interstadial (a) was comparable with that of the present with chemical weathering and soil horizon development. The mid-stadial climate (b) was cold and wet, particularly in summer months, and erosion under periglacial conditions caused cryoplanation on interfluves and solifluction on midslopes; locally-derived loess was also deposited. The late-stadial climate (c) was warm and moist and mass wasting and creep transported much of the hillslope mantle material which was redeposited as colluvium. Forest cover became established and persisted until the 1850s when most of the forest was removed by European settlers; mass wasting is now a commonly active process.

Hillslope form changes

Traverses along the contours of hillslopes may reveal evidence of past processes in deposits and soils, especially on hillslopes which have experienced markedly different environmental regimes from those of the present. The evidence may be visible at the groundsurface as channels and vegetated scars of old landslides (Plate 10.3) or it may be apparent in the faces of deep cuttings which have been cut along the contours (Plates 10.4, 10.5); of particular importance are those indications of ancient landslides, stream channels, and sites of preferentially greater weathering which are not conformable with the modern ground surface, for such sites present strong evidence not only of ancient environments but also of changes: in the loci of denudation; of drainage density; of the relative importance of various erosion processes; and in the position of spurs and depressions. There is clear evidence, then, that hillslopes may not evolve in a simple sequential form towards progressively lower slope angles, extending basal concavities, or lower relative relief.

An example in which spurs and depressions do not retain constant position through time, but undergo topographic inversion in which ridges become sites of depressions and depressions become the sites of ridges, has been provided by Mills (1981, 1990). This inversion seems to occur in parts of the Appalachians of USA where the floors of depressions become infilled by large (>1 m) boulders of orthoquartzite which armour the floors of channels and depressions and so prevent further erosion of the depression floors by running water. Erosion then becomes focused on the sideslopes of the depressions. The sideslopes are progressively

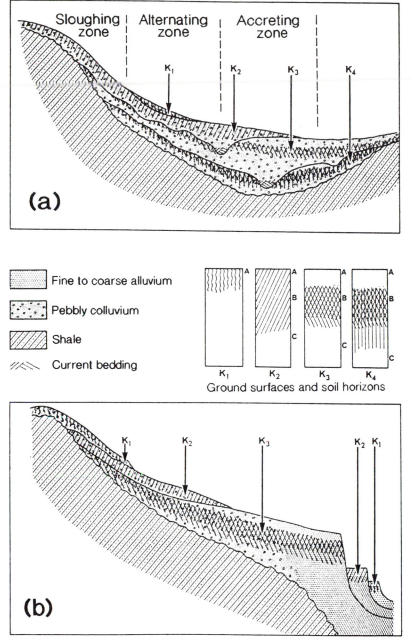

Fine to coarse alluvium

Pebbly colluvium

Shale

Current bedding

Ground surfaces and soil horizons

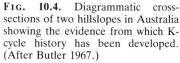

FIG. 10.4. Diagrammatic cross-sections of two hillslopes in Australia showing the evidence from which K-cycle history has been developed. (After Butler 1967.)

eroded until the old depression floor is at a higher elevation than the original ridge. The evidence that remains is of ridge tops with caps of bouldery colluvium and beneath them deeper saprolites than underlie the tops of many uncapped ridges. As the process of topographic inversion is still continuing, its various stages can be recognized in different parts of the landscape.

Infilled gullies, perhaps containing solifluction debris or thicker soils than those of surrounding areas, are widely recognized features of hill country in temperate landscapes which have experienced

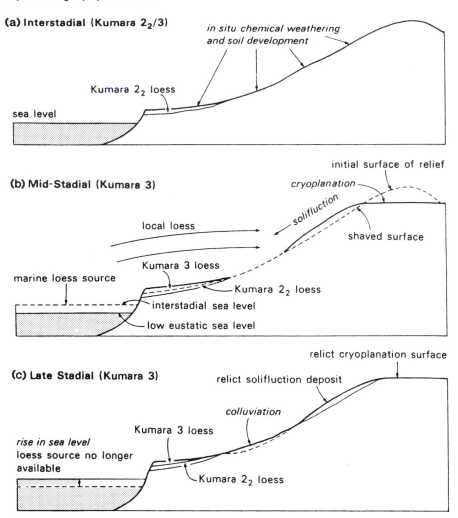

(a) Interstadial (Kumara $2_2/3$)

in situ chemical weathering and soil development

Kumara 2_2 loess

sea level

(b) Mid-Stadial (Kumara 3)

initial surface of relief

cryoplanation

solifluction

local loess

shaved surface

marine loess source

Kumara 3 loess

Kumara 2_2 loess

interstadial sea level

low eustatic sea level

(c) Late Stadial (Kumara 3)

relict cryoplanation surface

relict solifluction deposit

colluviation

Kumara 3 loess

rise in sea level
loess source no longer available

Kumara 2_2 loess

(d) Postglacial

initial surface of relief

zone of contemporary mass movement

relict cryoplanation surface

relict solifluction deposit

Kumara 3 loess

colluvium

sea level

Kumara 2_2 loess

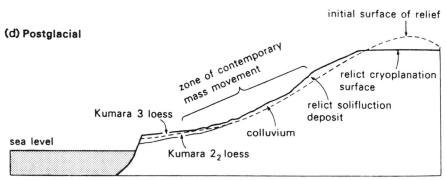

FIG. 10.5. Sequence of geomorphic events and landform development on the Otago Peninsula, New Zealand (after Leslie 1973). 'Kumara' is the local name for stadials of the Last Glacial. During periods of low sea-level, (b), exposed continental-shelf areas became sources of loess.

PLATE 10.3. Hillslopes with landslide scars forming the heads of the drainage channels. None of the channels has permanent streams and most do not have seasonal flows, indicating that these are forms relict from earlier phases of erosion.

colder climates during the Last Glacial. The infills may be bedded indicating wash processes, or unbedded indicating solifluction, debris flow, or other landsliding processes. Indications of former climate may be obtained from pollen or structure in the deposits.

A major problem in carrying out studies of regoliths is the absence of exposures through them. Seismic methods (e.g. Mills 1990) and ground-penetrating radar (Mellett 1990) are now readily applicable to regolith studies; the equipment can be moved in a small handcart.

Evidence of past erosional events

Many geomorphic processes occur too frequently for soil profiles to develop distinct horizons in the deposits derived from those events. The evidence for the events may then be recognized from slight horizon development, bedding in the deposit, separation of deposits by lenses of material from other sources, incorporation of datable material such as logs or tree stumps, survival of stone lines,

deposits of charcoal, and formation of sedimentary structures in the deposit. A few examples of such features are given below.

Landslides may be discrete, catastrophic events or, as with some large failures, very slow or episodically moving. Discrete events are commonly the result of major storms or periods with climatic regimes which are conducive to erosion. Such periods may be separated from each other by more stable conditions in which slope stability, with soil-horizon development, prevails. Slope instability (Fig. 10.6) with landsliding often creates conditions which are favourable for further erosion. A landslide scar may undercut the slope above and become a collecting area for water or of slope debris which is uncompacted and poorly drained and therefore likely to slide. Footslope debris may form a deposit of weak material which is prone to rill and gully erosion and it may cause the water-table in the lower slope to rise, thus destabilizing the midslope and promoting a further landslide. A period of more stable climatic conditions may

PLATE 10.4. Three bodies of sediment are evident in this exposure. At the base are nearly horizontal bedded tuffs which had a gully cut into them. The gully was subsequently infilled by material washed into it leaving bedded sediment dipping from top-right to bottom-left. An eolian deposit has subsequently been laid down as a mantle covering both older materials. The present groundsurface is accordant only with the most recent mantle.

follow in which soil may develop on landslide scars and footslope deposits.

The evidence for the period of instability is most obviously in the erosion scars which may be visible long after vegetation and soil have obscured the form of the scar; soil profile development is also likely to be more limited in the scars than on the surrounding slopes, and the long profile of the eroded slope may be distinctly steeper towards the upper slope and of lesser gradient at the footslope. Confirmatory evidence will be found primarily on the footslope where lobate forms, perhaps cut by gullies, may survive and in these deposits it may be possible to distinguish landslide events, if they were well separated in time and soil formation has occurred between them, or lenses of slope wash, tephra, loess, or other sediment has stratified the deposits.

The periodicity of events is often difficult to establish. Local records, maps, and air-photos of successive periods may aid recognition of very recent events. Older events may be recognizable from ^{14}C dating of paleosols (see Matthews 1985) or wood in the debris; vegetation established on the footslope deposit may be of a uniformly young age compared with that on most of the hillslopes and in forested areas trees may be datable by their growth rings (dendrochronology). Tree-growth rings may also yield evidence of the climate itself. If the landsliding event is part of a major and widespread erosional period then regional evi-

PLATE 10.5. An inclined hillslope on chalk in which solution hollows formed under cold climate conditions and were later infilled with unbedded soliflual clay containing flint and chalk clasts, northern France.

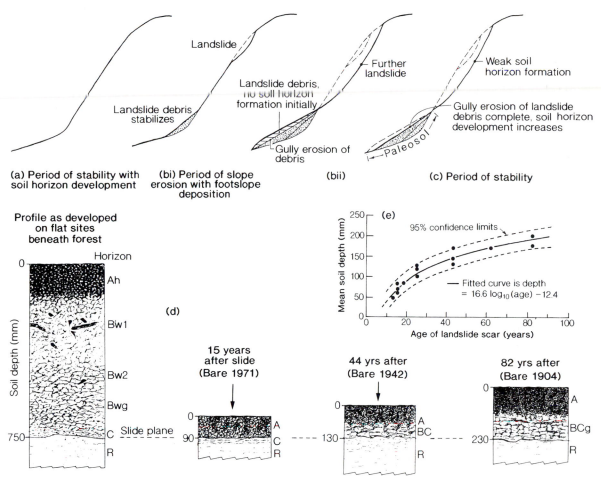

FIG. 10.6. (a–c) Development of a hillslope profile as a result of periodic instability and stability. (d) Soil profiles as developed under original forest on spurs, and in landslide scars after removal of forest and sowing of pasture grasses. The sharp break between the A and C horizon of the youngest soil indicates the position of the old shear plane. In the older soils the potential shear plane is likely to develop between the BC or C horizon and the bedrock (R), especially where the soil develops on impermeable rock. The soil depth chronofunction is shown in (e). (d, e after Trustrum and DeRose 1988.)

dence of valley-floor deposition may be present (see Chapter 16).

A detailed study of soil-horizon development in an area of periodic landsliding (Trustrum and De Rose 1988) has shown that chronosequences can be recognized in the soils (Fig. 10.6d) and that these sequences can be calibrated against evidence from old records and photographs. Soil formation in landslide scars is primarily a result of bedrock weathering and accumulation of colluvium derived from breakdown of exposed bedrock, and crumbling of soil of headscarps. Mean soil depth has increased from 50 mm on 15-year-old scars to

200 mm on 82-year-old scars. The rate of soil-depth increase averaged 3.5 mm/y over the first 40 years after landsliding but declined to 1.2 mm/y over the following 50 years, indicating a logarithmic chronofunction.

This study indicates not only the relationship between geomorphic and soil features but that the history of erosion can be interpreted from a study of the soil profiles and their pattern in the landscape.

Wind-throw of trees, with or without burning, has been known to induce periods of erosion in uplands (e.g. Grant 1963). The evidence for wind-

(a) Pit and mound topography

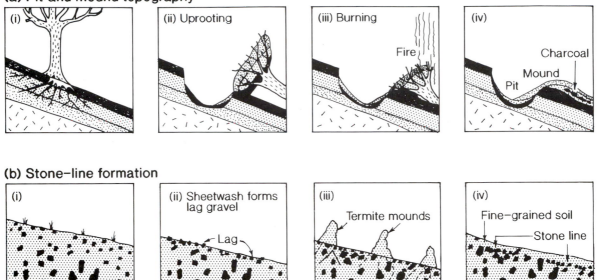

(b) Stone-line formation

FIG. 10.7. (a) The effect of wind-throw of trees on soil horizons and landsurfaces. (b) Formation of local stone lines. (Based, in part, on Johnson 1990.)

throw may be preserved as hummocky ground where trees have been wrenched from the soil, leaving a pit, and soil from the roots has created a mound alongside the pit (Fig. 10.7a). Soil profiles may retain a record of such events in the resulting variability in horizon thickness. Where forest burning has occurred, as a result of lightning strikes, volcanic or human activity, lenses, or patches of charcoal may be preserved in the soil; such material can be dated by ^{14}C. Pit and mound topography may be preserved in the landscape for more than 1000 years, although 200–500 y is more common (Schaetzl and Follmer 1990). If forest cover is removed, the evidence is less likely to survive than if the area remains forested.

Stone lines are relatively common features of soil profiles in semiarid savanna and humid tropical environments (Johnson 1990) and especially in Africa (Fig. 10.7b, Plate 10.6), they also occur in many other environments but usually as discontinuous features.

Stone lines occur in a variety of forms ranging from thick (>1 m) bodies of pebbles and blocks to single grains of bedrock, ferricrete, calcrete, or transported rock fragments forming distinct layers

or lenses in soils of large areas (some km^2) or local patches (a few m^2). Stone lines are distinct from the soil materials above and below, both by their concentration of large fragments and by the relatively fine-grained materials in which they lie. In a particular site there may be more than one line separated from others by layers of fine-textured soil.

Stone lines may be caused by several processes: extensive and uniformly thick lines are almost invariably relics of former lag-gravels which originally developed as groundsurfaces where surface wash or wind was a dominant process responsible for removing fine-grained soil. The fine-grained soil, now overlying the stone line, is a result of the activity of burrowing animals carrying that soil through the lag to the groundsurface and depositing it in mounds. Earthworms, ants, and termites (Plate 10.7) may be effective over large areas. Termite numbers typically average 1000–4000/m^2 in tropical savannas and rainforests, and in extreme cases their mounds may cover 30 per cent of the land area and more than 1 tonne/ha of fine earth may be deposited by them on the surface annually (Crossley 1986). The transfer of soil from

PLATE 10.6. The modern groundsurface is formed by a thick lag-gravel overlying a bed of fine-grained soil. A thin stone line (between the arrows) has been formed below the upper bed of fine soil and above the lower bed of fine soil. A paleosol below the stone line is indicated by the darker tone and structure of the material immediately below it. The area has a thorn-scrub savanna vegetation, South Africa.

depth to the surface may cause the soil above the buried stone line to thicken at average rates of 25–200 mm/ky.

Virtually all animal mounds are subject to erosion by wash and the fine material is gradually dispersed over the ground. Inevitably the greatest burial will occur on flat depositional sites with lag-gravels continuing to form the groundsurface on steeper slopes.

Local lensoid and discontinuous stone lines may be caused by tree-fall with blocks of bedrock or other stones carried to the surface by the roots, by gophers, rabbits, and other burrowing animals which live in warrens and other small local communities. Stone lines may also represent old groundsurfaces buried by eolian, colluvial, or other sediments. Whatever their cause, stone lines can be used to give indications of ground stability and instability.

Hillslope form

The shape of a groundsurface can be identified by consideration of its profile form with respect to a vertical plane and of its plan form normal to the profile and therefore in a horizontal plane parallel to the contours.

Profile and plan forms together provide a three-dimensional image and record of hillslope forms which are made up of a combination of planar, concave, and convex units (Fig. 10.8). Contour maps provide a very convenient source of information on the shape of large slopes, but few general-purpose contour maps are sufficiently accurate for the special purposes of geomorphological research. The techniques available for recording hillslope profiles are described in the very important bulletin by Young (1974), in Goudie (1981), and Dackombe and Gardiner (1983); Parsons (1988) has made

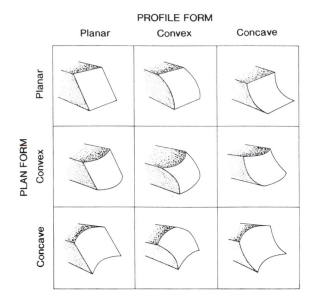

PLATE 10.7. A large termite mound, about 2.5 m high, in northern Australia, indicating the considerable volume of soil brought to the surface. Note the stony soil of its surrounds which will be buried when this mound is dispersed.

FIG. 10.8. The nine possible shapes for hillslope units. (Based on Ruhe 1975, as classified by Parsons 1988.)

critical comment on methodology and the use of statistical treatment of data.

Field-workers should be aware that substantial error can be introduced into hillslope surveys by: (1) failure to 'fix' a relocatable and mapped start-point and end-point of every profile; and (2) failure to record the profile along the true slope, rather than the apparent slope. True slope is the angle of surface gradient of the ground in the direction of maximum gradient. Following the true slope may mean that the profile changes direction at several points along its length.

The methods chosen should be compatible with the accuracy needed for the investigation. Survey by theodolite and stadi-altimeter is very accurate;

for other purposes simple levels, such as the Abney, provide adequate data. Automatic profile recorders—which are, in effect, wheelbarrow-mounted recorders of distance and inclination—are expensive, sometimes difficult to maintain, but ideal for repetitive work on slopes of modest inclination and roughness.

Profile recording is an important part of investigations of the relationships of hillslope form with: processes; soil characteristics; bedrock features such as weathering grade and jointing; slope stability; hydrology; vegetation cover; and soils. Separate profiles may be compared as a way of assessing hillslope response to varying degrees of undercutting or, where the age and initial form of a hillslope is known—as it sometimes is where the hillslope was formed on a spoil-heap—the effect of erosional and depositional processes on hillslope form. Frequency distributions of hillslope units of a specified gradient can be presented as histograms in order to identify characteristic gradients (Young 1961), but such studies require a careful assessment of the statistical treatment of the data if valid conclusions are to be drawn from profile data. Sampling procedures, representativeness of those samples, and unbiased definition of the slope units which are recorded, must be given critical consideration.

If hillslope forms result from the activity of a single, or limited number, of processes it becomes important to classify hillslope units so that correlations between form and process can be explored. Both Blong (1975) and Parsons (1978) have attempted to classify hillslope morphological attributes; both studies subdivided profile form into four components: size, shape, gradient, and roughness; but little success has been achieved in producing a universally applicable set of hillslope attributes, nor has much success been achieved in the more limited objective of establishing significant correlations between landslide attributes and hillslope attributes (Blong 1974). The study of hillslope forms for analytical purposes appears to lag behind advances in understanding the mechanics of processes and the causes of soil and rock resistance to those processes.

Field-investigation methods

Introductions to methods of field investigations for geomorphological purposes are given in Goudie (1981) and Dackombe and Gardiner (1983). The International Association of Engineering Geology (IAEG) set up a Commission on mapping and the classification of rock and soil for mapping purposes; its reports (Matula 1981*a,b*) are useful guides which overlap in content with the manuals mentioned above, and for some purposes extend them. A Working Party of the Geological Society (London) has produced a report (1982) on 'Land Surface Evaluation for Engineering Practice' which is a comprehensive guide to mapping for many geomorphological and engineering purposes. Many countries have national codes or government agencies which have published manuals on field procedures; one of the most comprehensive of these is the US Department of the Interior, Bureau of Reclamation, *Engineering Geology Field Manual*.

11

Water in Soils and Hillslope Hydrology

Hillslopes with a soil cover undergo a great variety of weathering, soil formation, soil erosion, mass wasting, and deposition processes. The energy for these processes is that of gravity and received solar radiation and the agent for change is, almost invariably, water. The action of water and ways in which its action is modified by vegetation, soils, slope angle, and surface relief are thus the focal points of studies of slope processes.

Water in soils

The physical properties of water result from its asymmetrical arrangement of hydrogen and oxygen atoms (Fig. 2.3) and its resulting dipolar molecules. The importance of these properties for the action of water as a solvent (Fig. 8.10a), as a contributor to cohesion through the attraction of water molecules to each other (Figs. 2.3, 5.1) and to adsorption (also called adhesion) through the bonding of water molecules to surfaces of solids (Figs. 3.12, 5.1) have been discussed in earlier chapters. The development of positive and negative pore-water pressures has also been discussed (Fig. 5.11).

The movement and retention of water in soil is influenced by cohesion and adhesion acting to create surface tension and capillarity.

Surface tension and capillarity

Surface tension is a phenomenon developed at liquid–air interfaces and results from the greater cohesive attraction of water molecules for each other than for the air (Fig. 11.1a). The net effect is that water molecules close to the interface have a higher energy than those in the bulk of the water and an inward force at the water surface causes the water to behave as if it were part of a stretched elastic membrane. This surface tension is greater in water than in most other liquids. It is an important property in the phenomena of capillarity and water retention in soils.

Capillarity is commonly observed in the 'wick effect' in which water moves up the wick against gravity or up a narrow tube which has been dipped into water. If a series of tubes of increasing diameter is dipped into water the greatest rise will occur in the tube of least diameter (Fig. 11.1b). This phenomenon of capillarity is due to the adhesion of water to the walls of the tube, creating a meniscus (Fig. 11.1c), and to the cohesion of water molecules to each other. Water will continue to rise in the tube until the weight of water in the tube counterbalances the cohesive and adhesive forces.

The height (h_c) of rise in a capillary tube is inversely proportional to the internal tube radius (r) and weight of the water ($\rho_w.g$) and directly proportional to the surface tension (T):

$$h_c = \frac{2T}{r.\rho_w.g} \qquad (11.1)$$

As the unit weight of water is a constant, this equation approximates, when h_c and r are expressed in centimetres, to:

$$h_c = \frac{0.15}{r}. \qquad (11.2)$$

The rise is greatest in a tube with the smallest radius of curvature of the meniscus.

The phenomenon of capillarity explains both the effect of suction, or negative pore-water pressures, in a moist, but unsaturated, soil (see Figs. 5.5, 5.11) and the capacity of fine-grained soils to hold water in micropores while coarse-grained and macroporous soils drain freely. Soil, however,

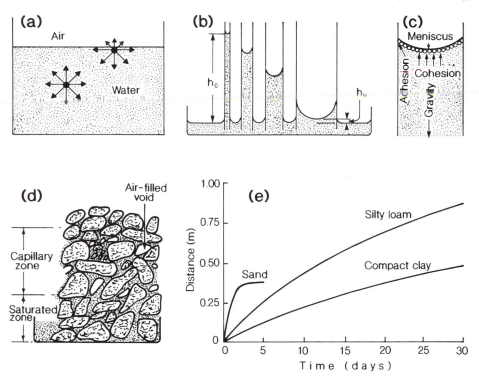

FIG. 11.1. (a) Forces acting on a water molecule within the body of the fluid are equal in all directions as each molecule is attracted equally by neighbouring water molecules. A molecule at the water surface is attracted more strongly to neighbouring water molecules than to those of the air. Consequently, there is a net downward force on surface molecules of water, creating surface tension. (b) Capillary rise is greatest in the narrow tube, in which the radius of curvature of the meniscus is least. (c) Forces acting between the wall of a tube and water molecules create adhesion, and the mutual attraction of water molecules creates cohesion. (d) Irregularities in pore sizes and distributions create irregular boundaries of the capillary. (e) The rate and height of capillary rise above a saturated zone depends upon soil pore size. (Based on Brady 1990.)

does not behave as a set of capillary tubes. Its channels are tortuous, variable in diameter, contain entrapped air and are discontinuous. The rate of movement of capillary water is therefore variable and often slow (Fig. 11.1e). The rate of capillary rise from a water-table is greatest in coarse-grained soils, but the maximum height attained is greatest in medium-grained silts and loams over periods of a few weeks. If many months are available, the greatest rise will occur in dense clays (Table 11.1). In soils, capillary movements occur in all directions, not just vertically.

Soil-water potential

The retention and movement of water in soils is governed by the energy available. That energy may be potential, kinetic, or electrical. The term 'free energy' may be used to indicate the summation of all available forms of energy to do work. The

free-energy level in a substance is then a general measure of the tendency of that substance to change. All substances tend to move from a state of higher to one of lower free energy. In soils this trend is commonly recognized as a tendency for water to move from wet to drier zones. If the *differences* in energy level between two contiguous zones are known, it is possible to predict the direction of water movement.

For practical purposes in soil science, the differences in energy levels between one site and another are more important than the absolute free-energy levels. This is perhaps best understood by reference to a hydro-electricity generating plant: water in a reservoir above a power station has a potential energy which is proportional to the difference between the height of the water surface of the reservoir and the height of the turbines. It is not the absolute height of the reservoir and

TABLE 11.1. *Typical height of capillary rise above a water-table given unlimited time*

Soil	Pore radius (mm)	Capillary height (mm)
Coarse gravel	0.4	0.038
Coarse sand	0.05	0.3
Fine sand	0.02	0.7
Silt	0.001	1500
Clay	0.0005	3000

Source: Fetter (1980), *Applied Hydrogeology* (Columbus, Ohio: Merrill).

TABLE 11.2. *Matric potential in common units for selected soil-water conditions*

Condition	Matric potential			
	(kPa)	(m)	(bar)	pF
Saturation	-10^{-1}	-10^{-2}	-10^{-3}	0.0
Field capacity	-10	-1	-10^{-1}	2.0
Wilting point	-1.5×10^3	-1.5×10^2	-1.5×10	4.2
Air dry	-10^5	-10^4	-10^3	6.0

Source: Marshall and Holmes (1988), *Soil Physics* (2nd edn.) (Cambridge University Press).

turbines above a datum level which is of concern. In soils the potential energy, known as the soil-water potential (ψ), has four major components:

$$\psi = \psi_g + \psi_m + \psi_p + \psi_o, \qquad (11.3)$$

where ψ_g is the gravitational potential resulting from the elevation of a point of interest above another point of interest in the soil ($\psi_g = gh$); ψ_m is the matric potential of the soil matrix (i.e. solids) due to adsorption and capillarity; ψ_p is the pressure potential developed as a result of the pressure head due to the pressure of water in the saturated zone above a point of interest (in an unsaturated soil $\psi_p = 0$, and in saturated soil it has a positive value); and ψ_o is the osmotic potential due to the attraction of ions, in solutions, for water. Osmotic forces tend to reduce the free energy of water in a solution as 'pure water' is drawn towards the ions in that solution. Osmotic potential does not greatly affect water movement in non-saline soils, but it is extremely important in water uptake by plants.

A common means of expressing soil-water potential is in terms of the height (in metres) of a unit water column whose weight just equals the potential under consideration. Alternatively it may be expressed in terms of standard atmospheric pressure at sea-level and, in the SI system, in kPa, (10 metre column of water $\sim$ 1 bar $\sim$ 1 atmosphere ~ 100 kPa). In unsaturated soils the work done in transferring water from one point to another is against a suction and values are therefore negative. The term pF, meaning the logarithm to the base 10 of the suction expressed in cm, also remains in use. The matric potential for important soil-water conditions is given in Table 11.2 (Fig. 11.2a).

Soil-water content and tension

The *amount of water* in a soil is commonly measured as a mass or volume of water within a mass or volume of soil solids. The simplest procedure is to collect a soil core in the field, with a coring tube of known volume and mass; the core is weighed and then dried in an oven at 100–110 °C for a standard time, and then weighed again. The water lost by the soil represents the soil moisture in the moist sample. This is known as the gravimetric method.

Various methods exist for indirect measurements in the field. The resistance method is based on the recognition that certain porous materials, such as gypsum, nylon, and fibreglass, have an electrical resistance which is related to their water-content. Blocks of these materials can be placed in the soil over long periods and their absorbed water content, which is in equilibrium with the soil-water content, read from an attached meter or data-logger as required. Neutron moisture probes contain a source of fast neutrons and a slow neutron detector. The probe is lowered into the soil through an access tube. The emitted fast neutrons collide with the hydrogen atoms of water and lose part of their energy. These slowed neutrons are measured with a detector tube and a scalar. The reading is related to moisture content.

The *tension* or suction with which water is held in soils is an expression of soil-water potential (ψ) except that it may be expressed in positive rather than negative terms. Field tensiometers are tubes with a porous cup at the lower end. The tube is filled with water and the tension recorded results from the drawing of water through the cup into the soil.

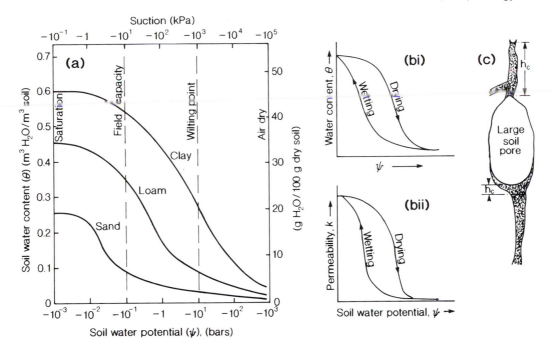

FIG. 11.2. (a) Soilwater potentials for soils of different textures derived from slowly drying saturated soils. Whatever the texture, the field capacity, defined as the moisture content of a freely drained soil which has not lost water by evaporation or to plants, and the wilting point of plants, are best defined in terms of soilwater potential, as they differ in water-content for different soils (based on Brady 1990). (b) Hysteresis effects on (i) water-content, and (ii) permeability. (c) Variation in pore diameter causes variations in height of capillary rise on the wetting curve; note that the upper pore is drawing its water from a narrow lateral pore, not from the large pore below it.

As the water-content (θ) of the soil decreases, the matric tension increases (Fig. 11.2a). The re-wetting curve, however, does not follow the drying curve; the drying soil has a higher water-content than the wetting soil at the same suction. This hysteresis effect is commonly explained by the fact that soil pores are variable in diameter with many containing voids larger than their openings. The so-called ink-bottle effect is illustrated in Fig. 11.2c. It is shown in Fig. 11.1b that capillary rise in a wide tube is small and much higher in a narrow tube. As water-content declines, the meniscus in a large tube will fall rapidly to a level which is adjusted to the radius of the meniscus, and will be held against gravity in a narrow part of a pore where the radius of curvature of the meniscus is smaller. Where water is re-wetting a soil it cannot rise through a wide pore by capillarity; the rise is delayed until the water is forced into the wide pore by hydraulic pressure. The need for hydraulic pressure delays the re-wetting in contrast with the

relatively unimpeded drainage possible during drying.

A rather similar effect occurs at the boundary between soil layers of different pore diameter and particle size: water will not move by capillarity alone from a fine-grained microporous soil into a coarse-grained macroporous soil; an hydraulic pressure is required to induce such flow and in unsaturated conditions this is usually unavailable and the texture boundary forms a barrier to soil-water flow.

Soil-water movement

Three types of water movement occur in soil pores: saturated flow, unsaturated flow, and vapour movement. Vapour movement is of little importance for geomorphic processes and will not be discussed further.

The flow of liquid water is due to a gradient in the total water potential (ψ) from one soil zone to

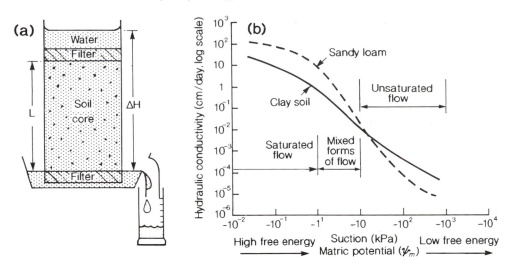

FIG. 11.3. (a) Definition of hydraulic gradient. The arrangement shown here represents the principles used in the design of a simple permeameter. (b) A generalized relationship between matric potential and hydraulic conductivity for two soils (note the logarithmic scales). Saturated flow occurs at near zero potential and unsaturated flow at potentials (suctions) of tens to hundreds of kPa. Because of variations of pore size in most soils, saturated and unsaturated flow may occur at the same time over a narrow range of average potential. (Based, in part, on Brady 1990.)

another. The direction of flow is from a zone of higher potential to a zone of lower potential.

Saturated flow takes place where the soil pores are completely filled with water. Saturation is common in the lower horizons of poorly drained soils, in portions of well-drained soils above stratified clays, and in the upper horizons of most soils during or immediately after heavy rain.

The flow of water under saturated conditions is determined by the hydraulic gradient and the hydraulic conductivity of the soil. The hydraulic gradient (i) is indicated in Fig. 11.3a as:

i = head loss/drainage path length, or $\Delta H/L$.

Hydraulic conductivity is a measure of the ease with which soil pores permit water movement through the soil. In saturated soils, hydraulic conductivity (k) is equivalent to permeability and k is called the coefficient of permeability (Table 11.3). The coefficient of permeability is the rate of flow of water per unit area of soil when under a unit hydraulic gradient; its dimensions are those of velocity (m/s).

If Q is the quantity of flow passing through an area A in time t (with units of m^3/s), then the average velocity v is given by Darcy's law:

$$v = \frac{Q}{A}, \quad \text{or,} \quad v = ki,$$

or

$$Q = Aki,$$

or

$$k = \frac{Q}{Ai} = \frac{Q}{A \cdot \Delta H/L}, \qquad (11.4)$$

then, for saturated soils:

$$k = \frac{QL}{A \cdot \Delta H}. \qquad (11.5)$$

In the absence of a hydraulic-pressure gradient the flow would be substantially less. The presence of a pressure gradient allows saturated flow to occur horizontally and upwards, as well as downwards.

It has been shown that in clay soils of very low permeability, Darcy's law is not strictly applicable because of adsorption effects. None the less it is used for most practical purposes.

The permeability of a soil varies with effective pressure and void ratio and will be influenced by soil stratification. In many soils, permeability is greater in a horizontal direction through one soil horizon than it is in a vertical direction through several horizons, which may vary in texture and structure.

A commonly accepted approximation is that

TABLE 11.3. *Coefficient of permeability (k, m/s)*

Gravels	$>1 \times 10^{-2}$
Coarse sands	1×10^{-5} to 1×10^{-2}
Fine sands	1×10^{-7} to 1×10^{-5}
Silts	1×10^{-9} to 1×10^{-7}
Clays	1×10^{-9} to 1×10^{-11}

void ratio is linearly related to the logarithm of the coefficient of permeability. The total flow rate (q) in soil pores is proportional to the fourth-power of the pore radius:

$$q = \left(\frac{\pi r^4}{8\eta}\right)(\rho_w g . \Delta H/L), \qquad (11.6)$$

where η is the kinematic viscosity and ρ_w is the density of water. Thus the size of pores has a very large effect on soil hydraulic conductivity. For example, the flow rate through a pore 1 mm in radius is equivalent to that in 10 000 pores of radius 0.1 mm, for the same hydraulic gradient, even though 100 pores of radius 0.1 mm have the same total cross-sectional area as a 1 mm pore. The macropores clearly account for most saturated flow in soils. Open granular structures in soils are far more conducive to flow than compact platy structures and earthworm and termite burrows form relatively large channels. Straight pores aligned parallel to the direction of flow are also more conducive to flow than tortuous channels and those aligned across the direction of flow.

Unsaturated flow is normal for most soils for most of the time. It is characterized by a lack of hydraulic gradient, absence of water in large soil pores, and the presence of water in small pores where it is held by adhesion and cohesion as adsorbed and capillary water. In an unsaturated soil, therefore, it is the matric potential (ψ_m) gradient from one zone to another which acts as the driving force for water movement. Flow is through the films of adsorbed water and through capillary pores. As water tends to seek an equilibrium, it flows from a zone of high potential to a zone of lower potential, but very slowly.

For any particular soil there is a relationship between the hydraulic conductivity and the matric potential (Fig. 11.3b). Near saturation the conductivity is high, but as the potential decreases the size of the pores which are filled with water decreases, thereby reducing the conductivity. Hydraulic conductivity is therefore a function of both water-content and matric potential.

Sandy soils have high conductivities near saturation because of their large pores. Near the air-dry state they have very low conductivities because they lack fine pores. Clays near saturation have lower hydraulic conductivity than sands but higher conductivities when unsaturated.

A modified form of Darcy's law can be applied to unsaturated soils (see standard texts on soil physics such as Marshall and Holmes 1988; Hillel 1971); permeability measurements are best done in the field where there is less chance of modifying soil structure and compaction in the testing process. Frequent monitoring of changes in soil-water content by neutron probe, and water potential by tensiometers, during drainage is a useful method. In general, the higher is the initial water-content of a soil, the greater is the potential and the more rapid is the water movement from wetter to drier zones. This point is of great importance for an understanding of infiltration (see below).

Hillslopes in the hydrological cycle

Water reaches the ground by falling from clouds as rain, snow, hail, or by the condensation of dew and fog—these forms of water are collectively called precipitation. Some of the falling precipitation may directly hit soil or rock or fall into the streams, but in a humid region much hits the vegetation and is intercepted. Part of the intercepted water is evaporated back into the atmosphere, both during and after rainstorms, but in prolonged or intensive rain some water falls between the plants, or drips from the leaves to the ground, some flows down the plant stems, and some may be absorbed by the plants. Of the water which reaches the ground part may flow off directly into the streams but, except on frozen or bare rock surfaces, most of the surface water infiltrates into the soil or is temporarily ponded in depressions on the surface. The water which infiltrates into the soil first fills the voids in the soil and, if the soil is very impermeable, any excess water must flow off the soil surface—as overland flow—but, if the properties of the soil profile permit, subsurface water may be taken up by plants, or may flow laterally as throughflow, or may move down vertically to become part of the

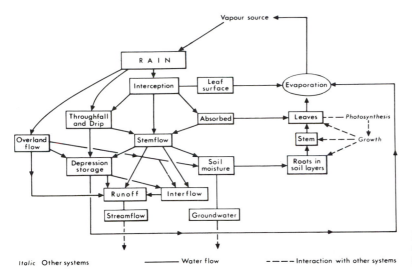

FIG. 11.4. A schematic representation of the links between components of the hydrological cycle.

groundwater. Both throughflow and groundwater may eventually reach the streams occupying valley floors. The water of streams flows into the oceans from where it may again be evaporated and then condensed into cloud droplets, and be precipitated on the land. This system of circulation of water between the atmosphere, the land, and the oceans is called the hydrological cycle. The main elements relevant to hillslopes are indicated in Fig. 11.4.

Interception

As rainfall is intercepted on its path to the soil the water may drip off the leaves of the intercepting tree or shrub, or it may run down the trunks of the plants as stemflow. When interception occurs at the beginning of a storm the leaves are dry and water may be directly absorbed into small depressions in the bark and the crotches of trees, or evaporated from plant surfaces. The capacity of plants to store water, and lose it by evaporation, declines with increasing duration of a storm, but evaporation losses from trees may be significant even during prolonged storms. Evaporation losses from water stored on plants will continue after the storm.

It is impossible to draw quantitative universally applicable conclusions about interception losses because they depend upon various meteorological factors: the duration, amount, intensity, and frequency of rainfall; windspeed; pre-existing water on plant surfaces; air temperature; and the type of vegetation. The bulk of the experimental evidence indicates that interception losses are generally greater beneath evergreen than beneath deciduous trees. Winter and summer losses beneath evergreens are usually about the same but winter losses beneath deciduous trees are much lower (about 4–7 per cent) than the summer losses (10–15 per cent) (Lull 1964).

A high initial loss from initially dry plant surfaces, with a decline to a much lower loss, is normal for all reported studies (see Figs. 11.5 and 11.6). In New Zealand, Aldridge and Jackson (1968) found that manuka intercepted 30 per cent of the total gross rainfall. Stemflow accounted for about 23 per cent of the gross rainfall and throughfall about 39 per cent. The rainfall measured at the groundsurface increased with gross rainfall amount and intensity. Interception losses beneath grasses and crops are usually in the range of 10 to 20 per cent and beneath forest 5 to 50 per cent. Initially wet plant surfaces reduce initial and total interception losses.

Not all effects of vegetation on moisture balance are negative; in cloudy and foggy areas condensation on leaves and the resulting drip can account for a large proportion of the moisture supply; some so-called cloud forests get nearly all their moisture this way. Methods of measuring stemflow and throughfall, are shown in Plate 11.1.

Infiltration

Infiltration is the process by which water enters the surface horizon of the soil. Percolation down to the

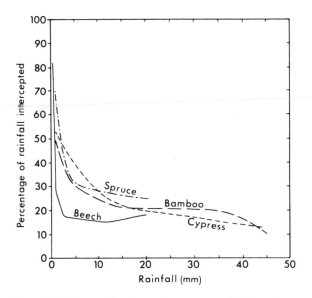

FIG. 11.5. Interception by various tree species (data from Lull 1964).

tion; (f) cracks; (g) slope; (h) freezing; (i) cultivation; (j) swelling clays which may block pores; (k) low soil pH (where the soil cation-exchange complex is dominated by H^+ ions, soil particles disperse, but divalent cations such as Ca^{2+} encourage flocculation and soil structure development); (l) animal trampling; (m) vegetation; (n) litter; (o) soil organisms; and (p) soil moisture. The actual controls on infiltration into a particular soil depend upon the local operation of these factors in various combinations and magnitudes.

At any site the rate at which water infiltrates is determined by: (1) the amount of water available at the soil surface; (2) the nature of the soil surface; and (3) the ability of the soil to conduct water away from the surface. Water will normally only enter the soil body when water films exist on the surfaces of grains and peds at the ground surface. Some soils have high resistance to wetting, that is they are said to be hydrophobic, because of the presence of organic resins or they have lost adsorbed water and surface electrical changes as a result of extreme heating, especially by burning of vegetation. Very deeply cracked soils may receive water into the cracks irrespective of any influence of adsorbed water. At the other extreme, soils which are already saturated may not be able to transmit further water because of low permeability or maintenance of saturation by throughflow from upslope. Rain falling onto saturated soils, which cannot drain, will therefore become overland flow.

water-table occurs beneath the surface horizons. Infiltration may be controlled by a variety of factors including: (a) the type of precipitation, especially its intensity; (b) surface-soil compaction; (c) soil porosity, especially at the surface; (d) splashing of fines on a bare surface to form a crust which blocks soil pores; (e) depth of surface deten-

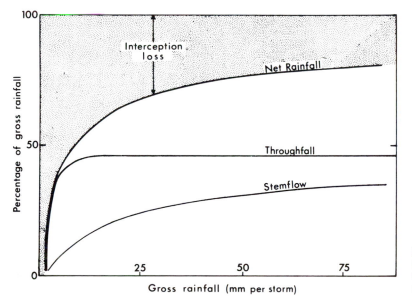

FIG. 11.6. Distribution of rainfall after it has reached the canopy of 5 m tall manuka (*Leptospermum scoparium*). (After Aldridge and Jackson 1968.)

PLATE 11.1. Experimental plots with long troughs for catching throughfall and collars for catching stemflow. The segmented collar in the inset is for measuring fall and drip close to the trunk.

The most widely used expression for infiltration is that of Philip (1957):

$$f = k + (B/2t)^{1/2},$$

where: f is the maximum infiltration rate, or infiltration capacity at time t after water is impounded on the soil surface; B is the sorbtivity of the soil (which indicates the capacity of the soil to absorb water), and k is the hydraulic conductivity. The parameter B is obtained from the rate of downward penetration of the wetting front.

In the early stage of infiltration into an unsaturated soil, sorbtivity (B) is the dominant term and can be described as the initial infiltration rate which is that of rapid flow into empty cracks and macropores in response to a gravitational and

matric potential gradient; it is therefore influenced by the initial water-content and particularly films of adsorbed water on the walls of macropores. Sudden drops in infiltration rate can occur where potential gradient falls rapidly as unsaturated soil in peds is wetted (Davidson 1985).

At later stages the parameter k becomes the increasingly dominant term until at saturation it is equal to the saturated hydraulic conductivity. The value of k can be approximated from the measured rate of infiltration once this has reached a steady state (Figs. 11.7a, 11.8).

After initial wetting the surface horizons form a transmission zone above a wetting front. Existing soil water is displaced downwards out of the macropores by newly infiltrating water which

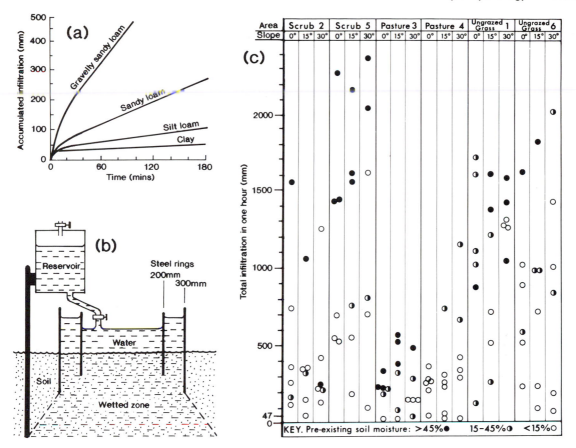

FIG. 11.7. (a) Curves indicating initial high rates of infiltration and then reaching steady rates, for different soil types. (b) A double-ring infiltrometer. A constant head is maintained in the inner ring and the amount of water lost by infiltration is recorded. The outer ring creates a wetted zone, around the soil being wetted from the inner ring, in order to prevent lateral flow from the inner ring. (c) Total infiltration in one hour into pumice soils with three types of vegetation, three set angles of ground-slope angle and three levels of antecedent soil moisture. The variability is characteristic of soils with weakly developed profiles, such as occur on many hillslopes. Note the higher infiltration into moist soils. (After Selby 1970.)

moves as a pressure wave through the profile. In multilayered soils, hydraulic conductivities vary so that saturated layers may form above each less permeable horizon, and lateral throughflow may develop in the more permeable horizons.

A steady rate of infiltration is achieved when the entire profile is transmitting water at the maximum rate permitted by the least permeable horizon. This may occur within ten minutes of the onset of a storm or only after several hours. Excess water which cannot infiltrate is stored initially in surface depressions (depression storage), and once these are filled the excess spills downslope as overland flow.

In a soil which is very uniform in its properties,

slope, vegetation, and land use, a single curve may give a reasonable representation of the infiltration occurring during a storm, but for a soil with variable properties a family of curves gives a better indication of infiltration (Fig. 11.8). The extreme variability of hillslope soil infiltration rates both in space and over time has been illustrated by the work of Loague and Gander (1990). Variations over time are particularly common where clay peds swell and impede infiltration during storms, but do so at irregular rates (Dunne and Dietrich 1980). Seasonal differences in infiltration rates and processes may be illustrated by an example from South Australia (Smettem *et al.* 1991): in summer, soils under pasture grasses become hydrophobic

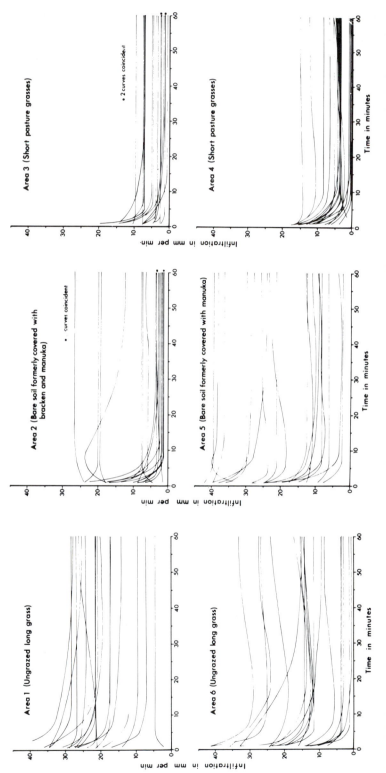

Fig. 11.8. Families of curves showing rates of infiltration into yellow-brown pumice soils beneath ungrazed long grass, short grazed grass, and a shrub vegetation of bracken and manuka. (From Selby 1970.)

as a result of extreme heating and drying of vegetation and overland flow is the dominant hydrological process, with little infiltration; winter is the season in which soil moisture is recharged but macropores are so widely open that water movement down them is rapid and effectively bypasses the soil matrix. The water passes through a clay-rich B horizon to the soil–rock interface and only when all macropores are filled is water drawn into the soil matrix and excess water becomes overland flow. Where the infiltration rate is highly variable from one soil plot to another a few centimetres away, the runoff from one plot of low infiltration capacity may well be absorbed by another plot of higher capacity lower down the slope. Surface runoff from one part of a slope does, therefore, not necessarily reach a stream as overland flow.

Land use practices and vegetation cover greatly modify the ability of a soil to absorb water. One pass of a tractor wheel, for example, has been known to reduce non-capillary pore space by half, and infiltration rate by 80 per cent (Steinbrenner 1955). Compaction by machinery or by trampling animals may be so great that infiltration into cultivated fields is markedly less than beneath nearby woodland. Wise land use is, therefore, aimed at reducing compaction, especially in critical areas in valley bottoms and near gullies where increased runoff could start rapid rill and gully erosion.

Reported infiltration capacities vary from 2 to 2500 mm/hour. The range of infiltration in relation to vegetation, slope, and pre-existing moisture content of a variable soil is indicated in Fig. 11.7c.

Methods of measuring infiltration rates are varied (see Hills 1970). The simplest device for use on hillslopes is a ring infiltrometer (Fig. 11.7b) consisting of a steel ring inserted into the soil and a small reservoir from which water is fed to pond on the soil surface. A constant head of water is maintained and the water drained from the reservoir in a known time gives the infiltration rate. A double ring may be used to reduce the amount of subsurface lateral flow. Sprinkling infiltrometers are used to simulate rainfall and have been used in varying sizes, especially in irrigation studies.

Concepts of runoff

The history of modern hillslope hydrology can be outlined by reference to a few research papers which identified the mechanisms by which water is transmitted from hillslopes to stream channels.

Overland flow generation was first described by Horton (1933) who regarded the soil surface as a site at which rainfall is divided into two components: some infiltrates, that which cannot infiltrate into the soil is ponded and then flows down the slope as overland flow. This infiltration-excess concept of overland flow was extraordinarily influential because it accorded well with the contemporary methods of predicting the generation of storm and baseflow discharges from drainage basins. It is still commonly termed 'Hortonian overland flow'. In modern terminology, the mechanism operates when the rainfall rate exceeds the infiltration capacity of the soil; because as the soil saturated-hydraulic conductivity is exceeded the soil surface horizon becomes saturated and the saturated zone propagates downwards; the soil infiltration capacity then decreases to a constant rate. The rainfall duration must be greater than the time required for ponding; this will be greatly influenced by the initial water-content of the soil.

Field observations indicate that Hortonian overland flow is a rare phenomenon, especially in areas with undisturbed vegetation cover and deep permeable soils. Overland flow is most readily generated in semiarid environments with thin, impermeable soils with low water-storage capacity, and in any environments where loss of soil structure (and therefore macropores) by compaction, removal of vegetation, freezing, and blocking of pores are associated with prolonged and/or high intensity rainfalls.

Field observations have been confirmed on theoretical grounds by Freeze (1972) who analysed soil hydraulic conductivities and rainfall frequency, intensity, and duration. He concluded that infiltration rates will be exceeded by rainfall rates only rarely in undisturbed, vegetated soils.

A modification of the Hortonian model of runoff generation, known as the 'partial area' concept (Betson 1964), drew the attention of researchers to the variability of infiltration rates and rainfall intensity across the landscape, and particularly to the evidence that only small areas of a landscape may contribute to overland flow.

A '*variable source area*' model came from recognition that in forested uplands, storm runoff may be generated by subsurface flows: or by surface flows from soils which are saturated to the soil surface (Hewlett 1961; Dunne and Black 1970*b*;

Dunne 1978*b*). Subsurface stormflow is now regarded as the major runoff-generating mechanism in most humid environments, both because of its influence on the development of saturated zones and as an important contributor to stormflow in its own right (Anderson and Burt 1978).

In areas of permeable soils where hydraulic conductivity decreases with soil depth, subsurface flow moves laterally as throughflow within the soil profile. When and where the profile becomes completely saturated, saturation-excess overland flow will occur. Both processes may occur at rainfall intensities and durations which are well below those required to produce Hortonian overland flow. Furthermore, both throughflow and saturation-excess flow may be generated from source areas which are variable in extent and different in location from source areas of Hortonian overland flow.

The influences on storm runoff and the recognized mechanisms are indicated in Fig. 11.9.

Runoff processes

Infiltration-excess overland flow

Overland flow was regarded by Horton as an explanation of sediment erosion and transport on hillslopes. He conceived it as developing a sheet of water, some way below the slope crest, and progressively increasing in thickness towards the slope base. The crest-slope area may therefore be subject to rainsplash erosion but not to sheetwash.

The difficulty with this concept is that sheets of water are virtually never seen on hillslopes; and even where they are observed, flow depths are varied, with water being concentrated in pathways which have the flow characteristics of sheetwash. In a series of experiments using simulated rainfall on sparsely vegetated slopes in a semiarid environment, Emmett (1978) was able to demonstrate the importance of pathways around vegetated mounds and pebbles as well as the washing of organic debris into small barriers. The failure of such barriers in succession may concentrate flow and permit entrainment and transport of debris and sediment. Natural slopes are usually so irregular that uniform sheetwash is impossible.

Locally, wash processes may be enhanced by overland flow on very impermeable surfaces of clay or bare rock and by stemflow from large trees during intense rainfalls (Herwitz 1986). Such

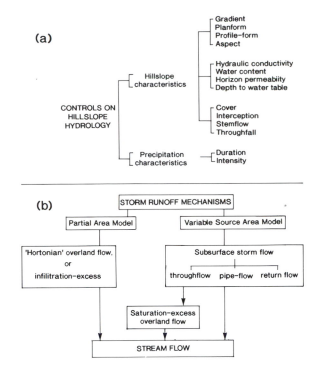

Fig. 11.9. (a) Controls on hillslope hydrology; (b) Storm runoff mechanisms.

phenomena emphasize the highly variable nature of infiltration and therefore of Hortonian overland flow on hillslopes. Notwithstanding the general validity of this comment, the importance of this process in small catchments in badlands and on cultivated slopes should be noted, but data from such sites should not be extrapolated to slopes with a full vegetation cover. Concentration of overland flow into rills and gullies enormously increases the power of water to entrain sediment. Rills form where slopes become sufficiently steep for sheetwash to be unstable and incise into loose sediment or where slopes are long enough for pathways to converge and form unstable concentrated flows.

Subsurface stormflow

Subsurface flow may be generated by two mechanisms: non-Darcian flow through macropores or pipes; and by Darcian flow through the soil matrix.

The importance of macropore flow is limited by its restriction to periods when all peds are filled by capillary water, i.e. they are at field capacity; it will therefore not occur at low rainfall intensities nor will it occur at low antecedent soil moisture

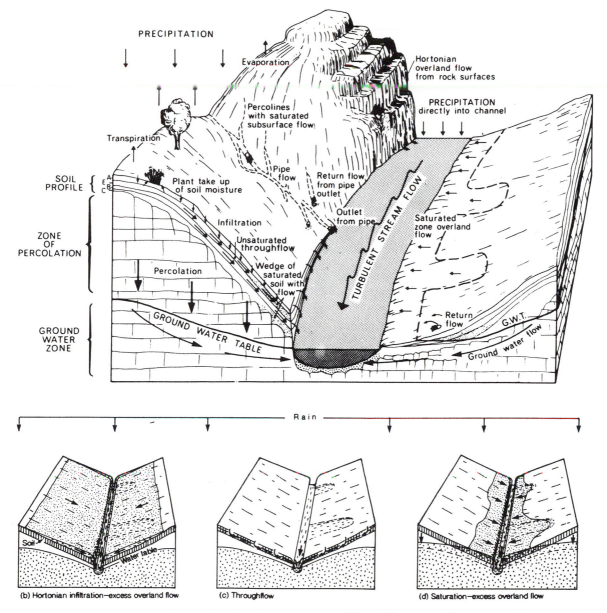

FIG. 11.10. (a) A schematic landscape with the various types of runoff from hillslopes and the sources and paths of runoff. (b–d) Detail of individual mechanisms. Note that the films of overland flow water are grossly exaggerated and unnaturally regular.

contents, for many soils this implies a water-content of about 30 per cent by volume (Coles and Trudgill 1985), and for allophanic soils about 80 per cent by volume. Studies using tracers are needed to confirm this theoretical view as the downslope connectivity of macropores is seldom known and some evidence suggests that little of the water draining from the bases of slopes during storms is from newly falling rain.

Flow through a soil matrix can occur most readily where upper soil horizons are very permeable and saturated, and where the water-table is

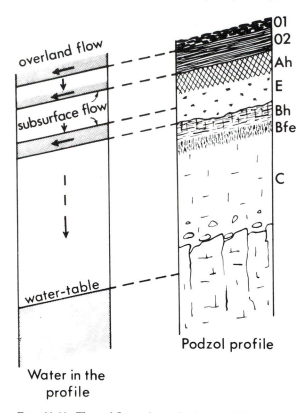

Water in the
profile

FIG. 11.11. Throughflow above horizons of lesser permeability, in a podzol.

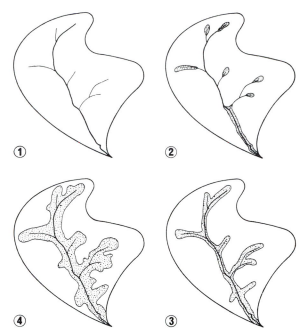

FIG. 11.12. An expanding source area for streamflow during a storm (dotted).

so high that the capillary zone (also called the capillary fringe) extends to the groundsurface. In the latter situation only a small amount of water is needed to cause the water-table to rise (Hewlett and Hibbert 1967; Anderson and Burt 1982).

Water-tables are likely to be near the groundsurface in concavities on the hillslope, in the concavities where midslopes and footslopes merge, and alongside stream channels. Such sites are also sites at which water from all sources converges. Geomorphic processes are therefore likely to be concentrated at such sites.

Note: the terms 'interflow' and 'throughflow' are both used to describe types of subsurface flow; they are not exact equivalents. Interflow was first used to indicate groundwater which re-emerges as overland flow or to indicate shallow groundwater flow; it has subsequently been used for subsurface flow of whatever origin. Throughflow is, specifically, that water which moves laterally through

relatively porous soil, above a less permeable body of soil or rock, or above a saturated zone (Fig. 11.11). 'Return flow' is a general term for re-emergence of subsurface water at the groundsurface. Methods of measuring subsurface flow are described by Atkinson (1978) and the process is analysed by Whipkey and Kirkby (1978).

Saturation-excess flow

Saturation-excess flow occurs on saturated sites into which water cannot infiltrate. Such sites vary in area seasonally and during storms (Dunne 1978*b*). The flow is commonly a mixture of return flow and direct runoff from rainfall. With very permeable soils the proportion of the catchment area yielding such flow will be small, but in humid areas with soils of low permeability the areas will be much greater and the proportion of water lost as stormflow may be large.

The incidence of severe floods arising rapidly in hill country catchments, which experience no Hortonian overland flow, emphasizes the importance of saturation-excess flow from catchment areas which expand during the period of a storm (Figs. 11.12, 11.13). Storm-runoff contributing

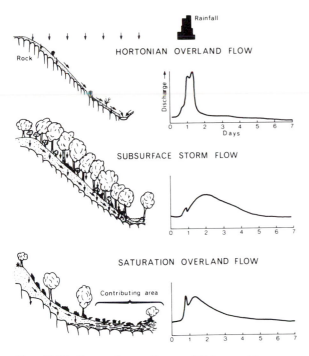

FIG. 11.13. Types of stormflow on hillslopes with stream hydrographs it generates. The initial rise in the hydrograph is produced by water falling directly into the channel. The shape of the recession curve indicates the storage and rate of drainage from the soil on the slopes.

areas commonly develop first alongside stream channels and in concavities and then expand as surface runoff occurs from operation of several processes. An extreme example is provided by Griffiths and McSaveney (1983) for a mountain catchment of 28.5 km^2 in the New Zealand Southern Alps. Slopes there have inclinations up to 70° with an average of about 30° and mean annual rainfalls exceed 10 000 mm/y. The average time between the beginning of effective rainfall and river rise during storms is only 0.5 h. Storm quick-flow durations are similar to rainfall durations and storm runoff accounts for 50–60 per cent of the storm rainfall. About half of the storm rainfall is stored, albeit for a short time. Five processes of runoff have been identified. (1) First overland flow occurs from bare rock surfaces; then (2) Hortonian overland flow occurs on soil-covered slopes. (3) Subsurface stormflow migrates through open rock joints, coarse talus, soil pipes, and permeable soils. (4) Saturation-excess overland flow takes place alongside channels and in swampy areas of valley floors. (5) Some rain falls directly into streams. The limited significance of Hortonian overland flow in mountainous regions, except where soils are frozen or large areas are of bare rock, is emphasized by the observations of Pathak *et al.* (1985) in the Himalayas, and of Cheng (1988) in British Columbia; in both areas subsurface flow is dominant in forested areas.

Pipe-flow

Flow in pipes has been greatly underestimated as a hydrological process, according to experimental work in a very small number of catchments (Jones 1987*a,b*; Bryan and Yair 1982; McCaig 1983). It is now recognized that subsurface natural pipes exist in many environments ranging from arid, through semiarid to humid temperate and humid tropical. They occur in many soil types and at various depths. Natural pipes are known with diameters ranging from 0.02 to >1 m and lengths of a few metres to >1 km; they may carry perennial or ephemeral flows. The major requirement for their existence appears to be a soil body which is strong enough to support the walls and roof of a pipe but not so strong that it inhibits pipe erosion by flows which, at least initially, are of low volume and velocity. The mechanics of pipe development are discussed in Chapter 12.

Pipe-flow may be derived from areas of saturated soil, areas of cracked surface soils or with many large, open macropores, or zones of converging saturation flow in macropores. Some pipe-flow may come from concentrated overland flow and channel flow which is diverted into a pipe. The velocity of pipe-flow has been variously estimated as being in the range of that of overland flow to being (at 0.1 m/s) an order of magnitude more rapid. It can therefore be a major contributor to storm runoff and especially to peak flows. Furthermore networks of pipes extend the areas of a catchment which contribute to storm runoff and they may be major contributors of water to saturated zones from which saturation-excess overland flow occurs. In some catchments pipe-flow has been assessed as contributing up to 50 per cent of the total storm discharge.

The total significance of pipe-flow in both catchment hydrology and in geomorphic development of hillslopes is, however, not well understood. The proportion of large regions in which pipes occur is usually regarded as being small; but as they are difficult to detect, unless their roofs collapse, they may be underestimated. Research into pipe-flows and the effects of pipes on delivering water to erodible sites, such as hollows and those with unstable soil masses, is rather neglected.

Conclusion

The nature of hydrological processes is now fairly well understood (Kirkby 1978, 1988; Burt 1989).

The significance of these processes for hillslope erosion and development is still being assessed but it is clear that advances in hillslope geomorphology must be based upon a sound appreciation of soil physics and hillslope hydrology.

12 Erosion of Hillslopes by Raindrops and Flowing Water

Controls on erosion

Erosion is an inclusive term for the detachment and removal of soil and rock by the action of running water, wind, waves, flowing ice, and mass movement. On hillslopes in most parts of the world the dominant processes are action by raindrops, running water, subsurface water, and mass wasting. The activity of waves, ice, or wind may be regarded as special cases restricted to particular environments.

Climate and geology are the most important influences on erosion with soil character and vegetation being dependent upon them and interrelated with each other (Fig. 12.1). The web of relationships between the factors which influence erosion is extremely complex. Vegetation, for example, is dependent upon climate, especially rainfall and temperature, and upon the soil which is derived from the weathered rock forming the topography. Vegetation in its turn influences the soil through the action of roots, take-up of nutrients, and provision of organic matter, and it protects the soil from erosion. The importance of this feedback is most obvious when the vegetation cover is inadequate to protect the soil, for eroded soil cannot support a close vegetation cover. The operation of the factors which influence erosion is most readily seen in their effect upon the disposition of storm rainfall (Fig. 12.2). By comparison with the high runoff from an eroded catchment a well-vegetated catchment with a permeable soil will experience higher infiltration, lower surface runoff, and less surface erosion.

Erosion is a function of the eroding power (i.e. the erosivity) of raindrops, running water, and sliding or flowing earth masses, and the erodibility of the soil, or:

$$\text{Erosion} = f \, (\text{Erosivity, Erodibility}).$$

Erosivity is the potential ability of a process to cause erosion, and for given soil and vegetation conditions one storm can be compared quantitatively with another and a numerical scale of values of erosivity can be created. Erodibility is the vulnerability of a soil to erosion and for given rainfall conditions one soil can be compared quantitatively with another, and a numerical scale of erodibility created. Erodibility of the soil can be divided into two parts—first, the characteristics of the soil, such as its physical and chemical composition, and second, the manner of treatment of the soil beneath land use (i.e. cropping, forestry, or grazing, etc.) and management (i.e. fertilizer applications, cropping, harvesting, etc.). All these factors operate together and are expressed in the Universal Soil Loss Equation (Fig. 12.3). The equation is widely used in soil erosion studies in croplands with Hortonian overland flow, but it has not been generally applied to areas with a complete grass or tree cover and it does not apply to soils being eroded by mass wasting. It does, however, demonstrate the interrelatedness of the various factors which influence the rate of soil erosion (Wischmeier and Smith 1965; Foster *et al.* 1977a,b).

The USLE is expressed as:

$$A = RKLSCP, \qquad (12.1)$$

where: A = the soil loss;
R = the rainfall erosivity factor;
K = the soil erodibility factor;
L = the slope length factor;
S = the slope gradient factor;
C = the cropping management factor;
P = the erosion control practice factor.

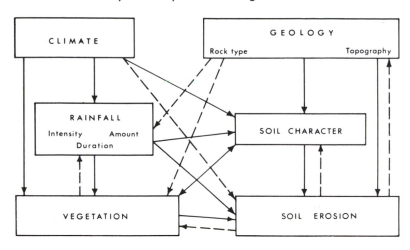

Fig. 12.1. The interrelationships between the main factors influencing soil erosion. (After Morisawa 1968.)

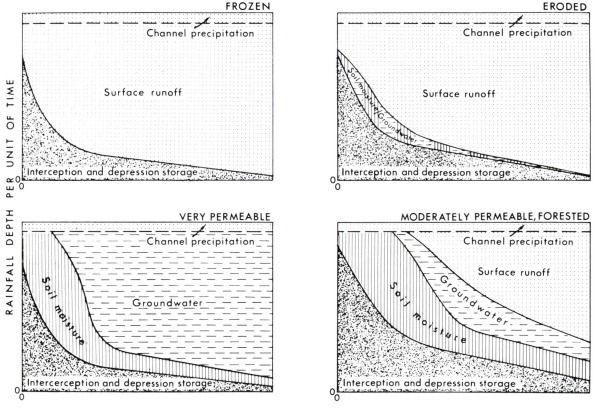

Fig. 12.2. Proportions of water in various parts of the drainage basin vary with time from the onset of rain and with the characteristics of the drainage basin. Throughflow and saturated overland flow are included in soil moisture. The greater the Hortonian surface runoff the greater is the probability of soil erosion.

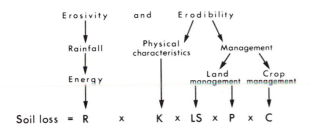

FIG. 12.3. Components of the Universal Soil Loss Equation.

The USLE was devised for practical conservation purposes and, as stated by Wischmeier (1976), may be used to:

(1) predict average annual loss of soil from a cultivated field with specific land-use conditions;

(2) guide the selection of cropping and management systems, and conservation practices for specific soils and slopes;

(3) predict the change in soil loss that would result from a change in crops or land use on a specific field;

(4) determine how conservation practices should be adjusted to allow higher crop yields;

(5) estimate soil losses from areas which are not in agricultural use; and

(6) provide estimates of soil losses for planners of conservation works.

An understanding of the importance of the USLE factors was gained from study of erosion on standard plots using cultivation practices appropriate to the USA. In order to apply its concepts more widely, modified versions have been developed and applied in other parts of the world: for example, Hudson (1961) developed a method for use in the African subtropics, Elwell (1977) a method for southern Africa, Stehlik (1975) a method for Czechoslovakia, and Morgan *et al.* (1984) a general method with refinements beyond those of the USLE. All of these methods are for use on fields or slopes of limited area and are concerned with annual sediment-loss estimates; they cannot be used for studies of drainage basins, nor for sediment yields from individual storms. Charts demonstrating how the USLE and Morgan *et al.* methods can be applied are presented in Morgan (1986).

In a less specific manner, not necessarily related to agriculture or other land uses, the factors affecting soil erosion—climate, topography, rock type, vegetation, and soil character—can be summarized into a descriptive equation:

$$E = f(C, T, R, V, S, \ldots [H], \ldots), \quad (12.2)$$

which has to be left open for additional factors, such as human interference [H], which may be significant at a particular site. This equation has seldom been expressed in quantified terms because of the extreme complexity of the variables, but a few attempts have been made to isolate some of the most significant factors and to express their relationship.

The human factor can exert its influence in a great variety of ways, especially by modifying the other factors, as when land is cleared of forest for pasture or cultivation; topography is modified by terraces or drainage ditches; and soils are changed by effects on vegetation, by cultivation, compaction by machinery, or application of fertilizers. Human disturbance has frequently disrupted the approximate balance between soil formation and soil erosion. It is, of course, possible for people to have the opposite influence and to protect soil from erosion by such activities as afforestation.

The climatic factor and raindrop erosion

The major climatic factors which influence runoff and erosion are precipitation, temperature, and wind. Precipitation is by far the most important. Temperature affects runoff by contributing to changes in soil moisture between rains, it determines whether the precipitation will be in the form of rain or snow, and it changes the absorptive properties of the soil for water by causing the soil to freeze. Ice in the soil, particularly needle ice, can be very effective in raising part of the surface of bare soil and thus making it more easily removed by runoff or wind. The wind effect includes the power to pick up and carry fine soil particles, the influence it exerts on the angle and impact of raindrops and, more rarely, its effect on vegetation, especially by wind-throw of trees.

Many reports of soil erosion phenomena have their value limited by uncertainties in the terminology used, consequently the key terms are defined here.

Raindrop erosion is recognized as being responsible for four effects: (1) disaggregation of soil aggregates as a result of impact; (2) minor lateral displacement of soil particles (a process sometimes referred to as creep); (3) splashing of soil particles into the air (sometimes called saltation); (4) selection or sorting of soil particles by raindrop impact which may occur as a result of two effects—(a) the forcing of fine-grained particles into soil voids causing the infiltration rate to be reduced and (b) selective splashing of detached grains. Wash is the process in which soil particles are entrained and transported by shallow sheet flows (overland flow). Rainwash is the combined effect from raindrops falling into a sheet flow.

Raindrop erosion is controlled by the resistance of the soil and the amount, intensity, and duration of the rainfall. A large total rainfall may not cause much erosion if the intensity is low, and likewise an intense rainfall of short duration may cause little erosion because the amount is small. Where both intensity and amount of rain are high, erosion is likely to be rapid. The erosive power of rain may, of course, be reduced to nothing if a complete vegetation cover prevents raindrops from hitting the soil.

In most temperate climates rainfall rate seldom exceeds 75 mm per hour, and then only in summer thunderstorms. In many tropical countries intensities of 150 mm per hour are experienced regularly. A maximum rate, sustained for only a few minutes, was recorded in Africa at 340 mm per hour (Hudson 1971).

Raindrop sizes may be measured by photographing them or, more commonly, by using an absorbent paper with a dusting of very finely powdered water-soluble dye on its surface. When dry the dye is invisible, but on exposure to rain each raindrop makes a roughly circular stain which can be measured. Stain sizes can be calibrated with drops of known size produced in the laboratory.

The largest natural raindrops appear to be about 5 mm in diameter and drops bigger than about 4.6 mm diameter break up into smaller drops. Only rarely are very large drops (6 mm or more) produced, and then only by the collision of two smaller drops.

During a storm the rain is made up of drops of all sizes and the distribution of these sizes depends upon the type of rain. Low-intensity rain is made up of small drops, but high-intensity rainfalls have a greater proportion of medium and large drops (Fig. 12.4).

A body falling under the influence of gravity only will accelerate until the frictional resistance of the air is equal to the gravitational force, and it will then continue to fall at that speed. This is known as the terminal velocity and depends upon the size,

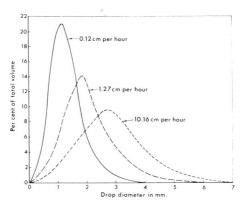

FIG. 12.4. A distribution of raindrop sizes in storms of given intensities. (Data from Laws and Parsons 1943.)

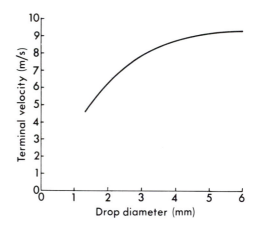

FIG. 12.5. The terminal velocities of raindrops of various sizes. (Data from Gunn and Kinzer 1949.)

density, and shape of the body. The terminal velocity of raindrops increases as the size increases; large drops with diameters of about 5 mm have a terminal velocity of about 9 m/s (Fig. 12.5, Table 12.1).

In a series of experiments, Laws (1941) found that the terminal velocity of raindrops in the open is affected by wind velocity and turbulence, but in spite of these effects most raindrops reach the ground at 95 per cent of their 'still air' terminal velocity. During rain, wind may 'drive' raindrops, and the resulting vector may be greater than the 'still air' velocity.

TABLE 12.1. *Raindrop velocity and kinetic energy*

Rainfall type	Median diameter (mm)	Velocity of fall (m/s)	Drops per (m²/s)	Intensity (mm/h)	Kinetic energy (J/m² per mm of rain)
Fog	0.01	0.0	67 000 000	0.05	0.02
Mist	0.10	0.2	27 000	0.13	4.14
Drizzle	0.96	4.1	150	0.25	6.61
Light rain	1.24	4.8	280	1.02	11.95
Moderate rain	1.60	5.7	500	3.81	16.94
Heavy rain	2.05	6.7	500	15.24	22.17
Very heavy rain	2.40	7.3	820	40.64	25.92
Torrential rain—1	2.85	7.9	1215	101.60	29.42
Torrential rain—2	4.00	8.9	440	101.60	29.42
Torrential rain—3	6.00	9.3	130	101.60	29.42

Sources: Basic data from Laws (1941); Laws and Parsons (1943); and Ellison (1947).

If the size of raindrops is known, and also their terminal velocity, it is possible to calculate the momentum of the falling rain, or its kinetic energy, by the summation of the values for individual raindrops. Kinetic energy, E_k, is given by the relationship:

$$E_k = \tfrac{1}{2}mv^2 \qquad (12.3)$$

where m = mass (kg) and v = velocity (m/s).

As the kinetic energy is closely related to rainfall intensity, it is possible to use the intensity value derived from automatic rain gauges (Fig. 12.6).

Experimental work by Wischmeier *et al.* (1958) showed that there is a strong statistical correlation between the soil eroded during a particular storm and the product of the kinetic energy of the storm and the 30-minute intensity. The 30-minute intensity is the greatest average intensity experienced in any 30-minute period during the storm. Wischmeier (1959) was thus able to propose that these factors could be used as an index of erosivity or an EI_{30} index. Development of this index has led to the idea that there is a threshold value of intensity at which rain starts to become erosive, because low-intensity rainfall may be observed to cause little or no erosion. Studies in Africa indicate that rain falling with an intensity of less than 25 mm/h is not erosive. By ignoring the rainfall with less intensity than 25 mm/h it has been possible to produce a more reliable index of erosivity called the KE > 25 index. A simplified calculation

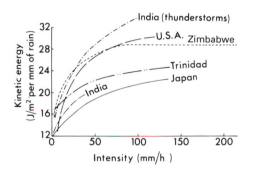

FIG. 12.6. Relationships, determined in a number of countries, between kinetic energy and rainfall intensity. Each curve extends to the highest intensity recorded.

of this index is given in Table 12.2, using only rainfall intensities during a given storm.

Indices of erosivity are most useful where they can be used to predict the erosive effects of rainfall and so influence the design of land management and soil-conservation plans. Rainfall erosivity can be mapped for any area for which there are sufficiently detailed rainfall records. Information can also be prepared on how erosivity varies during the year for any location. As an example (Hudson 1971) two locations may be compared.

At location A, 5 per cent of the rain is erosive, and the total annual rainfall is 750 mm, hence there is 37.5 mm of erosive rain.

TABLE 12.2. *Calculation of the* KE > 25 *index*

(1) Rainfall intensity (mm/h)	0–25	25–50	50–75	>75	
(2) Rainfall amount (mm)	30	20	10	5	Total 65
(3) Energy of rain (J/m² per mm of rain, derived from Fig. 12.6)	ignore	26	28	29	
(4) Total: line 2 × line 3	ignore	520	280	145	Total 945

The erosive power of the storm is thus 945 J/m².

Note: A joule (J) is a measure of the work done when the point of application of a force of one newton is displaced through a distance of one metre.

Source: Hudson (1971).

At location B, 40 per cent of the rain is erosive, and the total annual rainfall is 1500 mm, hence there is 600 mm of erosive rain.

Furthermore locality B has a higher average rainfall intensity—say 60 mm/h compared with 35 mm/h for locality A. It has been shown that the kinetic energy per mm of rain increases as the intensity increases so that for A the average kinetic energy may be 24 J/m² per mm of rain; and 28 J/m² per mm of rain at B.

The annual erosivity values are thus:

location A: 37.5 × 24 = 900 J/m²
location B: 600 × 28 = 16 800 J/m².

The erosivity of rain falling on B is thus 18 times greater than that falling on A.

The value of the index is thus evident, but the actual erosion which occurs is also dependent upon the characteristics of the soil, vegetation, and slope on which the raindrops fall.

For a bare soil the transport of soil by splash is likely to be expressed by an equation with the general form (Poesen 1985; De Ploey and Poesen 1987):

$$q_s = f(E_k, R, D_{50}, \beta, B), \qquad (12.4)$$

where:

q_s = the net splash transport on a bare, smooth, unchannelled surface (m³/m² per year);
E_k = kinetic energy of rainfall (J/m² per year);
R = the shearing resistance of soil (N/m²);
D_{50} = the median grain size of the soil material (m);
β = the slope angle (°);
B = the rainfall obliquity (°).

Raindrops driven by wind strike the soil surface at an angle from the vertical and so affect the splash-detachment angle. Wind therefore affects both the proportions of upslope and downslope splash and increases the impact energy of raindrops. On hillslopes into which raindrops are driven by wind, low angle slopes may experience net upslope splash and the maximum downslope transport will occur at slope angles >20° (Moeyersons 1983).

A generalized and approximate formula for the calculation of sediment discharge by splash from vertically falling raindrops could take the form:

$$q_s = \frac{E_k \cdot \cos\beta}{R}(\sin\beta + D_{50}), \qquad (12.5)$$

and for wind-driven raindrops:

$$q_s = \frac{E_k \cdot \cos(\beta \pm B)}{R}(\sin\beta \pm B) + D_{50}, \qquad (12.6)$$

where the terms are as given for equation 12.4 and:

B = the angle between a vertical line and a mean trajectory of raindrops, then:
$(\beta + B)$ = rain directed to produce increased splash;
$(\beta - B)$ = rain directed to produce decreased splash.

A specific form of these equations has been presented by Poesen (1985) in which the resistance to splash is represented as a function of soil-bulk density and the kinetic energy needed to detach 1 kg of material. The forms of the equations given above assume that R is a measure of total shearing resistance. It must be stressed that rainfall energy varies rapidly over time and the median particle size is a convenient, rather than fundamental, characterization of soil properties. Furthermore, soil from A horizons is often aggregated so that the effective particle size is different from that measured for individual grains. Because many aggregates break down under the impact of raindrops, by slaking, agitation, and splashing, the effective particle size on a surface changes during a storm. Changes due to aggregate instability enhance those due to selection by splashing, washing into voids, and transport in overland flow. The practical value of the generalized formulae of equations 12.5 and 12.6 may therefore be limited. Further discussion of these topics is presented by De Ploey and Poesen (1985).

The effect of raindrop impact both compacts the soil surface and disperses water from the crater of impact in lateral flow jets which have local velocities which are nearly double those of the raindrops at impact (Huang *et al.* 1982). These jets form water droplets which may contain soil particles.

Soil-surface compaction by impact is further enhanced by selective dispersal of fine grains of soil which are moved into the soil pores to create a sealed crust about 0.1–3 mm thick (Farres 1978). Aggregates at the soil surface are consequently broken down, but those below the crust are usually protected. A major effect of the crust is to reduce infiltration capacity. The percentage reduction is variable, depending upon the soil physical characteristics and the diameter, and therefore the energy, of the raindrops; recorded reductions are as high as 50 per cent for a single storm and 1000 per cent for a succession of storms (Morin *et al.* 1981; Hoogmoed and Stroosnijder 1984). Soils with high clay contents and high organic matter contents are generally less prone to crusting than loams and sandy loams. The overall detachment of grains, as measured by mean weight of soil displaced, usually follows the order sand > silt > clay loam > clay (Quansah 1985). Under most conditions gravel-size particles would be the least detachable and remain as a lag on the soil surface.

The topographic factor

This factor is evident in the steepness and length of slopes. Nearly all of the experimental work on the slope effect has assumed that the slopes are under cultivation. In such conditions raindrop splash will move material further down steep slopes than down gentle ones, there is likely to be more runoff, and runoff velocities will be faster. Because of this combination of factors the amount of erosion is not just proportional to the steepness of the slope,

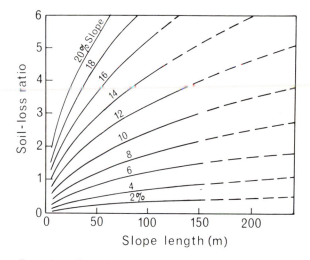

FIG. 12.7. The relationship between soil losses from bare soils and slope angle (in per cent) and slope length. (After Wischmeier and Smith 1965.)

but rises rapidly with increasing angle (Fig. 12.7). Mathematically the relationship is:

$$E \propto S^a, \qquad (12.7)$$

where E is the erosion, S the slope in per cent, and a is an exponent. Values of a derived experimentally range from 1.35 (Musgrave 1947) to 2 (Hudson and Jackson 1959).

The length of slope has a similar effect upon soil loss, because on a long slope there can be a greater depth and velocity of overland flow, and rills can develop more readily than on short slopes. Because there is a greater area of land on long than on short slope facets of the same width, it is necessary to distinguish between total soil loss and soil loss per unit area. The relationship between soil loss and slope length may be expressed as:

$$E \propto L^b, \qquad (12.8)$$

where E is the soil loss per unit area, L is the length of slope, and b is an exponent. In a series of experiments Zingg (1940) found that the values of b are around 0.6 but experiments elsewhere indicated that a rather higher value is more representative.

The vegetation factor

Vegetation offsets the effects on erosion of the other factors—climate, topography, and soil characteristics. The major effects of vegetation fall into at least seven main categories: they are (1) the interception of rainfall by the vegetation canopy; (2) the decreasing of velocity of runoff, and hence the cutting action of water and its capacity to entrain sediment; (3) root effects in increasing soil strength, granulation, and porosity; (4) biological activities associated with vegetative growth and their influence on soil porosity; (5) the transpiration of water, leading to the subsequent drying out of the soil; (6) insulation of the soil against high and low temperatures which cause cracking or frost heaving and needle ice formation; and (7) compaction of underlying soil.

The interception of raindrops by the vegetation canopy affects soil erosion in two ways: (1) by preventing the drops from reaching the soil and allowing water to be evaporated directly from leaves and stems; and (2) by absorbing the impact of the raindrops and minimizing the harmful effects on soil structure.

A close vegetation cover not only reduces runoff velocities by friction with plant stems, but it also prevents the runoff becoming channelled and so made able to cut into the soil. The slowing down of the runoff increases the time for infiltration.

A close forest cover often gives a virtually complete protection to soil. Not only do the canopy and understorey prevent raindrops hitting the ground, but the accumulation of litter forms a complete protective blanket (Plates 12.1–12.5). Experimental work has shown, however, that if the understorey and ground vegetation is destroyed by grazing animals then the energy of falling drops may actually be greater under forest than in the open. Rain falling with a low intensity may collect on the leaves of the canopy and then drip from the leaves in large drops. From a height of 10 m or so these drops will reach their terminal velocity so that even though some water is lost by evaporation and stem-flow, the effect upon the soil may be considerable. This appears to be the situation in some tropical forests where leaf litter is rapidly decomposed and the soil left bare.

Within managed forests, careful siting of roads and logging tracks to avoid channelling of runoff water, the preservation of a ground layer of plants, the prevention of fire, and the avoidance of clear felling on steep slopes can provide for soil conservation and wood production. Both conservators and foresters would probably agree that the ideal protective combination is a regular stand of trees with a close ground-cover of grass or herbs.

A well-managed high-yielding grassland is unlikely to be greatly affected by rain splash or sheet erosion, and poor grazing land is unlikely to be sufficiently productive to warrant expensive mechanical methods of erosion control. The methods which are used to decrease erosion on grasslands do not attempt to control soil movement directly, but aim at improving the vegetation so reducing runoff and increasing infiltration.

The prevention of 'pugging' of the soil by animal trampling during wet periods, especially in gateways and other well-used places, and other farming practices which reduce soil compaction and exposure are obviously necessary on all soils. In some areas mechanical methods, such as pasture furrows and other soil depressions, are used to reduce runoff. Pasture furrows are shallow, follow the contours exactly, and are spaced only a few metres apart. The best methods, however, involve the maintenance of a dense grass sward.

General relationships are indicated in Table 12.3 which shows the effects of vegetation in inhibiting erosion in northwestern USA. Ursic and Dendy (1965) found similar relationships in midwestern USA. The erosion rate can clearly vary by factors of 5 to 100 000 depending upon the vegetation (Fig. 12.8).

The soil factor

The soil factor is expressed in the erodibility of the soil. Erodibility, unlike the determination of erosivity of rainfall, is difficult to measure and no universal method of measurement has been developed. The main reason for this deficiency is that erodibility depends upon many factors, which fall into two groups: those which are the actual physical features of the soil; and those which are the result of human use of the soil (i.e. the management of the soil).

The resistance of soil to detachment by raindrop impact depends upon its shear strength, that is its cohesion (c) and angle of friction (ϕ) (Cruse and Larson 1977). It is difficult, in practice, to measure the appropriate values of c and ϕ for grains at the surface of a soil or soil crust, partly because of variability in the size, packing, and shape of particles and partly because of the varying degrees of wetting and submergence of grains by water (see Poesen 1985). More success has been achieved with simple rotational shear vanes than with most other methods (Brunori *et al.* 1989; Crouch and Novruzi 1989).

PLATE 12.1. The complete sward of tussock grasses prevents splash or wash erosion.

PLATE 12.2. The broken tussock grass cover permits frost action, splash, and wash erosion to occur between the tussocks.

Many attempts have been made to relate the amount of erosion from a soil to its physical characteristics. Pioneer work in this field was done in North America in the 1930s. Bouyoucos (1935) suggested that erodibility is related to the sizes of the particles of the soil in the ratio:

$$\frac{\text{per cent sand} + \text{per cent silt}}{\text{per cent clay}},$$

and Middleton (1930) used a 'dispersion ratio' based upon the changes in silt and clay content of the soil after dispersion in water. Other methods

PLATE 12.3. A cedar forest in the Atlas Mountains of Morocco. Much of the forest has been destroyed and the soil in the foreground is severely eroded. Grazing by goats prevents regeneration of the forest.

TABLE 12.3. *The relative relationships between erosion and vegetation*

Crop or practice	Relative erosion
Forest ground layer and litter	0.001–1
Pastures, humid region, and irrigated	0.001–1
Range or poor pasture	5–10
Grass/legume, hay	5
Lucerne	10
Orchards, vineyards with cover crops	20
Wheat, fallow, stubble not burned	60
Wheat, fallow, stubble burned	75
Orchards, vineyards clean tilled	90
Row crops and fallow	100

PLATE 12.4. Removal of cedar forest has permitted severe sheet wash erosion on these slopes, Atlas Mountains, Morocco.

PLATE 12.5. Leaf litter on the floor of a deciduous forest protects the soil from splash and wash erosion. The amount of protection varies with the seasons.

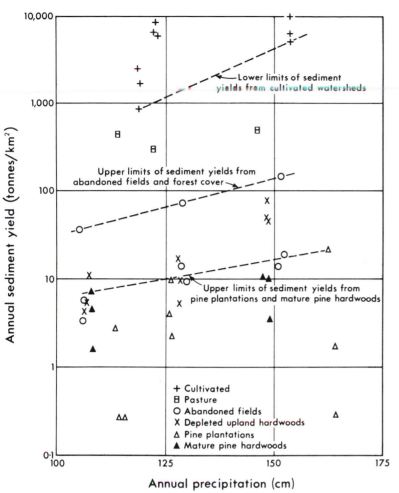

Inside the graph:

Lower limits of sediment yields from cultivated watersheds

Upper limits of sediment yields from abandoned fields and forest cover

Upper limits of sediment yields from pine plantations and mature pine hardwoods

Legend:
+ Cultivated
⊟ Pasture
O Abandoned fields
X Depleted upland hardwoods
△ Pine plantations
▲ Mature pine hardwoods

Y-axis: Annual sediment yield (tonnes/km²)
X-axis: Annual precipitation (cm)

Fig. 12.8. Variation in sediment yields from individual watersheds in northern Mississippi under different types of land use and various amounts of precipitation. (After Ursic and Dendy, 1965.)

involve measures of the stability of soil aggregates and the rates of water transmission through the soil. Stability of soil aggregates, resulting from the cohesive properties of clay or organic matter, is the main inhibitor of erosion by raindrops (Farres and Cousen 1985). Slaking of aggregates reduces them to smaller aggregates or single particles; slake-durability of the aggregates is therefore a useful measure of resistance.

The failure to find indices of erodibility has encouraged many research workers to subject soil samples to splash erosion in laboratory tests on trays of soil beneath simulated rainfall and to rank soil types in order of erodibility (Plate 12.6).

A second method is to lay out a series of plots in the field where natural rainfall can be measured, and the soil washed from the plots can be collected

in troughs at the downslope end of each plot. The sediment yield can then be related to the rainfall and, by using a number of plots with various forms of land management and vegetation cover, it is possible to compare the efficacy of various forms of land use for inhibiting erosion. A variation of this method is to use plots in the field and to measure the natural features (i.e. properties of the soil, climate, vegetation, etc.) and then to relate these properties statistically to the sediment yield in an attempt to determine which of the natural features influences the erosion (Plate 12.7).

Wischmeier and Mannering (1969) established a correlation between erodibility and a large number of variables which are soil physical properties and soil profile characteristics. From a statistical study of fifty-five US Corn Belt soils Wischmeier and

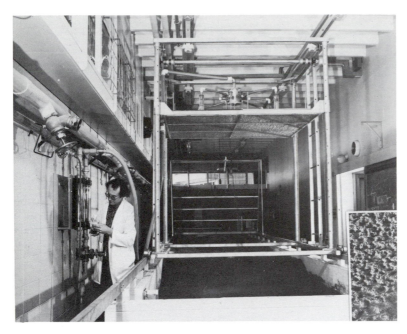

PLATE 12.6. A laboratory rainfall simulator, and the result of a raindrop impact upon bare soil (inset).

PLATE 12.7. An experimental plot with a trough to collect runoff and sediment. A rain gauge stands beside the trough.

Mannering derived an empirical equation to calculate the erodibility factor in the Universal Soil Loss Equation. It contains twenty-four variables representing soil properties which analysis has shown to be important. The equation predicted the erodibility of eleven bench-mark soils, whose erodibility has been previously established, by other means, very closely. It was found, however, that this equation was too complex for practical use, although valid for a broad range of medium-textured soils. A method using only five variables has been devised subsequently, and by use of a nomograph (Fig. 12.9) simply measured soil characteristics can be related to soil erodibility under rainfall impact and surface runoff.

In spite of the encouraging success of the Wischmeier and Mannering equation it is still true to say that none of the proposed measures conforms completely to the requirements of an erodibility index: that it be simple to measure, reliable in operation, capable of universal application, and capable of providing a quantitative measure of the erosion which will take place when the soil is subjected to rain of known erosive power.

One conclusion of overriding importance, which has been reached as a result of field and laboratory studies in many countries, is that soil erosion is largely controlled by the type of management soil receives, and this conclusion is the basis of nearly all soil conservation practice. This point may be illustrated by a hypothetical example in which one soil might lose say 300 tonnes/ha/year when it is used for the cultivation of row crops which run up and down the slope, whereas an identical soil under the same climatic regime, under well managed pasture, would lose only a few kilograms of soil per hectare per year. The erosion from two areas with the same soil but different management is commonly greater than the differences between two soils with the same management.

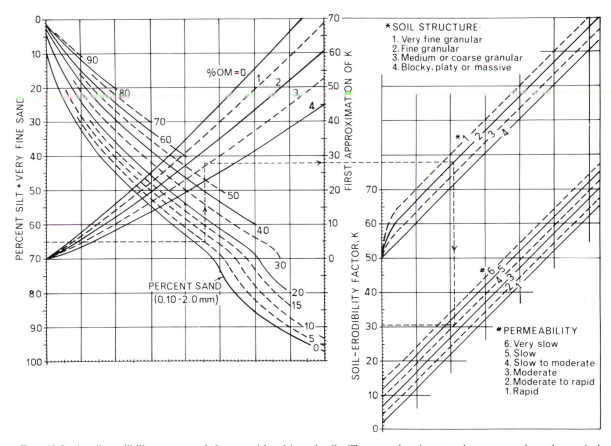

FIG. 12.9. A soil erodibility nomograph for use with cultivated soils. The procedure is: enter the nomograph on the vertical scale at the left with the appropriate percentage of silt plus very fine sand (0.1 to 2.0 mm); proceed horizontally to intersect the correct percentage sand curve, then vertically to the correct organic matter curve, and horizontally towards the right; (for many agricultural soils with a fine granular structure and moderate permeability the value of the erodibility factor K can be read directly from the first approximation-of-K scale on the right-hand edge of the first section of the nomograph and the procedure can terminate there); for soils with other than fine granular structure and moderate permeability continue the horizontal path to intersect the correct structure curve; proceed vertically downward to the correct permeability curve and move left to read the value of K. The dotted line illustrates the procedure for a soil with 65 per cent silt and very fine sand, 5 per cent sand, 2.8 per cent organic matter, 2 structure, 4 permeability, and hence erodibility of 0.31. (After Wischmeier *et al.* 1971.)

Management includes both the broad issues of land use and the details of crop management. The best management might be defined as the most intensive and productive use of which land is capable without causing any degradation (Fig. 12.10).

Wash, rill, gully, and piping processes

Running water removes soil from slopes by a variety of processes—sheet wash, rilling, gullying, and piping.

Wash erosion

Soil detachment by raindrop impact can occur anywhere that soil is exposed to the shear force of drops (Plates 12.8, 12.9). Detachment by runoff is generally limited to that small portion of the land area where runoff concentrates and flows at erosive velocities. The quantity and size of the particles that can be transported by runoff are a function of runoff velocity and turbulence and these increase as the slope steepens and the depth of flow increases. Once entrained the particle will remain in suspension until a lower, depositional velocity occurs. Recorded velocities for overland

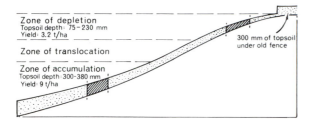

Fig. 12.10. Topsoil depth and yields of crops down a slope in Otago, New Zealand. (Data from New Zealand Soil Bureau.)

flow are commonly in the range of 0.015 to 0.3 m/s, which is great enough to move silts and fine sands.

The hydraulic conditions of flow vary greatly over short distances and groundsurfaces show patterns of miniature alternating scours, sediment fans, and debris dams (Moss and Walker 1978). Consequently, no single, simple, mathematical expression is likely to describe all relationships between sediment discharge, flow, and soil particles. Relationships have to be determined from experimental studies, but will take the general form of a ratio of driving forces/resisting forces:

$$q_s = \frac{k \; Q_w^a \; \sin\beta^b}{n^c D_{50}}, \qquad (12.9)$$

where: q_s = the net sediment transport;
  Q_w = the discharge of water;
  β = the slope angle;
  n = the resistance to particle entrainment;
  D_{50} = the median particle diameter.
  k, a, b, and c have empirically derived values.

Komura (1976) and Morgan (1980) used forms of the above equation and have determined values for the exponents.

Because overland-flow water depths are usually limited to a few millimetres, raindrops falling into the flow may create splash and enhanced entrainment. Moss and Green (1983), Palmer (1965), and Kinnell (1990) found that the rate of transport of soil by this mechanism increases with increasing depth of water to a maximum at a depth that is about one to three times the diameter of the raindrops, and is consequently at a maximum when flow depth is in the range of 3–10 mm. It is possible, however, that at depths of flow greater than one diameter the increased soil detachment is

due to the turbulence in the flow created by the raindrops. At depths greater than three diameters water is likely to be turbulent anyway and the raindrop impact energy is dissipated in the flow. In areas with bare, cultivated soil, overland flow in association with splash can account for large proportions of the total soil loss, and up to 95 per cent (Van Asch 1983; Morgan *et al.* 1984). In semiarid areas of New Mexico, surface erosion on unrilled slopes has been shown to account for 98 per cent of sediment production from all sources (Emmett 1978).

In some soils sediment may be transported through macropores (Pilgrim and Huff 1983) but the general importance of this process is unknown.

Rill erosion

Most studies of rills have been carried out either on agricultural soils or on field or laboratory plots established in agricultural soils. Studies related to geomorphic development are fewer and have been concentrated in semiarid environments or specific environments such as those with calcareous and smectite clay-dominated soils (see Bryan 1987, for a review). The agricultural importance of rills is indicated in the most commonly used definition: rills are 'microchannels . . . small enough to be removed by normal tillage operations' (FAO 1965). As rills exist in areas which are not cultivated, a more general definition is needed.

Rills are small channels with cross-sectional dimensions of a few centimetres to a few tens of centimetres. They are usually discontinuous; may have no connection to a stream-channel system; are often obliterated between one storm and the next, or even during a storm when the supply of sediment from splash on inter-rill areas, or collapse of rill walls, or liquefaction of the bed and walls, exceeds the transporting capacity of the rill-flow; they occur on slopes steeper than 2–3° (De Ploey 1983; Dunne and Aubry 1986; Bryan and Poesen 1989).

Parallel rills on a fresh surface become integrated into a drainage net by the breaking down of divides between rills with diversion of the water into the deeper rill, and the overtopping of rills and diversion of the water towards the lowest elevation (Plates 12.10 and 12.11). These two processes Horton (1945) called micropiracy and cross-grading. Their effect is to cause wider spacing of rills downslope.

It is commonly stated that rills do not become

PLATE 12.8. The interior of a forest showing damage to the floor by browsing animals. All tree and shrub regeneration has been prevented and wash erosion is now active. (J. H. Johns: Crown copyright, by permission of the New Zealand Forest Service.)

PLATE 12.9. Sheet-wash erosion.

progressively enlarged downslope, but instances are recognized of enlargement (Schumm 1956) even to the size of gullies (Plate 12.11), and rills may form the heads of natural drainage systems. Alternatively rills may become wider, shallower, and feed into a braided wash downslope. Braiding usually occurs where slope angles decline and is common in sandy deposits.

Smectite-rich rocks and soils of some badlands (Plate 12.12) have a high shrink–swell capacity and develop surfaces dominated by desiccation cracking. At the onset of rainfall, water fills the

PLATE 12.10. Rill erosion showing the development of a network.

PLATE 12.11. Rill erosion expanding downslope into gully erosion. The man in the gully gives scale.

cracks before overland flow can commence. Flow may start in depressions where subsurface water, which has moved through micropipes and tunnels, returns to the surface. The return flow may also carry silt which is then deposited in the depressions (Bryan *et al*. 1978).

Rill formation in smectite-rich rocks and soils may occur in four major ways: (1) return flow; (2) collapse of the roofs of micropipes and tunnels (Hodges and Bryan 1982; Gerits *et al*. 1987); (3) as a response to the hydraulic conditions of surface flow on micropediments (Hodges 1982); and (4) in badlands, by the subsurface flow of water through macropores and cracks to a site where a critical moisture content is reached creating rill formation in soils which may be highly dispersive (Imeson and Verstraten 1988). In this last mechanism, the development of critical moisture contents is seen as occurring at spacings which will be influenced by slope angle and the hydraulic conductivity, so controlling a pattern of evenly spaced parallel rills on uniform slopes. Variations in the properties of the rocks and soils may create conditions of extreme local variability in rill development and spacing (Gerits *et al*. 1987).

Because of the importance of erosion on agricultural soils a major debate has, at various times, taken place on the relative erosive power of raindrops, wash, and rill processes. Rills are both collection areas for inter-rill sediment and transporting agencies for those sediments removed from

PLATE 12.12. Badland relief developed in Quaternary lake deposits on the margin of the Salar de Atacama, Chile. Note the cracked surface, the rills, and unroofed pipe in the left foreground. The rills feed into a steep gully in the centre of the view.

rill walls and floors (Foster and Meyer 1975; Moss *et al.* 1982). Rill erosion is then the primary agent for sediment transport on slopes with little vegetation (Morgan 1977). Estimates of the proportion of sediment yield from slopes which is removed by rills range from 50 to 90 per cent.

The loss of soil by raindrop, wash, and rill erosion appears, from observational evidence, to be related to two types of rainfall event: (1) short-lived storms of high rainfall intensity in which the infiltration capacity of soil is exceeded, and (2) prolonged rainfall events of low intensity which saturate the soil. Conditions of saturation may also be produced by snowmelt, and either rapid melting, or rainfall on to snow covers, may produce wash and rill erosion without splash.

Soil-conservation measures are based on the knowledge that runoff increases as slope length and slope steepness increase. The aim of good soil-conservation measures is then to reduce the effective length of slopes by dividing them into sections using grass strips, hedges, walls, shallow drains, terraces, or furrows along the line of the contour. This inhibits both sheet wash and rilling as long as the barrier is wide enough and designed so that runoff cannot become channelled (Plates 12.13 and 12.14).

Slope steepness is reduced also by terracing which divides the slope into short, gently sloping sections separated by a terrace wall. As with all soil conservation measures the breakdown of a terrace can cause channelling of runoff and be the cause of very serious gully erosion.

Rainsplash erosion is best inhibited by the preservation of a complete vegetation cover and where this is not possible by ensuring that soil is not bare in seasons when rainfall intensities are greatest. Mulches of straw, cut weeds, or plastic sheeting may be used when intensive cultivation can justify the use of such methods.

PLATE 12.13. Terraces for conservation purposes at an altitude of 3000 m on the equator in Rwanda, central Africa.

Gully erosion

Gullies are erosional features found in many parts of the world, and variously known by regional names, for example, 'arroyo' in southwestern United States (Schumm and Hadley 1957; Cooke and Reeves 1976), 'lavaka' in Madagascar (Riquier 1958; Hurault 1971), and 'donga' in Southern Africa (Dardis *et al.* 1988).

A master rill may so deepen and widen its channel that it is classed as a gully—arbitrarily defined as a recently extended drainage channel that transmits ephemeral flow, has steep sides, a steeply sloping or vertical head scarp, a width greater than 0.3 m, and a depth greater than about 0.6 m (Brice 1966: 290). Gullies may also form at any break of slope or break in the vegetation cover

when the underlying material is mechanically weak or unconsolidated. Gullies are therefore most common in such materials as deep loess, volcanic ejecta, alluvium, colluvium, gravels, partly consolidated sands, and debris from mass movements (Blong 1970; Stocking 1980).

A second general class of gullies is of those which develop in weak argillic rocks—such as argillites, phyllites, mudrocks, and shales. These gullies are often channels through which debris flows and mud flows, with very high sediment discharges, drain unstable catchments (Schouten and Rang 1984; Sidle *et al.* 1985; Li Jian and Wang Jingrung 1986).

A third class of gullies is of those which are initiated by subsurface flows in natural pipes, with subsequent collapse of the roof of the pipe. This condition is being recognized more commonly. It is discussed in the next section under 'pipes'.

Because they are very rapidly developed erosional forms, gullies are usually not regarded as features of normal erosion, but the result of changes in the environment, such as faulting, burning of vegetation, overgrazing, climatic change affecting vegetation, extreme storms, or any other cause of a break in vegetation which will bare the soil.

Gully erosion nearly always starts for one of two reasons: either there is an increase in the amount of flood runoff, or the flood runoff remains the same but the capacity of water courses to carry the flood waters is reduced. The most common causes of increases in runoff or deterioration in channel stability are changes in vegetation cover—

PLATE 12.14. Conservation strips on the contour. The strips are well-grassed furrows below short-graded slopes in pasture, Transkei, southern Africa.

TABLE 12.4. *Values of roughness coefficient* n *for different channel conditions*

Description of channel	Range of values		
	Minimum	Normal	Maximum
Concrete, trowel finished	0.011	0.013	0.015
Concrete, shuttering	0.012	0.014	0.017
Brickwork	0.012	0.015	0.018
Excavated channels:			
earth, clean	0.016	0.022	0.030
gravel	0.022	0.025	0.030
rock cut, smooth	0.025	0.035	0.040
rock cut, jagged	0.035	0.040	0.060
Natural channels:			
clean, regular section	0.025	0.030	0.040
some stones and weeds	0.030	0.035	0.045
some rocks and/or brushwood	0.050	0.070	0.080
very rocky or with standing timber	0.075	0.100	0.150

especially removal of trees, increases in the proportion of arable land in catchments, excessive burning of vegetation, or overgrazing—or a climatic change with accompanying variations in rainfall periodicity and intensity.

The capacity of a stream channel depends on the cross-sectional area, the slope, gradient, and roughness. Changes in these factors can easily disturb an equilibrium between the channel geometry and the processes of erosion and deposition which have moulded it.

The relationship between the velocity of water in a channel and the geometry of that channel are expressed in Manning's equation:

$$V = \frac{1}{n} R^{2/3} S^{1/2}$$

in which: V is the average velocity of flow (m/s);
R is the hydraulic radius (m);
S is the average gradient of the channel (m/m);
n is a coefficient, known as Manning's n or Manning's roughness coefficient. (Typical values of n are given in Table 12.4).

(Note: the hydraulic radius is given by $R = A/P$ in which A is the area of a transverse section of a stream, and P is the wetted perimeter or length of the boundary along which the water and channel-bed are in contact in that transverse section.)

An increase in the resistance to flow by the growth of vegetation in or at the edge of a channel will cause an increase in the value of n and hence be accompanied by a decrease in the velocity of flow. This reduces the capacity of the channel to accommodate flood flows and increases the chances of flood-waters spilling over channel banks and starting new erosion patterns. By contrast a reduction in vegetation along a waterway may cause a decrease in resistance to flow, and therefore a decrease in the value of n, and increased velocity with the possibility of the development of channel scouring.

Any local effect can upset the equilibrium which already exists. Such minor starting points could be cattle tracks, pot-holes, or the diversion of drainage by a new road so that, instead of spreading over a whole valley floor, flood-waters are routed into confined areas with consequent smaller wetted perimeters and less resistance to flow.

Once gullying starts the gullied channel has a more angular and deep shape than the original bed (i.e. R increases). The gullied channel is rough and irregular so the value of n increases. For the velocity to remain constant the gradient must, therefore, decrease and this is what usually happens. The gradient of the floor of the gully is flatter than that of the original stream-bed—S decreases—(see Fig. 12.11). The head of the gully then works back upstream and the height of the

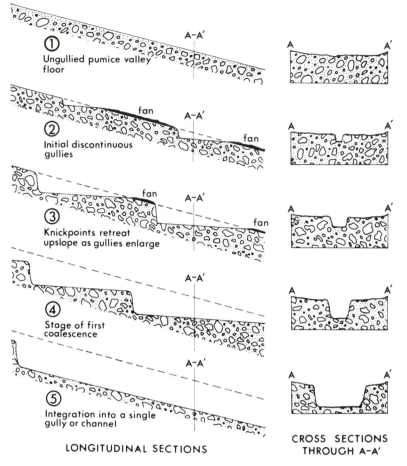

① Ungullied pumice valley floor

② Initial discontinuous gullies

③ Knickpoints retreat upslope as gullies enlarge

④ Stage of first coalescence

⑤ Integration into a single gully or channel

LONGITUDINAL SECTIONS

CROSS SECTIONS THROUGH A-A'

FIG. 12.11. Stages of development of discontinuous gullies. (Modified from Leopold *et al.* 1964.)

gully head progressively increases. As the head gets higher, or the velocity increases, gullying is likely to become more rapid; once it has started gully erosion thus becomes increasingly difficult to control. It is always easier and cheaper to prevent than to stop once it has started.

There are two main types of gully, although compound types are also common: (1) continuous gullies, and (2) discontinuous gullies. There are also three main processes operating to form gullies, although these usually occur in combination: (a) surface flow, (b) mass movement, and (c) piping.

A continuous gully which forms by enlargement of a rill may have no headscarp because it forms in non-cohesive materials without a resistant capping. This type of gully is particularly common where erosion of shattered rock and talus gravels has resulted from a deterioration in the vegetation

cover. Such gullies usually increase in width and depth downslope as the runoff increases and tributary gullies contribute their flow to the master gully, but in their lowest reaches where the mountain slope decreases, the gullies have zones of deposition which extend, as coalescing fans, beyond the gully mouth. The long profile of a typical gully therefore shows a slope near its head steeper than the mountain slope, but less steep near the base. Gullies may be cut by storm runoff and snow meltwater and therefore have ephemeral streams. Wind and frost action cause crumbling of the valley walls during dry periods.

Gullies without headscarps also develop within the scars and deposits of large mass movement features. Such features have a disrupted vegetation cover, and disturbed materials are then readily moved by surface flow or by minor mass move-

PLATE 12.15. Discontinuous gullies in valley floor infills, central North Island, New Zealand.

ments occurring at the head and sides of the gully (Fig. 12.12).

Gullies with headscarps will maintain that headscarp if (a) the gully originates at a scarp or steep face and the gully is cutting into cohesive materials and (b) if non-cohesive materials are capped by resistant materials such as root-bound soil. This is because the material making up the bed at the knickpoint has a resistance to shear stress greater than the stress provided by the flow. It is also essential for the flow to transport the eroded material away from the base of the headscarp if undercutting processes are to continue. Retreat results mainly from sapping as seepage water loosens material on the face of the headcut, or water under pressure flows out of the face after rain which has raised the water-table above the foot of the face. A plunge pool resulting from surface flow may contribute both to erosion at the base of the headscarp and to downstream removal of the sediments.

Where the slope of the gully floor does not exceed more than a few degrees the ephemeral stream usually has a wide channel because it can easily remove sediments from the base of the gully sides. Such gullies therefore have a rectangular cross profile, with the side walls being kept steep by undercutting.

In many dry valleys discontinuous gully erosion may begin at several points where vegetation cover is broken. At each point a small headscarp forms and as this retreats headwards a trench is left down-valley with a debris fan spreading out from its toe to form an alluvial fan.

It can be seen from Fig. 12.11 that the newly developed gully floor has a lower angle of slope than the original valley floor, but as the gully deepens and retreats headwards the slope increases again so that when two gullies coalesce their floor is usually of nearly the same slope as the original valley floor (Plates 12.15 and 12.16). In the original valley rainfall is distributed over the whole valley floor where it infiltrates into the soil, but as the gully develops, the storage capacity of the valley floor sediments is reduced and the subsurface water moves towards the gully. Peak discharges in

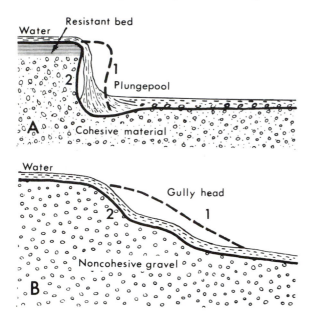

Fig. 12.12. Types of gully head. In A the materials are cohesive and have a resistant cap of root-bound soil supporting a headscarp. In B the non-cohesive material does not support a headscarp.

the gully therefore far exceed the previous discharges over the unchannelled valley floor. The floor of a discontinuous gully usually has a plunge pool at the foot of the headscarp, and downvalley from that a veneer of sediment, left by the falling flood of the last storm, over the original valley floor materials. The depth of the gully trenching is usually limited by the existence of a resistant bed in the valley materials.

Attempts have frequently been made to control gully erosion by building dams of concrete, stone, and wood (Heede 1977) (Plates 12.17 and 12.18). But most dams are liable to decay, to be bypassed or undermined, and in many cases they do not modify the basic cause of gully erosion.

In the terms of the Manning's equation: hydraulic radius is approximately equal to flow depth, so reducing the concentration of flow by reducing the contributing area for runoff, or increasing water losses by infiltration, or take-up by plants, could reduce flow depths.

Surface roughness in valley floors is largely caused by vegetation, so increasing vegetation cover by appropriate species will decrease velocities. In the southwestern USA, flow velocities

which are 'permissible' (i.e. will not be great enough to cause incision) on erosion-resistant soils are in the range 2.5–0.9 m/s, and on easily eroded soils are 1.8–0.7 m/s (Ogrosky and Mockus 1964). It may be necessary for mechanical methods such as wire-netting, permeable rock, brushwood, or timber dams to be used to hold back silt and control drainage before vegetation can be established. Planting a catchment with densely rooting trees and developing a close ground-cover of grasses, herbs, and shrubs are the usual conservation methods used (Heede 1976). Such methods are only effective if the climate permits a close ground-cover, and if grazing animals can be controlled. Because the discharge of a gully is controlled by catchment area, large gullies can sometimes be controlled by diverting water out of the gully along a controlled channel.

Gully widening by undercutting and bank collapse may be a major source of sediment discharge (Crouch and Blong 1989). Protection of the bases of gully walls can usually only be effected by decreasing flows and use of vegetation.

Gully erosion is particularly severe in semiarid countries where there is pressure on soil and vegetation resources. Where wood is the only available fuel for heating and cooking, and where people cannot afford adequate fences for animal control it is very difficult to maintain a strong vegetation cover and to apply conservation measures, hence soil erosion is often found around settlements.

Pipe erosion

Subsurface pipe erosion has been described by a number of terms including pothole erosion, suffosion, subcutaneous erosion, tunnelling, and tunnel-gullying, but the most widely used term is 'piping' (Parker and Jenne 1967; Crouch 1976; Jones 1987*a,b*). Natural pipes, and their role in slope hydrology, were described in the previous chapter.

Among the factors which dispose a soil to piping are: a seasonal or highly variable rainfall; a soil subject to cracking in dry periods; a reduction in vegetation cover; a relatively impermeable layer in the soil profile; the existence of a hydraulic gradient in the soil; and a dispersible soil layer.

Examples of piping are particularly common in semiarid badlands formed on smectite clays which have strong swelling and shrinkage properties and may also have high exchangeable sodium

PLATE 12.16. A fan formed at the lower end of a discontinuous gully.

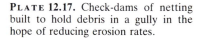

PLATE 12.17. Check-dams of netting built to hold debris in a gully in the hope of reducing erosion rates.

percentages (Heede 1971; Gutiérrez *et al.* 1988; López-Bermúdez and Romero-Diaz 1989; Swanson *et al.* 1989). Loess and loessic colluvium with high sodium content are also subject to piping (Laffan and Sutherland 1988).

The most commonly reported situation in which pipes develop is one in which a surface soil cracks as a result of desiccation. In a rainstorm water then infiltrates rapidly down the cracks and super-saturates a relatively permeable horizon in the

PLATE 12.18. This gully has cut down to bedrock, Transkei, southern Africa. Attempts are being made to reduce sediment transport by the erection of a porous dam of rock-filled wire baskets.

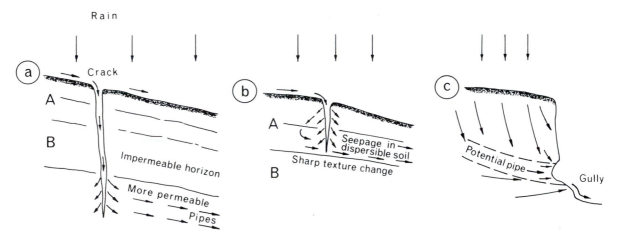

FIG. 12.13. Conditions favouring the formation of pipes: (a) cracking and a permeable horizon below an impermeable horizon; (b) an horizon of dispersible clay; (c) a gully head.

subsoil. Lateral seepage may be fast enough to move soil particles and develop a channel, or, if the soil has dispersible clays, these may lose aggregation. Movement of water through subsurface cracks and voids is slow until water breaks through the soil surface further down the slope, and rapid flow can then work headwards within the soil and form a gully or enlarge a pipe (Figs. 12.13 and 12.14).

Dispersion in soils is the deflocculation of the clay fraction when water is added. The dispersion may be a result of weakening of chemical bonds between particles by ion exchange or by leaching.

Piping is particularly common in the walls of gullies and in the heads of landslides where the confined flow in soils is suddenly accelerated and seepage pressures permit particles to be washed out of the soil. Pipes can extend headwards very rapidly once they start to grow. Rates of 45 m/hour have been reported in pumice soils in New Zealand.

Reclamation procedures include the destruction of existing pipes, the establishment of vegetation, infilling of cracks, and may include chemical treat-

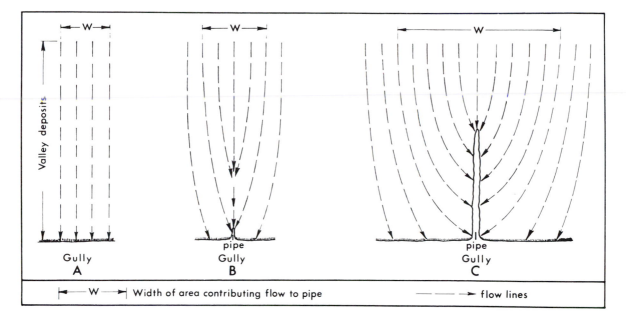

FIG. 12.14. Direction of flow of subsurface water showing an increase of catchment area (W) as a pipe erodes headwards. A, gully wall with no pipe; B, incipient; C, after a pipe has extended. These are plan views. (Modified from Terzaghi and Peck 1948.)

ment, such as lime or gypsum applications on soils with a high sodium content, to reduce soil dispersion.

Distribution of erosional and depositional sites on a hillslope

The nine-unit landscape model (see Fig. 10.2) was developed from field observations: it indicates that flat interfluves are zones of little erosion, transportational slopes are prone to erosion, and eroded material is carried across them to footslopes which are sites of predominant accumulation. This model has received general support from many observations and, on cultivated soils, explicit confirmation in the USLE slope factors for erosional sites.

Footslope colluvial deposits with their increases in organic matter and bases, with greater fertility (Plate 12.9, Fig. 12.10), and lenses of sands and silts are well recognized. In semiarid areas, colluvium may contain interbedded lenses of gravels and sands. Fan deposits, of colluvial footslopes, are usually of small dimensions—a few tens of centimetres to a few metres in length—because

their catchments and rills are also of limited dimensions. Particle sizes are commonly in the silt to sand range with some organic material and remnants of aggregates which may be dominantly of clay particles.

Whatever its texture, colluvium is usually deposited in such thin units, and at such low rates, that bioturbation by plants and animals, such as earthworms, causes bedding to be destroyed as pedological processes incorporate the sediment into the thickening soil. Rates of colluvial soil thickening over the last 1000 years in Iceland, where soil erosion has been periodically very active and where tephra is a readily eroded material, have been assessed as being in the range of 0.7–5.7 mm/y for slope-wash and 1.3–10 mm/y for fans (Gerrard 1985); most colluvial sites have rates of soil thickening which are of much smaller magnitude.

In spite of the general validity of the above statements, it has not been possible to assess quantitatively the rates of soil loss and accumulation or map sites of loss and gain over large areas. Work carried out in the 1980s, however, has changed this situation and provided a means of assessing sediment budgets for individual hillslopes and whole catchments through the use of caesium-137.

Caesium-137 tracer

Caesium-137 is a major component of the fallout from atmospheric testing of nuclear weapons; this radionuclide is currently present in soils and sediments around the Earth as a result of weapons-testing during the 1950s and 1960s. Fallout of ^{137}Cs was first detected in 1954 and fallout rates peaked in 1963–4, and then declined markedly as a result of the 1963 Nuclear Test Ban Treaty. The total amount of fallout, from weapons-testing, received at the land surface varies in response to the magnitude of annual rainfall and the pattern of atmospheric circulation responsible for dispersing the fallout.

In most situations, ^{137}Cs reaching the soil surface as fallout is strongly adsorbed by clay minerals in the upper soil horizons, and little is translocated to lower horizons. Subsequent movement of this radioisotope is generally associated with erosion, transport, and deposition of soil particles. Because of its relatively long half-life of thirty years, approximately 60 per cent of the total input since 1954 may still remain in catchments where it may be used as a tracer (Campbell 1983; Ritchie 1987).

The concentration of ^{137}Cs in soils and sediments can be measured with a germanium detector, although counting time may be as long as ten hours and require a sample with a mass of at least 20 g.

Tests of a hillslope model of soil depletion and gain have been carried out by Loughran *et al.* (1987, 1989, 1990). They have shown convincingly and quantitatively that: (1) ^{137}Cs is found in greater quantities in surface horizons than lower horizons of soils under grass and trees; under cultivation the ^{137}Cs is distributed fairly evenly through the profile down to the level of soil disturbance; (2) sites on drainage divides suffer little or no erosion; (3) sites at the top of cultivated slopes are most depleted in ^{137}Cs and are most eroded, compared with (4) sites at the bases of eroded slopes, which have greater amounts of ^{137}Cs, which is distributed fairly evenly down the profile; (5) hillslope erosion is indicated by the isotope and related to slope length and inclination; (6) it is possible to distinguish between (a) colluvial sediments from channel banks (with very low levels of isotope), (b) those eroded since the 1950s and 1960s (with high levels of ^{137}Cs), and (c) those from sites eroded before the 1950s (zero levels of ^{137}Cs). In general, (7) the conclusions

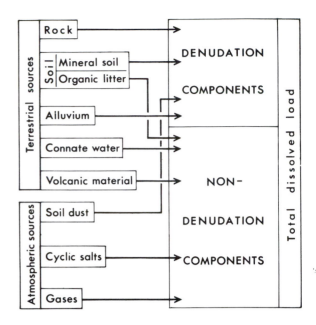

Fig. 12.15. Natural sources of material in the total dissolved load of rivers. (After Janda 1971.)

support the experience from erosion plot work. More importantly (8) they show that it is possible to trace sediment from sites of erosion to sites of deposition, and (9) maps of ^{137}Cs distribution (mBq/cm^2) in cultivated fields (Walling and Bradley 1990) show that spurs and convex slope units are depleted of soil, and concavities accumulate soil, even where these features are minor relief units. Furthermore, (10) the loss of soil from hillslopes is variable depending on both the extent of erosion and the size of the sediment traps within any landscape.

Solution

Solution is a much more important process on slopes than is commonly recognized. In many catchments more than half of the material removed in erosional processes is carried in solution, and it is the transport of solutes from one part of the soil profile to another which permits continuation of weathering processes, the synthesis of secondary minerals, and the formation of pans of iron oxides or carbonates within the profile. With the exception of measurements of chemical erosion of limestones few studies have been made of the effect of solution on slope forms: most studies have been concerned with the contribution of solution to total denudation and with the composition of drainage waters.

The geochemical budget

Solutes derived from a slope or drainage basin may come from a number of sources (Fig. 12.15): only solutes derived from rock and soil weathering are truly part of the denudational component, but they may be exceeded in quantity by the non-denudational component from rainfall, windblown dust and salt, or deep-seated migration of groundwater. If a catchment is virtually watertight so that there is no input or output in groundwater, no inputs from fertilizer or sewage, and no outputs from cropping or water abstraction, then additions which are not denudational can usually be accounted for as inputs of dust or in precipitation and measured in collecting devices. Where vegetation is stable so that the incorporation of nutrients into plant tissue should, over a year, be equalled by the return from biological decay, then:

solutes derived from weathering =
    output as stream solutes −
    input of solutes in precipitation

The precipitation component can be very important, for in areas close to an ocean, or desert salt-flats, catchments may obtain most of their sodium and chloride from the atmosphere. Net gains of potassium are also common, and nitrogen may be greatly increased in abundance by disturbance of vegetation, such as by the ploughing-up of pasture or burning of forest; some nitrogen may also be added from the ammonium (NH_4^+) generated in the electrical activity of thunderstorms (Douglas 1968).

In studies of soil water or river loads the total solutes are measured by evaporation of the water, by the electrical conductivity (which is a measure of the ionic activity), or by chemical analyses made of individual constituents. The results are usually expressed as concentrations in parts per million or milligrammes per litre (1 ppm = 1 mg/l). The distinction between solutes and suspended loads is, however, made on the basis of filtration, with all material passing a 2μm filter being classed as solute. A proportion of solutes is adsorbed on to colloids and may be included in the suspended load. Consequently, and particularly in areas with a high output of colloids with high base exchange capacities, such as montmorillonite or humus, a considerable proportion of the cations may be

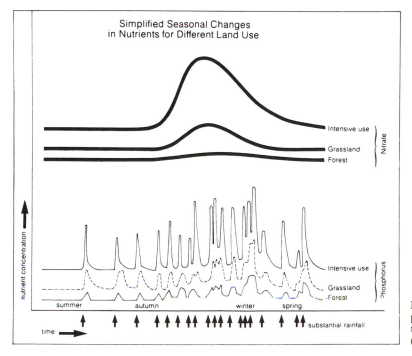

FIG. 12.16. Discharges of nitrate and phosphorus, during a year, in relation to rainfall. The curves are simplified. (After Haughey 1979.)

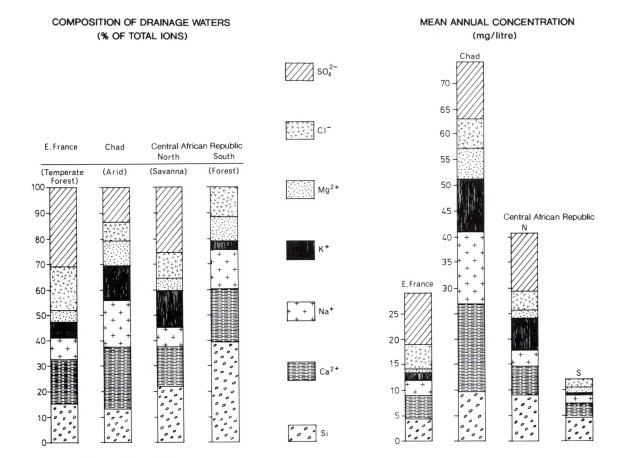

FIG. 12.17. Composition of drainage waters and concentration of ions in them from four areas of sandstone bedrock. (After Mainguet 1972.)

classed with solid loads and the significance of solutes thus underestimated.

Variations in solute discharge

Concentrations and total outputs of solutes vary through the year, especially in areas with seasonal climates. During the growing season decay of soil organic matter and the production of CO_2 by plant roots increases the bicarbonate content of soil water in contact with mineral matter. Chemical weathering is thus more vigorous, and silica and bicarbonate ions show a strong peak in discharge, in the productive season. Plant nutrients such as phosphorus, nitrate, calcium, and potassium show a marked peak in the period of leaf-fall. The most soluble compounds are readily transported and show least fluctuation in output with runoff, but the more insoluble compounds are usually asso-

ciated with solids in suspension and show marked association of concentrations with water discharge: Fig. 12.16 illustrates this point although both nitrate and phosphorus used in the example are probably not produced by rock weathering.

Discharges of the soil solutes may be very closely related, in time, to discharges of water because each storm may produce water which expels older soil water in the profile (where it has gained ions) and replaces it by freshly infiltrating water. Each storm is thus accompanied by the discharge of a mass of enriched water.

Solution and rocks

The ions, occurring in drainage waters, which are the result of chemical weathering of primary or secondary minerals may be only part of the product of weathering. The less soluble products of

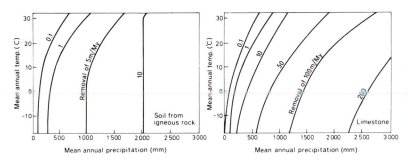

Fig. 12.18. Estimated chemical removal from an igneous rock soil and a pure limestone in metres per million years. Calculated for cations only. As this is a theoretical study it should be regarded as a working hypothesis rather than an established fact. (After Carson and Kirkby 1972.)

weathering remain within the regolith, and in arid climates drainage or soil waters may evaporate and leave pans of iron oxides, calcium carbonate, or other products. Interpretation of solute production from different rock types or the same rock type in different climates has, therefore, to be approached with care and variations in inputs in precipitation recognized. Some general trends do, however, appear to be detectable.

In an important study of sandstones (all consisting of about 92 per cent silica) Mainguet (1972) compared solutes from those rocks in four contrasting environments: the temperate forested area of the Vosges, eastern France; the tropical desert and semi-desert of Chad; the savanna woodlands of northern Central African Republic; and the tropical forests further south. The mobility of the ions has strong seasonal variations, but silica is generally more soluble in the humid tropics and less soluble in arid and temperate areas. Concentrations of solutes are greatest in semiarid areas where water has prolonged contact with rock minerals between rainfall events (Fig. 12.17). Total solute discharges are, however, normally greatest in humid climates where solutions are not saturated, and the flow of water through the soil is high. Total solutional denudation, therefore, tends to increase primarily with rainfall and discharge (Dunne 1978a), and to a lesser extent with temperature—the latter in response to greater rates of chemical reactions (Fig. 12.18).

The conversion of solute output data to estimates of rates of groundsurface lowering can be misleading, for the removal of solutes may be compensated for by a decrease in soil bulk density. Furthermore not all data are representative of long periods nor corrected for precipitation inputs, changes in land use, or variations within a catchment, and some solutes may come from depth where they have no effect upon slope profiles. The

TABLE 12.5. *Estimates of ground lowering by solution alone*

Rock type	Ground lowering mm/1000 year
Precambrian igneous and metamorphic	0.5–7.0
Precambrian micaceous schist	2.0–3.0
Ancient sandstones	1.5–22.0
Mesozoic and Tertiary sandstones	16–34
Glacial till	14–50
Chalk	22
Carboniferous limestone	22–100

TABLE 12.6. *Estimates of ground lowering in central Europe where annual rainfall is 800 mm/year*

Rock type	Ground lowering mm/1000 year
Crystalline rocks	0–2
Sandstone	2–14
Shales and slates	6–18
Limestones and dolomite	24–42
Gypsum and anhydrite	c.400
Sand and gravel	0–36

Source: Data from Hohberger and Einsele (1979).

data from published reports (Tables 12.5 and 12.6) suggest that there is a general increase in solute discharges from Precambrian igneous and metamorphic rocks such as granite, quartzite, and schist, to ancient and Mesozoic or Tertiary sedimentary rocks, with limestones being by far the most soluble common rocks.

In their study of data from 80 catchments in

Central Europe Hohberger and Einsele (1979) emphasize that groundwater carried the greatest part of the chemical load supplied to rivers, so that areas with a high relative yield of surface runoff have low chemical denudation rates. In Central Europe rivers from the Alps transport nearly similar quantities of solid and dissolved material, but rivers of hills and lowlands carry more solutes than solids.

Very few comments can be made on the effect of solute production on slope forms. It might be expected that solution would be at a maximum in areas of rapid throughflow and high soil moisture contents, and at a minimum in areas of low permeability and moisture retention. The contributing areas of an expanded saturated zone would thus lose mass by solution more rapidly than spurs and upper slopes, while footslope areas in semiarid zones of Hortonian overland flow would lose least, and may gain from deposition of salts. Much more field measurement will have to be done before such conclusions are verified or refuted.

13

Mass Wasting of Soils

Mass wasting of soil and rock is a major process in the development of hillslopes and especially those of steep and mountainous regions. It is also a process of enormous economic and social importance—usually in a negative sense. Among the most devastating landslides of the twentieth century are the loess dry-flows in Kansu Province, China, in 1920, which were triggered by a magnitude 8.6 earthquake, and which caused the deaths of 100 000 to 200 000 people (Close and McCormick 1922). The Mt. Huascaran landslide in the Peruvian Andes, 1970, destroyed the city of Yungay and killed 18 000 people: it was triggered by an earthquake and formed a debris avalanche–debris flow (Plafker and Ericksen 1978). The failure of the slopes of the reservoir behind the Vaiont dam in the Italian Alps in 1963 caused a flood wave which overtopped the dam and washed away villages with the loss of 2000–3000 people; the slope failure developed as heavy and prolonged rain caused water-tables to rise and reactivated an ancient landslide (Kiersch 1965). In 1985, lahars killed about 22 000 people at Armero, Colombia.

Tables listing many more devastating slope failures are given in Sidle *et al.* (1985). These tables indicate that the circum-Pacific region, which includes the Andes, Rockies, Japanese Alps, and New Zealand Southern Alps, are particularly prone to large-scale slope failures. These mountains have extreme relief, are tectonically, volcanically, and seismically very active; they have geological and climatic conditions conducive to slope failure, and have suffered widespread changes in vegetation and land use in the last 300 years. Other parts of the world have similar hazardous conditions but not in such concentration. The economic significance of landslides for most countries and regions is described in Brabb and Harrod (1989).

Classification of mass wasting

Mass wasting is the downslope movement of soil or rock material under the influence of gravity without the direct aid of other media such as water, air, or ice. Water and ice, however, are frequently involved in mass wasting by reducing the strength of slope materials and by contributing to plastic and fluid behaviour of soils. The term 'mass wasting' is more inclusive than 'landslide' because the latter does not include falls, topples, and creep as these do not have distinct planes or zones of sliding.

A great variety of materials and processes is involved in mass wasting with a consequent variety of types of movement. Distinguishing between these types requires considerations of at least the following criteria: velocity and mechanism of movement; material; mode of deformation; geometry of the moving mass; and water-content. With so many criteria available, it is perhaps not surprising that there are many classifications in use and conflicts in applications of terms. The earliest widely used classification is that of Sharpe (1938) and most workers since then owe some debt to him for his pioneer effort. More recent classifications are those of Varnes (1958, 1975), Hutchinson (1988), Nemčok *et al.* (1972), and Sassa (1989).

Sharpe's scheme is shown in Table 13.1, that of Varnes (1958) in Table 13.2, and that of Hutchinson in Table 13.3. The classification of Varnes (1958) is of landslides only and excludes creep and frozen ground phenomena, while Hutchinson's is the most complete scheme.

Hutchinson's classification (1988) has two parts: the first part is of all movements on hillslopes, whether they are in soils or rocks, and is based primarily on morphology of the moving mass, with

TABLE 13.1. *Classification of Sharpe (1938)*

	Nature and Rate of Movement	GLACIAL TRANSPORT	With increasing Ice Content ←	Rock or Soil →	With increasing Water-Content	FLUVIAL TRANSPORT
FLOW	imperceptible		SOLIFLUCTION	CREEP (ROCK CREEP SOIL CREEP)	SOLIFLUCTION	
FLOW	slow to rapid		DEBRIS AVALANCHE		EARTH FLOW MUD FLOW DEBRIS AVALANCHE	
SLIDE	slow to rapid			SLUMP DEBRIS-SLIDE DEBRIS-FALL ROCKSLIDE ROCK FALL		

TABLE 13.2. *Classification of Varnes (1958)*

Type of Movement		Type of Material			
		Bedrock		Soils	
FALLS		ROCKFALL		SOILFALL	
SLIDES	few units	rotational SLUMP	planar BLOCK SLUMP	planar BLOCK GLIDE	rotational BLOCK SLUMP
SLIDES	many units		ROCKSLIDE	DEBRIS SLIDE	FAILURE BY LATERAL SPREADING

		All Unconsolidated			
		rock fragments	sand or silt	mixed	mostly plastic
FLOWS	dry	ROCK FRAGMENT FLOW	SAND RUN	LOESS FLOW	
FLOWS	wet			RAPID EARTHFLOW SAND OR SILT FLOW	DEBRIS AVALANCHE DEBRIS FLOW — SLOW EARTHFLOW MUDFLOW
COMPLEX		Combinations of Materials or Type of Movement			

some consideration of mechanisms, material, and rate of movement; the second part is for geotechnical purposes in which the objective is design of stabilizing works, and it is therefore based on recognition of the conditions which control soil strength. Sassa's scheme is based on type of shear and grain size of material. Nemčok emphasizes the velocity of movement. The proliferation of classifi-

TABLE 13.3. *Classifications of slope movements by Hutchinson (1988)*

I. CLASSIFICATION BY MORPHOLOGY WITH SOME CONSIDERATION OF MECHANISM, MATERIAL, AND RATE OF MOVEMENT.

A. Rebound

Movement associated with:
1. Excavations, from human activity
2. Naturally eroded valleys

B. Creep
1. Superficial, predominantly seasonal creep; mantle creep:
   (a) Soil creep, talus creep (non-periglacial)
   (b) Frost creep and gelifluction of granular debris (periglacial)
2. Deep-seated, continuous creep; mass creep
3. Pre-failure creep; progressive creep
4. Post-failure creep

C. Sagging of Mountain Slopes
1. Single-sided sagging associated with the initial stages of landsliding:
   (a) of rotational (essentially circular) type (R-sagging)
   (b) of compound (markedly non-circular) type (C-sagging):
       (i) listric (CL); (ii) bi-planar (CB)
2. Double-sided sagging associated with the initial stages of double landsliding, leading to ridge spreading:
   (a) of rotational (essentially circular) type (DR-sagging)
   (b) of compound (markedly non-circular) type (DC-sagging):
       (i) listric (DCL); (ii) bi-planar (DCB)
3. Sagging associated with multiple toppling (T-sagging)

D. Landslides
1. Confined failures:
   (a) in natural slopes
   (b) in excavated slopes
2. Rotational slips:
   (a) Single rotational slips
   (b) Successive rotational slips
   (c) Multiple rotational slips
3. Compound Slides (markedly non-circular with listric or bi-planar slip surfaces):
   (a) released by internal shearing toward rear:
       (i) in slide mass of low to moderate brittleness
       (ii) in slide mass of high brittleness
   (b) progressive compound slides, involving rotational slip at rear and fronted by subsequent translational slide
4. Translational slides
   (a) Sheet slides
   (b) Slab slides; flake slides
   (c) Peat slides
   (d) Rock slides:
       (i) Planar slides; block slides
       (ii) Stepped slides
       (iii) Wedge failures
   (e) Slides of debris:
       (i) Debris-slides; debris avalanches (non-periglacial)
       (ii) Active layer slides (periglacial)
   (f) Sudden spreading failures

E. Debris Movements of Flow-Like Form
1. Mudslides (non-periglacial):
   (a) Sheets
   (b) Lobes (lobate or elongate)
2. Periglacial mudslides (gelifluction of clays):
   (a) Sheets
   (b) Lobes (lobate or elongate, active and relict)
3. Flow slides:
   (a) in loose, cohesionless materials
   (b) in lightly cemented, high porosity silts
   (c) in high porosity, weak rocks
4. Debris flows, very to extremely rapid flows of wet debris:
   (a) involving weathered rock debris (except on volcanoes):
       (i) Hillslope debris flows
       (ii) Channellized debris flows; mud flows; mudrock flows
   (b) involving peat; bog flows, bog bursts
   (c) associated with volcanoes; lahars:
       (i) Hot lahars; (ii) Cold lahars
5. Sturzstroms, extremely rapid flows of dry debris

F. Topples
1. Topples bounded by pre-existing discontinuities:
   (a) Single topples
   (b) Multiple topples
2. Topples released by tension failure at rear of mass

G. Falls
1. Primary, involving fresh detachment of material; rock and soil falls
2. Secondary, involving loose material, detached earlier; stone falls

H. Complex Slope Movements
1. Cambering and valley bulging
2. Block-type slope movements
3. Abandoned clay cliffs
4. Landslides breaking down into mudslides or flows at the toe:
   (a) Slump-earthflows
   (b) Multiple rotational quick-clay slides
   (c) Thaw slumps
5. Slides caused by seepage erosion
6. Multi-tiered slides
7. Multi-storied slides

TABLE 13.3. (*Contd.*)

II. GEOTECHNICAL CLASSIFICATION OF SLOPE MOVEMENTS BY SHEARING BASED ON SOIL FABRIC AND PORE-WATER PRESSURE CONDITIONS.

A. Soil Fabric (*effects on c', ϕ'*)

1. FIRST-TIME SLIDES IN PREVIOUSLY UNSHEARED GROUND: soil fabric tends to be random (or partly orientated as a result of depositional history) and shear strength parameters are at peak or between peak and residual values.

2. SLIDES ON PRE-EXISTING SHEARS associated with:
2.1 Re-activation of earlier landslides.
2.2 Initiation of landsliding on pre-existing shears produced by processes other than earlier landsliding, i.e.:
   (a) Tectonics
   (b) Glacitectonics
   (c) Gelifluction of clays
   (d) Other periglacial processes
   (e) Rebound
   (f) Non-uniform swelling

In these cases the soil fabric at the slip surface is highly orientated in the slip direction, and shear strength parameters are at, or about, residual value.

B. Pore-Water Pressure (*conditions* on shear surface, effects on *u*)

1. SHORT-TERM (undrained)—no equalization of excess pore-water pressure set up by the changes in total stress.

2. INTERMEDIATE—partial equalization of excess pore-water pressures. Delayed failures of cuttings in stiff clay are usually in this category.

3. LONG-TERM (drained)—complete equalization of excess pore-water pressures to steady seepage values.

Note that combinations of drainage conditions 1, 2, 3 can occur at different times in the same landslide. A particularly dangerous type of slide is that in which long-term, steady seepage conditions (3) exist up to failure but during failure undrained conditions (1) apply, i.e. a drained/ undrained failure.

cations has tended to defeat the objectives of all classifications—achievement of clear, unambiguous communication. Because Varnes' terminology for landslides is commonly used it will be followed here, but it has to be realized that this, like all other classifications, can be difficult to apply. Recognizing the nature of a process from the debris it leaves behind is not always easy and the group of processes—debris slide, debris avalanche, debris flow (Fig. 13.1)—collectively called translational slides—are not always distinguishable from each other. Slides frequently break up into avalanches and may become flows towards the base of a slope if sufficient water is available. A second source of confusion in classifications is the varied uses attached to the words debris, soil, earth, and mud. The 1975 classification of Varnes tries to overcome this problem by using 'debris' to include only coarse material, and 'earth' to denote sand, silt, and clay (Table 13.4). Even this distinction is not always possible because many failed regoliths include a range of materials from boulders to clay.

Types of mass wasting in soils

Creep

The term 'creep' has been used in several senses in hillslope studies, and distinctions between these uses are not always recognized. Here a distinction will be made between: (1) creep of single particles over an exposed bedrock or soil surface; (2) depth creep in soils and rock in which an overburden is transported as a mass above the zone of deforming material; and (3) soil creep.

Particle creep is most readily discerned when painted stones are monitored and observed to move downslope. Most observations have been carried out in semiarid and montane environments. Leopold *et al.* (1966) recorded particle movements on a hillslope in New Mexico over a four-year period and suggested that gravity aided by freeze–thaw and wet–dry cycles was largely responsible for transport of the stones. Schumm (1967) placed sandstone fragments on shale slopes, over seven years, and found the downslope movement to be directly proportional to the sine of the

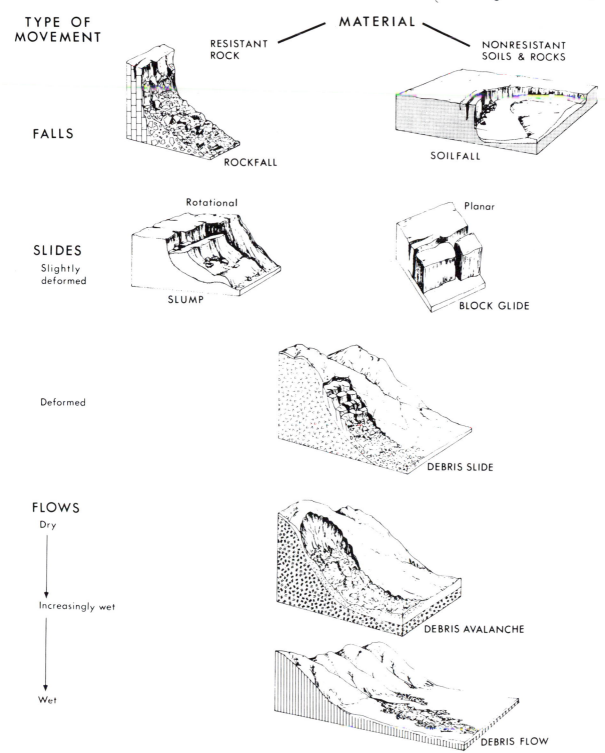

TYPE OF MOVEMENT

MATERIAL

RESISTANT ROCK

NONRESISTANT SOILS & ROCKS

FALLS

ROCKFALL

SOILFALL

SLIDES
Slightly deformed

Rotational

SLUMP

Planar

BLOCK GLIDE

Deformed

DEBRIS SLIDE

FLOWS
Dry

Increasingly wet

Wet

DEBRIS AVALANCHE

DEBRIS FLOW

Fig. 13.1. The main mass-wasting types according to the classification of Varnes (1958).

TABLE 13.4. *Classification of Varnes (1975)*

Type of Movement			Bedrock	Soils coarse	fine
FALLS			ROCKFALL	DEBRIS FALL	EARTH FALL
TOPPLES			ROCK TOPPLE	" TOPPLE	" TOPPLE
SLIDES	rotational	few units	" SLUMP / " BLOCK GLIDE	" SLUMP / " BLOCK GLIDE	" SLUMP / " BLOCK GLIDE
	translational	many units	" SLIDE	" SLIDE	" SLIDE
LATERAL SPREAD			" SPREAD	" SPREAD	" SPREAD
FLOWS			" FLOW (deep creep)	" FLOW (soil creep)	" FLOW
COMPLEX			Combination of 2 or more types		

hillslope gradient which is, in turn, proportional to the component of gravity acting parallel to the hillslope (see Fig. 13.8(1)); he inferred that the movement was creep induced by frost action. Kirkby and Kirkby (1974) studied movements on hillslopes in southern Arizona, during a two-month period, of all particles with diameters >1 mm; observations confirmed that rainsplash caused the movements and that the distance moved was directly related to hillslope gradient, inversely related to particle size, and unrelated to the distance from the divide. Abrahams *et al.* (1984) studied stones with a minimum diameter of 8 mm, on hillslopes in the central Mojave Desert, and found the distance moved to be directly related to length of overland flow and hillslope gradient, and inversely related to particle size; they concluded that hydraulic action is responsible. Movements of particles on talus slopes (see Chapter 15) are far more complex than those on slopes with lower inclinations, and may involve creep from action of freeze–thaw, but also movement from impacts of rolling particles as well as a range of mass movements caused by debris flows and snow avalanches.

The main conclusion is that particle creep from freeze–thaw may be dominant in areas where climate is suitable and that rainsplash and wash may have similar effects in semiarid environments where frost action is limited.

Depth creep is a process which has been noticed mainly by engineers involved in construction of deep cuttings and tunnels and in stabilizing slopes on which very slow, but continuous, movements are occurring. Some such movements may be due to sliding along distinct failure planes—a process which is, by definition, not creep. Continuous creep can occur in a zone where no distinct failure planes are evident and where the shearing stresses have values which are much less than the peak strength of the material (Terzaghi 1936). Such slow creep movements can gradually close tunnels and cuttings but, more catastrophically, can lead to accelerated creep and shear failure (see Fig. 4.4a).

Models used to describe depth creep are usually based on Bingham rheological behaviour described by the relationship (Van Asch *et al.* 1989):

$$\tau = \tau_o + \eta\dot{\varepsilon},$$

where: τ = shear stress;
τ_o = yield strength (creep threshold shear strength);
η = a flow factor;
$\dot{\varepsilon}$ = shear strain rate.

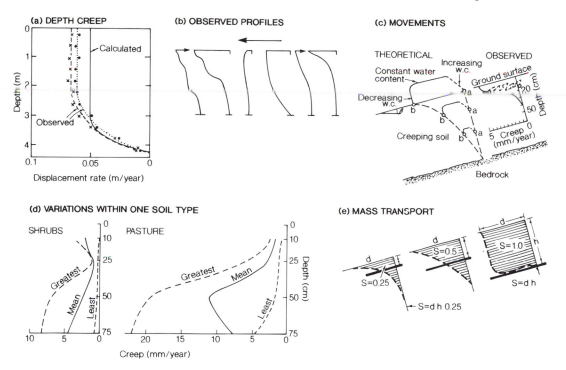

Fig. 13.2. Creep profiles: (a) depth creep as calculated and as observed in two profiles (after Van Asch and Van Genuchten 1990); (b) types of profile movement as recorded by columns of blocks and by plastic pipes (after Jahn 1981); (c) a theoretical profile of soil particles moving from a to b in a soil with seasonal moisture changes, and an observed profile with varied directions of movement (the theoretical profile is after Fleming and Johnson 1975); (d) rates of creep in pumiceous soils (after Selby 1974*c*); (e) the mass transport of soil calculated in accordance with the observed creep profile (after Jahn 1981).

The yield strength, τ_o, can be described by the Coulomb equation; for most soils this will describe the residual strength, and in most clay soils $c'_r \approx 0$, so the appropriate equation is:

$$\tau_o = \sigma' \tan\phi'_r.$$

The flow factor η is a viscosity coefficient if the soil behaves as a fluid but, if the soil has a structure which modifies flow behaviour, the flow factor value may increase with the effective stress and a modified flow factor is required (Van Asch and Van Genuchten 1990). In a study of varved clays, these two authors found that the combination of a residual strength value for overconsolidated clays and a flow factor which depends on effective stress gave the best prediction of creep velocity as measured in the field (Fig. 13.2a). This figure illustrates the essential feature of deep creep—that is the transport of a rigid body of material above the zone of deformation.

Soil creep is usually defined as the slow down-slope movement of superficial soil or rock debris which is usually imperceptible except to observations of long duration. Early geomorphologists attributed to it a significant part in the development of hillslopes. W. M. Davis (1892) invoked creep to explain the summit convexities of badland divides (see Plate 12.12) and low-angle summit convexities of hillslopes in humid regions. He also suggested that creep rates are fastest near the groundsurface and increase with slope angle. G. K. Gilbert (1909) attributed the convexity of hilltops to creep, and demonstrated how creep can produce a convex slope profile by assuming that creep rate is a function of slope angle. Sharpe (1938) made a distinction between creep in temperate and tropical climates and creep and solifluction of cold climates.

Davis and Snyder (1898) attributed creep to freeze–thaw and the action of soil animals and plant roots; other workers have emphasized wetting and drying, and heating and cooling, with resulting expansion and contraction of the soil. Soil creep is therefore seen to be periodic or seasonal in areas with variable climate.

The mechanics of creep have been investigated experimentally and theoretically (Terzaghi 1953; Goldstein and Ter-Stepanian 1957; Saito and Uezawa 1961; Culling 1963, 1983; Haefli 1965; Bjerrum 1967; Carson and Kirkby 1972; Van Asch et al. 1989). Four principal mechanisms have been postulated as producing creep on hillslopes: pure shear, viscous laminar flow, expansion and contraction, and particulate diffusion (Donohue 1986). Shear and flow should be detectable in the profiles of displacement of pins, blocks, and tubes, placed in the soil. Diffusion is the movement of single particles, with many possible causes of displacement; it is necessarily very difficult to detect and would require the insertion of marked or chemically 'labelled' particles into the soil. Expansion and contraction can be more readily detected and has been identified (Fig. 13.2c).

Any theory should be testable by laboratory experiment and field measurement. Herein lie two problems: laboratory experiments are carried out on samples which cannot readily contain all of the processes of natural conditions, and in the field all measurements disturb the soil under study; they are carried out over relatively brief periods and by a range of methods of varying precision. The most common techniques are described in Goudie (1981). Rates of creep are usually studied by placing pins, or acrylic rods, in the walls of trenches which are then refilled; by inserting columns of beads, blocks, or tubes into the soil; by attaching cones to piano wire which is then led to the surface; by inserting sensitive tilt bars into the surface soil; or by using strain gauges. All methods have errors associated with them, but there are now sufficient measurements available to indicate that common creep rates downslope are between 0.1 and 15 mm/year in vegetation-covered soil, but on exposed talus slopes, or in cold climates where freeze–thaw processes are common, higher rates up to 0.5 m/year have been recorded. (See Young 1972; Carson and Kirkby 1972; Swanson and Swanston 1977; Jahn 1981, 1989.)

A severe limitation on most sets of data is that they have been collected over relatively short periods of 2 to 3 years. The studies of Young (1978) and Jahn (1981, 1989) are exceptions to this. Young made observations over a 12-year period, in northern England, and showed that the linear downslope component of movement within the top 20 cm of mineral soil was 0.25 mm/year, but this was exceeded by a component of 0.31 mm/year inwards towards the bedrock and perpendicular to the groundsurface. The inward movement is interpreted as being caused by loss of weathered material in solution with rearrangement and settling of the remaining particles. The greatest loss at the surface, with progressively smaller losses with depth in the profile, is consistent with the suggestion that chemically undersaturated rainwater enters the soil from the surface, and takes ionic materials into solution until it reaches chemical equilibrium at depth. As solution represents ground loss, and presumably results in rearrangement of soil particles, it is interpreted as being more important than creep. There are, however, few comparable records of this process.

Work on creep has been undertaken in the Sudetes Mountains of Poland by Jahn and his colleagues over the period 1960–88 using plastic tubes and columns of wood blocks inserted 1 m into the soil. The tubes and columns of blocks were excavated after intervals of 5, 10, 12, and 17 years. Fifty columns were established and 35 survived to be excavated, making this a unique programme of study. The shapes of the profiles which were excavated are varied (Fig. 13.2b) and do not support any of the theoretical models which have been published. Three ecological zones were sampled: in low altitude pastures, up to 800 m, surface creep averaged 2–3 mm/y over a 10-year period; in mature forest, at 800–1200 m, no creep was observed over 12 years; in the high alpine grasslands, above 1200 m, soil movement averaged 8–9 mm/y over periods of 5 and 10 years. Human activities of farming at low altitudes, pasture animals, tourism, and winter sports at high altitudes, are regarded as at least doubling the natural rate of creep.

A feature of some observations of creep is the resistance to downslope movement near the groundsurface resulting from the binding effect of plant roots. This is very clear in the measurements of Selby (1974c) made over a 3.5-year period in pumiceous soils at sixteen sites; those sites under pasture grasses showed the greatest retardation at the groundsurface and those under a close shrub cover showed a maximum retardation at a depth of 25 cm where root density is greatest (Fig. 13.2d). The high rates of movement at depths below 75 cm may be a result of the very high permeability, low bulk density and low particle density of the soil, with high rates of water movement, displacement of particles, and formation of pipes.

A study of Moeyersons (1988) on hillslopes in Rwanda has produced some instructive evidence. The hillslopes have terracettes, bent trees, and an old landslide which indicate the probability of creep. Young pits were installed, with lines of blades set into the soil normal to the sides of an excavated pit. Re-excavations over several years revealed patterns of movement which were exceedingly complex, with variable rates of movement with depth and from point to point and, more notably, indications of cells of apparent turbulent flow at depths of about 0.5 to 1.5 m. The site is in pasture grasses with mean annual temperatures around 20 °C with no frost, so cryoturbation is not the cause. The rises and falls of displacement lines indicating the paths of the blades suggest that zones of change of soil volume or soil density are developed and that the detailed ground surface form may be changed. Soil displacement appears to result from turbulent displacements in the solum as a result of biogenic and soil moisture changes being superimposed on a continuous deep creep. Deep creep is indicated by accurate triangulation of big eucalyptus trees which 'float' on a thick creeping soil mantle. Such information is important for the design and stability of buildings, roads, sewers, and other utilities.

Some measurement methods permit regular or seasonal measurements. Rashidian (1986) reported greatest movement in spring after snowmelt but observations were limited to 1.5 years. In Selby's observations rates of movement were high but showed no seasonal variation, no relationship to soil moisture change, to slope gradient, or aspect. The work of Finlayson (1985) fits even less with conventional views on soil creep. He used flexible tubes with a light source at their base and a travelling microscope on a tripod to sight on cross-wires in the tubes. The resulting profiles of the tubes show that movements are very irregular, often have a substantial uphill component, and are weakly influenced by slope gradient.

The mass transport of soil by creep is dependent partly on the creep rate, but even more on the creep depth and profile (Fig. 13.2e). If creep is occurring by shearing at a constant depth and all soil above that shear plane is moving as a rigid body, the cross-sectional area of the moving mass will be rectangular and the mass transport for a 1

metre segment measured along the contour can be represented by:

$$S = dh,$$

where: S is the volume of soil transported down the slope in one year, d is the annual creep rate at the groundsurface, and h is the depth into the soil over which that creep rate can be apportioned. For creep profiles which are concave downslope, the value of h will be reduced by some fraction.

The mass transport by creep is consequently very limited at all sites where there is a marked decrease in creep rate with depth and the depth to which creep occurs is shallow. The importance of creep has commonly been attributed to its theoretical occurrence at any point on a hillslope irrespective of distance from a divide or of slope length, but its total and relative importance for soil transport will depend upon the activity of other slope processes.

Many types of field evidence have been held to demonstrate the existence of soil creep including outcrop curvature, tree curvature, tilting of structures, soil accumulations upslope of retaining structures, turf rolls, and cracks in the soil. Most of these phenomena can, however, be produced by other processes such as wind, slope wash, depth creep, sliding, or tilting under the weight of the object. It is better to recognize that movement has occurred on the slope than to ascribe it to a particular process without unequivocal evidence.

In brief then, downslope soil creep does occur on slopes, but irregularly in direction and rate, probably by several mechanisms of which the activity of soil animals and plant roots may be important, but none of the data sets supports a single model of pure shear or viscous laminar flow, and diffusion is very difficult to study. Expansion and contraction by one or more of several mechanisms has greater support, but by implication rather than experimental proof.

The observation by Jahn, that creep does not occur beneath old forest, even on gradients of 30°, casts doubt on the general validity of the hypothesis that convex upper slopes, in humid temperate environments, are due to soil creep. Other hypotheses such as solifluction and other processes under pre-forest environments may need further examination.

The observations of Moeyersons, however, indicate that much further work is needed also in humid and subhumid tropical areas with deep soils,

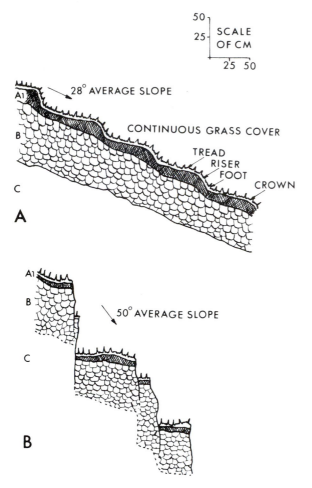

50⌐
25⌐ SCALE
     OF CM
      ⌐————
       25 50

28° AVERAGE SLOPE

CONTINUOUS GRASS COVER

TREAD
RISER
FOOT
CROWN

A

A1
B
C

50° AVERAGE SLOPE

B

Fig. 13.3. Idealized terracettes on gentle and steep slopes.

for these may be areas where creep is a major influence on hillslope form.

The uncertainties will only be resolved by many more carefully designed long-term programmes of measurement, which are not biased towards the assumption that the only direction of soil movement is downslope, and which permit repeated measurement of the precise location of indicators without disturbing the soil.

Terracettes are, perhaps, the most prominent surface features attributed to soil creep. They are interlacing networks of low but long tracks (see Vincent and Clarke 1976, for a review). The networks may form continuous staircases up some hillslopes, or they may occur intermittently. It has been estimated that on a 24° slope, on soils from

loess, they may occupy as much as 11 per cent of the groundsurface and on a 37° inclination on steep mudstone-based hillslopes 40 per cent.

It is evident that many short discontinuous terracettes form beneath forest by the washing of soil and accumulation of leaf litter behind tree roots. When forest is cleared, and the land sown in pasture grasses, an incipient terracette pattern is therefore already in place. It is equally clear, however, that animal treading integrates several different kinds of slope irregularities into continuous networks which roughly follow the contours. There is also evidence that animal treading can develop terracettes on previously smooth slopes in a few weeks (Higgins 1982). The habit of sheep to follow-the-leader is particularly conducive to this process. The spacing of terracettes is influenced by the size of the animals and the slope angle (Howard and Higgins 1987). Treads become repositories for animal dung and urine which encourage plant growth, particularly along the edge of a terracette, and this factor may also assist in the preservation of a sharp edge to the feature. In periglacial environments, frost action and flow, as solifluction above permafrost, can cause lobate terracettes.

In morphology, terracettes appear to belong to two main classes (Fig. 13.3). On low-angle slopes with slow rates of soil movement, the vegetation cover is not broken and the terracette form is subdued. On steeper slopes and with more rapid soil movement, soil cracking increases, especially at the back of each tread, and a minor process of slumping sometimes appears to be occurring.

Falls

Falls in soil or soft rocks usually involve only small quantities of material because steep slopes in weak materials are necessarily very short. These falls are usually the result of undercutting of the toe or face of a slope by a river or by wave action (Plate 13.1). They are facilitated also by weathering and the opening of fissures near a cliff top as a result of freeze–thaw, wetting and drying, earthquake shocks, or tension.

Slumps

Slumps have curved failure planes and involve rotational movement of the soil mass. They have received much attention in the engineering literature because they are common in the walls of cuttings in soft rocks such as shales, mudstones,

PLATE 13.1. A stream-bank showing, from left to right: a block near to failure, as indicated by the undercut and propagating joint; undercut and cambered blocks with evidence of recent falls; and blocks without cambering with slip debris at the base.

PLATE 13.2. The crown of a rotational landslide.

and over-consolidated clays. They occur also under entirely natural conditions, especially where the toe of a slope has been undercut by river or wave action. Slumps may be single rotational failures with shearing occurring rather rapidly on a well-defined curved surface. Backward rotation of the sliding mass often causes fracturing of further blocks at the crown of the slump (Plate 13.2) and sometimes is accompanied by ponding of water in the depressed, or backwards-tilted, area of the crown. Such water may promote continuing movement. In many materials, especially with high water-content, the toe of a slump may be so remoulded that it becomes a flow (Fig. 13.4; Plate 13.3).

Some slumps, and especially those in clays underlain by hard impervious strata, and overlain by porous caprocks which form water reservoirs, develop multiple rotational slides with two or more blocks tilting backwards and moving on the same slide plane at the sole of the slumps. In stiff fissured clays forming relatively high relief the face of a slope may be formed of numerous shallow rotational slides. They develop from the base of a slope where the lowest slump removes lateral support from the slope immediately above it, which then fails. This type of retrogressive failure may gradually extend up the slope. Deep-seated multiple retrogressive failures appear to be most common in sea-cliffs experiencing continuing removal of the toe, and shallow multiple failures are most common in the soil forming in a stiff clay as weathering reduces the shear strength of the superficial materials (e.g. Brunsden and Jones 1976).

Slumps with failure planes at depth occur in cohesive soils in which the shear stress is zero at the groundsurface and increases linearly with depth due to the weight of the overburden, but the increase in shear strength with depth occurs at a slower rate. At some critical thickness of the overburden, therefore, the shear stresses exceed the shear strength and failure is possible.

Lateral spreads

Failures by lateral spreading are special classes of slumps which are virtually confined to clay-rich sediment deposited in the shallow seas and lakes around the edges of ice-sheets. Nearly all known examples come from southern Norway, the St. Lawrence lowlands of eastern Canada, and the Alaskan coast. The failures usually begin with a single rotational slide in a bank undercut by a stream or the sea. The slide movements remould the clay along the slide plane very rapidly, and it turns to a dense liquid which supports moving blocks of overlying clay or sands to leave a chaotic topography of small horsts and grabens (Fig. 13.5; Plate 13.4). The movements are retrogressive as each block removes the lateral support of the soil upslope of it. The soil close to the river is sometimes more weathered than that further inland and the plan of the failure then becomes bottle-necked in shape with the narrow neck in the river bank.

Most failures by lateral spreading occur in sensitive soils (see Fig. 7.4). Many of the St. Lawrence basin marine glacial deposits are lightly overconsolidated and have exceptionally high undrained shear strengths. The clays contain high proportions of unweathered primary minerals, of less than $2\,\mu m$ size, as quartz and feldspar glacial flour, with few expanding lattice clay minerals (Cabrera and Smalley 1973; Smalley and Taylor 1972; Smalley 1976). In their undisturbed state the soils are naturally well-cemented, but once remoulded become fissured and reduced to gravel-sized aggregates with low strengths.

Block glides

Block glides are slides in which material remains largely undeformed as it moves over a planar slide plane. These are uncommon features in soils, except at a very small scale.

Translational slides

Translational slides are by far the most common form of landslide occurring in soils. They are always shallow features and have essentially straight slide planes which usually develop along a boundary between soil materials of different density or permeability (Plate 13.5). On many steep slopes the boundary between the solum and the saprolite becomes the slide plane and on mudrocks the solum-rock boundary is the slide plane. Depths to the failure plane are usually in the range 1–4 m and the length of the slide is commonly great compared with the depth.

The distinction between a slump and a translational slide is usually clear, but one failure can undercut a slope and lead to a failure of another kind (Plate 13.6).

The distinction between debris slides, debris avalanches, and debris flows is based upon the

PLATE 13.3. A large slump in mudstones showing the hummocky lower area with disturbed drainage pattern and ponding. The debris has dammed the river and continuing erosion of the toe is causing repeated movements.

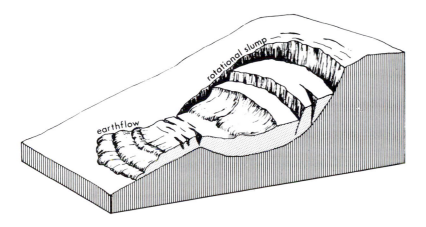

FIG. 13.4. Features of a slump.

PLATE 13.4. A large spreading failure in sensitive soils; the Nicolet landslide of November 1955, Quebec, Canada (from a paper by Jacques Béland, courtesy of Roger Bédard, 'Nicolet Landslide, November 1955', *Proceedings of the Geological Association of Canada*, 8, pt. 1 (Nov. 1956). By permission of M. Béland.)

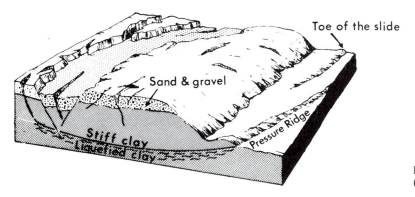

FIG. 13.5. A lateral spreading failure. (After Hansen 1965.)

degree of deformation of the soil material and the water-content of the sliding mass. As both deformation and water-content frequently increase downslope, what may be a debris slide at the crown of the landslide—with relatively large undeformed blocks of soil sliding downhill—may become an avalanche of small blocks and wet debris in mid-slope and a thoroughly liquefied flow at the base of

PLATE 13.5. The crown of a translational slide showing the failure plane at the sharp boundary between saprolite, which now forms the surface of the scar, and the solum. Note the macropores in the crown scarps, which increase infiltration capacity.

the slope, especially if the flow mass descends into a river (Plate 13.7).

Unlike falls, slumps, and glides which may occur as a result of deep percolation of water, and hence at a considerable time after a rainfall, translational slides nearly always occur during heavy rain. Rainstorms with sufficient intensity or duration are required to raise the water-table to near the soil surface or fill pre-existing tension cracks. In low-intensity rainfalls the removal of water from the soil by throughflow can keep pace with infiltration, and in short-duration falls the field capacity of the soil may not be exceeded. Only when the capacity of the soil to drain is exceeded for long enough for water pressures to rise substantially can the soil lose sufficient strength to fail.

Accounts of extreme storms causing many translational slides in the Appalachians are given by Jacobson *et al.* (1989) and Gryta and Bartholomew (1989), in Puerto Rico by Simon *et al.* (1990), in New Zealand by Salter *et al.* (1981), and in California by Ellen and Wieczorek (1988).

Flows

Flows occur when coarse debris, fine-grained soil, or clay are liquefied (Plate 13.8). The terms 'debris flow', 'earthflow', and 'mudflow' are used to distinguish between these three classes of material. The tendency for flows to develop may be encouraged by a number of factors: remoulding of soils during landsliding; the presence of clays with high liquid limits in areas where rainfalls are high; the presence of soils with low liquid limits in areas of low rainfalls—in such soils little water is required to make the soil behave as a liquid; the presence of soils with open fabrics resulting from

PLATE 13.6. A slump formed towards the base of the slope where a stream undercut the toe and water-tables are high. It has progressively extended up the slope. Another failure is developing to the right of the photograph where cracking and slight dropping of the head and bulging of the toe of a failure can be seen. Note the terracettes on the remoulded soil.

PLATE 13.7. Numerous translational landslides formed on these slopes during an abnormally wet winter, Wairarapa, New Zealand. The soils are in loess overlying mudstones. Note the debris which has flowed into the valley floors.

flocculation during deposition; or the thawing of soil ice. It has been suggested by Hutchinson and Bhandari (1971) that some flows may be promoted by the collapse of soil from surrouding cliffs or steep slopes on to the upper part of a concave moving mass, thus raising pore-water pressures and promoting flow which, in turn, causes loading on the debris further down the flow and rapid movement. They called this mechanism 'undrained loading'. A feature of many mudflows is that they can occur on slopes of very low angle. Mudflows may be relatively slow-moving (1–20 m/year) or very fast, as in the catastrophic flows broadly grouped as debris flows (Costa 1984). This topic overlaps with processes occurring on rock slopes and is discussed with them in Chapter 14.

The term 'solifluction' literally means soil flow and is usually used to refer to the flow failures of high latitudes and high elevations, although some

PLATE 13.8. A small debris flow. Note the levees.

authors prefer to use the term 'gelifluction' for the slow flows occurring in thawing soils.

Dry flows range in size from large catastrophic failures of thick deposits of loess (wind-blown silt-sized deposits) in China to small flows of dry sand and silt in quarries and sandpits or down the faces of dunes.

Field-study of landslides

Most geomorphic studies of landslides attempt to assess the quantitative significance of landslides in total denudation and in modifying the ground-surface, or attempt to determine the causes and mechanics of the process. Studies usually begin with the preparation of a map of the area relating the sites of landslides to lithology, slope angle, pre-existing slope disturbances, hydrological conditions, and locations on the slope. A study is also made of the climatic or seismic conditions before and at the time of the slide. Where an attempt is to be made to understand the mechanisms of the failure the geometry of the slide has to be determined and the appropriate method of shear strength testing adopted. The following notes are intended as a guide to appropriate procedures and the use of a common terminology.

One of the difficulties of attempting to discover the significance of erosional events is that many different forms of reporting are used and the information given for each case study is varied. In Table 13.5 four landslide-producing storms and their effects are compared. If this method were used more commonly a better understanding could be obtained of mass-wasting processes and their importance.

Morphometry

The shape, or morphometry, of landslides has its own terminology which is outlined in Fig. 13.6 in which a translational slide is displayed. If the terms were being applied to a slump the head of the displaced mass would be much closer to the crown of the slide and the length of the exposed slide plane would be small. Also shown in Fig. 13.6 are the measurements which are made to determine the area, depth, and volume of the scar and the displaced mass. This information is needed for quantitative studies of the stability of hillslopes against landsliding.

Attempts have been made to use the ratio between the depth and length of landslides as a means of interpreting the processes which gave rise to the scar (Table 17.1). Crozier's (1973) compilation of data for landslides in New Zealand showed that the ratios could be used to distinguish types of process but, by contrast, Blong (1973) concluded that simple morphometric indices are not of value for distinguishing the processes which have given rise to a landslide, as similar ratios may be associated with widely differing behaviour among failures. Because the process of erosion affects both the shape of the scar and the fabric and strength of the deposited material it is advisable to

TABLE 13.5. *Landslide records*

	Mgeta, Tanzania	Tarfala, Lappland	Longyear Valley, Spitsbergen	Mangawhara Valley, New Zealand
1. Latitude, Longitude	7° S; 37° E	68° N; 19° E	78° N; 16° E	37° S; 175° E
2. Altitude (m.a.s.l.)	1100	1130	37	150–350
3. Mean annual temperature (°C)	24.3	−4.3	−6.2	14
4. Annual precipitation (mm)	1058	950	203	1800
5. Rainfall records since	1951	1947	1915	1956
6. Relief range (m)	1100–2000	1200–2000	50–500	50–500
7. Bedrock	Gneiss/granulite	Amphibolite schists	Sandstone, schists	Sandstone/siltstone
8. Regolith type, depth (m)	Sandy/silty regolith	Bouldery debris	Bouldery debris on permafrost	Clay to sandy, with volcanic ash 1.5–2.0
9. Vegetation	Cultivation, grass fallows	Mountain tundra	Mountain tundra	Pasture grasses
10. Rainstorm date	23.II.1970	6.VII.1972	10-11.VII.1972	28.II.1966
11. Rainstorm total (mm)	101	45	31	150–230
12. Rainstorm duration (hours)	2	2	10	24
13. Rainstorm type	Convective	Convective	Cyclonic	Cyclonic with local convection
14. Rainstorm area[a] (km^2)	50	11	>30	250
15. Rainstorm return period (years)	10	>27	>53	10 or >10
16. Slope erosion type[b]	Debris slides	Gullies	Debris slides	Debris avalanches/debris slides
17. Slope gradient (degrees)	33–44	12–30	30–34	32
18. Depth of scars (m)	0.8–3	0.5–4	0.6–1	0.9
19. Width of scars (m)	5–30	1–13	5–20	12
20. Denudation rate/catchment area[c] (mm/km^2)	14 mm/20 km^2	5 mm/11 km^2	1 mm/4.5 km^2	80 mm/0.5 km^2 to 10 mm/20 km^2
21. Mass transport form	Mudflow	Debris flow	Debris flow	Debris flow
22. Mass deposition form	Total fluvial removal	Colluvial fan	Colluvial fan	Colluvium reworked in streambed
23. Mass max. depth (m)	—	3	1.5	1.5
24. Mass gradient (degrees)	0	10–15	8–15	5–25

[a] Area in which mass movement occurred.
[b] Classification of Varnes (1958).
[c] Vertical downwearing of the landscape. (Denudation rate is calculated from volume of erosion scars.)
Source: Data from Rapp (1975); and Selby (1976*a*).

study all available evidence before categorizing a landslide.

Shear-strength testing

In all strength tests an attempt is made to simulate the magnitude of loads, the loading rates, and drainage conditions which occur in the slope failures being studied. The situations and suggested parameters given in Table 13.6 should be regarded as a guide only. For 'first-time' landslides in which soil has not been remoulded peak shear-strength values (ϕ'_p) are usually most appropriate, but where prolonged creep or flow has occurred, or soils have been involved in an earlier period of instability, residual values (ϕ'_r) are appropriate. For shallow landslides, or those in very variable soils, pedo-

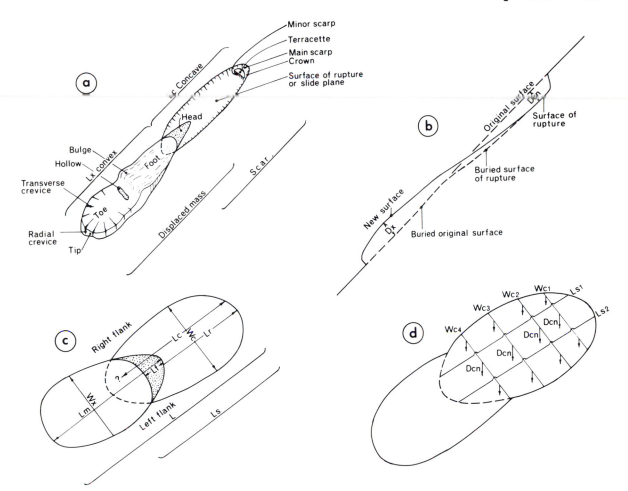

Fig. 13.6. The terminology of a landslide and the indices of landslide size and volume. (Adapted in part from Crozier 1973.)

TABLE 13.6. *Shear-strength test chosen for selected field problems*

Problem	Parameter	Test
Shallow translational landslide formed in prolonged wet season	$c'_p \, \phi'_p$	Consolidated undrained laboratory test, or field shear box test in saturated conditions
Shallow translational landslide formed in dry season rainstorm	$c'_p \, \phi'_p$	Consolidated undrained laboratory test, or field shear box test at natural moisture content
Long-term stability of clay slope subject to gradual toe removal, progressive failure, or first time failure in fissured clays	$c'_r \, \phi'_r$	Drained test, but for fissured over-consolidated clays $c'_r = 0$ and ϕ'_r parameters are appropriate
Long-term creep of clay slope or slow mud flow	$c'_r \, \phi'_r$	Ring shear test
Failure of slope in sand	$c_d \, \phi_d$	Drained test

logical structures and root materials may have a major effect upon shear strength. Field shear box tests may then be useful and attempts should be made to simulate moisture conditions and rates of strain which occurred during the landslide. For discussion of test procedures and landslides see Bishop and Bjerrum (1960); Bishop (1971); Skempton (1970); and Bromhead (1986).

Stability analyses

Hillslopes upon which landslides are active, or slopes which are being deeply undercut at the base, are oversteepened with respect to an angle for long-term stability. In the investigation of slope instabilities it is desirable to know what causes the instability and what effect various remedial measures, such as draining a slope or planting trees on it, may have.

Factor of safety

The stability of a slope is usually expressed in terms of a *factor of safety, F*, where

$$F = \frac{\text{sum of resisting forces}}{\text{sum of driving forces}}.$$

Where the forces promoting stability are exactly equal to the forces promoting instability $F = 1$; where $F < 1$ the slope is in a condition for failure; where $F > 1$ the slope is likely to be stable. There is no such thing as absolute stability, only an increasing probability of stability as the value of F becomes larger. Most natural hillslopes upon which landsliding can occur have F values between about 1 and 1.3, but such estimates depend upon an accurate knowledge of all the forces involved and for practical purposes design engineers always adopt very conservative estimates of stability (Table 13.7). It can be seen that the greatest uncertainties are usually associated with soil water, especially with its local variability of pressure and seepage.

Hillslopes which have too low an angle for landsliding to occur have F values much greater than 1.0 and are theoretically being modified entirely without the operation of landslides. Such slopes, however, may occur only at very low angles—less than 8° or so in many clays—and even on very gentle slopes mass wasting may occur in periglacial environments where soil ice and soil saturation in an active zone above permafrost can be effective.

TABLE 13.7. *Values of minimum overall safety factors*

Failure type	Item	F
Shearing	Earthworks	1.3 to 1.5
	Earth-retaining structure	1.5 to 2.0
	Foundation structures	2 to 3
Seepage	Uplift, heave, slides	1.5 to 2.5
	Piping	3 to 5

Source: After Meyerhof (1969).

Stability analyses of shallow translational slides

Translational slides are usually analysed by the infinite slope method which is a two-dimensional analysis of a slice on the sides of which the forces are taken as being equal and opposite in direction and magnitude. It is assumed that the mobile slice is uniform in thickness and rests on a slope of constant angle and infinite extent. This dispenses with the need to consider side and end effects, and is justified as translational slides are long in relation to their depth and width and are often uniform in cross-section. This mode of analysis was employed by Skempton and De Lory (1957).

The forces acting at a point on a shear plane of a potential shallow slide are illustrated in Fig. 13.7. The gravitational stress acts vertically, the normal stress is normal to the shear plane and is partly opposed by the upthrust or buoyancy effect of pore-water pressure; the shear stress acts down the shear plane and is resisted by the shear strength of the soil.

Where a rectangular block of soil rests on a slope (1 in Fig. 13.8), the resisting force holding the block in place is given by W (which is the mass × the gravitational force) multiplied by the cosine of the angle of slope. The driving force is $W\sin\beta$.

For an infinite slope analysis the soil block is within the regolith and the value of W has to be determined indirectly. The easiest method is to make measurements of the vertical thickness of the block. This is also realistic as cracks in the soil are approximately vertical. The block of soil therefore has the sectional form of a parallelogram ABCD (2 in Fig. 13.8). For computation this is converted to the equivalent rectangle so that the parallelogram ABCD is represented by the rectangle AEFD. The

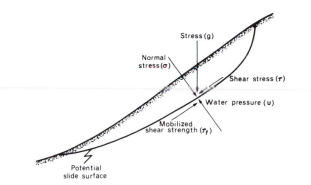

FIG. 13.7. Forces acting at a point on a potential failure plane.

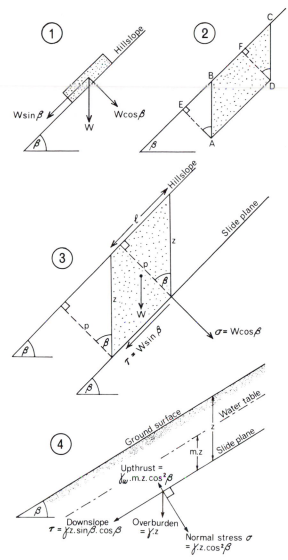

angles EAB, and FDC in (2) are equal to β and in (3)

$$\cos\beta = \frac{p}{z} \text{ thus } p = z\cos\beta.$$

Assuming a unit width of the soil block so that we are dealing only with a two-dimensional problem:

$$W = l\gamma p \text{ (substituting } p = z\cos\beta) = l\gamma z\cos\beta,$$

where γ is the unit weight of the soil at natural moisture content.

The normal stress on the shear plane is:

$$\sigma_n = W\cos\beta, \text{ hence } \sigma_n = l\gamma z\cos\beta\cos\beta.$$

The shearing stress on the shear plane is:

$$\tau = W\sin\beta, \text{ hence } \tau = l\gamma z\cos\beta\sin\beta.$$

(Note: (a) stress must be expressed in force per unit area, so a two-dimensional analysis is only realistic when it includes the assumption of a unit width to the soil block; (b) because the length of the block (l) is not relevant in an infinite slope analysis it can now be dropped from the equations.)

Effective shear strength at any point in the soil is given by the Coulomb equation as:

$$\tau_f = c' + (\sigma_n - u)\tan\phi',$$

where τ_f Is the effective shear strength at any point in the soil;

c' is the effective cohesion, as reduced by loss of surface tension;

σ_n is the normal stress imposed by the weight of solids and water above the point in the soil;

FIG. 13.8. Stresses acting on a slope, which are identified in an infinite slope analysis of a translational landslide. In (1) a block sits on a slope; in (2) the block is part of the soil profile and the area of a block with a parallelogram section is shown to be equal to that with a rectangular section; (3) the stresses acting on a slide plane; in (4) these stresses are analysed in a form suitable for inclusion in an infinite slope analysis.

u is the pore-water pressure derived from the unit weight of water (γ_w) and the piezometric head ($\gamma_w mz$) (see Fig. 13.8, 4);

ϕ' is the angle of friction with respect to effective stresses.

Substituting for σ_n the Coulomb equation can be rewritten as:

$$\tau_f = c' + (\gamma z \cos^2\beta - u)\tan\phi'.$$

Because $F = \dfrac{\text{sum of resisting forces}}{\text{sum of driving forces}} = \dfrac{\tau_f}{\tau}$

then $\quad F = \dfrac{c' + (\gamma z \cos^2\beta - u)\tan\phi'}{\gamma z \sin\beta\cos\beta}.$

It is convenient to express the vertical height of the water-table above the slide plane as a fraction of the soil thickness above the plane and this is denoted by m. Then if the water-table is at the ground surface $m = 1.0$, and if it is just below the slide plane $m = 0$. Pore-water pressure on the slide plane, assuming seepage parallel to the slope, is then given by:

$$u = \gamma_w m z \cos^2\beta$$

thus

$$F = \frac{c' + (\gamma - m\gamma_w)z\cos^2\beta\tan\phi'}{\gamma z \sin\beta\cos\beta}.$$

For example if:

$\phi' = 12°$, $c' = 11.9\,\text{kN/m}^2$, $\quad \gamma = 17\,\text{kN/m}^3$, $\beta = 15°$, $z = 6$ metres, $m = 0.8$, $\gamma_w = 9.81\,\text{kN/m}^3$, then:

$$F = \frac{11.9 + (17 - 0.8 \times 9.81)\,6 \times 0.92 \times 0.2}{17 \times 6 \times 0.25 \times 0.96}$$

$$= 0.9.$$

Thus the slope is prone to failure when the water-table is about a metre below the groundsurface. If the water-table can be lowered, by drainage, to just below the slide plane (when $m = 0$) then $F = 1.3$ and the slope will be stable in the longer term.

Where there is not a continuous water-table with flow parallel to the soil surface an alternative form of analysis must be used.

Under most conditions the free soilwater-level cannot rise above the soil surface and the pore pressure (u) is given by:

$$u = \gamma_w h,$$

where h is the piezometric height or the height to which water will rise in a stand pipe, inserted in the soil to the depth of the failure plane. Where seepage is not uniform, and directed out of the slope, it is convenient to use the ratio r_u between

pore pressure and the weight of a vertical column of soil:

$$r_u = u/\gamma z, \quad \text{thus} \quad u = r_u \cdot \gamma z.$$

The equation for determining the factor of safety then becomes (Haefli 1948):

$$F = \frac{c' + (\gamma z \cos^2\beta - r_u \cdot \gamma z)\tan\phi'}{\gamma z \sin\beta\cos\beta}.$$

substituting $\gamma_w h/\gamma z$ for r_u this becomes:

$$F = \frac{c' + \left(\gamma z \cos^2\beta - \dfrac{\gamma_w h}{\gamma z} \cdot \gamma z\right)\tan\phi'}{\gamma z \sin\beta\cos\beta}.$$

By dividing through by γz this may be rearranged to give:

$$F = \frac{\dfrac{c'}{\gamma z} + \left(\cos^2\beta - \dfrac{\gamma_w h}{\gamma z}\right)\tan\phi'}{\sin\beta\cos\beta}.$$

Under most conditions the highest pore pressures will exist when groundwater-level is at the surface. As γ is usually about $2\gamma_w$ the corresponding value of r_u is approximately 0.5. The extreme upper limit is the 'geostatic' value due to weight of soil and the pore-pressure ratio $r_u = 1.0$, which implies a piezometric level rising above the slope to a height approximately equal to a depth z—a state clearly impossible under ordinary conditions, but possible if frozen soil is thawed so rapidly that the entire weight of overburden is transferred to the pore water, without any of the water being able to escape. This condition occurs in periglacial environments, and fossil landslides in southern England have been explained in this way (Skempton and Weeks 1976; Chandler *et al.* 1976).

Very high pore-water pressures may also be produced where water fills a deep tension crack during a dry season rainstorm and moves laterally at depth below more impervious soil. The piezometric head downslope of the tension crack may then tend towards the geostatic value. Such a condition is relatively common in parts of California and New Zealand where severe summer storms have often been responsible for landsliding (Selby 1967a,b, 1976; Campbell 1975, Ellen and Wieczorek 1988).

In one case study Rogers and Selby (1980) demonstrated that the piezometric surface rose 0.2 m

The type of analysis given above can be of great value to engineers, foresters, and planners who wish to determine what effect altering land use, draining a slope, or planting trees may have upon stability.

Spring flows from highly fractured and closely jointed rocks may also cause local very high pore-water pressures at the base of the soil and trigger landslides during or after storms. Shallow translational landslides in much of the Appalachian region, USA, in 1972 are attributed to such a process (Everett 1979). Very high local water pressures may cause a burst or 'blow out' of the soil which may become the initiating event leading to a larger landslide.

Submarine translational landslides are very common in the weakly consolidated sediments of submarine deltas and they may be studied in the same way as terrestrial landslides. Such a study of slides up to 10 km long and 33 m deep (z) has been made by Prior and Suhayda (1979). They have shown that failure of sediment slopes, under gravitational stresses alone, can occur at angles as low as 0.5° where very high internal pore-water pressures exist. Such pore pressures have been measured directly using piezometers. Using the assumption that $F = 1$, and hence that the slope is in a condition for failure, the pore pressure (u) needed to initiate failure can be calculated from the rearranged stability equation:

$$u = \frac{c' - F\,(\gamma z \sin\beta \cos\beta)}{\tan\phi'} + \gamma z \cos^2\beta.$$

In the Mississippi delta front measured and calculated values have been shown to correspond closely. The pore-water pressure is so high that it constitutes a strong artesian pressure. Calculated values in some places approach, or even slightly exceed, geostatic pressures and indicate a condition of almost zero effective stress.

Stability analyses of rotational landslides

Deep rotational landslides are confined to clays and clay-rich soil, and do not occur in sands, because the strength of a soil due to cohesion only is not controlled by overburden pressure. Values of frictional strength for both clay and sand increase in proportion to the normal stress acting on a potential failure plane within a soil, thus for a frictional soil strength increases with depth. In frictional materials the rate of strength increase

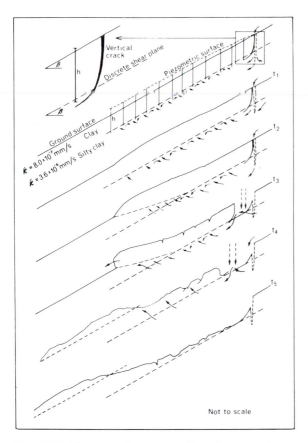

FIG. 13.9. Diagrammatic representation of an eye-witness account of a shallow landslide in Matahuru Valley, New Zealand. The slide began with a bulging of the toe and the formation of a tension crack at the crown. Over a period of 15 minutes the moving mass slid over the failure plane and burst through the sod at the toe as a saturated flow of debris. The upper part of the displaced mass then slid rapidly, but discrete soil blocks stabilized in the failure zone once the saturated material drained. *k* is the coefficient of permeability and *h* the piezometric head. (After Rogers and Selby 1980.)

above the groundsurface causing high uplift pressures at the failure plane where a silty clay, with relatively high permeability, underlies a less permeable clay (Fig. 13.9). Solutions of the stability equation are given in Fig. 13.10. They illustrate how the factor of safety against landsliding varies with changes in the value of the parameters. *F* is very sensitive to changes in values of *c'* and *h*, moderately sensitive to values of *z* and *β* and rather insensitive to values of *φ'* and *γ*. The value of *h* may be reduced by better soil drainage but *c'* may be modified only by increasing apparent cohesion through a denser plant root network.

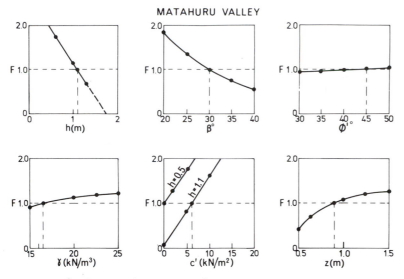

MATAHURU VALLEY

FIG. 13.10. Variations in the factor of safety against shallow landsliding as one parameter only varies. Parameter values are as given except where one parameter is varying.

$z = 0.9$; $\phi' = 45°$; $c' = 6 kN/m^2$; $\gamma = 17 kN/m^3$; $\gamma_w = 9.8 kN/m^3$; $h = 1.1$; $\beta = 30°$

with depth exceeds the rate at which shear stresses increase and deep failures cannot occur. For clays shear stresses may increase more than strength for each increment in depth, hence deep-seated failures are possible, especially for clays in which $\phi_u = 0$.

Rotational failures may be treated as a series of vertical slices for each of which a modified infinite slope analysis is carried out and the values for each slice are then summed. A toe of the slope which resists failure is treated as providing a negative driving force and because the length of the base of each slice may vary the value of l_A etc. (Fig. 13.11) is included in the formula thus:

$$F = \frac{\sum\limits_{O}^{A}[c'l + (W\cos\alpha - ul)\tan\phi']}{\sum\limits_{O}^{A}[W\sin\alpha]}.$$

This method of analysis by slices was first proposed by Fellenius (1936) and is named after him; alternatively it is called the Swedish Method of Slices or Conventional Method. A worked example using the method is provided in Fig. 12.12. The original profile is reconstructed and the shape and slope of the slide plane are plotted on graph paper from field measurements. The centre of rotation of the slide can then be established. The failed mass is

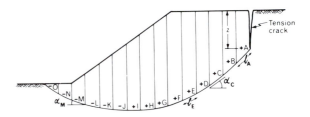

FIG. 13.11. Division of a rotational failure into vertical slices. Those resisting failure are treated as having negative values.

divided into slices (five or six are sufficient for a simple failure) and these are drawn on the graph paper so that the centre of gravity of each slice and its area can be obtained from the drawing. The length of the base of the slice is taken from the drawing and the angle of inclination of the base of the slice is found from an extended radius of the circle and a vertical through the centre of gravity of the slice. All slices are assumed to have a thickness of 1 m and thus a weight given by the product of the area (A in m^2), the unit weight of soil, and the thickness. Samples are collected from the unfailed soil in the sides of the landslide scar for strength testing and determination of unit weight of the soil.

The Fellenius Method is widely used because of its simplicity and because the calculations are easy.

BY THE SWEDISH METHOD OF SLICES

From field measurement angle of slope: $\beta = 34°$
From shear strength testing: $c' = 45\,\text{kN/m}^2$,
$$\phi' = 27°$$
Soil unit weight at natural moisture content:
$$\gamma = 20\,\text{kN/m}^3$$
Unit weight of water: $\gamma_w = 9.81\,\text{kN/m}^3$
From scale drawing, and assuming a water- table at the ground surface:

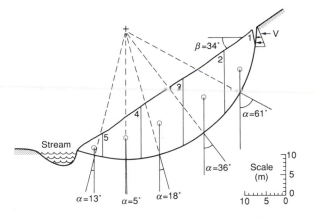

Slice	Area (m²)	h (m)	α°	l (m)
1	82	13.6	61	17.5
2	154	16.0	36	13.1
3	151	15.0	18	10.9
4	97	11.2	5	10.0
5	18	4.3	13	6.1

Slice	c'	l	$c'l$	W	$\cos\alpha$	$W\cos\alpha$	u	ul	$\tan\phi'$	$c'l + (W\cos\alpha - ul)\tan\phi'$	$\sin\alpha$	$W\sin\alpha$
1	45	17.5	787	1640	0.48	787	92	1610	0.51	367	0.87	1427
2	45	13.1	589	3080	0.81	2495	108	1415	0.51	1140	0.59	1817
3	45	10.9	490	3020	0.95	2869	101	1101	0.51	1392	0.31	936
4	45	10.0	450	1940	0.99	1921	76	760	0.51	1042	0.09	175
5	45	6.1	274	360	0.97	349	29	177	0.51	362	0.22	− 79
										$\Sigma4303$		$\Sigma4276$

$$F = \frac{c'l + (W\cos\alpha - ul)\tan\phi'}{W\sin\alpha} = \frac{4303}{4276} = 1.00$$

(by the Simplified Bishop Method $F = 0.99$)
Consequently the slope is in equilibrium with the water-table at the surface.
BUT if the tension crack is filled with water to a depth of 4.5 m

$$V = \tfrac{1}{2}\gamma_w h^2 = 99\,\text{kN/m}^2$$

Then: $F = \dfrac{4303}{4276 + 99} = 0.98$, hence the slope is liable to failure.

[Notes: W = Area × 1 m × γ (assuming a 1 m thick slice).
$u = \gamma_w h\,\cos^2\beta$ (because seepage is downslope, parallel to the groundsurface (see Fig. 13.8 (4)).
A negative value is assigned to $W\sin\alpha$ for slice 5 because it resists sliding.
The water-table was assumed to be at the groundsurface from the field observations that tension cracks were filled and seepage was occurring from the base of the slope.
If the water-table is not at the surface a flow net should be used to determine pore-water pressure for each slice (see Lambe and Whitman 1979: 359).]

FIG. 13.12. A worked example of a post-mortem stability analysis of a rotational failure.

Unfortunately the value of F derived this way is often 10 to 15 per cent below the value derived by more rigorous methods and may be even more in error in certain cases (Whitman and Bailey 1967). Most of the error occurs in the treatment of pore-water pressures; some error occurs because of the method's assumption that all side forces on each slice act in a direction parallel to the failure plane and that normal forces are assumed to act at right angles to the failure plane (Fig. 13.13). The Fellenius Method treats each slice as though it were nearly rectangular, but with increasing curvature of the failure plane this becomes an untenable assumption.

For a slice with a curved base and an upper surface which is not parallel to the failure plane corrections have to be made. As a result an alternative method was proposed by Bishop (1955) and this was simplified by Janbu *et al.* (1956). The Simplified Bishop Method of Slices assumes that forces acting on each slice are in a horizontal and vertical direction. This assumption is not entirely valid but the method has been shown to provide values of F which are in the range of values derived by more rigorous methods and are seldom more than 2 per cent in error. The correction factor is:

$$\cos\alpha\left(1 + \frac{\tan\alpha\tan\phi'}{F}\right),$$

and the formula for the Simplified Bishop Method is then:

$$F = \frac{\sum\limits_{i=1}^{i=n}\left[c'b + (W - ub)\tan\phi'\right]\left[1\bigg/\left[\cos\alpha\left(1 + \frac{\tan\alpha\tan\phi'}{F}\right)\right]\right]}{\sum\limits_{i=1}^{i=n}[W\sin\alpha]},$$

where b is the horizontal width of the slice.

It will be seen from the equation that F appears on both sides, and a trial-and-error procedure is required with values being assigned to F and the equation solved repeatedly until it balances. This disadvantage is overcome by using a computer. A number of other more rigorous methods of slices are in use. They differ in their handling of inter-slice forces, but the established methods give closely corresponding values of F (Fredlund and Krahn 1977).

Tension cracks frequently develop on slopes as a result of desiccation of the soil. For analysing a natural failure it is reasonable to assume that a

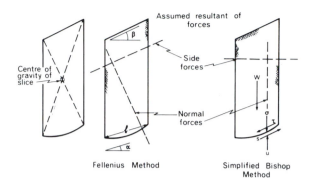

FIG. 13.13. Location of the centre of gravity of a slice, and the resultants of side and normal forces assumed in the Fellenius and Simplified Bishop methods.

crack developed at the head scarp of the failure and extended down to the failure plane. Where this cannot be determined the maximum possible depth of the crack (h_c) may be estimated from:

for cohesive soils ($\phi_u = 0$) h_c (metres) $= \dfrac{2c_u}{\gamma}$

in a soil with both frictional and cohesive properties

$$h_c = \frac{2c'}{\gamma}\left(\tan 45° + \frac{\phi'}{2}\right).$$

Tension cracks have the effect of decreasing the length of a failure plane over which cohesive resistance can be mobilized; where cracks are filled with water a hydrostatic pressure (V) may be exerted related to the depth of water (h) in the crack then:

$$V = \tfrac{1}{2}\gamma_w h^2.$$

V is an addition to the driving forces acting against the slide downslope of the tension crack.

Limitations and alternative forms of analysis

The infinite slope analysis and the Swedish Method of Slices indicate the principles employed in limit equilibrium methods. These methods postulate that the slope will fail by a mass of soil sliding on a failure surface and that, at the moment of failure, the shear strength is fully mobilized along the entire length of the failure plane, and the overall slope and each part of it are in static equilibrium. Uncertainties in analysis can arise for many reasons, but particularly the following: (1) the physical properties of the soil may vary from point to point along the failure plane; (2) the shape of the failure plane may not be known with certainty and it may vary from a simple planar or circular form; (3) pore-water pressures may vary in unknown or unpredicted ways; (4) the forces acting between slices may be significantly large; and (5) the assumption that stresses at the lateral margins of the slide can be ignored may be invalid.

Methods for overcoming some of these problems have been developed and are in general use. The availability of computer programs has greatly extended the capacity for using more complex forms of analysis. The complexities will not be discussed here but some references will be given to the major methods in use for overcoming them.

Variable soil properties have already been discussed briefly in the sensitivity analysis shown in Fig. 13.10. This technique shows the effect of varying only one of the parameters at a time. An alternative approach is to adopt a Monte Carlo simulation procedure which allows each of the variables to be changed simultaneously according to some well-defined rules. The probability of failure can then be estimated as the proportion of occurrence of the factor of safety being less than or equal to unity. Examples of this approach are given by Tobutt (1982); Priest and Brown (1983); and Nguyen and Chowdhury (1984).

The shape of non-circular curved failure planes is considered in several analytical methods including the Janbu Simplified Method, Spencer's Method, and Morgenstern and Price's Method. The main differences between these methods are in the way they assess the interslice forces; there is little to chose between them in their accuracy. These three methods are described by Bromhead (1986) and Nash (1987); Nash also gives an analytical account of the assumptions and limitations of the methods.

The driving force in a landslide is provided by the weight of the sliding mass acting down the slope. The average driving force across the shear plane may be reduced by the 'drag' along the margins of the slide where the sliding mass is in frictional contact with the stationary soil. Where this effect is significant the driving forces can be multiplied by a reduction factor to account for this effect. This was found to be appropriate in an analysis of the stability of the Mam Tor landslide in England (Skempton *et al.* 1989), and the 'drag' force was shown to increase the factor of safety by 2.3 per cent. The reduction factor is: $1/(1 + KD/B)$ where, D and B are the average depth and width of the sliding mass, and K is an earth pressure coefficient with a value which is usually in the range 0.5–1.0.

The problems of deriving a reliable value for pore-water pressure are discussed later in this chapter.

Where uncertainties in stability analyses exist, better estimates of slope stability may be obtained from mapping the susceptibility of slopes to landsliding than from an approach based upon a study of the forces involved in instability. Landslide susceptibility maps seek to identify potential failure areas by mapping old landslide features and factors likely to cause failure in the future—such as slope undercutting, drainage ponding, weak rocks, and recently deforested slopes. The most probable sites for new landslides are areas where sliding has occurred already, or areas which are similar to those where landslides have occurred. Old landslide debris is in a remoulded state, and hence of lower strength than the original soil, and is weakened further by weathering. It is also likely to be a site of impeded or disturbed drainage. Modification of such a site by drains cutting across a toe and removing support can therefore lead to renewed movement, as can periods with unusually high water-tables.

Factors influencing landsliding in soils

Many factors influence the development of landslides, and a particular slide can seldom be attributed to a single definite cause although it may be possible to identify a dominant or a triggering effect.

Possible causes and contributing factors are listed in Table 13.8 but, as many of these have been discussed already, mention here is confined to the effects of vegetation, vibration and ground acceleration, water, soil ice, weathering, and the form and orientation of hillslopes.

Vegetation

Vegetation change is a very important influence upon slope stability in areas where forests are being removed from hillslopes as part of a pro-

TABLE 13.8. *Factors contributing to mass movement in soils*

A. Factors contributing to high shear stress

Types	Major mechanisms
1. Removal of lateral support	(i) Stream, water, or glacial erosion (ii) Subaerial weathering, wetting, drying, and frost action (iii) Slope steepness increased by mass movement (iv) Quarries and pits, or removal of toeslopes by human activity
2. Overloading by	(i) Weight of rain, snow, talus (ii) Fills, waste piles, structures
3. Transitory stresses	(i) Earthquakes—ground motions and tilt (ii) Vibrations from human activity—blasting, traffic, machinery
4. Removal of underlying support	(i) Undercutting by running water (ii) Subaerial weathering, wetting, drying, and frost action (iii) Subterranean erosion (eluviation of fines or solution of salts), squeezing out of underlying plastic soils (iv) Mining activities, creation of lakes, reservoirs
5. Lateral pressure	(i) Water in interstices (ii) Freezing of water (iii) Swelling by hydration of clay (iv) Mobilization of residual stress
6. Increase of slope angle	(i) Regional tectonic tilting (ii) Volcanic processes

B. Factors contributing to low shear strength

Types	Major mechanisms
1. Composition and texture	(i) Weak materials such as volcanic tuff and sedimentary clays (ii) Loosely packed materials (iii) Smooth grain shape (iv) Uniform grain sizes
2. Physico-chemical reactions	(i) Cation (base) exchange (ii) Hydration of clay (iii) Drying of clays (iv) Solution of cements
3. Effects of pore water	(i) Buoyancy effects (ii) Reduction of capillary tension (iii) Viscous drag of moving water on soil grains, piping
4. Changes in structure	(i) Spontaneous liquefaction (ii) Progressive creep with reorientation of clays (iii) Reactivation of earlier shear planes
5. Vegetation	(i) Removal of trees (a) reducing normal loads (b) removing apparent cohesion of tree roots (c) raising of water tables (d) increased soil cracking
6. Relict structures	(i) Joints and other planes of weakness (ii) Beds of plastic and impermeable soils

PLATE 13.9. Cutover and burned forest five years after clearance.

gramme of agricultural expansion or as part of a regular cycle of cropping a forest before replanting the slopes with trees. Deforestation of slopes has frequently been followed by severe shallow land-sliding in areas such as New Zealand, Alaska, British Columbia, the Himalayan foothills, and Japan (Plate 13.9).

At least four effects of trees upon slope stability can be identified (Gray 1970; O'Loughlin 1974; Brown and Sheu 1975).

(1) Wind throwing and root wedging occur as trees are overthrown by strong winds and under heavy snowfalls. This effect is probably most noticeable on very steep slopes such as the walls of formerly glaciated valleys where soils are very shallow and root penetration into rock is limited. The wind effect on most forests is very small and adds less than 1 kPa to the shearing stresses even when a 90 km/h wind is blowing down the slope (Hsi and Nath 1970).

(2) Trees have the effect of increasing the surcharge, and hence the shearing stresses, on a slope by $T\sin\beta$ while the normal stress is increased by $T\cos\beta$ (where T is the weight of the trees). The effect of the surcharge is thus expressed by:

$$\frac{T\cos\beta\tan\phi'}{T\sin\beta},$$

where, for example, $\phi' = 36°$ and tree weight is 2 kPa (Fig. 13.14a), the trees increase stability on slopes of less than 34°, but where slopes are greater than this angle they may be detrimental to stability if the increase in shearing stress produced

PLATE 13.10. The shallow but laterally spreading root system of a fallen tree.

TABLE 13.9. *Strength added to soil by plant roots*

Plant	Soil	Increase in apparent cohesion (kPa)
Conifers (pine, fir)	Glacial till	0.9–4.4
Alder	Silt loam	2.0–12.0
Birch	Silt loam	1.5–9.0
Podocarps	Silty gravel	6.0–12.0
Poplar	Silt loam	2.0–9.0
Alfalfa (lucerne)	Silty-clay loam	4.9–9.8
Barley	Silty-clay loam	1.0–2.5
Clover	Silty-clay loam	0.1–2.0

Note: In any soil the strengthening effect of roots varies greatly with depth and laterally.

Source: Data from unpublished sources and O'Loughlin (1974); and Waldron (1977).

by the weight is not offset by root strength. Large, closely spaced trees on low-angle slopes can increase the normal stress by 5 kPa while increasing the shearing stress by about half that (Bishop and Stevens 1964).

(3) Mechanical reinforcement of the soil is provided by root networks. Some roots grow downwards through the potential failure zone into the underlying soil or rock but in shallow soils most trees have shallow root networks which interlock (Plate 13.10) and provide an apparent cohesion to the soil which usually falls in the range of 1.0 to 12.0 kPa (Table 13.9). Lateral reinforcing effects are particularly important around the perimeter of a potential failure.

(4) Trees modify soil moisture by lowering the water-table as a result of transpiration and interception and they thus delay or prevent soil saturation. Transpiration draws water into the tree at the root level and therefore creates a suction in the soil. A minimum value for this effect of 7 kPa, during the wet season, has been recorded in Hong Kong (Greenway 1987). Much higher values of soil suction have been recorded but it is not clear how much of this is caused by trees.

Trees also cover soil with leaf litter and may prevent the soil drying out and cracking in dry seasons. By contrast, trees provide conditions for rapid infiltration and interflow along root channels and cracks: the latter mechanism has been identified as promoting landsliding on steep slopes with shallow soils in Brazil (De Ploey and Cruz 1979).

In studies of landsliding on forested slopes soil-shear strength is measured with a large field shear box in which pedological structures and tree roots are retained in the sample. The effect of roots is incorporated as apparent cohesion so the infinite slope equation then becomes:

$$F = \frac{C'_{s+t} + [(\gamma z + T) - (mz\gamma_w)]\cos^2\beta\tan\phi'}{(\gamma z + T)\sin\beta\cos\beta},$$

where: C'_{s+t} is the total cohesion derived from soil effective cohesion plus the apparent cohesion due to tree roots (kPa);

T is the tree weight (kPa).

Upon deforestation many of the beneficial effects of trees upon slope stability may persist for a few years because the tree roots decay gradually (Fig. 13.14b), with about half the root strength being lost in two to five years. The decay of roots usually leaves many root channels and the soil is then more permeable and the water-table can respond more rapidly to storms. Any actions which impede soil drainage, such as diverting runoff from forestry roads into gullies or depositing spoil into depressions, and thereby raising the water-table in depressions, then become more likely to cause instability.

The overall effect of trees on hillslopes with shallow soils is to increase soil-shear strength by 60 per cent or more. The effects are summarized in Figs. 13.14 and 13.15. These effects are in accordance with the common experience that deforestation is followed by shallow landsliding in many upland areas, but only after an interval of some two to ten years. It is also in accordance with the observation that landsliding is far less common on forested slopes than on adjacent slopes under cultivation or grass (Selby 1967a,b; Swanston 1970; Pain 1971).

More detailed studies of the relationship between slope stability and vegetation are provided by Gray and Leiser (1982) and Greenway (1987); a specific example is the report of Yim *et al.* (1988).

Earthquakes

The most damaging effect of most earthquakes is the vibration of the ground, at frequencies of 0.1 to 30 Hertz, as shock waves, especially surface waves, pass through it. In the California earthquake of 1906, surface waves up to 1 metre high with a wavelength of 20 m moved across the landsurface at velocities of a few m/s. Surface movements normally last for less than one minute, but in the

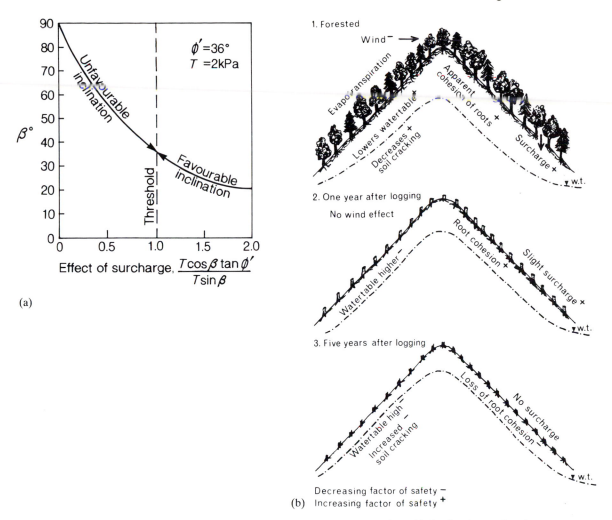

(a)

(b)

FIG. 13.14. (a) The effect of tree weight on slope stability. On slopes steeper than 34°, for the given soil and tree cover, tree weight decreases stability; (b) The effect of trees upon soil shear strength, and its decline after the forest is cut down. Note that the effect of surcharge may be detrimental on steep slopes.

Alaska earthquake of 1964 they lasted over five minutes and were followed by twenty-eight after-shocks in 24 hours, of these ten were strong quakes. The magnitude of a quake is measured on a scale named after Richter, who developed it. Richter magnitude (M) is defined as the logarithm of the amplitude (in 1/1000 mm) of the largest wave recorded on a standard seismograph 100 km from the epicentre. Each division on the nine-point scale therefore has ten times the energy of the next lower division. The intensity of an earthquake is recorded on the modified Mercalli (MMI) scale which has twelve divisions numbered in Roman numerals, I to XII, and the divisions are based on observed movement and the extent of damage. The intensity falls off with distance from the epicentre. An intensity of VII will be experienced near the epicentre of a magnitude 5, which is the weakest earthquake to cause extensive damage to buildings. Major earthquakes are those of magnitude 6.5 or greater.

Earthquake waves, like any other wave form, can be damped or amplified. Weak soils, especially those with a high water-content, commonly amplify earthquake waves. Topography can also cause amplification especially where waves are

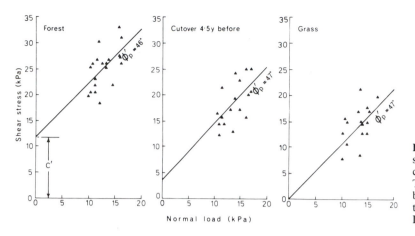

FIG. 13.15. The decline of shear strength of a soil after the forest is cut is caused by a loss of apparent cohesion. The scatter in the data points is caused by the high variability of structure and texture in steepland soils. (Supplied by D. Parker.)

focused in convexities; ridges and convex hills consequently become sites of greater shaking and are likely to suffer greater landsliding than concavities. This happened in the magnitude 7.5 quake in Guatemala, 1976, which produced more than 10 000 landslides over about 16 000 km², mostly in Pleistocene tephra and pumiceous pyroclastics (Harp *et al.* 1981).

In a study by Keefer (1984) it has been shown that on most soils the maximum area likely to be affected by landslides varies from a few km² at M = 4.0 to 500 000 km² at M = 9.2; but even at values of M = 2.9 landslides have occurred in loess in China (Xue-Cai and An-Ning 1986). The smallest earthquake to have caused landslides in USA had a local M = 4.0; such quakes have caused rock and soil falls and some slides. At higher magnitudes the record shows: at M = 4.5 soil slumps and block slides; at M = 5.0, rock slumps, soil lateral spreads, rapid soil flows and subaqueous slides; at M = 6.0, rock avalanches; and at M = 6.5, soil avalanches. M = 5 is the minimum magnitude for soil liquefaction and lateral spreads and flows. The reports of Kingdon-Ward (1952, 1955) [extracts were also published in *Terra Nova*, 2: 187–90 (1990)] indicate the damage to hillslopes caused by the largest continental earthquake of the twentieth century. This, the Assam earthquake of 1950, caused landsliding in the eastern Himalayas over an area of about 15 000 km², moving an estimated 5 × 10¹⁰ m³ of debris; this is thirty times the annual sediment discharge of the Brahmaputra River, the tributaries of which drain the area affected.

The Alaska earthquake of 1964 caused many landslides but none has been studied more intensely than the large spread-failure of Turnagain Heights, a suburb of Anchorage. Anchorage is built on the lowland bordering Cook Inlet. The lowland is about 20 m above sea-level at its lowest part and consists of late Last Glacial alluvial and glacio-estuarine sediments overlying Tertiary rocks. During the earthquake, which had a surface wave magnitude of 8.5 and a duration variously estimated at 4.5 to 7.5 minutes, the slide extended across a width of 2800 m, regressed inland from the coast a distance of 2100 to 400 m, and material from the original coastal cliff zone moved up to 660 m into the Inlet. Some of the houses moved laterally up to 200 m (Seed and Wilson 1967). Fig. 13.16 gives an indication of the development of the final profile of the slide along one transect.

The soils underlying Anchorage are predominantly silty clays or clayey silts with some bodies of fine to medium sands with traces of silt and gravel and some units of clay. The glacial deposits are overlain by a 1 m thick unit of alluvial fan deposits. Initial research carried out after the earthquake concluded that sands near the middle of the deposit liquefied with the weakest sediments being those near to sea-level. Some deposits were also recognized as being sensitive. Liquefaction was consequently believed to be the process which caused the lateral spread.

More recent work, using drilling, seismic and electric cone-penetrometer tests, indicates that there is a zone in the sediments of extremely sensitive clays and silty clays (up to $S_t = 1000$) that are both more abundant and more widely distributed than was indicated in earlier investigations (Updike

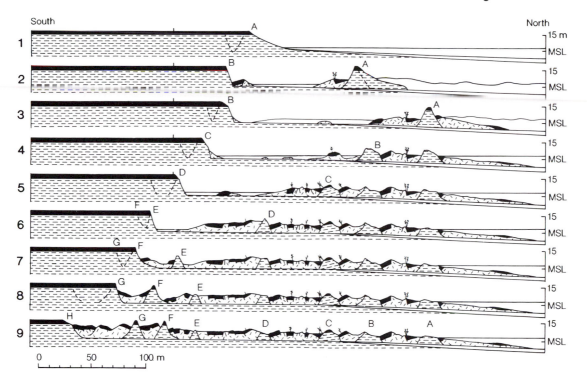

Fig. 13.16. Sections through the east lobe of Turnagain Heights lateral spread. This sequence of events is based partly on an eye-witness account. The development occurred during a period of about three minutes. (After Voight 1973b.)

et al. 1988a,b). The significance of these findings is that the sensitive clays may be more critical than the sands during future large earthquakes. Strength degradation of the silty clays probably resulted from long-duration accelerations between 0.2 and 0.4g.

In soils, most hillslope failures caused by earthquakes are shallow translational landslides. Many of these failures occur where the soils have a large factor of safety against failure under conditions of high pore-water pressures. Application of the infinite slope analysis (Matsukura and Maekado 1984) to a pumiceous soil, with the properties $c = 22\,kN/m^2$, $\phi = 21°$, and $\gamma = 10\,kN/m^3$, and calculating F for a range of accelerations (a) with:

$$F = \frac{c + \gamma z \cos\beta(\cos\beta - a/g \sin\beta)\tan\phi}{\gamma z \cos\beta(\sin\beta + a/g \cos\beta)}$$

demonstrates (Fig. 13.17) that, where total stresses apply, the slope has a factor of safety of 4, and that it was liable to failure only when the horizontal acceleration due to the earthquake had a value of

about $a = 0.8g$ ($g = 9.8\,m/s^2$). The earthquake had a magnitude, M = 6.7. This analysis, however, does not take account of the duration of shaking.

Examples of earthquake-triggered landslides have been reported from rapidly rising and forested ranges of Papua New Guinea (Simonett 1967; Pain and Bowler 1973; Pain 1975). The area affected by each earthquake is commonly a few hundred km^2 and the average downwearing of the landscape 70 to 400 mm. It seems probable that such earthquakes occur about every 200 years so the landscape is being lowered on average 0.35 to 2.0 mm/year by this process alone.

Some areas of frequent earthquakes such as Turkey and southern Italy have landscapes which have evolved largely as a result of landsliding. In Calabria, Italy, a combination of factors has contributed to massive and widespread terrain failures (Cotecchia and Melidoro 1974). Seismic activity there is among the highest in the world: the rocks are argillaceous marine formations of late Cenozoic age which have been uplifted more than 1000 m during the last two million years. As a

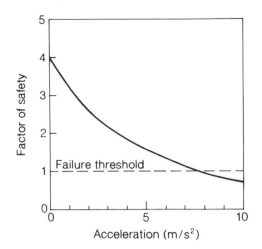

Fig. 13.17. The relationship between factor of safety, for one landslide, and earthquake acceleration. (Based on Matsukura and Maekado 1984.)

result of uplift stream incision is very active with each wave of erosion working up the valleys from the coast. The slopes are thus being steadily undercut. Furthermore the winter rainfall is often intense and prolonged. These factors acting in combination produce severe remoulding of soils and of rocks along fault planes, so that landsliding is frequent but at its peak during winters after each major earthquake (Cotecchia 1987).

Water

Water is by far the most important contributor to slope failure. Its operation may be through hydration of clay, undercutting of slopes, weight of rain, as an agent in weathering, as soil ice, in spontaneous liquefaction; but more than these by buoyancy effects, reduction of capillary tension, decrease in aggregation, and by viscous drag on soil grains.

Aggregation of soil particles commonly increases the internal friction of soil. In the rather extreme case quoted by Yee and Harr (1977) soils under forest in western Oregon (USA) had friction angles of 40°. During severe rainstorms the soils were saturated and lost aggregation so that there was a reduction of about 10° in the friction angle. Such a reduction may help to produce instability in wet conditions. This mechanism does, however, apply primarily to humic horizons and to soils with dispersible clays and promotes very shallow landsliding only.

Seepage pressure is the drag of moving water on soil particles (Terzaghi and Peck 1948). It results from friction effects created by molecular attraction between water and the solid particles between which it passes. Where the pressure gradient is steep, seepage pressures may become great enough to trigger landslides.

The effectiveness of seepage pressures depends on the particle size of the material on which it acts. In coarse gravel and other fast-draining materials, resistance to flow is slight and large seepage pressures seldom develop. Pressures caused by seepage in clays are also usually minor because of the impermeability of the clay, but materials in the size range of silt and fine sand are most affected by seepage pressures. Rapid large drawdowns in reservoirs, or along river banks or lake shores, can create steep pressure gradients in the groundwater and seepage pressures large enough to trigger bank collapses. As the water level falls in a stream or reservoir the friction between the alluvial particles of the bank and the escaping water creates an internal stress which acts towards the free face of the alluvium, at the same time pore pressure in the alluvium is no longer balanced by water pressure in the adjacent water body, and the unbalanced pressure promotes the bank collapse (Plate 13.11).

Pore-water pressure has been discussed already (see Fig. 5.12 and discussion under 'stability analyses'). It is necessary to emphasize here that pore-water pressure is not as simply distributed as is implied in most standard forms of analysis.

Theoretical flow lines indicating water movement usually show the flow to be nearly parallel to the groundsurface (Fig. 13.18a) but for many hillslopes subject to landsliding this can be a false assumption. Groundwater rise at the base of a slope due either to internal drainage of the slope and flow convergence into the basal concavity or, perhaps, to passage of a flood from higher up the catchment, can cause much higher temporary groundwater-tables than would be produced by flow parallel to the groundsurface. Debris from an old landslide may also form a mantle of low permeability on the groundsurface that will impede return flow and so raise the water-table.

Of even greater general importance is the convergence of flow—shown most readily as occurring on a groundsurface (Fig. 13.18d)—but of major importance also for subsurface flow. Hollows, such as old landslide scars become sites with higher groundwater levels and, therefore, pore-water

PLATE 13.11. Seepage pressures have caused small failures in the banks of this lake.

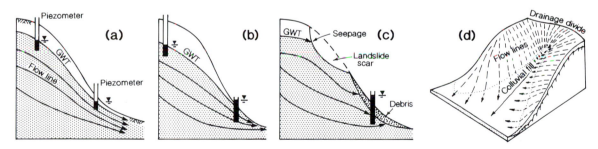

FIG. 13.18. Flow lines and groundwater tables with associated piezometric levels: (a) for flow nearly parallel to the hillslope surface; (b) for convergent flow into the base of the slope; (c) the influence of a body of low-permeability landslide debris; (d) flow lines indicating convergent flow into concavities and divergent flow from spurs.

pressures than surrounding areas. If they are being infilled by creep, wash, or other processes, they are then liable to failure of the infill. Furthermore, solution and other weathering processes are likely to be more effective there. There is thus a reinforcing effect, or feedback, in operation with spurs shedding water being least affected by erosion, and concavities collecting water and being most affected by erosion. Drainage network extension

thus concentrates in hollows and valley heads. Hollows are discussed further later in this chapter.

In stability analyses there is an implicit assumption that pore-water pressure remains unaffected by slight movements before the main failure event. It is inherently unlikely that pore pressures, in such heterogeneous materials as soils, do behave uniformly and also unlikely that a slight movement will always continue without an additional input of energy other than that of the driving force provided by the soil mass itself.

Experimental work by Iverson and LaHusen (1989) has demonstrated that dynamic pore-pressure fluctuations can be generated as a result of grain rearrangements during rapid shear, and that fluctuations can be large enough to modify grain contact stresses significantly and promote efficient deformation. The implication is, then, that the sliding process itself decreases internal resistance to shear and can enhance the possibility of continuing failure once movement has been initiated. The experimental work is significant for all slides of granular material but may be even more relevant to an understanding of how debris avalanches, debris flows, industrial slurries, and saturated sediments can remain liquefied and flow much farther (sometimes tens of kilometres across slopes of a few degrees) than similar masses of dry debris, and much further than anticipated from the material's initial potential energy and coefficient of internal friction.

The experiments were carried out using an idealized granular medium of water-saturated, close-packed, fibreglass rods. Rods can only move in one plane and this simplifies observation and measurement. Miniature transducers were used to sample water pressure near the centre of selected pores. Shearing was induced along a selected plane (Fig. 13.19a). The slide surface was formed of a series of hummocks which caused one line of rods to deform over the underlying line of rods. These vertical motions caused pore-volume changes and water-pressure changes along the slide surface (Fig. 13.19b). The pore-pressure fluctuations exhibited a series of high-pressure plateaux, formed as rods immediately above the slip surface floated on a cushion of water. Intervening troughs of low pore pressure occurred as rods recontacted and the pores between them dilated. At the scale of this experiment, the pore-pressure fluctuation at the shear plane was considerable at 17 kPa (1 kPa is approximately equivalent to the hydrostatic pressure at the base of a column of water 0.1 m high) and rose to +5 kPa.

Field tests using sandy soil and artificial rainfall have confirmed the existence of pore-pressure fluctuations although in soils the pattern is far more variable (Fig. 13.19c). The pressure fluctuation near the shear plane, at a depth of 0.9 m shows initial soil dilation and low pressure, then rapid rises of pressure with sharp fluctuations over a range of 3 to 7 kPa. These vertical pore pressures were large enough to support the entire overburden weight and locally may have reduced grain-contact stresses to zero.

These experiments not only suggest the mechanism responsible for long-distance movements of flows, but also the possibility that pore-pressure fluctuations may act as catalysts for propagating failure by transiently reducing grain-contact stresses in adjacent zones. Blockage of soil drainage with transition from drained to undrained conditions, and increases in pore-water pressures, may also occur as a result of initial movements in a landslide (see part IIB of Table 13.3). Loading the surface of a landslide with fallen, or deliberately placed debris, or unloading it by erosion, or excavation, can also have major effects on pore-water pressures, and therefore on the factor of safety (see Fig. 17.7).

Suction due to capillary stresses has a major influence on soil strength and therefore on the factor of safety. An example of a railway embankment of lacustrine silt, which began to fail several years after construction, has been analysed by Krahn *et al.* (1989). The embankment soil initially had high cohesive strength, but this decreased as negative pore-water pressures (i.e. suctions) dissipated. As the slopes were constructed at the optimum moisture content for stability, the embankment probably had an initial suction of about 50 kPa and therefore a factor of safety in excess of 3.0. Midsummer suctions, in soil near the surface, four years after construction, were in the range 10 to 17 kPa. It is shown in Fig. 13.19d that a suction of about 7 kPa is equivalent to a cohesion of about 2 kPa. Adding a mere 2 kPa of cohesion increases the factor of safety from unity to 1.35, that is, by about 30 per cent.

Field and laboratory testing procedures are available (Anderson and Kemp 1987; Anderson *et al.* 1990) which should make it possible for suction effects to be studied more closely. Losses of capillary stresses upon saturation have been recognized as a cause of landsliding at many sites, including areas around the Black Sea (Onalp 1988). Decreases in shear strength, as the water-content of soil increases towards saturation, are in the range of 20 to 30 per cent for many soils; much of this loss is in the suction effect but some may be a reduction in friction angle (Fig. 13.20).

Macropore flow and piping have been observed immediately prior to failure in a few instrumented

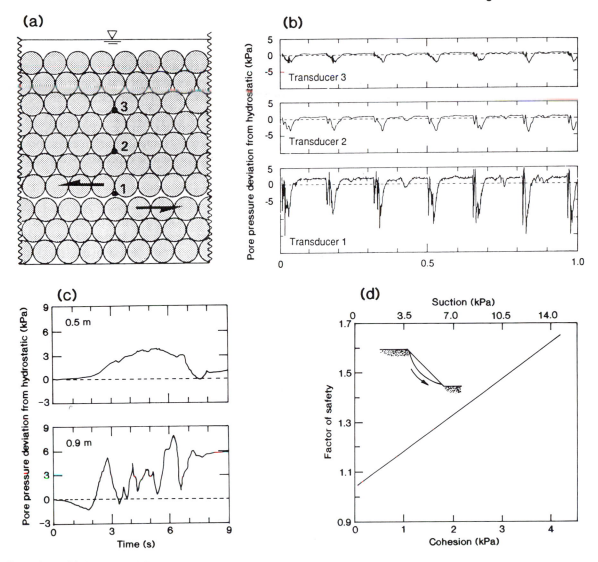

Fig. 13.19. (a) An array of fibreglass rods with three transducers in pores; the direction of shear is shown by arrows. (b) A time series of pore-water pressures measured at the transducers shown in (a); the dashed lines show a hydrostatic reference pressure. (c) Pore-pressure measurements made during an artificial landslide event; dashed lines show a hydrostatic reference pressure (a–c after Iverson and LaHusen 1989). (d) Effect of cohesion and suction on factor of safety for a near-surface slide with a profile as indicated (after Krahn *et al.* 1989).

experiments on the initiation of landslides (Harp *et al.* 1990). Such flow suggests that subsurface detachment and transport of fine-grained particles, immediately before failure, can both cause weakening of the soil and, through the escape of subsurface water, a sudden drop in pore pressures immediately prior to failure. If this is found to be a general condition, the sudden drop could be

detected by electronic piezometers and used as a warning, especially in critical urban areas.

Ice

The periglacial zones have a distinctive suite of mass movement processes and landforms associated with the thawing of frozen ground. Initiation of movements, and their subsequent velocities is

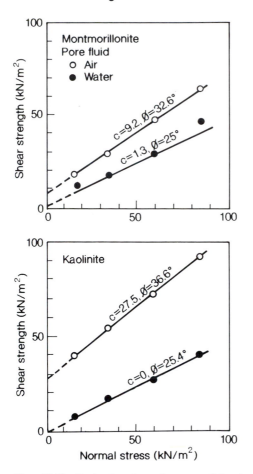

Fig. 13.20. Coulomb plots for consolidated drained shear box tests of statically compacted montmorillonite and kaolinite, showing the effect of water on shear strength. (Data from Sridharan and Venkatappa Rao 1979.)

more dependent on ice content of the soil and the rate at which it thaws than upon the slope gradient. In the freezing process, water is drawn to the freezing front by cohesive effects between water molecules (see Fig. 8.7b); consequently ice volumes in the soil can exceed, by far, the volumes of water which exist in unfrozen soils. Under rapid thawing, therefore, this water is involved in processes which are nearly all of the flow type. Three main classes of flow-dominated mass wasting are recognized in periglacial zones: solifluction, skinflows, and bimodal flows (McRoberts and Morgenstern 1974; Harris 1987).

Solifluction is the slow, downslope movement of saturated, non-frozen earth material, behaving apparently as a viscous mass, over an impermeable surface. It involves both frost creep in which particles are lifted by growing ice crystals and then moved downslope when the ice thaws, and gelifluction in which saturated soil flows over a frozen surface.

Gelifluction occurs in the active layer of seasonally thawed ground in which excess pore-water pressures are generated as thawing soil consolidates under its own weight. The excess pore pressures of thaw consolidation reduce frictional strength of the soil and these pressures cannot be reduced by infiltration into a frozen substratum (Morgenstern and Nixon 1971). Flow movements extend to depths of a few centimetres to about 2 m and flow-velocity profile patterns are as varied as those of soil creep. Surface movement is typically in the range 10–20 mm/y but in alpine areas 50–500 mm/y have been reported. These high rates in alpine areas may owe much to the action of needle ice lifting particles, and subsequent thawing allowing large displacements down steep slopes. Lobes of solif: lucted material (Plate 13.12) usually move faster than sheet-like flows. Most soil movement occurs in soil layers with ice contents in excess of 150 per cent by weight.

Plug-like movements of saturated soil are particularly, and perhaps uniquely, associated with permafrost, as this supplies large quantities of water at depths where flow can raft along the upper soil, including its vegetation cover.

Skinflows are the rapid detachment of thin veneers of soil and vegetation and their movement downslope over impermeable layers, usually over the permafrost table. They have shallow ribbon-like tracks and arcuate head scarps.

Skinflows develop where there is a sudden thawing of the active layer. This may be induced by such phenomena as heavy rain, fires, or an unusually warm day. In some areas, they involve sufficient soil for it to be necessary to protect roads and utilities against their development—usually by placing an insulating blanket of sand and gravel which also reduces the rate of pore-water pressure buildup and creates a normal stress which, on low angle slopes, is greater than the added shear stress.

Bimodal flows, unlike solifluction and skinflows, are not restricted to active layers, but result from deep and rapid thawing of ice-rich permafrost. They develop particularly on river-banks, coastal sites, and below or around thermokarst depressions where rapid thawing is possible. The term 'bimodal'

PLATE 13.12. Lobate terracettes of soliflucted debris at 1600 m in Central Otago, New Zealand.

PLATE 13.13. A bimodal periglacial landslide with an 8 m-high scarp at the rear which fails by repeated slumping, and retreats at 2–6 m/y; in the foreground is an earthflow of ice-rich sediment derived from the slump blocks (photo by J. Ross Mackay).

refers to the formation of a steep headscarp by undercutting, followed by slower thawing of permafrost with creation of a mudflow tongue (Plate 13.13). French (1976) reported such features with diameters of 300 m. Exposure of headscarps can lead to repeated retrogressive failures or, if the mudflow material blankets the feature, to stabilization.

Fossil periglacial forms are now found in some temperate zones on slopes of less than 5° which would not normally be attained under landsliding processes. It is believed that many landslides in Western Europe are relict from periglacial conditions and that they account for the very low angle failures in the Lias Clay of central England (Chandler 1971), and in the Weald Clay where solifluction occurred on slopes as low as 1.5° (Skempton and Weeks 1976). Under present climates the lowest slope angles at which slides occur in these areas is about 8°.

Weathering

Weathering is obviously of great importance in the long-term conversion of strong rock to weaker soils. It is also of importance in the progressive weakening of marine clays, mudstones, shales, and related argillaceous rocks.

Because weathering is a slow and undramatic process its importance in the progressive reduction of resistance is easily overlooked. The unexpected failure of a slope that is known to have been stable for hundreds of years may be attributable to weathering through indirect effects, such as creation of a more permeable material which will allow greater pore-water pressures than could occur in less weathered material.

The importance of fissuring in reducing strength of over-consolidated mudrocks is emphasized below.

Hillslope form and orientation

Hillslope form and orientation are passive factors which may influence the nature and effect of active processes.

Hillslope form is unlikely to have universally applicable relationships with landslide type and density because of the effect of combinations of the active processes—with the one exception of hillslope gradient, which is likely to be correlated with increasing density of landsliding. In one geomorphic zone, however, it is reasonable to expect some correlations.

In a study of a rainstorm in New Zealand, in April 1981, with a duration of three days and rainfalls in excess of 900 mm in some areas, and over 800 km[2] receiving 500 mm, Salter *et al.* (1981) found that no landslides occurred on hillslopes of less than 8°, and over 97 per cent of failures occurred on slopes steeper than 20°. This gradient appears to have been the threshold angle for

TABLE 13.10. *Hillslope gradient with landslide frequency and density for one storm*

Gradient group (°)	Frequency of slides	Slide density (slides/km^2)
0–3	0	0
4–7	0	0
8–15	9	0.13
16–20	218	2.12
21–25	2214	14.35
26–35	4111	12.42
>35	618	9.22
TOTAL	7170	—

Source: Data from Salter *et al.* (1981).

instability (Table 13.10). The slope group with the highest landslide density had gradients of 21–25°, but steeper slope units had a lower landslide density: the latter effect was probably due to two factors: (1) forest and shrubs are more prevalent on steeper slopes; and (2) the landslide-prone soils had already been eroded from the steep slopes.

In a separate study, Blong (1974) found that landslide type, landslide depositional characteristics, and shear-plane size are quite unrelated to hillslope size, shape, or surface irregularity.

Hillslope hollows cut in bedrock are common in soil-mantled hillslopes. Some of these hollows date from the Last Glacial and were produced by solifluction and other processes which are no longer active (Cotton and Te Punga 1955; Dietrich and Dorn 1984; Mills 1987). Many of these hollows are now filled with colluvium, from which the evidence for the formative processes and their age has been derived. Other hollows are much younger and were formed by landsliding under processes of the kind which are still active.

Three general forms of bedrock hollows can be identified: (1) those filled with colluvium; (2) those without colluvium and with a sharp headcut; and (3) those with a gradually sloping head (Fig. 13.21).

Infilled bedrock hollows in the Wellington region of New Zealand (Plate 13.14) contain angular coarse-grained debris and have a range of infill depths of 0.6 to 6 m. Analyses using a rearranged form of the infinite slope equation to calculate the critical depth, Z_c, in excess of which failure is likely to occur have been carried out by Crozier *et al.* (1990):

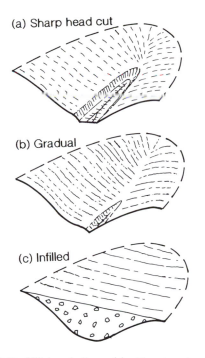

(a) Sharp head cut

(b) Gradual

(c) Infilled

FIG. 13.21. Hillslope hollows: (a) with a steep head cut; (b) with a gradual head; (c) infilled.

$$Z_c = \frac{(c' \sec^2\!\beta)/\gamma}{\tan\beta - [(\gamma - \gamma_w)/\gamma]\tan\phi'}.$$

This analysis shows that the thin (<0.5 m) regolith on the hillslope between hollows is potentially stable, that some shallow infills are stable, but deeper ones have infills near the critical depth for stability when the water-table reaches the surface.

The analysis raises the question of why all the infills have not failed, as it is known that water-tables do reach the surface. The answer appears to be that the strength of tree roots has added an apparent cohesion of 1 to 20 kPa and so stabilized many infills. Much of the region has been de-forested in the last 150 years and infills now fail during extreme storms.

The formation of hollows by landsliding, fol-lowed by quiescent periods in which there is gradual infill by creep, wash, and other processes, and then an episode of fill removal by landsliding, has been suggested for other areas by Dietrich and Dunne (1978), and Marron (1985).

Hollows with a head-cut in a colluvial infill commonly have forms and locations controlled by seepage erosion, whereas gradual channel

head locations in hollows appear to be governed by saturation overland flow (Montgomery and Dietrich 1989). Seepage and landsliding are more important on steep slopes and overland flow on gentle slopes. Landslide scars on steep slopes commonly form the heads of the drainage net-work, and the length and areas of basins in which hollows lie vary inversely with the local valley gradient at the channel head. The larger the source basins the more widely spaced must be the hollows and therefore the lower is the drainage density. There is, consequently, an intricate relationship between the hydrological network and landsliding, and hence between hillslope planform and drain-age, which also links to the stability of the regolith (Montgomery and Dietrich 1988). It follows from this that if a change of environment, such as greater rainfall or deforestation, induces a greater density of landslides it will consequently mark the beginning of an increase in drainage density. If stream-head drainage basin areas increase as hollows are infilled, the drainage density will decrease.

Hillslope orientation has been found to influence hillslope processes in a number of studies in various environments. Salter *et al.* (1981) in the study mentioned above, found that 48 per cent of all landslides occurred on equator-facing slopes and only 10 per cent on pole-facing slopes. The greater number of failures on equator-facing slopes has been noted in a number of studies (e.g. Crozier *et al.* 1980). A possible cause is that the greater insolation received causes more wetting and drying cycles with greater soil cracking, and there-fore more macropore development, higher infiltra-tion, and higher pore-water pressures. The deeper cracking may also increase the water holding capacity of the soil so that soils on sunny slopes may exceed or hold their liquid limit volume of water, whereas soils on shady aspects do not reach this limit. On sliding and remoulding, flow be-haviour is therefore more likely on equator-facing slopes.

In a study of badland erosion in South Dakota, Churchill (1982) found that the dry conditions of equator-facing slopes reduce weathering and fluvial erosion so that the pole-facing slopes have a drainage density which is several orders of mag-nitude greater than that of equator-facing slopes.

Many other studies (reviewed by Parsons 1988) have identified hillslope asymmetry, but few are based on critical examination of all slope units

PLATE 13.14. A colluvium-filled depression in Pliocene marine sediments, Taranaki, New Zealand (photo by M. J. Crozier).

in an area, and even in those that are, it has proved impossible to relate hillslope asymmetry to processes, because hillslopes develop over long periods. Only where a single storm event, or a rapidly evolving landscape is being considered can such correlations be well established.

Hillslopes on soils with distinctive properties

Cliffs in brittle granular soils

Many sedimentary soils (in contrast to residual soils) are composed of grains with a limited range of particle sizes, as for example, the silt and fine-sand sizes of loess and tephra; others have a wide range of particle sizes, as for example, glacial tills and alluvial deposits of mountain torrents. Many of these sedimentary soils support steep cliffs, sometimes cliffs of considerable height (Plate 13.15). Many highway cuts, steeper than 80° and higher than 30 m, in Japan, Guatemala, New Zealand, and China occur in volcaniclastic and loess deposits. In California and Oregon natural slopes in marine terrace sands are steeper than 70° and reach heights of 30–45 m and locally slopes steeper than 40° exceed 150 m (Sitar and Clough 1983). Tills of lateral moraines may support slopes of 100 m at angles of 60–70°.

The unconfined compressive strength of the materials may be low (70–700 kPa) and, more importantly, they may lose most of their strength after failure (i.e. they have very brittle behaviour) and much of it on saturation. Assessments of the stable angles of hillslopes in relation to the height of the slope are critical for many practical purposes, especially as many of the areas listed above are prone to both severe earthquakes and rainstorms. In a more general sense, many river-banks, coastal cliffs, and agricultural terraces formed in granular soils have steep slopes which are stable at low heights, yet are unstable above some critical height.

The form of stability analysis which is relevant to very steep cliffs, in soils which fail by the development of tension cracks along the cliff top, is the

PLATE 13.15. Cut slopes, each about 7 m high and nearly vertical, in a pyroclastic flow of pumiceous ignimbrite. The material can be readily dug with a spade but is stable because of its high internal friction and high permeability. However, on remoulding under saturated conditions it is very sensitive and has very low slake durability.

Culmann wedge analysis. The block which fails may have the form of a column with an inclined base dipping to the cliff foot (Fig. 13.22), or it may be a simple triangular wedge inclined from a point behind the cliff top to the cliff base. In both cases the position of a tension crack on the cliff top is considered.

Culmann (1866) has shown that:

$$H_c = \frac{4c}{\gamma} \cdot \frac{\sin\beta\cos\phi}{[1 - \cos(\beta - \phi)]}.$$

As the slope gets steeper the critical height for stability will decrease, or as a valley is deepened the critical slope angle for stability decreases. At higher angles of surface slope (β) the potential failure plane is inclined at a higher angle (α) so that a smaller depth of slope is necessary to produce a landslide.

It is common for a tension crack to develop behind the slope face and if this extends down to a potential failure plane the critical height is reduced by that depth (z). The theoretical maximum possible depth of a crack is about half of the critical height of the slope, but because the actual depth is related to its position behind the cliff the usual procedure is to solve the equation for H_c and then to determine the critical height of the slope with a crack as $H_c - z$.

Analyses of Culmann wedge failure in loess have been carried out by Lohnes and Handy (1968), in glacial tills by McGreal (1979), and in pyroclastic deposits (unwelded ignimbrites) known in Japan as 'Shirasu', by Matsukura *et al.* (1984). The relationship between critical height and slope angle can be plotted for the soil and the plot used as a planning tool as well as a means of understanding the controls on the inclinations of natural slopes (Fig. 13.22b).

It is essential in the Culmann analysis, as in any stability analysis, that appropriate soil properties are used. In very free-draining soils the results of drained shear tests, with the soil at partial saturation, may be appropriate, but the worst case condition of undrained shear testing at saturation may be appropriate for poorly drained soils with low permeabilities; such may be the case for agricultural terraces in steep hill country, like that of Nepal where few terraces higher than 2 m are stable—as is shown by experience and confirmed by the Culmann analysis (Caine and Mool 1982).

Mass wasting on over-consolidated mudrocks

Over-consolidated mudrocks outcrop over large areas of Earth's landsurface. In the UK they occupy over 20 per cent of the surface area, in North America they underlie large areas of the Great Plains. In North Island, New Zealand, they underlie much of the eastern and western parts of the Island. In considering these materials, it should be remembered that the term 'mudrock' includes large bodies of sand- and silt-size grains, clay minerals of several species, a range of fabrics (see Chapter 3), various degrees of cementation

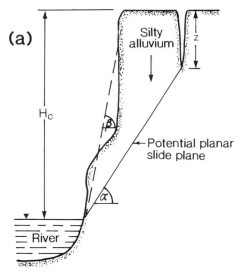

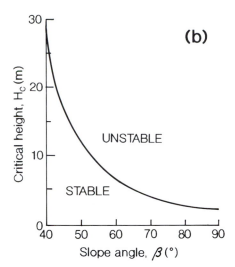

BY THE CULMANN METHOD

From shear strength testing: c = 8.0 kN/m^2, ϕ = 25°   Where β = 90° and ϕ = 0°, because sin90° = 1
Soil unit weight at natural moisture content: γ = 22 kN/m^3 and cos0° = 1
From ground survey: β = 80°, z = 0.8 m

Then H$_c$ = $\dfrac{32}{22}$. $\dfrac{0.98 \times 0.91}{(1-0.57)}$ − 0.8   then H$_c$ = $\dfrac{4c}{\gamma}$

H$_c$ = 2.2 m

FIG. 13.22. (a) Culmann analysis of a wedge failure in stiff soil. (b) Relationship between slope height H_c and slope angle β, for the same soil, and prediction of failure conditions. As the slope steepens the height for stability declines.

by calcite derived from marine shellfish, and inclusions of such minerals as pyrite. These materials occur in varied amounts and combinations so that there may be many departures from the general comments given below.

Mudrocks have low compressibility, moisture content, permeability and sensitivity as a result of burial and overburden pressure of younger sedimentary deposits. The reduction of void space, expulsion of pore water, and diagenetic bonding during burial, give mudrocks high effective shear strengths when compared with the strength of normally consolidated soils of similar composition.

Erosion of overburdens, of 1000–2000 m thickness in many cases, leaves the surface mudrocks exposed to low normal stresses and to weathering processes. Stress-relaxation causes dilation (unloading) joints to develop normal to the ground surface (see Fig. 4.14b) and water can then penetrate the near-surface mudrock. Swelling clay

minerals can expand, slaking occurs at the ground-surface and chemical weathering can cause chemical and mineralogical changes. The weakening of diagenetic bonds causes a reduction in cohesion, especially where calcite is removed in solution.

An increase in water-content with very small displacements along joints and fissures causes strength to decline to the 'fully softened' or 'critical state' value. After much larger displacements the strength falls to a residual value (see Fig. 5.26b). This two-stage process of softening can be illustrated by reference to two examples from Britain.

The Upper Lias Clay of the East Midlands has, as a result of weathering, a surface zone to a depth of 10 m, or more, in which the clay is so fissured that it has the characteristics of a breccia (see Fig. 5.4); the clay mass consists of clay lumps in a matrix of remoulded clay and the water-content has risen to about the plastic limit. At failure of a 'first-time' landslide the weathered material is

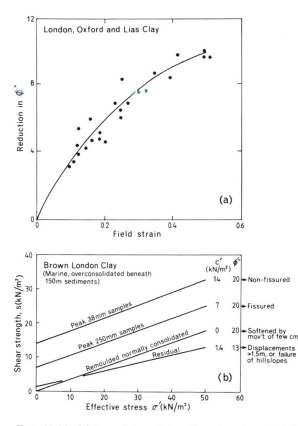

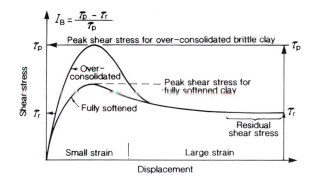

Fig. 13.24. Definition of the brittleness index.

Fig. 13.23. (a) Loss of strength in sedimentary clays related to field strain (after James 1971); (b) Coulomb plots for direct shear box strength tests on weathered London Clay showing the effect of sample size and degree of remoulding: small samples lack fissures (data from Skempton 1977).

well below peak strength and the 'fully softened' clay has the same strength values as remoulded, normally consolidated clay soil (Chandler 1972).

The slide planes of old landslides or ancient soliflucted material, as in shear joints, bedding planes, and faults, have much lower strengths which are very close to residual values.

The London Clay is of Eocene age and was mostly deposited in a moderately deep marine environment with an overburden of 150–400 m of sediment. Many deep landslides occur in this material because of numerous excavations for railways, roads, and canals in the London area. All slides occur in the brown weathered clay and none penetrates the underlying unweathered blue clay to any appreciable depth (Skempton 1977). Unweathered and unfissured London Clay exhibits

peak strength at failure, but progressive slight movements during formation of fissures cause work softening along parts of potential slide planes so that there is a non-uniform mobilization of shear strength along the failure plane and a reduction in the apparent cohesion. Large field strains (defined as the ratio of the amount of slide movement occurring to the length of the slide plane) may also cause a reduction in ϕ', but most of the loss of strength at the fully softened condition is from loss of apparent cohesion (James 1971; see Fig. 13.23). Residual shear strengths are reached most commonly in movements along bedding-planes and in shallow failures on natural slopes where final failure has been preceded by large displacements along the shear plane.

The large percentage drop in strength displayed by many over-consolidated mudrocks and clay shales can be indicated by a Brittleness Index, I_B:

$$I_B = (\tau_p - \tau_r)/\tau_p \quad \text{(Fig. 13.24).}$$

Brittleness in such materials is largely due to dilation and forced alignment of clay minerals along the shear plane.

Similar weathering effects of argillaceous rocks are common on the Great Plains of central USA and Canada. The rocks were over-consolidated beneath about 600 m of sediment, which has been removed since Eocene times, and subsequently suffered rebound, swelling, and fissuring. In early postglacial times meltwater cut deep channels into the weathering rocks and many landslides occurred along the flanks of the channels. More recently water-tables have dropped and the slopes have stabilized in many areas, but where water levels are still high translational slides are common and often associated with bentonite-rich beds (Mollard

1977). Strength reductions similar to those of the British mudrocks are reported by Silvestri (1980) and Wu *et al.* (1987).

In New Zealand two contiguous bodies of mudrocks illustrate differences in compositions and their effects on geomorphic processes (Pearce *et al.* 1981).

The Tikihore Formation underlies a landscape of long (900–2800 m), rectilinear slopes with inclinations of about 14°. Most of the land surface is formed of intersecting earthslide-flow complexes. The rock is dominated by highly crystalline illite (60 per cent) and kaolinite (40 per cent) which have a remoulded strength which controls the slope processes and inclination.

The nearby Mangatu Formation underlies a landscape of shorter (350–1400 m), convexo-concave slopes with mean ridge crest to valley bottom inclinations of 17°, but slope angles are variable and up to 50°. The mass-wasting forms include some major amphitheatre-shaped features which are probably scars of old rotational failures and many small mass movements, of which few are flow failures. The Mangatu mudrocks contain approximately equal amounts of clay minerals and quartz grains with plagioclase feldspar and calcite each forming 5 per cent of the rock, but ranging up to 15 per cent. Of the clay minerals 60–80 per cent is montmorillonite. In view of a total rock composition which is 20–30 per cent montmorillonite it is initially surprising that the rock, and the soil on it, display relatively high strength, little swelling, support steeper slopes, and have fewer flow failures than the Tikihore Formation. However, the calcite acts as a cement and gives the rocks added strength stability, except in a few notable areas with large, gullied, complex slide-flow features.

These large, complex failures, including that shown in Plate 13.16, coincide with areas where fine-grained, disseminated pyrite is widespread and also found in veins and blebs. The weathering of pyrite yields up to four moles of sulphuric acid per mole of pyrite and it is thought that acid-sulphate weathering has dissolved out the calcite cement, thus decreasing the rock strength and leaving a montmorillonite-rich regolith of very low remoulded strength. Bacterially-aided acid sulphate weathering of pyrite, sometimes followed by precipitation of gypsum, has also caused slope stability and foundation problems in clay shales in Canada (Gillott *et al.* 1974).

Slope failure in saprolites and residual soils

Soil mechanics as a discipline developed because of a need to understand the nature of sedimentary soils, and their behaviour when used for foundations and as earth dams and other forms of embankment. The application of this knowledge to saprolite materials, and residual soils developed by weathering *in situ*, has not always been successful. It is largely because of the problems of slope stability in expanding cities such as Rio de Janeiro and Hong Kong that major advances have been made in recognition of the distinctive properties of saprolitic and residual soils. Much of this knowledge, and recognition of remaining problems, has been summarized in review papers by Irfan and Woods (1988); Massey and Pang (1988); and Vaughan (1988).

The major difference between sedimentary soils and saprolitic soils is the inheritance in saprolites of relict joints and other structural features of the bedrock, even where weathering has completely altered the mineral composition from that of the bedrock and formed material of weathering grades IV and V (see Fig. 9.4). Discontinuities are of importance for the following reason: they form planes of separation along which weathering has penetrated more effectively than into the unjointed saprolite mass, and this may cause the relict joint to be either a zone of preferential weakness or, if precipitation of iron and manganese oxides has occurred, a zone of great resistance and less permeability than the mass of saprolite. In both cases the relict joint is likely to be a plane of preferential shear or different pore-water pressure. Quartz veins, soft clay, fault gouge, in-washed joint fills, and other bands of material can also cause discontinuities in soil properties.

Slickensides, resulting from either tectonic displacements in the parent rock, or from differential movements within the saprolite, are commonly reported (e.g. Koo 1982). Within a saprolite, slickensides may result from stress relief as overburden is removed by erosion or settling if the weathering has not been isovolumetric.

Slope failure along relict joints is particularly common where: (a) one or more of the joint sets dip out of the hillslope; (b) randomly oriented joints intersect and form irregular paths along which shear movements can propagate; (c) high water pressures can develop along joints which

PLATE 13.16. A mass-wasting landscape on the Mangatu Formation mudrocks, see also Plate 17.1, p. 382. The deeper features are slump-earth flows which are also gullied. All slope surfaces are involved in shallow earthflows (photo by J. H. G. Johns, New Zealand Forest Service, Crown Copyright reserved).

have high permeability or, more commonly, where clay-rich relict joints form barriers to water movement and thus create perched water-tables or the development of pipes and seepage pressures; and (d) weathering along a joint reduces the strength of joint asperities and a reduction in the basic friction angle of rock surfaces through changes in mineralogy (Fig. 13.25a, b). Koo (1982), for example, obtained peak strength parameters of $c' = 0$, $\phi' = 20°$ from direct shear testing of smooth relict joint surfaces, in weathered volcanic rocks in Hong Kong, in contrast to the strength of intact saprolite of $c' = 6\,\text{kPa}$, $\phi' = 36°$.

The type of mass wasting occurring on a slope formed on saprolites is closely influenced by the relict joints. (1) Where joints in the bedrock were very closely spaced, the weathered saprolite is likely to behave as a granular material in which each granule is a block of saprolite. Rotational failures are then likely to occur. (2) Relict joints dipping nearly parallel to the hillslope may occur through inheritance of stress release joints parallel to the slope in bedrock, or to foliation and bedding in that plane. Translational landslides are then likely to occur with the relict joints forming the failure planes. (3) Intersection of two joints at close to 90° and with an orientation close to that of the hillslope can form wedge failures; such failures have to be analysed in the same way as wedge failures in rock. (4) Relict joints with a near vertical

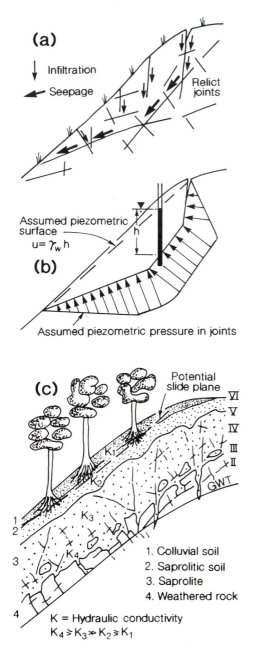

Fig. 13.25. (a) Relict joints in a saprolite with major joints providing a pre-determined failure plane. (b) The pattern of water pressure along the major joints and the assumed piezometric surface used in a stability analysis where pore-water pressure is a controlling factor (a, b, based on Massey and Pang 1988). (c) The hydraulic conductivity, K, and weathering grade of saprolitic soils on a hillslope. The groundwater-table is at some depth and the failure occurs in the solum and saprolitic soil as suction is lost during infiltration (based on Wolle and Hachich 1989).

dip, and due to foliation or bedding, can produce toppling failures; these also have to be treated as rock slope failures.

Residual soils forming weathering zone VI, the solum, have properties which are very different from those of sedimentary soils. They have a very small overburden and therefore little consolidation, they are highly, but variably, structured and have variable textures. The porosities of residual soils are generally high, when compared with those of sedimentary soils. Residual soils have varying proportions of clay minerals depending upon their parent material and degree of weathering. In general, a soil with less than about 15 per cent of clay content behaves much as a granular material. No orientation of the clay minerals during shearing is possible and a unique value of ϕ'_{cv} (the friction value during shear at constant volume) is attained at large strains. A soil with more than about 40 per cent clay content has properties dominated by the presence of the clay. In such a soil the platey clay minerals can orientate and a continuous shear surface can form. The value of ϕ'_{cv} is obtained only while the particles have random orientation. At large strains the strength drops to ϕ'_r. At clay contents between 15 and 40 per cent an intermediate situation occurs. The variable texture and composition of residual soils may cause uncertainty in choice of appropriate shear strength properties for a stability analysis.

The structure (i.e. fabric) of residual soils and the opening of cracks during dry periods creates considerable variability of infiltration from place to place on a hillslope. This may be part of the cause of uncertainty about pore-water pressures during the intense storms which cause, or at least trigger, most slope failures in the humid and seasonally humid tropics (Massey and Pang 1988). A second factor which is being recognized as being of great importance is that of matric suction.

Many hillslopes in tropical areas such as Tahiti and Papua New Guinea have inclinations of 35° to 55° and up to 70° (e.g. Simonett 1967) yet the soils appear to have lower strength at saturation than is needed to maintain stability. Several studies (e.g. Anderson *et al.* 1987) have suggested that failure occurs when the soil suction, which maintains stability under most conditions, is reduced by infiltration.

The situation which allows this process to operate has been described by Wolle and Hachich (1989) for the Serro do Mar escarpment in south-

PLATE 13.17. Shallow translational landslides resulting from a single storm. The failure planes are at the boundary between the solum and saprolite (see Plate 13.5).

eastern Brazil. On hillslopes of 40 to 45° the soils are relatively thin and consist of 1 to 1.5 m of colluvium overlying 1 to 2 m of saprolitic soil with features inherited from the basement granitic and gneissic migmatites. Below the saprolite is a zone of corestones and blocks on weathered bedrock with open fractures (Fig. 13.25c). The important feature is that the hydraulic conductivity is least at the surface and greatest at depth, with the weathered rock and corestone zone having much greater conductivity than the soil above. The weathered rock therefore acts as a drain to the upper horizons and water levels remain restricted to depths of 20 to 30 m below the surface.

The vegetation intercepts a significant proportion of the rain and protects the soil from erosion. However, shallow translational landslides result from major storms and also occur in many wet seasons. The landslide forms are very similar to those shown in Plate 13.17.

The slides are triggered by heavy rain, particularly when antecedent soil moisture is high, and infiltration is rapid and the wetting front can penetrate to depths of 1.0 to 1.5 m in four to twelve hours.

Degrees of saturation of surface soils in the rainy season vary between 37 to 86 per cent with 60 to 70 per cent being common. Apparent cohesion due to suction is eliminated at saturations of 92 to 95 per cent. Soil strength, during a wet season with prolonged rainfall, can be reduced from $c = 6.0\,\text{kPa}$ and $\phi = 34°$, to $c' = 1.0\,\text{kPa}$ and $\phi' = 34°$. The corresponding factor of safety falls from 1.66 to 0.92.

The reduction of strength occurs without any increase in pore-water pressure and is entirely the result of loss of apparent cohesion as capillary suction is eliminated by infiltrating water penetrating as a wetting front.

Suction levels are generally high enough to maintain stability except in high intensity rain which cannot drain quickly enough to maintain cohesive strength in the upper 1 m or so of the soil. Knowledge of pre-existing soil-moisture content from an automatic tensiometer, and of rainfall intensity from a recording rain gauge, can provide a ready warning system for critical urban areas.

Case studies

Much of what is known about mass wasting, and particularly about large failures has been obtained from detailed investigation of failures after the event. Case studies are particularly helpful in recognition of the range of geological conditions which may lead to instability and as guides to appropriate methods of study. Many case studies are reported in Zaruba and Mencl (1976), in Voight and Pariseau (1978) a two-volume collection, and in volumes 1 and 2 of Bonnard (1988). Important systematic studies of aspects of mass wasting are included in Schuster and Krizek (1978); Brunsden and Prior (1984); Sidle *et al*. (1985); Bromhead (1986); and Crozier (1986).

14

Flow Failures on Hillslopes

Flow failures are gravity-induced mass movements intermediate between sliding and water flows. They involve various forms of rapid mass movement in which the proportions of granular solids, water, and air, and the velocity with which they move, are intimately related to the rheology of the mixture. The proportions of solids range from total dominance by clays, silts, sand, gravels, or boulders, to any mixture of these; the pore fluid may be entirely air or water or any combination of the two, and the velocity of flow, and hence the internal shear rate, may range from that of solifluction (0.001–0.5 m/y) to about 360 km/h. The volume of material involved ranges from a few cubic metres to tens of Mm^3.

Because of both the large masses of material and the high velocities which are possible, flows are some of the most destructive natural processes known on Earth's surface, causing hundreds of deaths and losses of millions of dollars worth of property each year (Costa 1984).

Rheology of water—sediment mixtures

With the exception of clear water and air, natural fluids are multiphase involving mixtures of sediment, water, and air. The behaviour of the mix, in response to shear stress, is a function of: (1) the relative proportions of these components; (2) the grain-size distribution of the sediments; and (3) the physical and chemical properties of the sediment. In some circumstances temperature and solutes may also have an effect.

It was noted in Chapter 4 (see Fig. 4.2) that water is a Newtonian substance and that many clays near their liquid limit water-contents behave as either dilatant plastics or as Bingham plastics

(Figs. 7.3 and 14.1). The term 'dilatant' in this context indicates a fluid that exhibits an increase in viscosity with shear rate (shear thickening). A fluid with a viscosity that decreases with increasing rate of shear (shear thinning) is termed a pseudoplastic or a general plastic.

In field situations, velocity of flows is a surrogate for rate of shear, and flow processes have been identified with characteristic rates. The rheological response of sediment–water mixtures at a given strain rate is governed primarily by sediment concentration, or water-content, and is affected to a lesser extent by the grain-size distribution of the solids. Velocity of flows and their sediment concentration were consequently chosen as the primary criteria for classification of flows by Pierson and Costa (1987).

Classification by rheology

In Fig. 14.2 are shown the fundamental features of the classification. The boundary lines (A, B, C) are not fixed in position, but will move to the right or left, depending on the grain-size distribution. The positions as shown assume that the flowing sediment is a coarse, poorly sorted, non-cohesive mixture typical of most colluvial soils of mountains. For cohesive sediments and those with high proportions of fines, the boundary lines shift to the left; for coarser, better sorted, non-cohesive sediment they shift to the right. As more data become available the boundary positions may be defined more closely. The shaded zones indicate conditions thought not to occur in nature.

Major categories of flows

We may distinguish: streamflow; and two types of plastic flow—(1) slurry flow, and (2) granular flow.

Streamflows and floods are movements of multiphase liquids in which fine-grained sediment and

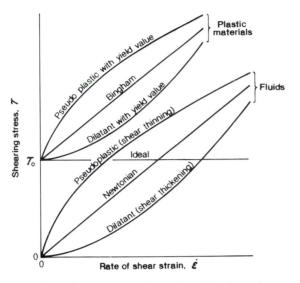

FIG. 14.1. Flow curves for idealized liquid and plastic substances. Liquids exhibit no initial shear strength; plastic materials may deform only after a yield stress, τ_o, is exceeded.

air bubbles are dispersed and suspended in water. As long as the dispersion is dilute, the particles and bubbles do not interact and the liquid behaves as a Newtonian fluid.

If the solids increase in concentration, a point will be reached at which the particles begin to interact; clays will flocculate and develop cohesive strength. The fluid then has a yield strength and becomes non-Newtonian. The concentration required to produce yield strength is about 3 per cent by volume for smectites, and 13 per cent for kaolinite (Hampton 1972); fine-grained non-cohesive sediment may reach 50 per cent by volume before frictional interactions create yield strength (Rodine 1974).

Hyperconcentrated streamflows are flowing mixtures of water and sediment which are sufficiently dense to damp turbulence and have low but readily measured yield strength, commonly about 30–40 N/m² (Beverage and Culbertson 1964).

Slurry flow occurs in sediment–water mixtures having sufficient yield strength to exhibit plastic behaviour. It develops as a sharp transition from hyperconcentrated flow (indicated by boundary B in Fig. 14.2) and is evident by the capacity of the flow to form steep, lobate fronts and lateral levees, and to carry gravel-sized particles in suspension

and yet to become partially liquefied as the material is remoulded. Such behaviour is evident in wet concrete mixtures and is a result of internal friction between grains and of cohesion in clay-rich mixtures.

The natural moisture content of a slurry flow exceeds a critical limit, which for clay-rich mixtures is the liquid limit and for other mixtures is the closest approximation to that limit. At that water-content part of the weight of the solids is carried by the pore water, which accounts for the liquefied state and the pore-fluid pressures which exceed the hydrostatic pressure (see Fig. 13.19). The mixture flows as a coherent mass when the yield strength is exceeded, consequently the mass may support boulders or roll along those too large to be carried. The mixture remains homogeneous as it comes to rest and there is, therefore, no sorting and no bedding in the deposits. This is in contrast to the deposits of hyperconcentrated flows in which differential settlement from suspension can occur.

Viscous behaviour may control the flow behaviour of slurries when the clay and silt content of the mixture is high, or the velocity and water-content are low (Bagnold 1954). Under such conditions either a Coulomb-viscous or a Bingham model is often used for predicting flow behaviour (Johnson 1970) (Figs. 7.3, 14.1).

Inertial forces dominate flow behaviour where momentum is transferred through particle collisions. Such forces operate where the viscosity of the pore fluid is low (e.g. where there are few fines in suspension and the fluid is mainly water, and particle sizes and velocity are high). The appropriate rheological model is then the dilatant fluid model (Takahashi 1978). An intuitive estimate of the critical velocity separating viscous and inertial mechanisms is about 2 m/s for a coarse-sediment mixture but this estimate would change with the composition (Pierson and Costa 1987).

Granular flows develop when the sediment concentration increases to the point where the material is not liquefied by remoulding; that is, the pore fluid does not support the granular mass and the full weight of that material is borne by grain-to-grain contacts or collisions (Keefer and Johnson 1983).

At low rates of shear, sliding friction, interlocking of grains, and any viscous effects of pore fluids, together determine the bulk stresses within the deforming mass. Such slow flow may be called frictional granular flow and this is operative at low

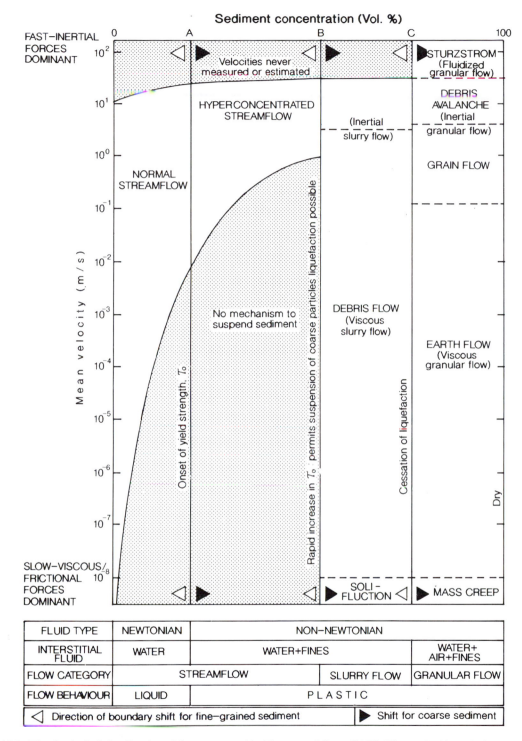

FIG. 14.2. The rheological classification of flows proposed by Pierson and Costa (1987). The vertical boundaries A, B, and C are rheological thresholds related to sediment composition; they can therefore move to left or right; their position in this figure assumes a coarse, poorly sorted mixture. Horizontal velocity boundaries are influenced by sediment concentration and by grain-size distribution and density. Shaded zones indicate conditions not known in nature.

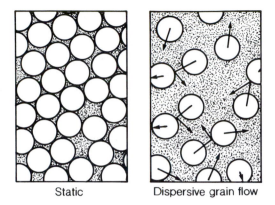

Static   Dispersive grain flow

FIG. 14.3. A representation of dispersive grain flow in which particle collisions force grains apart and reduce frictional retardation. (After Melosh 1987.)

velocities, which for a well-graded coarse material is about 0.1 m/s.

As velocities greater than 0.1 m/s develop, grain inertia increases and grain collisions transfer momentum resulting in dispersive stress (Bagnold 1954) which forces grains apart (Fig. 14.3), reducing frictional retardation. In such inertial granular flow, velocities are thought to be in the range of 0.1 to 35 m/s.

Massive granular flows, known as debris avalanches and sturzstroms (Hsü 1975), having volumes exceeding $1-5 \, Mm^3$, achieve very high mean velocities in the range 50–100 m/s (180–360 km/h). Field investigations of active sturzstroms are, inevitably, impossible, so there are still many hypotheses of the causes of their high velocities and extremely large distances of travel. The most probable explanation lies in momentum transfer by energetic interparticle collisions with negligible frictional resistance (see below for discussion of mechanisms).

Problems of classification

The literature is replete with terms designed to distinguish various forms of mass wasting. The most obvious categories lie in attempts to distinguish flows on the basis of their composition, thus occur clayey debris flow, coarse debris flow, mudflow, silt flow, sand flow, peat flow. The difficulty with all this is that there is no consistent terminology, for example, many so-called mudflows have large proportions of silt and sand in their composition, and more importantly science is not advanced by such terms because they are not related to flow mechanisms. It seems better, therefore, to use broad terms based on rheology and, if necessary, to add descriptors such as 'peaty' debris flow.

A second area of confusion lies in the use of the terms 'flow' and 'slide'. Slide is clearly appropriate where a block of material is moving over a clearly defined shear plane and the block has a nearly uniform velocity throughout its mass. The shear strength may then be modelled by the Coulomb equation and limit equilibrium forms of stability analysis applied. Many features classified as mudslides and earthflows (e.g. Brunsden 1984; Bovis 1985) appear to have the characteristics of slides and may change into flows in their terminal sections; 'flow' being used to describe units of the failure which move as viscous substances. Whether or not a slide which becomes a flow is termed mudslide or earthflow will depend on judgement as to which is the dominant process. It will be unfortunate for scientific advance, if traditional usage,

TABLE 14.1. *Physical properties of flows*

Flow	Sediment load by weight (%)	Bulk density (Mg/m^3)	Shear strength (N/m^2)	Flow rheology	Deposits
Stream flow	1–40	1.01–1.3	<10	Newtonian	Sorted, stratified with bedforms
Hyperconcentrated flow	40–70	1.3–1.8	10–20	Approximately Newtonian	Poorly sorted, weakly stratified
Debris flow	70–90	1.8–2.6	>20	Visco-plastic	Levees and lobes of largely unsorted material, largest clasts on top and at face of lobes

Source: Costa (1984).

which is not rheologically based, is allowed to control terminology.

The term debris torrent was devised to distinguish those types of debris flows, common in the Pacific Northwest of USA and Canada, which are channelized and have a large organic matter content, particularly of logs, and a limited fine-grained content. As Slaymaker (1988) has pointed out, the term 'torrent' in European languages refers to a fast-flowing mountain stream. The term 'debris torrent' is therefore linguistically confusing. No doubt local usage will continue, but definition in terms of rheology, which will place debris torrents at the high velocity end of debris flows in Fig. 14.2, would be useful. Terms such as 'channelized' can be added as descriptors.

Lahars are debris flows which are composed of volcaniclastic materials and drain from volcanoes. It is unnecessary for most purposes to distinguish them as a separate class of phenomena. Because of the abundance of debris available on volcanoes, together with sources of water, volcanoes are conducive to production of debris flows and many of these have been highly destructive (see, for example, Osterkamp *et al.* 1986; Palmer and Neall 1989; Arguden and Rodolfo 1990).

Debris flows

Conditions for development

Debris flows develop where there is an abundant source of material which can be mobilized by the addition of water. Such conditions exist on hillslopes with colluvial soils and saprolites, and especially in hollows where there are thick colluvial infills. On mountains, glacial, talus, and other deposits add to the possible source materials. In semiarid zones, alluvial deposits of earlier floods and flows may be available.

Rainstorms are probably the most common sources of water but, as an abundance of water is required, certain conditions are particularly favourable: high rainfall intensity of long duration; rainfall on thawing snow packs; small drainage basins with steep slopes in which rainfall intensity is uniformly high and runoff is rapid; and, more rarely, glacier and lake overflows into talus, glacial, and unconsolidated alluvial deposits.

Various criteria related to rainfall intensity and duration, and to combinations of these with antecedent rainfall and soil water status, have been proposed as predictors of threshold hydrological conditions under which debris flows are likely to occur. Some of these predictors are demonstrably useful for distinct terrains (e.g. Crozier 1986, ch. 6), but the range of relief types, sediment availability, and storages of water, as well as rainfall conditions, make the successful development of a universal predictor improbable. Caine (1980) has attempted to calculate a threshold of rainfall intensity and duration which is required before debris flows will develop on undisturbed slopes. The data were derived from 73 published records of events for which the total rainfall depth, d, its duration, D, and the mean intensity, I, are known. When intensity is plotted against duration the threshold curve is described by:

$$I = 14.82D^{-0.39},$$

and a plot of depth of rainfall against duration gives a threshold curve described by:

$$d = 14.82D^{0.61}.$$

This threshold seems to be reasonable for durations of ten minutes to ten days, although debris flows may be generated when durations are longer than ten days; it is probable that periods of particularly high rainfall intensity within storms are conducive to failure.

Reports of observers

Something of the variety of debris flows, and of their features in common, can be understood from reports of observers. The following statements are brief summaries of published reports.

(1) The scene is the Ghizar Valley of the Karakoram mountains of northern Pakistan (Nash *et al.* 1985). The valley is arid, 2500 m deep, and has steep walls with talus slopes and fans along their bases, and the valley floor consists of low-angle fans and flood plains. At the head of the tributary valley of the Jandorote Nala is a debris-covered relict glacier within which was a lake. Down-valley, a glacial moraine covers the valley floor and provided a source of debris.

On 27 July 1980, a wave of mud swept down the Nala destroying crops, houses, and livestock as well as the road; flowed across the alluvial plain and blocked the Ghizar River. Waves of debris followed. From their source about 3 km away, each wave, travelling at about 100 km/h (27 m/s), surged down the valley slewing from side to side of its incised channel as it went around bends, until with a deafening roar it reached the bottom of the Nala where it was 5 m high, occupied the 80 m wide channel, and threw up debris and boulders as it passed. The wave then spread out over the landslide dam surface, which 'rose and fell like a jelly. The scene closely resembled a breaking wave passing through the narrow entrance to a tidal inlet.' When movement had ceased the debris had the consistency of wet concrete and was firm enough for people to walk over it.

Waves of debris were roaring down the valley at four per hour for several hours. The formation of waves was not observed, but it is probable that they developed as water from the lake was released in bursts. Each burst caused the glacial debris to reach a critical moisture content so that it flowed down valley as a lobe which generated the pulsing waves of mobile debris as they burst through its bouldery snout.

(2) The scene is an unvegetated ravine on the flank of Mt. Thomas in Canterbury, New Zealand (Pierson 1980). The head of the ravine is cut in highly sheared sandstone and argillite undergoing severe erosion, and the lower reaches of the ravine are incised into an alluvial fan. During April 1978, three severe rainstorms in nine days produced at least 325 mm of rain, and debris flows were active during the last three days of this period.

The debris flows came down the entrenched channel in successive pulses about ten to twenty minutes apart. Between pulses, a fluid slurry flowed turbulently around boulders, with a flow depth of 0.1 to 0.3 m. A thick, submerged layer of gravel moved as bedload. A debris surge appeared usually as a low wave, much like a tidal bore, travelling at up to 5 m/s, or more than twice the velocity of the slurry. The highest surge front was roughly 3 m high. Slugs of coarse bedload material built up in the most viscous slurry flows, allowing debris to accumulate behind them that would initiate, or be flushed out by, the next surge.

A notable feature of wide sections of the channel was the development of confining levees of well sorted cobbles and boulders which were pushed out of the viscous slurry flows. The slurries were, therefore, periodically moving within their own levees at higher elevations than the channel floor.

Deposition on the lower fan occurred as the slurry of mud and rock emerged from the entrenched channel, lateral spreading permitted debris thickness to decrease below that critical for flow. At the distal edge of the fan most of the deposit was of particles smaller than gravel size and had a total thickness of 0.1 to 0.2 m.

(3) The scene is the Gissal Village area of Tajikistan, bordering Afghanistan (Ishihara 1989). On 23 January 1988, an earthquake of M = 5.5 triggered a series of landslides in the gently sloping hills formed on thick (20–60 m) loess deposits.

Slope failures originated as slides, two of which merged to produce a 15 m deep retrogressing failure with a front 850 m wide. Close to the crown of the slides, the loess formed multiple slightly rotated blocks, but a few tens of metres from the crown the loess developed an irregular hummocky surface. The material involved was about 20 Mm^3. The loess was probably close to saturation to a depth of about 15 m as a result of irrigation. Liquefaction of the loess during initial sliding produced a debris flow of mud that moved over 2 km, burying houses on the edge of a village.

(4) The scene is steep hill country (20–40° slopes) with a full vegetation cover and saprolitic and colluvial soils. It would fit descriptions of storm events in Puerto Rico (Jibson 1989); the southern Appalachians (Neary and Swift 1987), and the San Francisco Bay area of USA (Ellen and Wieczorek 1988); and the south Auckland area of New Zealand (Selby 1976a). All of these areas have suffered storms of about 200 mm total rainfall over 15–100 hours with maximum local rainfall intensities in the range 70–100 mm/h. Antecedent moisture content of soils is less well known but varies from dry summer to moist winter conditions. The extreme of these examples is the case of Puerto Rico where 4-day rainfall exceeded 750 mm and 24-hour rainfall exceeded 560 mm. In the San Francisco Bay event over 18 000 landslides transformed downslope into debris flows that killed fifteen people and did damage exceeding \$US280 million in about 32 hours.

The storms were generally severe, with return periods estimated variously as being in the range of ten to one hundred years. In all cases landsliding began after a regional threshold rainfall had been exceeded. Slope failures were shallow (0.8 to 2 m) and started as slides or avalanches which were quickly remoulded so that the failed material became a debris flow. The source areas include rectilinear side slopes, upper-slope concavities, and the heads of first-order drainage basins. Concentration of water promotes failure and this is evident in concavities, but on side slopes concen-

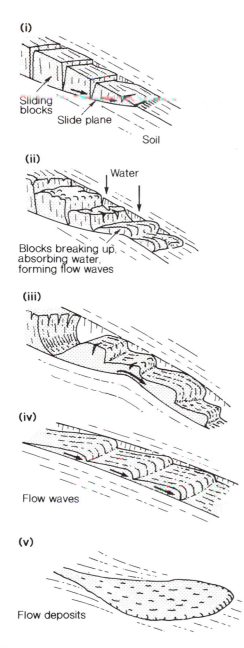

(i)

Sliding
blocks
Slide plane
Soil

(ii)

Water

Blocks breaking up,
absorbing water,
forming flow waves

(iii)

(iv)

Flow waves

(v)

Flow deposits

FIG. 14.4. A schematic sequence of events down a channel where an initial translational slide is converted first into a channelized debris flow and then spreads out into a thin flow at its terminus. (Based on Ellen and Wieczorek 1988.)

sliding slab, with tensile cracking as the slab moves over hummocks or over the lip of the slide scar. Intense rainfall, overland flow, and tributary flows cause incorporation of more water into the moving and disintegrating blocks of soil, so that remoulding, loss of structure and liquefaction occur concurrently in various parts of many failures (Fig. 14.4).

Viscous fluid behaviour develops as downslope velocity increases and shear resistance declines (see Fig. 7.3a). The debris flow may be very fluid, in which case it may move over grassed open hillsides without causing erosion and with leaving only a slick of soil and miniature levees as evidence of its passage (see Plate 13.8). More viscous flows, and those with greater volume, leave larger levees and may scour channels leading to the valley floor (Plate 14.1). In areas with many failures, valley floors become channels in which severe widening and deepening occur in very brief periods. The final channel form may be aggraded with marginal levees (Plate 14.2), or with lobate rock debris (Plate 14.3), depending on the source material, its rate of supply, and the runoff during the storm.

Morphology of debris flows

It is evident from the descriptions given above that not all debris flows have the same morphology. The loess and Karakoram flows may be near to being end-members of a set of morphologies. There are, however, some features which are common and from which it is possible to deduce information on the mechanics of flow.

Flow deposits with a lobe at the head and wave forms behind (Fig. 14.5) are common forms resulting from events such as those of Mt. Thomas and the Ghizar Valley. A typical flow of this kind is a wave of wet debris moving through a channel, with superimposed smaller waves travelling at higher velocities than those of the debris flow itself. Each lobe comprises a snout with lateral deposits at the side of, or immediately adjacent to, the channel, and medial deposits within the channel. Many alluvial fans of mountains, and of arid and semiarid regions, have deeply incised channels near the apex of the fan and such channels form routes followed by debris flows which may overtop the banks leaving concentrations of coarse clasts. The debris flow deposits typically contain large clasts that appear to be randomly distributed in a fine-grained matrix (Plate 14.4). Lenses of sorted and weakly bedded pebbly silts or sands

tration can occur in cracks, from pipe flows, or macropore flow above impermeable soil or rock units. Initial failure is usually by rotational failure or translational sliding followed by bending of the

PLATE 14.1. Debris flow derived from a translational slide in saprolitic soil. The flow has been sufficiently erosive to scour a channel in the hillside. Note the levees and the lobate terminus.

may occur within the main bodies of debris flow as a result of deposition from thinner more fluid pulses between the passages of lobate debris (Johnson and Rodine 1984, provide detailed descriptions). Levees formed by the expulsion of large clasts from the margins of slurries are usual features of coarse flow units (Plate 14.5).

Examples of debris flows which have, usually, 20–30 waves in each event, but which may involve 100–200 waves in an event, without any fluid flow between waves, are provided by Li Jian and Luo Defu (1981). These flows, which they call 'mud-flows', carry boulders with diameters of 3–5 m. Even the short duration event of the Aberfan spoil-heap failure in 1966 (Hutchinson 1986) moved

in successive debris sheets as different parts of the source spoil-heap fed into the flow-slide. Evidence for waves is not always evident in debris left after a storm, especially where the source area is a major slope failure of rather uniform physical properties (Plate 14.6), and subsequent streamflow has incised into the debris. The evidence for a lobate snout may then be confined to the presence of a group of unusually large boulders left on and in levee deposits.

Mechanics of failure and movement

Catastrophic flow follows upon reductions in viscosity, with increased shear rates, down to values associated with hyperconcentrated fluids.

PLATE 14.2. Deposition of debris-flow materials in a valley floor. The debris is derived from shallow saprolitic soils. The weakly developed levees are of rather coarser debris than the main body of material.

Slowly developing slope failures are unlikely to reach velocities at which viscosities are significantly reduced, and they will therefore stop. Catastrophic flow is most likely where an initial impetus is provided by a sudden event such as earthquakes, landslides, and dam bursts.

Landslides on steep slopes may fail suddenly and have sufficient energy to remould debris so that it can incorporate water beyond the liquid limit content, or its equivalent for coarse-grained soils (Fig. 14.4) or, less commonly, the collapse of a soil mass into, or on to, another may so raise pore-water pressures in the lower mass that the overburden weight is transferred to the fluid and the soils grains are forced apart by fluid pressure, leading to liquefaction (Fig. 14.6)—a mechanism of undrained loading proposed by Hutchinson and

Bhandari (1971) and confirmed experimentally by Sassa (1985). The change from slow creep to plastic flow in clay-rich soils follows on destruction of strong bonding, with a consequent decrease in the viscosity of the order of nine magnitudes (Culling 1988).

The ability of a body of flow material to maintain momentum is controlled by its viscosity and shear strength, which will also control the threshold thickness for cessation of movement. The relationships have been analysed by Johnson (1970). The cohesive part of shear strength is provided by the clays and the frictional part by the coarser grains. Flow becomes impossible if the friction angle is greater than or equals the slope angle of the channel or hillslope over which the debris flow is moving, that is if $\phi \geq \beta$. The existence of levees

PLATE 14.3. Debris-flow deposits of highly fractured sandstone left in a channel. The virtual absence of fines is a feature of this deposit, which has not had a channel cut through it after the initiating storm. Levees are absent.

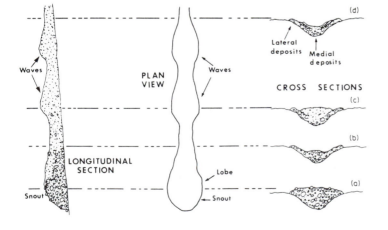

FIG. 14.5. Idealized representations of a debris flow showing waves of debris. (After Johnson and Rodine 1984.)

PLATE 14.4. A section through an alluvial fan in a valley of the Nepalese Himalayas. The bedded material on the right is a fluvial deposit. That on the left is a debris-flow deposit filling a nearly square channel cut in the alluvial deposits. Note the lack of sorting and bedding, and mixed grain sizes of the debris-flow deposits. The lenses of finer material near the top of the debris-flow deposit may be from a fluid pulse in a flow event.

PLATE 14.5. Channels with levees left by debris flow, Austrian Alps.

and lateral deposits thicker than the channel deposits clearly indicates that a flow cannot be treated as a rigid plug of material obeying the Coulomb shear strength model. To that model a viscosity term must be added as in a Coulomb-viscous model:

$$\tau = c + \sigma_n\tan\phi + \eta_c \,(du/dy), \qquad (14.1)$$

or in a Bingham model

$$\tau = s + \eta_B \,(du/dy), \qquad (14.2)$$

where: η_c and η_B are appropriate and Bingham viscosities respectively and s is shear strength.

The Coulomb shear strength can be measured in a large shear box, but the Bingham shear strength, s, can be determined more readily from field observation of the thickness, T, of deposits which overtop a channel then:

$$s = T\gamma_D\sin\beta, \qquad (14.3)$$

where γ_D is the unit weight of the debris, and β is the surface slope.

In places where debris flows have plugged channels with known cross-sections:

$$s = \frac{D\gamma_D\sin\beta}{(2D/W)^2 + 1}, \qquad (14.4)$$

where D is the depth (i.e. thickness) of the plug, and W is the width of the plug.

The Newtonian viscosity, η_N, of moving debris is seldom, if ever, relevant for debris flows, but is often quoted as a method of comparing viscosities

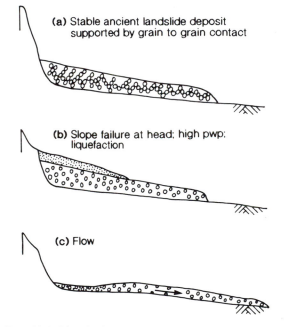

(a) Stable ancient landslide deposit supported by grain to grain contact

(b) Slope failure at head; high pwp; liquefaction

(c) Flow

FIG. 14.6. Liquefaction by high pore-water pressures resulting from loading by a collapsing soil mass on to a saturated body of material.

for different types of flow; values are in the range 10 to 500 Ns/m² (Johnson and Rodine 1984):

$$\eta_N = \frac{\gamma_D \sin\beta D^2}{2V_s}, \qquad (14.5)$$

where V_s is the surface velocity of the flow.

If the debris flows are Bingham, plastico-viscous, fluids, then the model requires a rigid plug of material moving down the centre of the flow. The Bingham viscosity, η_B, can be derived from:

$$\eta_B = (sW_p/4V_m)[W_c/W_p - 1]^2, \qquad (14.6)$$

where: s is the shear strength;
      W_p is the width of the plug;
      V_m is the maximum velocity of the plug;
      W_c is the width of the channel.

For a debris flow at Wrightwood, California, Johnson and Rodine calculated a Bingham viscosity of 130 Ns/m² and a Newtonian viscosity of 500 Ns/m², nearly four times greater. The differences between the Newtonian and Bingham viscosities are shown particularly in the velocity distributions in flows. A Newtonian fluid, such as water, produces a parabolic profile at the surface and the fluid cuts an open V-shaped channel. The Bingham plastico-viscous fluid moves as a rigid plug and cuts a U-shaped channel (Fig. 14.7). Field observations made during, or immediately after a debris flow event support the use of a Bingham model.

Solution of several of the above equations requires a knowledge of the surface velocity of the plug. It is well recognized that flow of a fluid round a bend results in elevation of the fluid surface on the outside of the bend in proportion to the

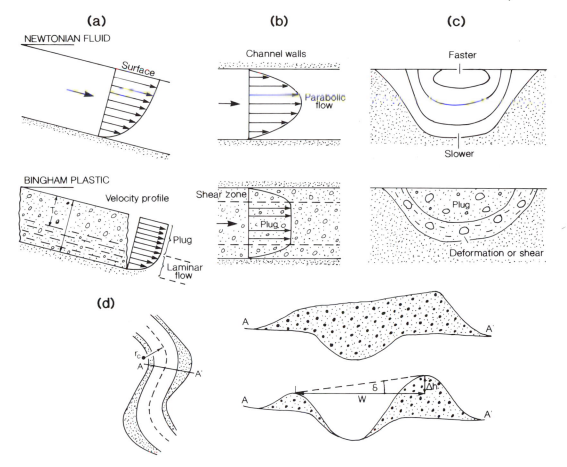

FIG. 14.7. Velocity profiles for Newtonian and Bingham rheologies: (a) vertical profiles; (b) horizontal-surface profiles; (c) channel cross-sections. In (d) are shown the parameters used in equation 14.7 (All figures are based on Johnson and Rodine 1984.)

velocity. The mean velocity, $\bar{V}$, of the surface can be derived from:

$$\bar{V} = (r_c\, g\, \cos\beta \tan\delta)^{0.5}, \qquad (14.7)$$

where: r_c is the radius of curvature of the bend;
  g is the acceleration due to gravity;
  β is the channel slope (if this is $<15°$, $\cos\beta$ can be dropped from the equation as the value is close to unity);
  Δh is the elevation difference between the inside and outside of the bend (Fig. 14.7d);
  W is the width ($\tan\delta = \Delta h/W$).

Transport of large clasts is characteristic of many debris flows. Isolated boulders tend to sink in a flow, due to gravity, but they remain supported in a fine-grained matrix; this leads to the question of what keeps them suspended and prevents separation. In a static body of material, (1) cohesion, (2) buoyancy, and (3) structural support may be effective; in a flowing material (4) turbulence, and (5) dispersive pressure may also be effective.

(1) Cohesion of clay particles can permit indefinitely a clay–water slurry, with a density of $1.17\,\mathrm{Mg/m^3}$, to suspend medium sands; and a slurry of density $1.26\,\mathrm{Mg/m^3}$ can support coarse sand (Kuenen 1951). Many debris flows have little clay (<8–10 per cent), yet have large boulders. Particles coarser than sand must be supported by other forces.

(2) Buoyancy is determined by the difference in

density between submerged solids and the fluid. If the fluid density is taken to be that of the whole mass (as it was for the slurries mentioned above), the differences between the fluid and particle densities may be small. For a boulder with a density, ρ_b, of $2.65\,\text{Mg/m}^3$, in a debris flow of density, ρ_f, $2.0\,\text{Mg/m}^3$, the submerged weight is:

$$\frac{\rho_b - \rho_f}{\rho_b} = \frac{2.65 - 2.0}{2.65} = 0.25\,\text{Mg/m}^3.$$

Buoyancy can thus support about 70–90 per cent of the particle weight in debris flows.

(3) Structural support is provided by a framework of grains in point-to-point contact transmitting overburden stresses to the channel floor. It will be most effective in static deposits although it may also operate within rigid plugs of moving material.

(4) Turbulence is a major process in streamflows, but is damped in progressively more viscous flows until it is eliminated. Plug flow is, by definition, not turbulent, but some eye-witness reports suggest that turbulence may develop in the fronts of some lobes and waves of debris flows.

(5) Dispersive pressure is the term Bagnold (1954) gave to the lift produced when stresses are transmitted between particles in collision, or near collision as one is sheared over another (Fig. 14.3). In high concentrations of poorly sorted grains, sheared by flow, it results in the drift of large particles to the flow surface. Dispersive stress, P, is given by:

$$P = 0.042\lambda D^2 \left(\frac{du}{dy}\right)^2 \cos\phi_D, \qquad (14.8)$$

where: λ is the linear grain concentration;
D is the particle diameter;
du/dy is the velocity gradient;
ϕ_D is the dynamic angle of internal friction.

The dispersive pressure on a given particle increases as the square of the diameter, so the largest particles are most affected and should migrate to the front and top of the flow; this is commonly observed to be the situation.

Deposition and cessation of flow occur as the slope angle of the channel or groundsurface declines. In a moving flow the plug is being transported on a basal layer of shearing debris; at decreasing angles of slope, and therefore

shearing rate, the plug thickens as the shearing basal layer becomes thinner, until the basal layer is too thin for further internal shearing; the flow then stops. Alternatively a flow may spread laterally and the plug thickens to the critical thickness, T_c:
$T_c = s/(\gamma_D \sin\beta)$ (from Eq. 14.3).

While it is true that the loss, or gain, of just 2 or 3 per cent water, by weight, can cause a change in debris flow shear strength by a factor of two or more, there is no evidence that debris flows halt primarily because of water loss.

Channel erosion by debris flows is common and can be very rapid. The total shear stress exerted on a streambed is:

$$\tau = \rho_f g R\sin\beta, \qquad (14.9)$$

where ρ_f is fluid density, R is the hydraulic radius (which is approximately flow depth in wide channels), and β is the channel slope.

During passage of a debris flow, the fluid density can be twice as great as that of a water flood (2.0 compared with $1.0\,\text{Mg/m}^3$) and the flow depth of a debris flow is commonly three times that of the water flows between debris surges. Debris flows can thus exert about six times the shear stress on channel beds than can water flows of the same event.

Impact by debris flows can be very destructive, compared with water floods. For a simple demonstration consider only the relative kinetic energy, E_k, as given by: $E_k = \frac{1}{2}mv^2$. The differences in density alone indicate that debris flows will have twice the impact pressure of water flows. Take into account the greater thickness of debris flows and their greater velocities (say $10\,\text{m/s}$ compared with say $4\,\text{m/s}$ for water floods) and it can be seen that debris flows can readily have 5 to 100 times the impact force of water flows.

Protective measures against debris flows can, consequently, be expensive. Where it is feasible the source areas of sediment can be stabilized more cheaply by afforestation than can debris-flow channels be diverted or barriers built. Energy dissipation devices are commonly necessary to limit the impact pressures of flows.

Debris flows, locally called debris torrents, in the Pacific Northwest of USA and Canada, are so destructive that considerable attention has been given to them. See, for example, Van Dine (1985) with corrections by Bovis *et al.* (1985); Bovis and Dagg (1988); and Hungr *et al.* (1984).

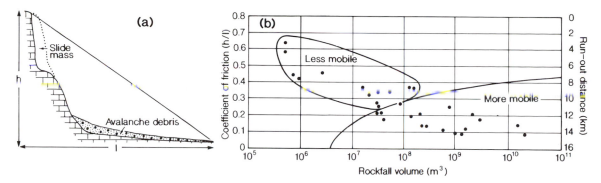

FIG. 14.8. (a) Calculation of the coefficient of internal friction of a rock avalanche (sturzstrom) from h/l. (b) Relationships between rockfall volume, the coefficient of internal friction, and the mobility of the flow. (Based on Hsü 1975, 1989.)

Denudation rates

Debris flows clearly have enormous power and capacity to transport debris. Pierson (1980) suggested that the Mt. Thomas debris flow had a sediment yield equivalent to that of several thousand years' worth of erosion at average sediment discharges of streams from small catchments in New Zealand mountains. It is very clear that, in some regions, debris flows are the dominant processes of erosion in small mountain catchments, see, for example Machida (1966) on Japan, the example from China quoted above (Li Jian and Luo Defu 1981), and the comments of Starkel (1976: 219) on 'sjels' in the Caucasian regions. Sjels are clearly debris flows which, on addition of water from tributaries, turn into turbulent debris floods.

Large rock avalanches

Rock avalanches result from the common condition in which well-jointed rock loses internal cohesion by joint propagation and becomes a mass of loosely fitting masonry with strength which is entirely frictional. Friction angles for discontinuous rock in a dry state vary from about 43–45° for rock in a loose aggregate state, to about 65° in a densely packed state (Silvestri 1961). Weathering will ultimately decrease this friction angle.

Large rock avalanches occur primarily in mountains where high, steep slopes, undercut by fluvial or glacial processes, allow large rock masses to descend at high velocities, by combinations of falling and sliding, into valley floors. If their descent is primarily a sliding process, they move as frictional materials and their coefficient of friction, as defined by the ratio h/l (where h is the height from the crown of the failure to the base of the descent, and l is the distance of total horizontal displacement to the toe (Fig. 14.8a) is greater than about 0.58 = tan 30°. Some avalanches fall on to glaciers and may travel further than predicted by the run-out line (or Fahrböschung) joining the top of h to the limit of l; others may follow a tortuous path of descent along pre-existing drainage lines and have a more limited horizontal displacement. Coefficients of friction for avalanches with volumes of <5 Mm^3 may vary therefore from about 0.4 to 0.65 (Fig. 14.8b).

Very large rock falls (>6 Mm^3) with large vertical drops, have run-out distances which are far greater than predicted for a frictional slide. They clearly move as flows with coefficients of friction which may be in the range 0.08 to 0.36 for avalanches of dry non-volcanic rock and 0.06 to 0.2 for dry volcanic debris (Ui 1983; Ui *et al.* 1986).

Rock avalanches with large volumes; low coefficients of internal friction and large travel distances have been called sturzstroms (Hsü 1975). Sturzstroms have attracted much attention because they are now recognized as major processes of erosion in high mountains (e.g. Whitehouse 1983, for the New Zealand Southern Alps; Hewitt 1988, for the Karakoram; Evans 1989, Evans *et al.* 1989, and Kaiser and Simmons 1990, for the Canadian Mountains). Volcanoes are of particular concern because of the potential danger around their bases (e.g. Voight *et al.* 1983, for Mt. St. Helens;

Crandell 1989, and Crandell *et al*. 1984, for Mt. Shasta; and Ui *et al*. 1986, for Japanese volcanoes). The work of Whitehouse indicates that in the Southern Alps there have been nineteen large rock avalanches in the last 2000 years, which have moved about $1\,Gm^3$ of debris, equivalent to an areally averaged erosion rate of $100\,t/km^2$ per year. Over the last 10 000 years in the central Southern Alps, forty-two avalanches, with deposits ranging in volume from $1\,Mm^3$ to $500\,Mm^3$, have occurred with a frequency of one per 244 years (Whitehouse and Griffiths 1983).

In mountains around the world, rock avalanches are probably triggered more by earthquakes than any other single cause. Volcanoes have distinct features which make them particularly prone to failure: (1) during the rise of magma beneath and within a volcano the whole massif dilates and its flanks tilt and this may be associated with earthquake tremors; (2) as magma sinks, the massif may subside forming ring faults; (3) injection of dykes also causes variations in stress in the host-rock mass; (4) hydrothermal fluids may be injected into rock bodies and promote rapid rock alteration and weakening; (5) gas pressure from within the vents may displace parts of the flanks; (6) the layers of lava and pyroclastics of many stratiform volcanoes permit many perched water-tables and high pore-water pressure zones to develop; (7) dacitic and rhyolitic lavas are very viscous and form particularly steep slopes; and (8) in semiarid and arid areas, such as the central Andes, the lack of flank erosion by water permits remarkably steep volcanoes to grow so that heights of fall may be very large and angles of descent steep. Combinations of these factors cause flank collapses, like that of the Mt. St Helens avalanche in 1980, which may be followed instantaneously by horizontally directed volcanic blasts and pyroclastic eruptions.

The velocity of large rock avalanches is commonly in excess of 90 km/h (25 m/s) and may reach 350 km/h (100 m/s) with individual boulders being propelled through the air at up to 1000 km/h (277 m/s). The high velocities attained permit not only long run-out distances (10–30 km), but also ascent of opposing slopes to heights of several hundred metres; the feature at Avalanche Lake, in the MacKenzie Mountains of Canada, may have the greatest rise, so far recognized, of 640 m (Evans 1989).

The material of avalanches is mostly from bedrock, although in some mountains snow and ice may be incorporated in the debris and water may be acquired from lakes and rivers as the mass travels. There is a general absence of gravity sorting of debris; normally the material of the avalanche is chaotic, and large blocks may either fit closely together or be separated in a matrix of finer material. The debris sheet is commonly thin, being a few tens of metres thick, but it usually has a well-defined distal rim, lateral ridges, and transverse surface patterns of hummocks and ridges. The surface pattern of the debris thus has many features in common with glacial moraines.

Examples of rock avalanches

The size of features and energy involved in large avalanches may be gauged from a brief description of what is probably the world's largest known landslide—the Saidmarreh in southwestern Iran (Harrison and Falcon 1937). The slide occurred on the northern flank of Kabir Kuh, an elongate anticlinal ridge of Asmari limestone dipping at about 20° and resting on thin-bedded Eocene marl and limestone. A segment of the ridge 15 km long, about 5 km wide, and about 300 m thick, slid off the mountain into an adjacent valley (Plate 14.7). Part of the moving mass had sufficient momentum to rise 600 m above the valley floor, cross the nose of a neighbouring anticlinal ridge, and come to rest in the next valley 20 km distant from its source. The dammed Saidmarreh River formed a lake 40 km long and, on average, 5 km wide. The mass involved was about $20\,km^3$ and covered an area of $166\,km^2$ to a maximum depth of over 300 m and an average thickness of about 130 m. The edges of the deposits are sharp fronts at least 50 m high.

A well-known example of a devastating rock avalanche is that from the east face of Turtle Mountain which destroyed the southern end of the town of Frank, in the Crowsnest Pass of Alberta, in 1903 (McConnell and Brock 1904). The Frank avalanche is variously estimated as involving about 30 million to 90 million tonnes of fissured limestone which moved as a fluidized mass for a distance of up to 4 km, and travelled 140 m up an opposing slope at speeds calculated as exceeding 160 km/h. Approximately $2.6\,km^2$ of the valley floor was covered with debris to an average depth of 20 m.

The crest of Turtle Mountain is anticlinal and the avalanche material came from a dipping limb of the anticline, so it probably took place along bedding surfaces. The surface of rupture close to

PLATE 14.7. Scar and debris of the Saidmarreh landslide, Zagros Mountains. The slide plane follows a bedding plane and the hummocky debris covers the valley floor. Some of the debris which rose up the opposing slope can be seen on the right of the anticline in the foreground. (Photo: by permission of Aerofilms.)

the toe of the avalanche followed a minor thrust above the Turtle Mountain Fault. The base of the mountain is formed in compressible and impervious shales, and underground coal-mining at the toe of the slope may have added to instability (Fig. 14.9).

Day and night temperatures preceding the avalanche were respectively about +21 °C and −17 °C, indicating that regular freeze–thaw processes were occurring and failure may also have been triggered by earthquake activity during the preceding months.

A recent devastating rock avalanche is that which fell from the crest of Mount Huascarán (6654 m) on 31 May 1970, and killed 18 000 to 20 000 people when it obliterated the towns

of Ranrahirca and Yungay (Fig. 14.10). Mount Huascarán is the highest mountain in the Peruvian Andes. The upper part of its peak, which failed, is composed of granodiorite and the avalanche debris included a mass of glacial ice and glacial till which it picked up at the base of the cliff (Browning 1973).

The avalanche was triggered by an earthquake which occurred off the coast about 125 km from Mount Huascarán. The same earthquake triggered many thousands of small shallow landslides in the mountains.

The cliff face has an angle of 70–90°: the rock mass of 50 to 100 Mm³ fell for approximately the first 600 m and then descended a further 2700 m along a valley with a slope of about 23° for 14.5 km.

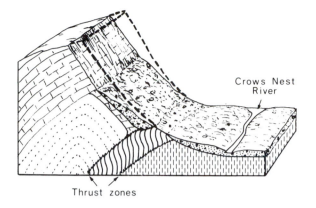

FIG. 14.9. The Turtle Mountain rock avalanche. The geology is from a study by Cruden and Krahn (1973). Shear tests on bedding planes in blocks from the slide debris gave for peak strength, $c = 220 \, kN/m^2$, $\phi = 32°$; for residual strength, $c_r = 124 \, kN/m^2$, $\phi_r = 16°$.

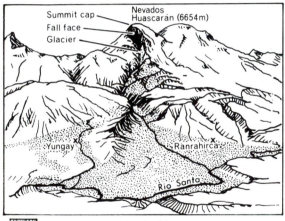

☐ Area covered by avalanche debris

FIG. 14.10. Sketch of the Huascarán rock avalanche, 1970. (Modified from Plafker and Ericksen 1978.)

This part of the journey took three minutes and was accomplished at average velocities of around 280 km/h. The avalanche had a low coefficient of friction of 0.22, and although it left virtually no debris in the valley it must have moved as a flow of at least 100 m thickness. Air escaping from the debris carried dust and mud high up the valley walls and boulders with a mass of several tonnes were hurled as much as 4 km through the air. The valley down which the avalanche moved was

incapable of containing all the materials and, at a narrow bend, the debris split with one part travelling up and over a 200 m high slope before plunging down on the town of Yungay.

From Yungay the avalanche continued down-valley for a further 50 km at an average velocity of 25 km/h. This second part of the movement was presumably a wet flow mobilized by water picked up from the river valley and from melting snow and ice.

The material which covered the two towns where the avalanche lost its velocity is composed of unsorted clay and boulders. Many boulders have a mass of 700 tonnes and a few exceed 14 000 tonnes. Some sorting occurred down-valley in the area of flow with the largest boulders being dropped highest up the valley.

The Huascarán rock avalanche appears to have been an event with a greater height of fall, velocity, and probable volume than any avalanche known to have occurred in historic times. Nevertheless there is clear geological evidence that a considerably larger avalanche from the north peak of Huascarán devastated the same area some time before the arrival of the Spaniards (Ericksen et al. 1970).

An example of a medium-sized rock avalanche which slid and fell from Ama Dablam (6856 m) in the Himalayas of Nepal is shown in Plates 14.8 and 14.9. The avalanche originated on the back wall of a cirque at an altitude of about 5600 m. The debris descended a 70° slope to about 5100 m on the Tsuro Glacier; from there it flowed about 2 km down a slope of 25° over ice and moraines, and then over a break of slope down a 50° incline into the Imja Khola which it dammed. There was thus a total vertical displacement of 1500 m and a horizontal displacement of 3.0 km. When the dam was breached, debris flowed down the valley for over 20 km, leaving an unsorted debris sheet with an average thickness exceeding 3 m. When photographed, five years after the failure, which occurred in 1979, at least 1 Mm³ of debris was still present in the floor of the Khumbu Valley (Selby 1988).

Mechanisms of flow

The mechanics of flow of rock avalanches and fluidized rock masses, called sturzstroms, has been explained in a number of ways. The first fundamental paper on the subject was published by Heim (1882) who reported the Elm rock avalanche in Switzerland (see Hsü 1978). Heim recognized that the debris flowed, that sliding was of little

PLATE 14.8. Ama Dablam showing the rock avalanche source on the triangular face, in deep shadow, above the right-hand cirque. The avalanche track can be seen sloping down from the cirque into the valley below. Himalayas of Nepal.

PLATE 14.9. Rock avalanche debris in the floor of the Khumbu Valley, about 25 km downstream from the source. Note the complete lack of bedding of the debris, the huge range in grain sizes and the very large boulders in and on the debris. The largest boulder visible is about 125 m^3 and over 300 tonnes. Fluvial erosion has removed part of the top of the deposit and cut a channel through it.

or no importance in the transport, that the large distance of travel was a result of low internal friction, and that the reduction of internal friction is velocity dependent. Thus the principles of fluid mechanics can be applied to the study of rock avalanches and the velocity of the slide is proportional, among other things, to the square root of the thickness of the debris wave. Velocities will, therefore, be highest in confined valleys and greatly reduced as the debris spreads out.

The idea of fluidization of debris was taken up by Kent (1966) who postulated that trapped air is the fluid medium, and by Shreve (1968) who suggested that the moving mass is carried on a cushion of air. Both of these ideas appeared to be supported by observations of a blast of displaced air which precedes the debris mass, but neither idea is now generally accepted because they both require implausibly low permeabilities of the sliding mass to trap the air. Furthermore, very similar rock avalanche deposits have been recognized on Mars and the Moon, where neither air nor water is present. Fluidization may be enhanced by the presence of air and water, but neither is required.

Goguel (1978) suggested that frictional heating in large landslides might vaporize small amounts of included water and that the resulting steam could fluidize the moving mass. The objections to this idea are the same as for those of Kent and Shreve. Habib (1975) proposed that frictional heat might melt rock debris near the sole of a slide; the slide mass would then glide on a molten layer. Glass from melts is, however, rare but has been reported from a few slides—notably those at Koefels by Erismann (1979) and that in the Langtang Valley of Nepal (Heuberger *et al.* 1984).

The idea of dispersive grain stresses (Bagnold 1954) has been taken up by several workers. Hsü (1975) has reformulated the original idea of Heim, who noted that an individual block travels in a zigzag bounding path through elastic impacts with its surroundings. A large aggregate of small blocks behaves quite differently because each block is confined to bouncing back and forth between its neighbours, and only the outer ones may fly away. Thus kinetic energy is exchanged between particles by elastic collisions, and the same energy keeps the particles separated during countless elastic contacts. The mass therefore behaves as a fluid with very low internal friction.

The presence of fine particles increases the frequency of collisions, and also the dispersion of

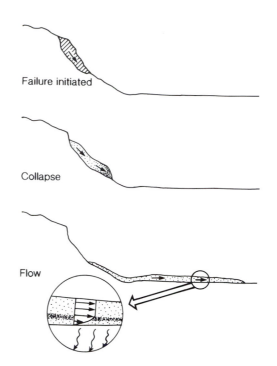

FIG. 14.11. Schematic illustration of events in a large rock avalanche with an inset illustrating the velocity profile expected for highly non-Newtonian flow. In the inset, stipple density represents acoustic energy density at a maximum in a basal layer where it is generated. Wavy arrows denote loss of acoustic energy to the bedrock. (After Melosh 1987.)

large blocks which may then pass one another. Rock avalanches thus have high fluid viscosities when the mass is thick (perhaps around 4 MPa/s at a thickness of two or three metres) but as little as 100 kPa/s when the flow is less than this in thickness. Such high fluid viscosities prevent internal turbulent mixing and internal deformation, so the mass moves as a thin flexible sheet of material with plastic behaviour and with shearing at the base of the debris.

Melosh (1979, 1987) has proposed that the random movement is not of individual rock fragments, but of groups of fragments organized into elastic waves. Since the rock fragments seldom lose contact with one another, this process dissipates energy at a far slower rate than does individual grain dispersion. The elastic waves are part of a strong acoustic field which creates fluctuating pressure in a small volume of debris. These fluctuations vary between being much less than the overburden pressure of the rock mass and much more. The average pressure must equal that of the overburden, but the energy of each high pressure event may be stored from the initial moment of collapse of a slope, with a concentration of acoustic energy in the zone of high shear strain near the base of the moving mass (Fig. 14.11). As the mass spreads out and thins, the available energy eventually reaches a stage at which the losses exceed that being generated, and the flow stops.

In a study using computer simulations, Campbell (1989) has examined the hypothesis that the bulk of avalanche material rides on a thin layer of highly agitated particles of low concentration. His calculations show that: (1) the process is feasible; (2) it can produce coefficients of friction low enough to explain the long run-out distances; and (3) that the observed forms of sturzstrom debris are compatible with the proposed flow mechanism.

These forms in the debris indicate that: (1) the debris has moved as a more-or-less undisturbed sheet which permits angular boulders to be carried on the surface; (2) some size-sorting has occurred as a result of agitation of the sheet as it passes over rough topography, consequently large blocks can rise to the surface of the sheet; (3) a fine-grained layer has been formed, near the base of an avalanche, where intense shear and brecciation

occurs; (4) the flow sheet has spread out from the base of steep slopes until the leading edges came to a halt and the rest of the material has piled up behind them, leaving a steep-fronted edge. The calculations and observations are in conformity with the assumption that the larger the mass, the larger the kinetic energy, the longer the time required for the agitated layer to dissipate that energy, and, therefore, the longer the run-out distance will be.

Hazards and uses of avalanches

Rock avalanches are sufficiently rare for their occurrence in one site to have a low probability. However, in large valleys avalanches may dam rivers and breached dams may become significant threats to all downstream areas. The capacity of avalanches to dam major rivers has been used in USSR where nuclear explosions have produced sturzstroms with volumes up to $80\,Mm^3$. Dams of large size, produced by controlled explosions under high cliffs with the intention of creating large rock avalanches that will block valleys, were planned for the Tien Shan mountains. The impounded reservoirs were to supply power and irrigation water (Melosh 1990).

15

Rock-Slope Processes

Bare rock slopes exist for a number of reasons. They may be too high, as a result of uplift or deep incision, for debris to accumulate and bury them; there may be active processes at their bases removing debris so that it cannot accumulate; or they may be too steep or the climate too severe, as a result of cold or aridity, for chemical weathering and vegetation to maintain a regolith. In many environments bare rock faces exist where slope angles are steeper than about 45°, for this is approximately the maximum angle maintained by rock debris, but in the humid tropics weathering and vegetation establishment may be so rapid, on rocks such as mudstones and basalts, that a regolith may form on slopes as steep as 80°. Examples of such slopes occur on Tahiti and in Papua New Guinea where bare rock may be exposed for only a few years after a landslide.

The form of rock slopes is determined both by the properties of their rocks and by the processes acting to modify them. An arbitrary distinction may be made between those rocks which have such high internal strength that they fail, almost invariably, along joints and fractures and those rocks of lower intact strength or intense fracturing which behave more like soils. The first class of rocks may be spoken of as 'hard' rocks and the second class as 'soft' rocks. In general bare rock slopes are formed on hard rocks, but this is not an invariable condition. Soft rocks such as mudstones and shales are often raised to high elevations by tectonic processes and steep slopes can be retained on them by regular undercutting of the slope. The rate of denudation on such slopes, however, is very rapid compared with that on hard rocks and they acquire a soil and vegetation cover far more rapidly than do hard rocks.

Factors in rock resistance and failure

Hard intact rocks have strengths controlled by their internal cohesive and frictional properties, but few rocks forming hillslopes are intact. Their strength is largely controlled by the size, spacing, and continuity of partings within the rock mass, and their stability by the angle at which partings dip with respect to the hillslope angle (see Chapter 6).

Weakening of hard rocks on hillslopes may lead to rock failures occurring by falling, toppling, or sliding of discrete joint blocks and rock fragments, or by the development of landslides. No matter what the scale of the failures the causes can usually be attributed to geological, climatic, weathering, or human factors.

Geological factors

Nearly all of the world's largest landslides occur in the tectonically and seismically active belts of rising mountain chains formed at the boundaries of the major crustal plates—particularly around the rim of the Pacific Ocean and along the Alpine–Himalayan chains (Kingdon-Ward 1955). High mountainous terrains, with plateaux and ranges towering above steeply incised river and glacial valleys, have multiple-layered, jointed, and fractured rock masses exposed on steep valley walls, groundwater-pressure fluctuations, severe physical weathering, and steeply dipping discontinuities.

Tectonic activity not only increases the relief but growing folds cause increases in hillslope angles, and faulting may produce growing scarps (Fig. 15.1). Many of the largest and most devastating landslides have followed earthquakes which

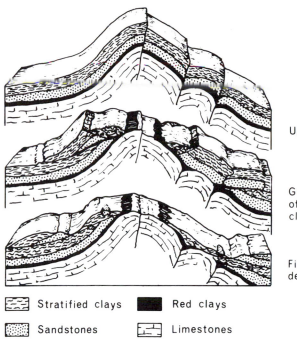

Upheaval

Gravitational slide
of apex of fold on
clay bed

Final state with
denudation

Stratified clays   Red clays

Sandstones   Limestones

FIG. 15.1. Development of a large gravitational landslide as a result of an increase in folding and uplift. (Modified from Zaruba and Mencl 1969.)

through ground accelerations, cause fracturing of joint blocks and loss of strength along joints. In Alaska, for example, twenty-four earthquakes of Richter magnitude 7.0 or greater occurred between 1898 and 1975, causing many thousands of landslides and large rock avalanches (Voight and Pariseau 1978). The distance at which earthquakes can trigger landslides depends upon many factors. These include the stability of the slopes, the orientation of the earthquake in relation to the slide mass, earthquake magnitude, focal depth, seismic attenuation, and after-shock distribution. Large earthquakes can be effective at a considerable distance. That which triggered the Mount Huascarán rock avalanche of 1970 in Peru was located 125 km from the mountain and had a focal depth of about 54 km (see also Chapter 13).

The total effect of earthquakes may well be underestimated for we commonly note only those which are the final 'trigger' events which produce failure. The long-term cumulative effect of many low-magnitude earthquakes may be as important as the rarer high-magnitude event.

Tectonic activity may have, and commonly does have in mountains, a major influence on the alignment of valleys (Scheidegger and Ai 1986) and the development of systematic joint orienta-

tions (see Fig. 6.1). Conjugate joint sets have a major influence upon the orientation and form of high alpine peaks, cirques, and first- to second-order channels (Plates 15.1, 15.2) (Gerber and Scheidegger 1973).

Dilation of rock masses and opening of joints can be a major cause of failures on rock slopes or surfaces. Dilation may be caused by three main processes: (1) the release of locked-in stresses; (2) reduction of the internal forces which make a rock competent (Lindner 1976); and (3) weathering.

(1) Overburden and tectonic stresses cause compression of the mineral fabric of rock so that intergranular stresses and strains gradually increase over a very long period and rock masses therefore acquire a considerable amount of 'locked-in' strain energy. A substantial proportion of this energy may be lost in the permanent deformation of mineral grains and can no longer be recovered, but much of the strain energy will be released when confining pressures are reduced by erosion of adjacent rock, or wasting of overlying glacial ice. The rocks most obviously affected are those, like laminated shales or closely bedded sedimentary rocks, which have inherent planes of parting which can open, but even massive rocks,

PLATE 15.1. A cirque in the Himalayas with its side walls developed precisely along the planes of two joints which intersect at the lowest point of the cirque. These joints form the upper half of a conjugate pair. The back wall of the cirque shows scars of wedge failures forming the notches on both sides of the central bastion.

PLATE 15.2. A serrated ridge developed on schists. The foliations dip from top-right to bottom-left and systematic cross-joints have controlled wedge-form valleys and chutes which terminate in debris cones at the base of the ridge. Himalayas of Nepal.

with no pre-existing planes of parting, can develop large-scale joints through stress release, and the greater the *in situ* stresses the more severe will be the effect of stress change caused by unloading (Lajtai and Alison 1979)—see Chapter 4.

In addition to joint development and opening, rock bodies which dilate can suffer general weakening by development of microfracturing with consequent decreases in shear and compressive strength of up to 25 per cent (Bauer 1987). This has obvious consequences for tunnelling and mining beneath valleys, but it also enhances valley erosion by fluvial processes (see also Fig. 4.14).

Valley deepening can occur at a much faster rate than general land-surface lowering on plains and plateaux. Consequently, rocks which have been deeply buried can have confining pressures reduced rather rapidly. In a valley the rock at the base of a slope is under the greatest stress and is in compression (Fig. 15.2) (Sturgul and Scheidegger 1967). Granite may expand its volume by as much as 1.5 per cent upon stress release without fracturing if the release occurs slowly, and the base of a slope may bulge as a consequence (Zaruba and Mencl 1976). More commonly, valley walls may bulge as a result of joints opening under tension and this may be followed by the linking of tension joints to form a shear plane along which a landslide may form (Fig. 15.3).

Stress release joints often open parallel to the hillslope or groundsurface and so may be straight, concave, or convex, although convex joints are probably most common (Plate 15.3). Consequently there is a tendency for the formation of dome-shaped hills or outcrops. These extensive joints can form in any massive rock and are particularly

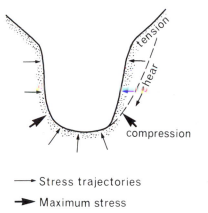

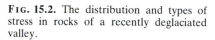
→ Stress trajectories
⇒ Maximum stress

Fig. 15.2. The distribution and types of stress in rocks of a recently deglaciated valley.

common in granite and sandstone. They normally terminate at intersections with weaker or laminated strata (Fig. 15.4). In addition to preparing rock slabs for spalling these joints permit water to penetrate the rock and so promote weathering processes.

Stress release may occur in some mountain valleys which were formerly glaciated. Retreat of glaciers has been known to permit the sudden fracturing and up-arching of rock on valley floors (e.g. Gage 1966) and the occurrence of slab-like joint blocks on valley walls is commonly thought to be evidence of stress release joint formation after the removal of laterally supporting ice (Kieslinger 1960). Persuasive evidence of stress release is seen in some walls of formerly glaciated valleys where

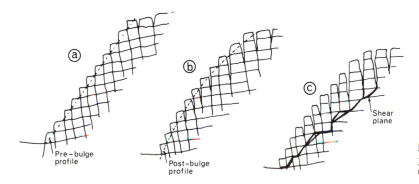

Fig. 15.3. The opening and linking of joints in a cliff to form a shear plane. (Modified from Whalley 1974.)

PLATE 15.3. Convex joints in well-bedded sandstone, Drakensberg, South Africa.

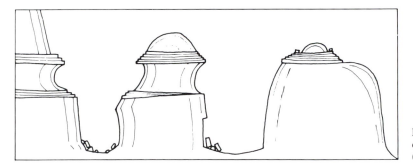

FIG. 15.4. Concave and convex joints developed in massive sandstones. (Modified from Bradley 1963.)

opening of joints exactly follows the change of hillslope from a cliff to an extending erosional slope of lesser angle below the cliff (Plate 15.4). Such joint patterns are particularly well developed in granites of formerly glaciated valleys in Antarctica. It has been pointed out by Whalley (1974), however, that tectonic stresses which have not been released by recrystallization of rock far exceed those which result from an overburden and so it is improbable that all slab jointing or rock failures are attributable to relief of overburden and lateral confinement.

(2) Internal forces holding rock intact may be substantially reduced by the addition of water to rocks which are not chemically altered in this process. The effect is most important in mudstones, shales, weakly cemented sandstones, and micaschists, but even strong igneous rocks may decrease in strength, from the dry to the saturated state, by up to 30 per cent (Broch 1979) (Fig. 5.26).

The addition of water reduces the attractive forces holding rock together and increases the internal repulsive forces. It involves cation exchange, hydration, the production of negative electrical force fields, the attraction of mineral surfaces for water, and capillary tensions. Natural cementation is strong enough to resist repulsive forces in many rocks, but those which contain high proportions of clay minerals may be readily disrupted. Atterberg limit tests on samples of swelling rock of this kind nearly always indicate high liquid and plastic limits and, commonly, the presence of montmorillonite.

(3) A third cause of rock swelling is weathering in which chemical alteration by hydration, oxidation, or carbonation creates by-products which may occupy a larger volume than the original rock. The conversion of sulphides to sulphates by the

PLATE 15.4. Granites in Antarctica. Major joints trending parallel to the slope surface even where there is a sharp change of slope angle.

addition of water is a common cause of swelling, and the slow processes by which olivine-bearing rocks are converted to serpentine are also accompanied by the development of high pressures in the rock mass.

Unfavourable angles of dip of bedding and joints are, perhaps, the most common cause of rock slope weakness. The angle at which a cliff will stand is controlled by the angle of dip of the joints compared with the friction angle between the blocks. Because of the effect of the overburden pressures the friction angle will decrease as the cliff height increases. This is probably because the increase in normal stress with depth is associated with an increase in the number of fractures across rock grains, which reduces the resistance to sliding due to interlocking between grains on opposite sides of a joint, and increases fracture intensities (Roš and Eichinger 1928). It follows then that stable slope angles decrease as the height of a slope in weathered and jointed rock increases. This relationship is well illustrated for New Zealand greywacke (indurated siliceous sandstones with beds of argillite) in Wellington where the highly sheared and shattered rock within 1 km of the Wellington Fault will support a stable slope lower in angle and height than the less shattered rock further from the fault (Fig. 15.5).

Stratified sedimentary rocks have bedding-

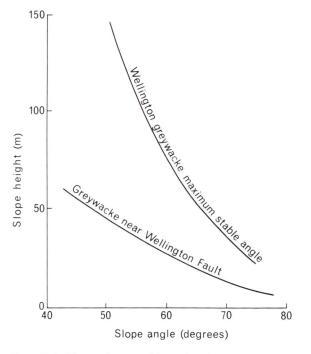

FIG. 15.5. The maximum stable angles of slope in relation to slope height and nearness to the Wellington Fault for greywacke, New Zealand. (After Grant-Taylor 1964.)

PLATE 15.5. A wedge-shaped block of mudstone, some 3 km long, has glided 1.5 km downslope from the range to the left. The failure plane is a curved surface and the rotation is indicated by the dip of the 25 m thick cap of sandstone towards the left. The block is sheared along its base. Hawke's Bay region, New Zealand. (Photo: J. R. Pettinga.)

planes, which are surfaces of minimum resistance, separating rock layers of various thickness. Bedding planes may become preferred surfaces of shearing because: internal erosion occurs along a sand–clay junction; tectonically induced bedding-plane creep or sliding produces a residual strength condition; stress relief by erosion permits shear deformation; shear strength of the rock below a bedding plane may be greater than that of the rock above a plane; or thin seams of weak material, such as clay, may occur at a bedding plane.

An example of the effect of a clay bed upon mass stability occurs in Natal (Sugden *et al.* 1977) where well-bedded sandstones and shales with a total thickness of 2000 m dip into the sea, at about 15°, from the flanks of a monocline. Along the bedding planes are occasional layers of clay 1–20 mm thick. Very large masses of rock move slowly above the lenses of clay and produce serious instability problems, especially during wet periods and in areas where the failure planes outcrop. The sedimentary rocks have strength parameters of $\phi = 21°$ and $c = 1.5 \, kN/m^2$, but the clay has $\phi'_r = 9.5°$ to $12.5°$ and is much weaker, but controls the stability of the whole mass. In this case the clay is at residual strength values because it was probably sheared during folding of the monocline, but a residual strength value has to be assumed in all such cases.

Similar conditions are reported for coastal landslides in southern England (Barton 1988) and for large rotational failures in New Zealand (Bell and Pettinga 1988). In both cases, creep and limited displacement have caused orientation of clay along bedding planes in mudrocks. Very large bodies of rock (Plate 15.5) may consequently slide on thin layers of remoulded clay (10–100 mm thick) along bedding planes dipping at angles as low as 5° especially where the clays are smectites. Trigger mechanisms, such as earthquakes and high porewater pressures, or tectonic tilting may be important in initiating movement but creep is the long-term preparatory process.

Joints dipping at critical angles for stability are commonly thought of as being continuous regular joints, such as are common in sedimentary rocks. In such rock masses the stability of a slope depends upon the orientation of the bedding planes with respect to the hillslope, because the rock mass will have an effective strength controlled entirely by friction along the bedding planes if the joints are continuous and lacking cohesion.

If the bedding planes are horizontal no landslide can occur and the critical hillslope inclination for long-term stability is theoretically vertical, although in reality weathering processes will reduce this. Some theoretical solutions with

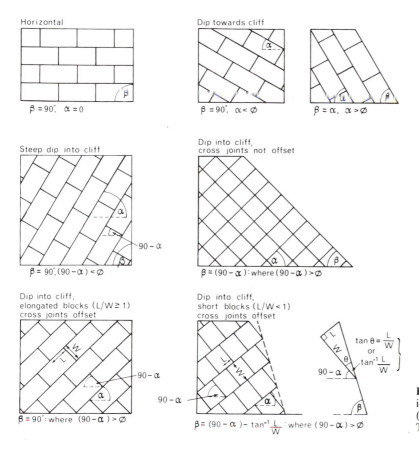

Fig. 15.6. The theory of critical angles in hard, bedded, and jointed rocks. (Modified by Young 1972, after Terzaghi 1962.)

respect to slope angle (β), friction angle (ϕ), the dip of the strata (α), and the relative spacing of the bedding joints (L) and cross-joints (W) are given in Fig. 15.6 (modified by Young 1972, from Terzaghi 1962).

These theoretical solutions are, however, greatly oversimplified because they ignore other important geological phenomena. The dip of strata and cross-joints together with the width of the joints (assuming that joints are not sealed by impermeable infill) largely determine the rate at which water and weathering processes can penetrate the joints. Suitably inclined permeable beds permit water to enter a slope and develop cleft-water pressures which promote instability and lower stable hill-slope angles (Fig. 15.7). Furthermore joint friction angles are not constant but change as joint roughness is modified by shearing of the asperities.

Investigations of over 300 slopes in the Rocky Mountains, by Patton (1966), showed that slopes are generally stable where the dip of the discontinuities ($\alpha°$) is less than the residual angle of sliding resistance (ϕ_{jr}) and that slopes are seldom unbuttressed where $\alpha > 45–50°$ (that is, α is much greater than ϕ_{jr}). Potentially unstable slopes are those for which values of i are high (i.e. ϕ_j is much greater than ϕ) as shearing through asperities can lead to failure.

On many rock slopes the toe is removed by erosion. The stability of the slope is then largely controlled by the depth of the cut. The height of a stable cliff is controlled by the cohesive and frictional strength along the bedding planes, the dip of those bedding planes out of the cliff and the unit weight of the rock (Fig. 15.8). The force, per unit area of the bedding plane, which tends to produce sliding along the discontinuity through the base of the cliff is:

$$\gamma H \cos\alpha \sin\alpha.$$

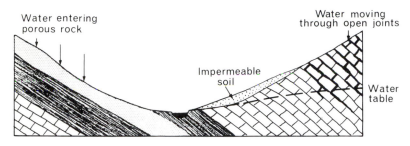

FIG. 15.7. The effect of bedding, jointing, and soil cover on the percolation of water into rock masses.

The force resisting this disturbing force is:

$$c_j + \gamma H \cos^2\alpha\tan\phi_j.$$

The cliff will be stable if resisting forces are equal to, or greater than, the disturbing forces:

$$H \leq \frac{2c_j}{\gamma\cos\alpha(\sin\alpha - \cos\alpha\tan\phi_j)}.$$

Thus the potential height of a stable cliff becomes very large as α approaches 90° (unless strata fail by buckling) or α is less than the value of ϕ_j. (It should be noted that this simple analysis ignores cleft-water pressure.) Any increase of H beyond a critical height would be followed by a rock slide with its foot at the base of the cliff (Terzaghi 1962). The importance of cohesion across the bedding joints is indicated in Fig. 15.8, where it is shown that a weak sandstone with strong cementation across the bedding planes is capable of supporting vertical slopes over 300 m high even when the bedding joints dip at 60° out of the slope, but a hard sandstone and a shale with the same dip but with little cohesion across the joints, can only support vertical slopes 10–20 m high: as the dip decreases so the stable vertical cliff height increases. Nearly vertical bedding is also theoretically stable, although this theoretical situation is usually modified by bulging or buckling of the strata.

The analysis given above assumes a vertical, or nearly vertical, cliff face; the analysis is suitable for undercut or over-steepened slopes. Where the dip of the critical joints, α, is less than the inclination of hillslope, β, which is not nearly vertical, the expression for, H, becomes:

$$H = \frac{2c}{\gamma} \cdot \frac{\sin\beta}{\sin(\beta - \alpha)(\sin\alpha - \cos\alpha\tan\phi)}.$$

Displacements along a shear plane can cause very large decreases in shear strength. The shear

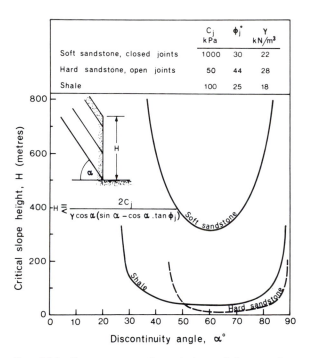

	C_j kPa	$\phi_j°$	γ kN/m³
Soft sandstone, closed joints	1000	30	22
Hard sandstone, open joints	50	44	28
Shale	100	25	18

FIG. 15.8. Curves representing solutions of the equation for stable cliff heights for rock slopes with bedding dipping out of the slope.

strength of intact rock is commonly 10 to 200 times that of soil, yet the residual strength after large displacements is commonly the same for a smooth joint surface in rock as for soil. The loss of strength on displacement is thus many times greater for rock than for soil, and consequently cumulative minor displacements along discontinuities can lead to unexpected catastrophic failures. Fig. 15.9 is a cross-section of a rock slope with an irregular joint plane with unfavourable dip and an uncemented fault plane. The shear-strength displacement figure for the joint indicates that initially the strength along the joint, with small displacements only, far

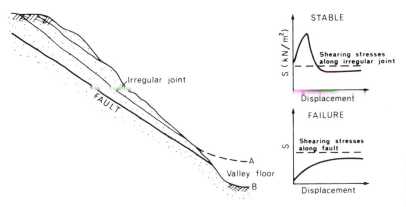

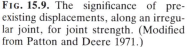

FIG. 15.9. The significance of pre-existing displacements, along an irregular joint, for joint strength. (Modified from Patton and Deere 1971.)

exceeds the shear stresses so that the slope is stable. Only when considerable displacement has occurred will shear strength fall to a level where failure can occur. If, however, the valley floor is lowered by erosion, failure can occur along the fault without prior small displacements. These examples assume no cleft-water pressures.

Displacements may have unexpected effects in some mudstones. Many clay-rich rocks have coatings of iron oxides, silicic acid gel, or manganese around joint blocks (Claridge 1960; Nankano 1967; Weaver 1978). These coatings prevent water readily penetrating the blocks but when mudstones and shales undergo creep, or other slow or limited displacements, the blocks may fracture, be ground against each other and slickensided, and the coatings broken. Water may then penetrate the rock mass. In some mudstones the water is adsorbed, and cannot be removed by drainage, and causes rock swelling. The most severe displacements may occur near the groundsurface where alternate wetting and desiccation causes deep cracking, slaking of the rock, and major losses of strength. Many of the severe landslides in Tertiary mud-stones in Japan and northeastern New Zealand have been attributed to such a mechanism.

The mode of slope failure is greatly influenced by both inclination and intensity of jointing (Fig. 15.10). Failures by toe bulging may occur where cross joints permit thrusting upwards and outwards to release stresses, even though the dominant joint set is at a potentially stable angle (Fig. 15.10a). The shape of failure planes is controlled not only by the dip of joints but also the strength across the strata compared with strength along the dominant joints (Fig. 15.10b, c). Where joints are very

closely spaced, but dipping very steeply into the slope, failure may be by rotational toppling of individual blocks, or, if jointing is sufficiently intense, the whole rock mass may behave as a soil and fail by deep rotational slumping (Fig. 15.10d, e).

Geological structures favourable and unfavourable to large-scale slope failure are difficult to classify because of their extreme diversity and the presence of many local conditions which reduce the validity of broad generalizations.

Certain structures which commonly inhibit the development of large gravitational failures and therefore ensure the predominance of superficial creep, fall, wash, solutional, and fluvial channel processes, include the following: (1) rock complexes of lower strength overlying those of greater strength such as occur in the normal stratigraphic sequence with older, more deeply buried and indurated, rocks underlying younger less indurated rocks; (2) very thick rock complexes that have approximately the same strength characteristics throughout, such as large intrusive bodies like granites, or thick strata of limestones, dolomite, sandstones, and greywacke; and (3) rock complexes only slightly affected by tectonic folding, faulting, or tilting.

Three general geological-tectonic conditions favourable for large-scale slope failures have been identified by Nemčok (1977).

(1) Rock complexes with high strength overlying weaker rock units are subject to undermining or failure along deep slide planes. The stronger rocks may be rigid and of constant volume, have high shear strength, be resistant to weathering, be impermeable, and capable of maintaining steep

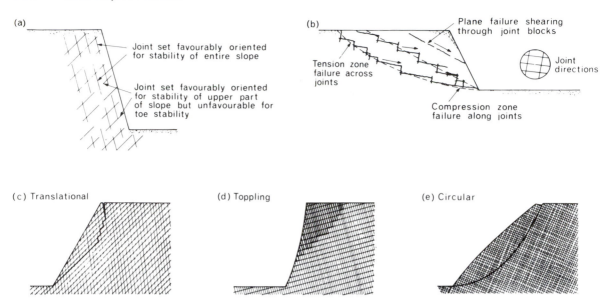

FIG. 15.10. Joint effects upon rock-slope failures.

slope faces. However, where they are underlain by soft, plastic, expandible, low strength, impermeable, easily weathered, crushed, or highly fissured beds the upper rocks may fail along very thin weak rock units, even when these have low angles of slope. Interbedded clays, silts, tuffs, fault pug, hydrothermally altered rock, or intruded bentonite layers provide examples of such weak units. If the overlying rocks are permeable or sufficiently porous to maintain a reservoir of water above the weak unit, failure along the toe of a slope can occur as earthflows rafting blocks of coherent strong rock. A very large example of such a failure has occurred on Samar Island in the Philippines (Wolfe 1977). Here a limestone block with an area of 18×25 km, and a vertical relief of 300 m or more, is underlain by argillized tuff which has been saturated by water percolating through the limestone. The whole block has glided about 5 km on a slope of only 0.6° during the last 10 000 years.

(2) Alternate strata, having higher and lower mass strength characteristics, are particularly prone to failure because of deformation, weathering, and water pressures along the weaker beds. Flysch complexes and their metamorphic equivalents characteristically have such variability.

(3) Rock units severely deformed by tectonism not only undergo progressive steepening of their major joint systems, but also seismic disturbance, the opening of joints, intensification of joints and formation of shatter zones, and the deepening of valleys where landmasses are rising.

Climatic factors

Climate influences slopes in hard rock mainly through the water pressure within the slope. In many high alpine areas cliff faces are frozen for much of the winter; in the European Alps the wave of winter cold may penetrate 10 to 20 m into the rock (Whalley 1974). This freezing of the face raises water-tables inside the cliff as well as causing freeze–thaw action on the face. In the succeeding spring and early summer the water-table will be lowered causing high seepage pressures acting towards the cliff face and thus reducing the strength within joints (Fig. 15.11). Heavy rainstorms can raise water-tables in well-jointed rocks. Seepage towards a cliff face may also be caused by rapid deglaciation of a valley during which water-tables in valley slopes will be lowered. The vertical height over which tension cracks may open is increased by these processes and this encourages shear planes to form.

Seasonal concentrations in the frequency of rockfalls, with maxima in the spring snowmelt period and the autumn (which is the wettest season) have long been recognized in the Norwegian mountains (Bjerrum and Jørstad 1957). The very

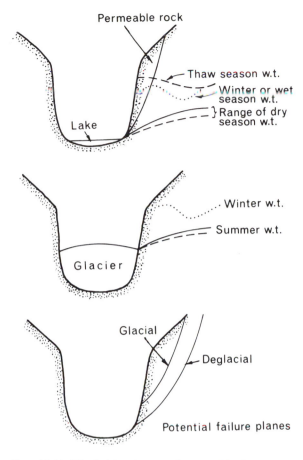

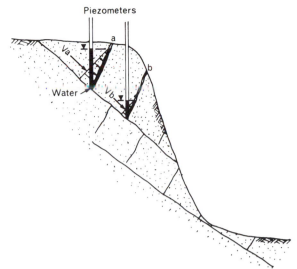

Fig. 15.12. Large differences in cleft-water pressures in adjacent joints. (Modified from Patton and Deere 1971.)

Fig. 15.11. The height of seasonal water-tables in rocks of a valley without and with a glacier, and the location of shear planes in the rock mass. Cleft-water pressures in the dry season are too low to affect stability, but in the wet season open joints may permit pressures to rise and failures to occur. (Modified from Terzaghi 1962.)

cold period of the seventeenth and eighteenth centuries, known as the Little Ice Age, was also a period of marked increase in rockfalls in European mountains (Grove 1972).

Groundwater conditions depend upon the spacing and openness of the joints. Where there are closely spaced, intersecting, open joints the water pressure within the rock mass can be treated in the same way as that used for soil slopes. The shearing stress (τ) along the joint is determined by the normal stress and coefficient of friction between the joint walls:

$$\tau = \sigma\tan\phi_j = \gamma z\tan\phi_j,$$

but cleft-water pressures reduce this to:

$$\tau = (\gamma - \gamma_w)z\tan\phi_j,$$

where: γ and γ_w are the unit weights of rock and water respectively, and z is the thickness of the rock mass above the shear plane. As rock is commonly three times as dense as water, rock-mass strength may be reduced by 30 per cent where clefts are filled with water to the groundsurface.

Where the joints are widely spaced, and of differing width and continuity, very variable distributions of water pressure may result and consequently the shearing forces also vary from one joint to the next. This is illustrated in Fig. 15.12 where the water level is much lower in joint b than in a. As a result the magnitude of the stress Va due to the hydrostatic water pressure along the joint a is several times the stress Vb. The figure also indicates the difficulty of obtaining reliable information on joint water pressures without an array of piezometers.

Groundwater levels fluctuate much more in rock slopes than in soils because the void space in rocks is much less. Fig. 15.13 shows the effect of 250 mm of rain which entirely infiltrates into a soil and into a slope of nearly non-porous rock. Where soil porosities vary from 33 to 10 per cent the rainfall may cause a rise in the groundwater-table of 750 to

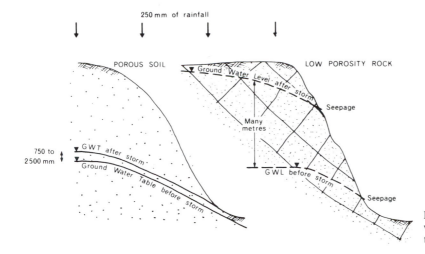

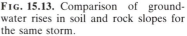

FIG. 15.13. Comparison of ground-water rises in soil and rock slopes for the same storm.

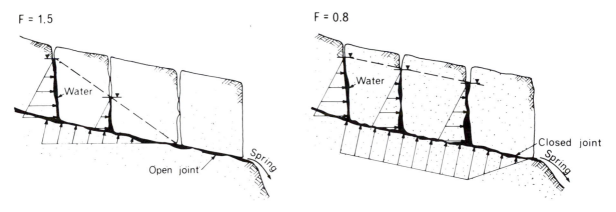

FIG. 15.14. Effects of joint openness on cleft-water pressures and slope stability.

2500 mm respectively. In the rock slope the joints may be filled and the rise of the groundwater level will depend upon their volume, it may be many metres and produce very high water pressures.

The effect of the openness of joints is illustrated in Fig. 15.14 where open joints drain readily and produce low water pressures near the cliff face, and hence a relatively high factor of safety against sliding, while a closed bedding joint permits high cleft-water pressures and a much lower factor of safety. The high water pressure may induce movement of the outer block and so lower pressures that failure may occur only after a series of small movements. Alternatively small movements along the sub-horizontal bedding joint could reduce the cleft

volume, so reducing seepage but increasing water pressure, thus leading to failure.

Weathering factor

Weathering is an important factor in all humid environments where the opening of joints, the weathering of asperities, and the formation of clay-rich joint infills can occur.

Human factor

Human interference with the stability of hillslopes in hard rock is still a minor factor compared with geological factors, but locally it can be of great importance. It usually results from the under-cutting of slopes or removal of their lateral support during quarrying or excavations for canals, roads, and railways. Its effect is similar to that of natural undercutting by waves or downcutting by rivers and glaciers, but it can also be enhanced by blasting which may open existing joints and cause further rock fracturing.

A more indirect influence is through the impoundment of lakes which then cause the rise of water-tables in the lower sections of rock slopes. The combined effects of reduction in strength and of buoyancy have been major factors in causing a number of very large, and costly, landslides into reservoirs (see for example discussion of the Vaiont failure, below). Remedial work to prevent such failures is also very expensive as it may involve drilling drainage tunnels into the hillslopes, grouting, rock bolting, and buttressing the bases of the slopes. In some cases investigations of slope stability have caused major projects to be abandoned because the hazards, and costs of remedial works, have been unacceptable (Schuster 1979).

Types of rock-slope failure

The types of failure which can occur on rock slopes range from very large features of tectonic-scale dimensions to movements of individual rock particles, with processes ranging from very fast falls, to slower slides and flows, and very slow creep. In each group phenomena can be further subdivided according to processes operating and the morphology of the moving material. The classification adopted here is based upon Nemčok *et al.* (1972) (Fig. 15.15).

Falls

Falls from cliff and rock faces are major contributors to talus and reworked debris which accumulates at the bases of steep slopes or is carried away by glaciers, rivers, or wave action. The distinguishing feature of all falls is that over much of their travel down a steep slope they are free-falling. It must also be recognized that this requires very steep slopes, or at least very steep segments, otherwise sliding processes preceding rock avalanching would be more appropriate forms of description.

The term 'rockfall' is commonly used to refer to a collection of processes which may involve the removal of material ranging in size from large rock masses through single joint blocks to particles ranging from boulder-size to gravel-size (Whalley 1984). It is improbable that the same suite of initiating processes operates across this enormous size range. Distinctions will be made here, therefore, between: (1) rock-mass falls; (2) rock slab and block falls; and (3) rock particle falls. That such a distinction is rarely made is an indication of our general ignorance of the processes, frequency, and significance for slope development of falls from rock slopes.

Rock-mass falls are, by definition, failures of large bodies of material which is likely to be internally jointed but which separates from its parent cliff along a single or stepped failure plane. For the failure to be a fall implies that the inclination of the whole cliff is likely to be in excess of 80°. Even in the highest mountains, individual large slopes rarely have such steep inclinations. Rock-mass falls are therefore likely to be most common on undercut slopes of sea cliffs, walls of deep glacial troughs and cirques, and sites undercut by rivers. An example of a mountain face, which is being shaped predominantly by rock-mass falls, is shown in Plate 15.6. There are some similarities between the forms seen in this example and those evident in the granitic gneisses involved in the Loen rockfalls of Norway.

Seven rock-mass falls, at Loen, came from Ramnefjell during the period 1905 to 1950 (Grimstad and Nesdal 1990). The debris slid or avalanched across a basal slope into Loen lake and, on two occasions, produced waves 74 m high which swept over villages and farms killing 134 people.

The rear scarp of the Loen failure after the last fall has an inclination of about 80° over a vertical distance of 450 m. The scarp is partly stepped and follows major joint surfaces. Large amounts of water are reported to have flushed out of areas of the scarp where there are boundaries between gneissic and schistose rocks. The water is reported to have come out as pulsating jets. The conclusion drawn by observers is that the cleft-water pressure against the vertical slabs, which were to fail, had two effects: (1) hydraulic jacking forcing joint blocks and rock masses away from the failure surface; and (2) hydraulic splitting of the vertical bridges of intact rock between pre-existing joints with the result that a continuous failure surface was produced. The rock mass had joint spacing exceeding 10 m so large rock volumes were involved. At

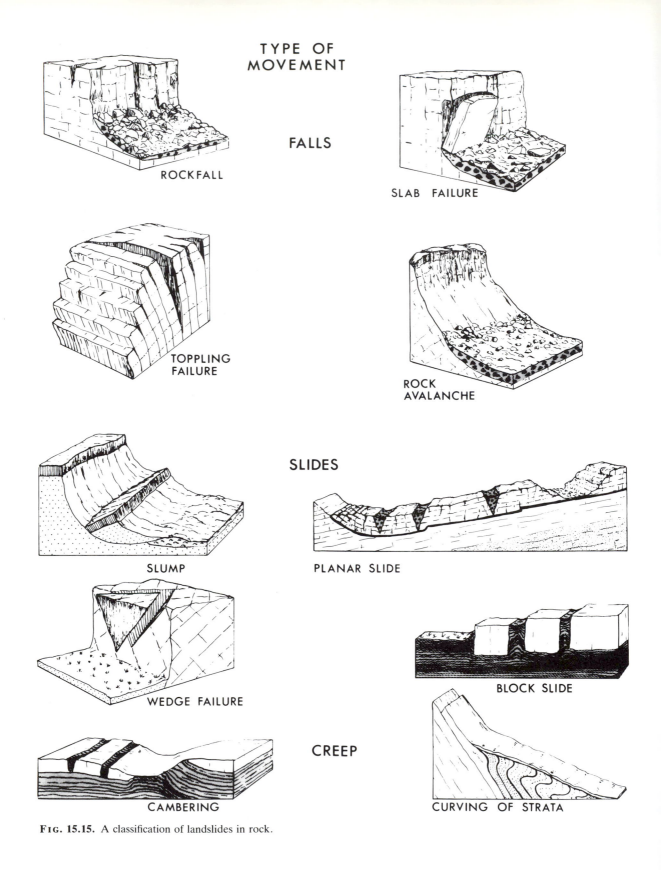

TYPE OF
MOVEMENT

FALLS

ROCKFALL

SLAB FAILURE

TOPPLING
FAILURE

ROCK
AVALANCHE

SLIDES

SLUMP

PLANAR SLIDE

WEDGE FAILURE

BLOCK SLIDE

CREEP

CAMBERING

CURVING OF STRATA

FIG. 15.15. A classification of landslides in rock.

PLATE 15.6. A cliff face up to 300 m high on Mount Boreas, Antarctica. To the right of centre can be seen the scar of a huge rockfall which has left an overhang below the high peak and massive debris on the slope below. The largest debris blocks are as big as a medium-size house.

least three of the seven failures are thought to have involved about 1 Mm³ of rock.

The Antarctic rock-mass failures shown in Plate 15.6 have occurred in an extremely arid environment and failure mechanisms did not involve water. It has been shown, however, that in nearby Mount Circe, the same sandstones suffer horizontal expansive stresses as a result of the gravitational load produced by their own weight (Augustinus and Selby 1990) (see Fig. 4.11). It is noticeable that the failures on the cliffs of Mount Boreas propagate upwards suggesting that failure is a response to stress release. The same mechanism has been shown to be effective at the base of glacial trough walls in the massive plutonic rocks of Fiordland, New Zealand (Augustinus 1988).

Two major mechanisms are clearly implicated in the separation of rock masses from their parent cliff: cleft-water pressures creating hydraulic jack-

ing, and hydraulic splitting of rock bridges; and (2) stress relief propagation of joints normal to the direction of stress—on cliffs this is usually parallel to the cliff face. Other processes may be involved at some sites, especially where tension joints develop at the top of cliffs or deep notches are cut into the base of the slope; the latter case is examined below under 'slab failures'. The paucity of studies of rock-mass falls does not permit informed discussion of other mechanisms.

The Loen failures are perhaps unusual in that a succession of failures has come from one site in a relatively brief period. The limited data suggest that large masses usually separate from a cliff face slowly, perhaps over periods of hundreds or even thousands of years, before a series of wet seasons (e.g. Schumm and Chorley 1964) or a sudden event, such as an earthquake, causes failure.

Slab and block failures are common forms of

PLATE 15.7. Face of a sandstone cliff. The blocks on the upper left and along the top have remained stable, as indicated by their weathered surface, but the faces below them have suffered slab failures, leaving overhangs above them. Grampians, Australia.

weathering on steep valley walls in hard rock. The release of lateral confining pressures permits the opening of tension joints which cut across geological structures and closed cooling joints or bedding planes. Slab failures in their initial stages of development are most obvious when they occur at the tops of cliffs where tension joints may be seen extending parallel to the cliff face. Over a period of time tension cracks will extend vertically, lengthening the slab, and sideways, widening it, until the tensile strength of the rock at the edges of the slab is exceeded and a fall occurs. Extension of the crack is frequently aided by water seepage, especially where this is under a high pressure head, by debris falling into the crack and acting as a wedge, and by ice. Slab failures on cliff faces frequently leave overhanging 'roofs' of rock above the scar from which they have fallen and this may become a site of further slab development (Plate 15.7).

The overhanging roof block or any coherent block of rock, or soil, can remain stable until its tensile strength along a vertical plane, extending upwards from the back of the underlying notch or rock face, is insufficient to support the mass extending from the face of the cliff. The overhanging block can be analysed as though it were the cantilevered end of a beam protruding from the cliff (Fig. 15.16).

A bending stress on a beam acts as a tensile stress on the top half of the cantilever beam and as a compressive stress on the bottom half of the

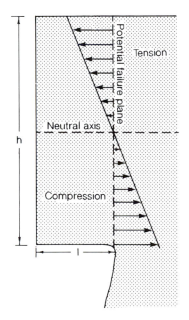

FIG. 15.16. Stresses acting due to self-weight on a protruding, rectangular, unjointed body of rock or soil. (Based on Maekado 1990.)

beam. The maximum bending stress, $\sigma_{b,max}$, is given by:

$$\sigma_{b,max} = M/Z,$$

where M is the bending moment and Z is the modulus of section;

$$M = \tfrac{1}{2}\gamma bhl^2 \quad \text{and} \quad Z = \tfrac{1}{6}bh^2,$$

where: γ is the unit weight of the beam material;
  b is the horizontal dimension of the beam parallel to the cliff face;
  h is the vertical thickness of the beam;
  l is the distance the beam protrudes from the face (or the distance of overhang).

The beam is stable until the bending stress becomes equal to the tensile strength σ_t, the critical value of overhang distance l_{crit} is then given by (Matsukura 1988; Maekado 1990):

$$l_{crit} = \sqrt{\frac{h\sigma_t}{3\gamma}}.$$

For a saturated rock, the tensile strength, as determined in a Brazil test (Brown 1981), and the unit weight, should be for saturated specimens. For conditions where a tension crack has already developed along the potential failure plane, the value of h should be reduced by the depth of the crack in determination of Z, but not for determination of M.

Where a slab is not undercut, but is separating from a cliff along a tension crack the Culmann analysis (Fig. 13.22) is relevant and has been used by Caine (1982).

Slab formation is a very slow process and even large vertical joints may exist for centuries before failure occurs. Many valley walls which have been deglaciated for 10 000 years, or more, still show no sign of slab failures developing. Once a failure has occurred a further slab may start to form behind the scar of the first failure. Sandstones, granites, and chalk are rocks which are most commonly reported as being subject to slab failure.

Rock particle falls are the most common types of rockfall. They are the result of weathering of cliff faces. It is commonly assumed that frost action is a dominant process in most cold, humid, alpine zones and it may be that wetting and drying with swelling and shrinking are effective in argillaceous rocks. In arid and coastal zones salt weathering may be important. These assumptions have, however, seldom been tested by detailed analyses and it is not clear whether chemical processes are significant in weathering of steep rock faces.

Two major studies which include observations of rockfalls are those of Rapp (1960*a,b*) at Tempelfjorden in Svalbard and Kärkevagge in Sweden. In the Swedish study, over the period 1952–60, measured yields of debris from rockfall were: pebble falls, $5\,m^3/y$; small boulder falls,

$10\,m^3/y$; large boulder falls, $35\,m^3/y$. The total of $50\,m^3/y$ from a vertical area of $0.9\,Mm^2$ of rock face corresponds to average retreat of the rock wall of $0.06\,mm/y$. Data, from many published sources, compiled by Ballantyne and Kirkbride (1987) suggest that Rapp's data from Sweden are from a site that is not as active as many others: cliff retreats, from sites in alpine environments, are more commonly close to 1.0 to 3.0 mm/y; in arctic and subarctic zones the range is 0.007 to 1.3.

There is, however, evidence that rates of cliff retreat by rockfall are variable over time and are notably affected by slope orientation. Ballantyne and Kirkbride calculated volumes of debris accumulated in protalus ramparts formed near the margins of glaciers of the Loch Lomond Stadial (*c*.11 000–10 000 BP); mean retreat rates of cliffs during the stadial were 1.69 or 3.38 mm/y depending on whether the stadial duration is taken as 800 or 400 years, these figures compare with modern rates in Scotland of 0.015 mm/y. A study by Gardner (1983*b*) in the Rocky Mountains of Canada demonstrated that greatest rockfall frequency and probability are in the hours around midday. Rockfalls there are most commonly associated with free faces associated with major fault scarps and north- to east-facing slopes. These observations provide some circumstantial evidence to support the view that the presence of snow and ice, which will provide meltwater, high solar radiation in summer periods at midday, and the presence of thinly bedded, closely jointed strata, are particularly favourable for frost weathering and particle fall.

It is evident from all of the available data that particle fall is by far the most widely distributed and most frequent type of fall. Slab failures and rock-mass falls may be productive of the largest volumes of debris at a single site, but the data do not permit comment on the relative contributions of the various size classes to slope denudation of whole mountain ranges over periods of thousands of years.

The relative significance of rockfall as a transport mechanism is also difficult to assess, partly because falls often involve sliding and avalanching, especially towards the bases of cliffs, but also because of the unknown contributions of snow avalanches, slush avalanches, and transport through chutes and gullies. From his observations at Kärkevagge, Rapp concluded that the order of significance is: running water > slides > avalanches

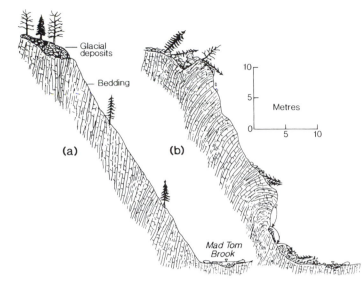

Fig. 15.17. Slope deformation in closely bedded quartzite. (a) An initial post-glacial condition. (b) The modern condition showing creep-induced structural changes: toppling is occurring at the top and towards the base as a result of overturning and dilation. A major displacement and sag-depression has formed at the crest, and flexure of beds is evident at all positions in the slope. (After Lee 1989.)

> falls > creep with solifluction. Rapp's study is of outstanding importance, and perhaps unrivalled in its contribution, but it is not clear how far its conclusions can be applied beyond its immediate area.

Toppling failures

Toppling is a process which may be treated as a distinct entity in a classification. Its primary characteristic is that it has a distinct component of rotation and sliding before a fall takes place. Many of the elements of toppling failures are evident in Fig. 15.17 in which originally nearly vertical beds of quartzite and schist have been deformed by creep and sliding along cross-joints and produced flexure of the main bedding units, fall of blocks from unstable sites, opening of tension cracks, and, near the base of the hillslope, folding as a result of drag and thrust from overlying beds.

Toppling is particularly common in slates and schists, but also occurs in thinly bedded sedimentary rocks, and in columnar-jointed igneous rocks such as basalt and dolerite. Toppling is a failure mode of slopes which involves overturning of columns. The distinction between toppling and sliding failures is based upon the ratio between the width of the joint block and its height (Fig. 15.18). Wide low blocks slide but tall narrow blocks overturn (topple).

(1) A block is stable where $\alpha < \phi$ and $b/h >$ tanα.

(2) A block will slide but not topple where $\alpha > \phi$ and $b/h >$ tanα.

(3) A block will not slide but will topple where $\alpha < \phi$ and $b/h <$ tanα.

(4) A block will both slide and topple where $\alpha > \phi$ and $b/h <$ tanα (Freitas and Watters 1973; Goodman and Bray 1976).

Recognition of the importance of toppling failures in natural cliffs, road cuts, and open-pit mines has promoted a number of studies on slope stability for such features (e.g. Wyllie 1980; Evans 1981; Brown 1982; Zanbak 1983; Cruden 1989). Toppling, may however, be of greater, but less obvious importance, in slopes which are undergoing rock-mass dilation and depth creep—see below.

Slides

Slides in rock or soil are characterized by movement above a sharply defined shear plane. In rocks such as slate, schist, and many sedimentary formations the shear plane commonly follows a structural plane—such as a plane of foliation or bedding—and it is often straight. The resulting failure is then a planar slide when movement is rapid and deformation of the rock mass occurs, or a rock glide when movement is slower and the hard rock mass is rafted on soft clay beds. Well-jointed blocks of hard rock may be tilted and deep fissures opened between them but the gliding blocks remain intact.

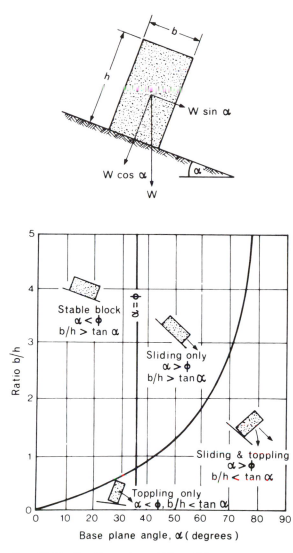

FIG. 15.18. Conditions for stability, sliding, and toppling failure of a block. (Modified from Hoek and Bray 1977.)

In clay-rich rock, such as mudstone, planes of failure are frequently curved and the resulting rotational movement is called a slump. Such failures are most frequently considered with soil failures but because they can occur beneath overlying hard rocks, and can cause tensional and shear failures in those rocks, slumps have to be recognized also as rock failures. Undercutting of basal slopes in clay by rivers, or the sea, frequently promotes slow but continuing slumping (Fig. 15.19).

Rock slides may be very large and catastrophic in mountain regions where the large available relief permits accelerations of rock debris to velocities as great as those of rockfalls and rock avalanches. An example is the huge landslide near Flims, Switzerland, which probably occurred during the last glacial (Heim 1933). This may be the largest Pleistocene landslide in Europe. It involved marly limestones dipping towards the Rhine Valley at 7–12° and may have been triggered by the retreat of a valley glacier which removed lateral support from the base of the slope. The scarp at the head of the slide is over 1000 m high: the volume of rock involved in the slide was 12 km³ and it blocked the Rhine to form a temporary lake over 15 km long and 400 m deep. The slipped rocks covered an area of about 49 km² and travelled over 150 m up the opposite valley wall.

A devastating planar slide into the lake impounded by the Vaiont Dam, Italy, occurred on 9 October 1963. The debris largely filled the lake and swept waves of water 100 m high over the dam. This surge of water carried away two high-level bridges in the valley below the dam and was still 70 m high at the confluence with the Piave Valley 1.6 km away. A compressive airblast preceded the flood down the valley and assisted in the destruction and death of 2600 people.

The Vaiont Valley is in the Italian Alps. It was

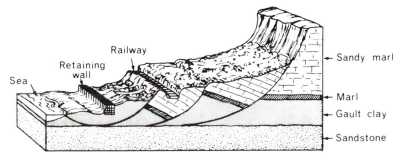

FIG. 15.19. Repeated slumping threatens the railway line near Folkestone, southern England. Marine abrasion has removed the toe of the slump and cylindrical failure planes have developed in the Gault clay which has been weakened by water seeping through the overlying glauconitic sandstones. Corrective measures include an extensive concrete apron to load the toe and protect it from further erosion.

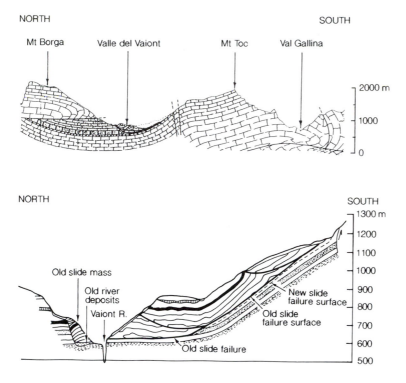

FIG. 15.20. (a) Cross-section showing the general geological structure of the Vaiont Valley (after Hendron and Patton 1985, based on the work of Semenza and Dal Cin). (b) A section across the Vaiont Valley, before the 1963 slide, showing the locations of failure planes and dip of the sedimentary rocks (after Hendron and Patton 1985, based on the work of Rossi and Semenza). In both sections the horizontal and vertical scales are the same.

occupied by a valley glacier until about 14 000 years ago, and since then a deep gorge 200–300 m deep has been cut below the glacial valley. The dam is in this gorge. The rocks of the valley are mostly limestones, with frequent clayey interbeds. The valley has been eroded along the axis of an east–west trending, asymmetrical syncline. The landslide occurred in beds which dip northwards towards the Vaiont gorge at 25° to 45° (Fig. 15.20).

There have been many published reports on the Vaiont failure, some by people who have never visited the site; the most comprehensive report is that of Hendron and Patton (1985) who had available to them detailed geological studies by Italian workers, especially D. Rossi and E. Semenza.

The area of the 1963 landslide had been affected by more than one previous period of shearing displacement; the earlier displacements were all associated with the weak clay interbeds within the lower rock units (Fig. 15.20b). The clay layers along the surface of sliding are commonly 10–20 mm thick but vary from 5 to 100 mm or more. They are montmorillonitic with 50–80 per cent of the clay being a Ca-montmorillonite which has

been sheared and slickensided by flexural-slips along the bedding planes during folding of the syncline. The shear strength along the failure plane was therefore, $\phi \approx 12°$, the range of values obtained in tests was 5–16°.

It is noteworthy, then, that large sections of the slope were dipping at angles far higher than the angle of shearing resistance of the critical rock units and also that the toe of the slope was unsupported because of the incision of the gorge. Detailed analysis showed that approximately 40 per cent of the total shearing resistance acting on the slide mass was supplied by near-vertical faces which formed the eastern (up-valley) boundary of the slide. The 1963 slide occurred because of the combined effects of a rising reservoir at the foot of the slope and increases in piezometric levels resulting from rainfall. The reduction in the factor of safety caused by reservoir filling alone was calculated to be approximately 12 per cent, and the reduction caused by pore-water pressures from rainfall and snowmelt was calculated to be at least 10 per cent. It is unlikely that the rainfall before the failure was uniquely high for the post-glacial period. The reservoir level that would cause failure without rainfall was about 710–720 m. The design reservoir level for full supply was 722.5 m. It was almost inevitable, therefore, that failure would occur at some time.

The geological conditions of the Vaiont failure—unbuttressed lower slopes and strata with dips which are close to the basic angle of shearing resistance of the rocks—are common. What is difficult to detect, without very detailed and therefore expensive explorative research, is the shearing resistance along critical beds which may not even be visible. Closely spaced bedding units, those with clay interbeds, zones of weathered and lamellar minerals such as biotites in schists, and crush zones and mylonites along faults, are all potential planes of weakness. The basic friction angle, ϕ_μ, for many rocks is close to 30°; decomposed schists and mylonites—even under overburdens of 40 to 60 m—may have strengths of $\phi_r = 5°$ (Brandl 1988). In closely bedded turbidites, with sandstone and mudrock alternating beds, the undisturbed bedding joints may have $c = 170$ kPa and $\phi = 35°$, but calcite veins with $\phi = 22°$, sheared joints with $\phi = 14°$, and fully softened mudrock interbeds with $\phi_r = 12–15°$ (Lembo Fazio *et al.* 1990). Even variations in joint spacing can result in variable shearing resistance where, for similar surface roughnesses, jointed rock masses of many small blocks show higher peak strengths than those with larger joint spacing. The scale effects are related to changing stiffness of the rock mass with changing block size or joint spacing (Bandis *et al.* 1986). Determining the threshold for catastrophic slope failures may therefore be a difficult process (Cruden 1988).

Velocities of planar failures on low angle slopes are seldom instantaneously great. It is common for long-term creep to be recognizable from the development of tension cracks, stretching of tree roots, cables etc., and for the creep phase to be followed by a period of steady displacement rates followed by accelerated rates of displacement until rapid failure occurs. Such a sequence is particularly common where movement is along clay beds, beds which have lost cohesion by weathering and solution, or along sheared joints in which asperities are sheared, crushed, and then infilled by debris. Even on very steep rock faces there may be a long period of opening of joints, minor rockfalls, and ejection of water before a catastrophic failure occurs. In open-pit mines it is common practice to place microphones in rock which is suspected of failure and to record the microseismic activity. As failure approaches the number and magnitude of microseisms increase as evidence of cracks propagating and asperities shearing.

Such warnings were recognized at Vaiont where, as the level of the reservoir rose, creep of the south slope accelerated from about 1 cm/week in April to 10 cm/day or more by early October. The creep gave warning of the slide and attempts were made to lower the lake level, but these largely failed because of the heavy rainfall and consequent large inflow of water. Creep reportedly reached a rate of 80 cm/day immediately before the slide which suddenly involved a mass of rock 1.8 km long, 1.6 km wide, with a volume of 250 Mm^3, moving across the valley in a minute or less. The most remarkable feature of the slide is that it crossed the 100 m wide gorge and pushed its front 140 m up the opposite side of the valley with the entire front retaining its structural integrity. The valley and lake were filled by material moving behind the coherent slide front.

It has been hypothesized by Müller (1964) that the catastrophic acceleration from a velocity of a few cm/day to 20–25 m/s was caused by a sudden decrease in internal friction of the slide mass so that a quasi-plastic creep behaviour was transformed into that of a viscous fluid. The enormous potential energy of the slide mass was transformed into kinetic energy in a few seconds. Such behaviour is analogous to that of some quick clays and snow avalanches in which particles lose contact almost instantaneously and flow but, as soon as a critically low velocity is reached again, bonding is reformed and the mass recovers frictional strength. Such behaviour in soil masses is said to be thixotropic but had, hitherto, never been recognized in rock masses.

Incipient failure of the Abbotsford landslide in a suburb of Dunedin, New Zealand, in 1979 provided a clear record of acceleration towards failure, because early movements were indicated by breaking of water pipes, then displacements of kerbs and then instrumental monitoring in the two months before final failure (Fig. 15.21). The failure involved soft Tertiary sandstone capped by Pleistocene colluvium and loess; the slide plane developed along a 50 mm thick bed of organic-rich, highly plastic montmorillonitic clay ($c' = 0$, $\phi' = 8–10°$). The strata are dipping at 6–10° and the total displacement of the 5.4 Mm^3 slide block was 100 m in about one minute. The residents had been evacuated from the suburb before the final failure

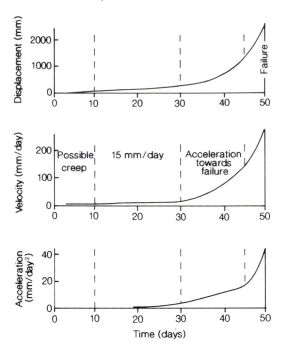

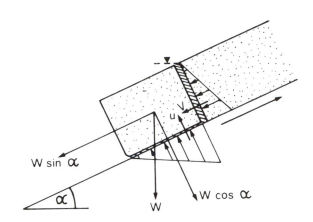

FIG. 15.22. Stresses acting on a joint block on a slope.

FIG. 15.21. Movement, and rates of movement, in the 50 days before failure of the Abbotsford landslide. (Data from Coombs and Norris 1981; and Salt 1985.)

occurred, because of the evidence of accelerating movement, but sixty-nine houses were destroyed (Coombs and Norris 1981; Salt 1985).

Stability analyses of slides

Stability analyses are usually more difficult to carry out, and the results are less reliable, for rock than for soil slopes. Complex geological conditions are the cause of this situation. It is the discontinuities in rock masses which largely control the stability and many features of such masses are difficult to measure. Cleft-water pressures are almost impossible to estimate unless a dense array of piezometers is in place and it is usually impossible to determine the cohesion along the joints forming a potential slide plane. As a result it is common practice to undertake a back analysis (i.e. a post-mortem) in which $F = 1$, cleft-water pressures are assumed to be at the worst case condition, ϕ_j is determined from shear box tests of the rock (to give ϕ), and i is estimated from joint surveys. A stability equation is then solved to determine possible values of c, and this value is used to estimate the stability of other slopes, in the area

of a landslide, which have similar geological conditions.

Where a large rock slide occurs with complex shearing patterns there is seldom much hope of accurately modelling all the parameters. A simple infinite slope analysis is then appropriate. Such an analysis was applied to the largest historical rock slide and debris flow, which occurred in the Peruvian Andes in April 1974. The Mayunmarca slide had an estimated volume of $1000 \, Mm^3$, a length of the debris mass of 8 km, a vertical difference of 1.9 km, and occurred on valley slopes in the range of 35 to 9° (Kojan and Hutchinson 1978).

Deep rotational failures occur in soft rocks, highly shattered bedrock, and in strongly weathered rock. Where there is no control on the slide plane exerted by a discontinuity, then the methods of analysing rotational failures, discussed in Chapter 13, for soil slopes are appropriate.

The simplest case for stability analysis is the condition of a block of rock on a potential slide plane (Fig. 15.22). The block will be on the point of sliding, or in a condition of limiting equilibrium, when the disturbing force acting down the plane is exactly equal to the resisting force on the basal area A of the block:

$$W\sin\alpha = c_j A + W\cos\alpha\tan(\phi + i).$$

The influence of water pressure in a tension crack is indicated in Fig. 15.22 where it is shown that the water pressure increases linearly with depth in the crack and produces a total force V acting on the rear face of the block, and down the inclined shear plane. Assuming that the water

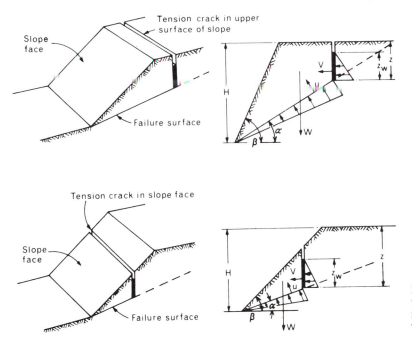

FIG. 15.23. Geometry of a rock slope with a planar slide with tension cracks in upper and lower parts of a sliding wedge.

seeps along the slide plane and flows out of the base of the block at atmospheric pressure, then the water pressure distribution results in an uplift force u which reduces the normal force acting across the plane at the base of the block.

The condition for limiting equilibrium is then defined by:

$$W\sin\alpha + V = c_j A + (W\cos\alpha - u)\tan\phi_j.$$

Not all planar slides are rectangular in section and tension cracks vary in their location. It is common for a failing block to have a triangular or trapezoidal section (Fig. 15.23).

For sliding to occur on a single plane, the plane must strike parallel or nearly parallel (within ±20°) to the slope face; the failure plane must outcrop in the slope face or at the toe; the dip of the failure plane must be greater than the angle of friction of this plane, i.e. $\alpha > \phi_j$; the lateral boundaries of the slide must provide negligible resistance.

Assuming that the tension cracks are vertical and filled with water to a depth z_w, that there is no rotation of the block, and that we are considering a slice of unit thickness then:

$$F = \frac{c_j A + (W\cos\alpha - u - V\sin\alpha)\tan\phi_j}{W\sin\alpha + V\cos\alpha},$$

where from Fig. 15.23

$$A = (H - z)\,\mathrm{cosec}\,\alpha$$
$$u = \tfrac{1}{2}\gamma_w . z_w(H - z)\,\mathrm{cosec}\,\alpha$$
$$V = \tfrac{1}{2}\gamma_w . z_w^2.$$

For a tension crack in the upper slope surface

$$W = \tfrac{1}{2}\gamma H^2\left[(1 - (z/H)^2)\cot\alpha - \cot\beta\right],$$

and for a tension crack in the slope face

$$W = \tfrac{1}{2}\gamma H^2\left[(1 - (z/H)^2\cot\alpha(\cot\alpha . \tan\beta - 1)\right].$$

Where the location, depth, and water height in a tension crack are known, solution of the equation for a factor of safety is relatively simple, but if a variety of groundwater conditions have to be considered then repeatedly solving the equation becomes tedious. Graphical plots, or stability charts, may then be used to solve components of these equations. Stability charts are given by Hoek and Bray (1977) and alternative forms of the analysis for various groundwater conditions by Stimpson (1979).

Limitations of stability analyses

It has been stated already that determining some of the parameters used in a stability analysis may be difficult. In mathematical analyses simplifying assumptions are fundamental, but if the assumptions are wrong, or the selection of parameters is wrong, then precise computation is in vain. It has to be accepted that detailed field observations may be more valuable in interpreting or predicting the causes or possibility of landslides in rock. Careful mapping of geological structures and hillslope forms may be a better guide than much study in the laboratory or much calculation. Unusual contour changes and breaks in slope, signs of large-scale creep, the opening of joints, frontal or toe bulges, lateral shearing or dislocation, remnant head-scarps, disturbed drainage, anomalous vegetation changes, tilted or displaced structures, or a history of instability are indicators which have to be sought and considered. It also has to be recognized that the larger the potential instability the more difficult it may be to detect. High altitude aerial photographs or satellite photographs may then be the most useful tools with which to start an investigation.

Wedge failures are slides produced by failure along two intersecting joint planes which dip out of a hillslope. They are most common in sedimentary rocks and foliated metamorphic rocks which have strongly developed sets of cross-joints which intersect the bedding and foliation at angles in the range of 45° to 135°. Wedge failures occur at a wide range of scales involving a few cubic metres of rock to those involving large volumes, such as the Ponui landslide in sedimentary rocks which involves $2.5\,Mm^3$ (Pettinga 1987).

Some of the most prominent results of multiple wedge failures occur in mountain ridges formed of schists with a strong direction of foliation intersected by a regular set of tectonic cross-joints. The result is a serrated, sawtooth ridge in which each angular depression is a site of wedge failures (e.g. Plate 15.2). Wedge failures are also evident as controls on gully form and alignment in Plate 15.13.

Small wedge failures are relatively common on natural cliffs, walls of open-pit mines and quarries, and on faces of cuts in saprolites where intersecting relict joints become the failure planes.

Gravitational spreading of ridges

Large-scale gravitational spreading (also known as 'sackung') of steep-sided ridges has been recognized in many parts of the world and in many geological structures; examples are provided in Mahr and Nemčok (1977), Mahr (1977), Radbruch-Hall (1978), Shimuzu *et al.* (1980), and Bovis (1982). An analysis of the mechanics of spreading has been published by Savage and Varnes (1987).

The field evidence for gravitational spreading is the formation of uphill-facing scarps aligned roughly parallel with the hillslope contours. Scarps near ridge tops are usually short and rather straight, but those further down the hillslope are longer and arcuate where they follow the contours across valleys and spurs. The alignments thus suggest that the scarps are produced by normal faulting with the faults dipping into the ridge top so that, where there are scarps on either side of a ridge, the ridge is an elevated graben. There is, however, clear evidence that the faults are short, usually less than 2 km, and also, because they follow the topography, they are not of tectonic origin. The displacements of the fault scarps are up to 40 m (Plate 15.8). A second type of common feature is elongated depressions which may be up to 100 m wide, hundreds of metres long, and 20 m deep.

At the bases of slopes there may be evidence of bulging which causes displacement of streams, damage to roads, and other deformations, but the evidence for basal bulging is usually far less clear than that of crestal sagging and displacement.

Some of the geological conditions under which gravitational spreading can occur are illustrated in Figs. 15.17 and 15.24; deformation of shales is shown in Plate 15.9. The conditions include: (1) deep-seated bending, folding, and plastic flow of rocks on slopes; (2) bulging, spreading, and fracturing of steep-sided ridges in mountains; (3) incremental movement along a dipping, irregularly surfaced plane, especially where foliated rocks like schists dip towards a valley and continued creep causes progressive decrease of frictional strength towards a residual condition (Huder 1976); (4) movement distributed over a thick zone in relatively uniform rock, as in mudstones or in schists and gneisses; (5) distortion and buckling of dipping interbedded strong and weak rocks (Fig. 15.17); (6) creeping of rigid over soft rocks without buckling; and (7) valleyward squeezing-out of weak ductile rocks overlain by, or interbedded with, more rigid rocks, causing tensional fracturing and outward movement of the more rigid rocks, sometimes with upward bulging in the centres of valleys.

The last type of cambering movement has

PLATE 15.8. Scarp produced by gravitational sagging of foliated schist. Scale is given by the man at the foot of the scarp, which is about 6 m high. Wanaka, New Zealand.

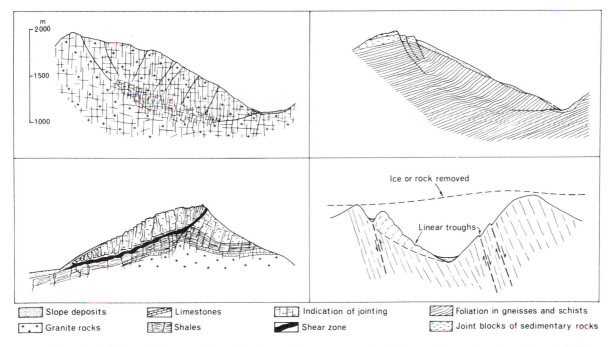

Slope deposits	Limestones	Indication of jointing	Foliation in gneisses and schists
Granite rocks	Shales	Shear zone	Joint blocks of sedimentary rocks

FIG. 15.24. Geological conditions and effects of rock-mass flow, creep, and local faulting. (Based, in part, on Nemčok 1977.)

PLATE 15.9. Creep in shales has caused bending of the strata. (Photo: by permission of The Institute of Geological Sciences.)

occurred in the iron-ore opencast mines of central England where valleys are cut through Jurassic limestones into underlying Lias clays. The bedding is nearly horizontal but the deformation of the clays, as they bulged towards the valleys, has caused fracturing of the limestone and, either cambering of jointed rock masses towards the valley (Hollingworth *et al.* 1944), or reverse tilting of the limestones. Deformation of some limestones has opened fissures, called gulls, parallel to the contours. In the Bath area disturbance of the strata occurs to depths of 30–40 m (Chandler *et al.* 1976). While the deformations can be interpreted as the squeezing out of plastic clays from a loaded zone to an unloaded one, the process may have been aided by saturation of the clay by water in colder climatic conditions and possibly by freeze–thaw effects (Horswill and Horton 1976).

Deformations in closely jointed rocks may involve toppling, folding, and dilation of the mobile rock masses. Toppling is evident in the type of situation shown in Fig. 15.17 and in the creep of joint blocks over plastic rocks (Fig. 15.24 bottom left). Folding can occur in a situation where finely laminated rocks, such as schist and flysch, have plastic interbeds, or in sedimentary rocks, such as sandstones and limestones, where bedding planes are closely spaced and the beds are nearly vertical. They can thus overturn and fold downhill, as has happened in the quartzites of Fig. 15.17.

Zones of slow, plastic flow deformation under gravity may extend to depths as great as 300 m below the groundsurface, and involve rates of movement ranging from 1 mm/year to 10 m/ year (Ter-Stepanian 1977). It is distinguished from soil creep by its great depth and isolation

PLATE 15.10. An avalanche from the western end of Nuptse, Khumbu Himal, Nepal. The avalanche started as a large slab failure, disintegrated as it fell into a rock basin, and poured onto the debris-covered surface of the Khumbu Glacier as a turbulent mixed avalanche with a leading cloud at least 200 m high. Note the talus cones to the right of the avalanche: they are formed of alternating layers of snow and rock debris.

from daily and seasonal climatic conditions, and from landsliding by the lack of a single, clearly defined failure plane and slow rate of deformation. Deformation at depth may, however, so weaken rock that it leads to landsliding (e.g. Giraud *et al.* 1990).

The fundamental cause of this form of gravitational failure appears to be oversteepening of the hillslopes by: (1) glacial trough formation, especially during the last glacial; (2) tectonic uplift of ranges accompanied by deep incision by rivers; and (3) by undercutting of valley slopes by rivers migrating laterally. The response of the uplands is therefore: (1) lateral spreading at the base which is compensated for at the crests of ridges by fault displacements; (2) a second form of response is applicable only in foliated and well-bedded rocks with near vertical dips and involves flexure, and may lead to folding and toppling; (3) another form of response is applicable only to geological conditions in which massive block-jointed rocks can creep, tilt, camber, or topple as a sub-stratum of plastic rock deforms beneath the overburden of the stronger block-jointed rock.

Snow avalanches

Snow avalanches have, in the past, been studied primarily because of the threat they pose to people and structures in mountains. They also have the capacity to entrain and transport rock debris and vegetation and leave these materials as deposits, sometimes of distinctive forms.

Three major forms of snow avalanche are recognized: slab, powder and slush avalanches; slabs may break up when they move, especially if they fall over steep faces, and a mixed form results (Plate 15.10).

In powder avalanches most of the snow swirls through the air as a snow cloud. In flowing mixed avalanches the snow moves in a turbulent tumbling motion. Large blocks and slabs of snow bounce over the ground, breaking up as they do so, and a cloud of snow dust travels above them. Avalanches may run over the snow ground layer or penetrate through the snow pack and run on the ground, entraining rock debris as they do so. Slush avalanches are wet, dense masses of snow which behave rather like small debris flows and they entrain much fine rock material and boulders. They commonly occur in spring melts or during rainstorms (Leaf and Martinelli 1977; Perla and Martinelli 1976).

Large avalanches usually occur on slopes of 30–50°, small avalanches on slopes of 50–65°, and minor shedding of small snow accumulations on steeper slopes where accumulation is most difficult. On slopes of less than about 30° snow avalanches are not common although dry snow can avalanche at angles as low as 25° and wet snow at angles as low as 10° (Mellor 1978). The optimum conditions for geomorphically significant avalanche snow accumulation are on slopes of 30–

PLATE 15.11. Avalanche tracks in the Coast Mountains, Alaska, cross areas vegetated by low-growing alder and willow shrubs. Areas with taller sitka spruce trees are less prone to avalanches. Note the snow fan at the base of the track on the right.

50° with an even surface, and hence above the tree-line. The even surfaces may be rectilinear, especially if they are talus slopes or in bowl-shaped depressions such as ice-free cirques.

Large snow accumulations are commonly formed on lee slopes, so where there is a dominant wind direction, and formation of cornices, snow accumulation and avalanche activity are likely to occur on slopes with a preferential aspect. Slab avalanches, however, may be most common on slopes facing the wind as wind-compaction of the snow increases snow density and the tendency for slab formation. Also favouring slab formation is a period of high insolation between accumulation events, with metamorphosis of the surface layer followed by hoar-frost. Such processes may cause layered accumulations to develop and slab release is then along one of the hoar-frost boundaries.

Dry-powder avalanches are most likely to occur where loose accumulations develop from drift and lee-side deposition without wind-packing.

Slush avalanches are released during rainfalls and melt periods when surface flow from gully walls is channelled into gully floors, causing saturated zones at the base of snow accumulations (McGregor 1989).

Avalanche tracks which terminate above the tree-line are likely to develop where the avalanches run down on to rough moraines or into the trees at the tree-line. Large avalanches may pass through forest zones along well-defined tracks which have been periodically followed by ava-

lanches for much of post-glacial time. Some may have been established at the end of the Last Glacial or in the 'Little Ice Age'. Some avalanche tracks are open talus slopes; some are over slopes on which shrubs and herbaceous plants are established (Plate 15.11), the plants being able to survive the passage of powder avalanches (Luckman, 1977; Butler, 1989). The run-out zones of avalanches which have passed through a forest zone, or over talus and through gullies, to valley floors are usually concave slopes with inclinations of 20° declining to about 8°. Run-out tracks may be concave transversely, indicating deposition but this may be from snow avalanches, debris flow, or fluvial processes.

Geomorphic activity of avalanches results from their high velocity and the rate at which they slow down. The terminal velocity of a dry flowing avalanche may be found from:

$$U^2 = \xi h(\sin\alpha - f\cos\alpha),$$

where: U is the velocity averaged over the depth of flow;

ξ is a turbulent friction coefficient;

h is the depth of flow;

α is the longitudinal inclination of the track;

f is a coefficient of sliding friction.

Values for ξ are difficult to estimate and vary with the roughness of the terrain traversed by the snow; they vary from 400 to 1800 m/s², for a channelled flow they are usually in the range 400–600 m/s². In

TABLE 15.1. *Typical impact stresses and velocities of avalanches*

Avalanches	Stress (kN/m²)	Velocity (m/s)
Small dust avalanches	1–10	20–40
Dust cloud of large avalanches	<100	30–70
Small, dry, surface avalanches	10–100	10–30
Dense, fast, confined surface avalanches	100–600	20–40
Extreme cases	<1000	<125

confined channels h is replaced by R, the hydraulic radius (Sommerhalder 1965).

The pressure (p) exerted on obstacles by snow avalanches depends upon the density of the snow mass (ρ) and its velocity (v):

$$p = \tfrac{1}{2}\rho v^2.$$

Dry slab avalanches release when densities are about 250 kg/m³ or less. Wet slabs and saturated slush masses release at densities of 450–600 kg/m³ (Ward *et al.* 1985). Maximum bed-shear stresses may range as high as 1 to 10 kN/m², but impact stresses can be far higher and the effect of an avalanche is, therefore, potentially greater on obstacles than upon even groundsurfaces. Table 15.1 shows typical impact stresses and velocities.

The greatest impacts occur where a large snow mass plunges over a near vertical face. In such conditions large basins may be formed in loose rock debris. In Fiordland, New Zealand, small lakes (tarns) with average areas of 11 000 m² are found at the base of steep tracks (38–59°) where there is an abrupt change of slope. The plunging avalanches may have masses up to 300 000 tonnes and impact pressures have been estimated of 600 kN/m² (Fitzharris and Owens 1984). Boulder rings around water-filled depressions are smaller scale features of similar origin.

Other microforms attributed to snow avalanches are avalanche boulder tongues, cones, mounds, ridges, gouges, pits, and modifications of talus slopes. Boulder tongues (Rapp 1959; Luckman 1978) are highly concave, elongated talus cones formed below chutes in the cliffs above. Such tongues are poorly sorted, or unsorted, piles of angular boulders heaped together in unstable forms. Perched boulders, bodies of unsorted, angular, and unstable debris forming low mounds,

and small ridges running up and down the slope, are also attributed to deposition from snow avalanches. Terminal ridges, looking like glacial moraines, may also be deposited from avalanches (Ward 1985).

Erosional forms include displaced boulders, which have left hollows, gouge, and scrape marks, and, more obviously, overturned trees which have left pits, and damaged trees. Buildings and engineering structures suffer primarily from impact pressures.

Slush avalanches and slush flows can occur on much lower angles of slope than snow avalanches. Their primary depositional forms are fans, spreads of debris, and boulder tongues. Many of the deposits they leave are difficult to distinguish from those of debris flows and flash floods (Nyberg 1989).

Overall, however, medium to small snow avalanches are not capable of moving large boulders or much debris. Only large avalanches falling over steep cliffs have impact pressures which can create significant new landforms. Minor modifications of deposits created by other processes are the main distinctive forms. Snow avalanches are more effective in modifying forests and human-built structures than creating erosional landforms. Large slush flows developed in prolonged and heavy rain in a melt season can be more effective than snow avalanches.

Deposits below rock slopes

The debris which has fallen from a mountain wall or a cliff may be carried away by streams, glaciers, or waves, or it may accumulate to form a talus deposit (also known as a scree). The rock debris making up a talus has characteristics related to the joint spacing and strength of the cliff material, and also to the degree of comminution it suffers in transport to the crown of the talus deposit. Travel down gullies and chutes may substantially break-down some material, whereas fall from a low cliff may leave it at the size of the rock-slope joint blocks (Plate 15.12).

Talus may take the form of a sheet of debris, known as a talus slope, where it has accumulated below a cliff which has weathered more or less uniformly along its face and shed its debris on to the slope below; or a talus cone may form below a chute or gully through which descending debris

PLATE 15.12. Left: a low cliff of dolerite is the source area for rockfall debris of joint-block size. Right: a rectilinear talus deposit formed entirely by rockfall. The arid zone of Victoria Land, Antarctica.

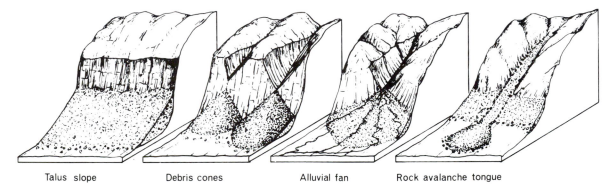

Talus slope   Debris cones   Alluvial fan   Rock avalanche tongue

FIG. 15.25. Debris accumulation below rock slopes.

is funnelled (Fig. 15.25; Plate 15.13). Rare but high energy events such as rock avalanches can produce long tongues of debris.

Talus deposits may form any combination of slope and cone forms depending upon the nature and distribution of weathering and erosion processes along a cliff face, and upon whether the joint pattern permits the development of chutes. Because streams may also be active on mountain slopes there is no simple distinction between cones

formed by rockfalls, landslides, snow avalanches, or debris flows and alluvial fans produced by stream deposition and debris flows. Thus a continuum may exist from steep-angled rock debris cones to lower-angled alluvial fans. Cones, however, are virtually always steeper than about 11° and more commonly have angles steeper than 20° and up to 46°; they may also lodge against a rock slope and be discontinuous with the chute which feeds them. Alluvial fans are nearly always less

PLATE 15.13. Debris cones below clefts formed in gneissic schist, Khumbu Himal, Nepal. The size of the cone is directly related to the size of the source area. Note evidence of debris-flow channels, which have levees. The medial ridges on the largest cone may be from avalanche activity. A small rampart of large blocks is forming at the foot of the middle cone. Talus on the right is from the whole cliff-face and has little cone development.

steep than 11° and are always continuous in long profile with the mountain valley which supplies their deposits.

Talus mantles and deposits

Many attempts have been made to develop simple models for the forms, development, and deposits of talus sheets and talus cones. Simplicity, however, is not a feature of talus accumulations for reasons which include the following. (1) Talus deposits found in formerly glaciated valleys have accumulated during the 10 ky of Holocene time, in which there have been many fluctuations of climate which have influenced both the supply of debris and the processes acting to modify the surfaces of talus deposits (e.g. Kotarba and Strömquist 1984; Francou 1988a,b). (2) In mountains in the tropics and subtropics, the dominant processes may have been, and may still be, substantially different from

those acting in glacial or periglacial zones (Francou 1988a). (3) In any locality, microclimatic variations may cause considerable variation in rates of supply to talus surfaces and in the importance of such processes as snow and slush avalanches, nivation, interstitial ice and creep created by that ice (Olyphant 1983; Akerman 1984). (4) The material supplied to a talus may be: of mixed size, or relatively equi-dimensional; dominantly cubic, or tabular; delivered as fall in single clasts, as blocks, or by catastrophic events; part of a dry avalanche of material, a wet slush avalanche, a powder snow avalanche, or part of a debris flow. (5) The processes acting to modify the talus surface may include one of, or a combination of: creep and rolling of particles caused by collisions; creep caused by needle ice; subsidence caused by melting of buried snow and ice; progressive weathering of talus materials; impact by rock and snow

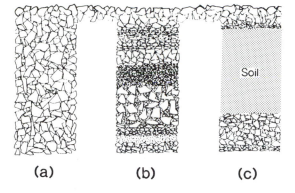

FIG. 15.26. Schematic profiles of three basic categories of talus-slope deposits encountered on the Craigieburn Range hillslopes: (a) open-work gravel; (b) stratified gravel and fines; (c) a paleosol within coarser deposits. (After Pierson 1982.)

PLATE 15.14. Bedded talus deposits of grèze-litée type. The more cohesive fine-grained beds stand out from the face.

avalanches; sliding of particles over snow patches; and rill and gully erosion.

Four main kinds of fabric are commonly recognized in talus deposits (Fig. 3.17). (1) An open-work fabric, in which there are few small clasts, usually results from fall of individual fragments or from small rockfalls in which the rock fragments are of similar size. An alternative source is from the accumulation of such materials in a chute, or high on the talus, to form a lobe which when over-steepened, or pushed by other debris or an avalanche, will form a dry flow which moves down the slope as a sheet. Such dry flows have been observed in the Karakoram (Wasson 1979; Brunsden *et al.* 1984). (2) A partly open-work fabric may result from infilling of some of the voids, in an open-work, and by washing down or fall of small grains. (3) A closed clast-supported fabric has all of the voids filled with fine-grained material; this is usually the result of washing of fines into an original open-work. (4) A matrix-supported fabric is usually the product of debris flows, solifluction, or wash.

Where there are several processes acting on a slope, a section through a talus may show varied fabrics (see Fig. 15.26): surface individual grain movements, caused by dry or wet sheet flows, create fine laminations or beds; channelized water creates distinct bedded units of limited width; debris flows create unsorted mixed-size and matrix-supported masses. Periods of slope stability followed by reactivation of talus formation may produce paleosols within the talus (e.g. Pierson 1982).

Stratified deposits, of different dominant fabrics and particle sizes, indicate alternations of processes in the accumulation of the talus mantle. A widely recognized form of such alternate beds is the type known as 'grèze-litée': this phenomenon consists of layers of angular rock fragments embedded in a finer matrix alternating with layers of equally angular material with an open-work structure (Plate 15.14). Such materials have been regarded as being produced by alternations of gelifluction, which is responsible for the material with a matrix, and sliding of coarse angular clasts over snow, which gives the open-work fabric (Van Steijn *et al.* 1984). Alternative hypotheses involve slope wash, eluviation to create the open-work fabrics, snow creep, rockfall, nivation, and miniature debris flows (see Van Steijn *et al.* 1984, for discussion and references).

Development of talus-slope forms

Three models of talus accumulation have been proposed.

(1) *Talus creep* moves downslope material which fell on the top of the slope; the talus surface is thus behaving as a conveyor belt (Sharpe 1938). Particle creep does occur on talus slopes but such a simple model fails to account for the observed fabrics of materials and known variety of processes.

(2) *Rockfall* has been regarded as the dominant process and control on talus form (Kirkby and Statham 1975). Their model emphasizes the accumulation of individual particles that roll, bounce, and slide downwards until they reach a zone where particles of similar size to their own form the surface. Large particles will roll over smaller ones and small particles will be trapped in the spaces between large ones. Consequently small particles will come to dominate upper slopes while large particles will dominate lower slopes. The lower concavity will be created by three effects: (a) the proportion of stones travelling a specific distance down the talus follows a Poisson distribution, hence most stones do not travel far; (b) some particles will travel further than others, creating a thinning tail; and (c) large blocks falling from a high cliff will have high kinetic energy which will carry them across the toe of the talus mantle. As the cliff height diminishes, this last effect will cease, the talus should become straighter (Plate 15.12), and the talus as a whole will have an inclination controlled by the angle of repose of the material. This model is relevant to rockfall mechanisms only.

(3) *Slush avalanching* has been observed to dominate accumulation processes on some talus slopes (Caine 1969). The slush flows redistribute debris from the head of the talus slope and carry it downslope so that the depositional layer increases in thickness with distance away from the source of slush flow, thus producing a concave form. Such a model is more specific to certain environments than the rockfall model, and it certainly cannot apply to hot arid or extremely cold Antarctic environments.

Three conditions of talus slopes are of great importance in considering the relevance of any model.

(1) *The long profiles* of talus mantles may take many forms: rectilinear; upper concavity → rectilinear → basal concavity; upper convexity → rectilinear → basal concavity; upper concavity → greater basal concavity.

(2) *Talus surfaces are commonly diachronous*. That is, surfaces are older and least disturbed by recent events towards the base of the slope. This has been demonstrated clearly by Whitehouse and McSaveney (1983) for some talus slopes in the Southern Alps of New Zealand, by using weathering rind thicknesses on clasts and colour of clasts as indices of time since deposition (Chinn 1981). The latter observations are also compatible with the findings of Gardner (1983) that lower parts of talus slopes in the Canadian Rocky Mountains have low accretion rates. The model of development is one of rockfall accumulations at the top of the talus, with movement of particles downslope until they are trapped where the surface roughness approaches the particle's own dimensions. There is thus some degree of sorting with small particles accumulating at the crest and the larger particles, with their greater energy, reaching the slope foot. The crest of the talus slope may therefore become convex until dry flow (= dry avalanching) redistributes the rock debris downslope. The dry avalanches have low initial momentum and are unlikely to travel to the slope foot, and will not cover a wide lateral surface. Hence the talus surface is less likely to be buried at the foot than nearer to the crest.

(3) *Frictional characteristics* of the materials forming a talus will influence the talus form, but only under certain conditions.

First it is necessary to define terms. If grains are poured on to a pile from a low height a cone develops with sides at an angle with the horizontal. If the platform upon which the pile stands is then tilted a steeper angle is attained, up to as much as 10° steeper, before failure occurs. The angle at which the failed material comes to rest is usually lower than either of the other angles already mentioned. Unfortunately all three angles have at various times been called the angle of repose.

The difference in the angles of slope of the pile are partly related to the differences between the dynamic friction angles which are relevant to the grains falling on the pile and the static friction angles, ϕ, which are relevant to the tilted pile at rest. The lowest angle achieved by grains after failure has occurred is related to the residual strength, ϕ_r, of the materials. The angles of slope of a talus are seldom, perhaps never, as steep as the static friction angle achieved by the tilted pile, and the angle of slope achieved by grains with high energy is related to dynamic friction and is lower than ϕ

PLATE 15.15. Virtually inactive talus cones and sheets flank the lower slopes of the Langtang Valley near the Nepal–Tibet border. The talus has developed during the Holocene and is now covered by vegetation, except at the apex which has some activity.

and ϕ_r. Thus although ϕ for clean sand is about 44°, a pile of loose clean sand will form a talus slope of about 33°.

Talus slopes are seldom composed of materials of a single-grain size and they may be affected by the shape of their particles. In general, massive, densely packed, coarsely grained, angular, or rough fragments form steeper slopes, while loosely packed, rounded, slaty, schistose, fine-grained, or smooth rounded particles form gentler slopes. High pore-water pressures may also reduce stable talus angles; for clean sands the angle of rest of saturated material with internal seepage is about half that of dry material. Processes such as slush avalanches or surface wash also have the effect of reducing talus-slope angles to below those attributable to the angle of rest or the strength of talus materials alone.

For most talus materials the minimum angle of shearing resistance is in the range of 39 to 40°

(Chandler 1973), but many talus slopes have rectilinear units with inclinations of 29 to 37° with 33 to 36° being very common. This lower slope inclination than the friction angle has several possible causes. (1) In a shear box test the specimen is under a normal load, and it has been shown by Barton and Kjaernsli (1981) that rock aggregate materials have peak, drained friction angles, ϕ, as low as 35° and in the range 35–45° where normal stresses are about 3 MPa, but the same materials have $\phi = 45$–55° under normal stresses of 0.06 MPa; talus materials on the surface of a deposit have little or no overburden and the material can readily dilate, so any comparison between talus inclination and the result of a shear test should be with material sheared under a negligible normal stress. There is a difference of up to 10° between the common talus inclination and the shear test results for ϕ. (2) Values of the residual friction angle ϕ_r may range from 15–35°, but are commonly in the range 25–35°; there are, however, considerable problems in undertaking a residual strength test because of the large displacements required and few large ring-shear machines are available (Barton and Kjaernsli 1981). A comparison between talus inclinations and friction angles suggests that the controlling factor is residual shear achieved after sliding in a loose state and the lower inclinations of talus may relate to transport of particles with high kinetic energy. (3) Any assessment of talus and frictional strength has to take into account the processes of deposition. It has already been pointed out that there are many possible contributing processes and on largely inactive talus slopes the processes acting during the talus formation may not now operate in the region (Plate 15.15). There is some evidence that many talus forms in Europe, for example, developed in the early Holocene and later periods of cold and wet climate (Kotarba and Strömquist 1984; Brazier *et al.* 1988; Brazier and Ballantyne 1989). (4) Studies comparing talus forms and inclinations in small areas indicate variations related to aspect and, with it different intensities of contributing processes (Akerman 1984).

Studies carried out in low mountains and in warm climatic zones may come to very different conclusions from those investigations centred on high mountains and very cold climates. In the latter, many taluses are of mixed layers of snow and rock debris. Some of the snow is converted to interstitial ice and the talus may become permafrosted (see Plate 15.10). In such circumstances the talus mass may behave like a rock glacier with mass deformation and flow at the surface (Johnson 1984); rock slides over ice-crusted snow may be a major process of transport (see Francou for work in the Andes). Rainfall on permafrosted, and therefore impermeable, talus may be a major factor in development of debris flows.

The diversity of talus-forming processes, diversity of talus materials, and diversity in age of individual talus accumulations, as well as a range of ages of the surface of a single talus accumulation, make any model based on a single mechanism inadequate. It may be appropriate to terminate the search for simple models of formation and to start using talus deposits as sources of information on past environments and processes. It should also be recognized that talus slopes in many areas are now inactive; their inclinations are likely to be stable for a long time, but if weathering has modified the talus material or the toe is undercut, processes unrelated to those of original formation may be dominant. By contrast, a forest vegetation cover may essentially fossilize an inactive form.

16 Models and Hillslope Development

Hillslopes may be considered as *systems*, that is, arrangements of materials which are acted upon by processes which rearrange the materials into newly formed bodies or units of rock and soil. In attempting to describe, analyse, simplify, and display understanding of hillslope systems, earth scientists develop and use *models*. Examples of the arrangements of hillslope systems are presented in Figs. 1.1 and 1.2. An example of a simple statistical model of the relationship between hillslope form and the mass strength of rock is displayed in Fig. 6.13.

Study of hillslope development has been at the heart of geomorphology since the subject developed in the nineteenth century. It now has many practical applications as people seek to understand, predict, and control the rate at which open-pit walls, deep cuttings, and natural cliffs and hillslopes will develop within the time during which a specific activity or land use will continue. The way in which hypotheses, concepts, and predictions of hillslope change can be presented and tested is in the form of models.

Types of models

There are three major classes of models: (1) conceptual models, (2) scale models, and (3) mathematical models (Huggett 1985).

(1) A conceptual model is a mental image of a natural phenomenon in which the essential features of the phenomenon are retained and the details of which are regarded as extraneous are excluded. A map, or block diagram, is a conceptual model in which major forms of an original landscape are selected for a particular purpose (e.g. Figs. 10.2, 10.3). Ideas of landform development are commonly expressed as schematic diagrams (e.g. Fig. 10.5) which are also conceptual models. The same ideas can be expressed in words, 'box and arrow' diagrams (e.g. Fig. 1.3), or equations such as the infinite slope model.

(2) Scale models are physical models in which both materials and the stresses acting on them should be scaled in proportion. Blocks of cement or plaster, which are used to represent joint blocks involved in planar sliding, for example, are acting only under Earth's gravitational field. For scaling, a ratio of plaster block size to rock joint-block size should be applied to the strength of the plaster, thus requiring it to be very weak. Similarly, scale models of soils should use smaller-size soil particles, but here the scaling method does not work, for clays and silts have distinctive properties of cohesion, permeability, and capillary suction which ensure that they do not behave as small models of sand and gravel.

In spite of the problems of scaling, some physical models have been very successful. Flume and wind-tunnel studies of sediment transport using sands have greatly assisted the search for understanding processes of entrainment, transport, and deposition. Models using plaster bricks have also aided understanding of the transmission of stresses in the rock masses of dam sites.

(3) In mathematical models the features of a system are represented by abstract symbols and their interactions are replaced by expressions containing mathematical variables, parameters, and constants.

The chief classes of mathematical models used by earth scientists are:

(i) stochastic models,
(ii) statistical models, and
(iii) deterministic models.

Stochastic models have a random process incorporated within them which describes a system, or some part of it, on the basis of probability. Random-walk and Markov-chain models come into this category.

Statistical models, like stochastic models, have random components, but the random components are the unpredictable data from naturally variable phenomena such as rainfall, soils, and earthquakes, together with variations caused by experimental and sampling error.

Deterministic models are conceptual models in mathematical form, but without random components. They may be derived from physical or chemical principles or from observational and experimental data.

Deterministic models are now in relatively common use in studies of hillslope evolution, as they can be expressed as a deductive argument as, for example, of how a rock slope will change as debris falls from a free face and accumulates as talus. The development of a hillslope profile as a continuous curve can be simulated by an analytical solution of an equation in which are incorporated assumptions about the relationship between specific slope processes and the slope form, or by an analytical solution to a hillslope development equation derived from the continuity condition of mass transfer of material on slopes.

All conclusions drawn from models should be tested against the field condition. Because hillslopes develop slowly compared with, say, beaches and sand-dunes, testing is difficult, but there are some sources of field evidence which can provide a degree of validation. In consideration of all conclusions it has to be recognized that models are not only simplifications of reality but their components are selected by scientists. If the selection of data and analytical procedures are not valid representations of nature, the model may produce results which are logically consistent, and even represent a natural form, but the conclusions may still be misleading.

Evolution of soil-covered hillslopes

Soil is fundamentally different from rock in that it is a particulate material with rates of infiltration, soilwater movement, solutional denudation, and responses to creep, rainsplash, and wash which are different from responses of rock.

Modelling the response of soil to erosive processes began with the efforts of pioneer workers such as W. M. Davis (1909), W. Penck (1924), and L. C. King (1967). Modern modelling also has its pioneers amongst whom Ahnert (1964, 1976a,b, 1988), and Kirkby (1971, 1985, 1987) are major contributors.

Deterministic models

The necessary basis for any process–response model is the continuity equation, which is a statement that, if more material is brought into a slope section than is taken out then the difference must be represented by accumulation. Conversely, if less material is brought into a slope section than is removed the difference must come from net erosion of the section. The rate of debris transport is thus a major term in the continuity equation, and the variation of the rate of transport with relief largely controls the slope form and rate of change. For a satisfactory statement of the continuity equation we also need to specify the initial form of the profile, the conditions at the crest (usually regarded as fixed), and at the base of the slope (where constant removal of material is the simplest condition).

The equation has the form:

$$\frac{\delta y}{\delta t} = -\frac{\delta S}{\delta x},$$

where: y is the elevation of a point;
t is the time elapsed;
S is the sediment transport rate;
x is the horizontal distance of a point from the crest.

Where there is no limitation on the supply of material then:

$$S = f(x)^m \cdot \left(\frac{\delta y}{\delta x}\right)^n,$$

where $f(x)^m$ represents a function proportional to distance from the watershed (roughly the distance over which overland flow is ineffective) and $(\delta y/\delta x)^n$ represents processes in which sediment transport is proportional to hillslope inclination. The values of exponents used by Kirkby (1971) are: for soil creep $m = 0$, $n = 1.0$; for rainsplash $m = 0$, $n = 1.0-2.0$; for soil wash $m = 1.3-1.7$, $n = 1.3-2.0$.

Solutions for the continuity equation take the general form of $y = f(x, t)$ which is a description of hillslope inclination, β.

As a first-order approximation for slopes of less than 30°:

$$S \propto \sin\beta \propto \tan\beta \propto \left(\frac{\delta y}{\delta x}\right)^n.$$

This model of transport rate as a function of slope angle (β) is confirmed from experimental data on soil erosion (see Chapter 12) and is used in analytical models such as those of Kirkby (1971) and Gossman (1976). By using values assigned to an initial condition (y, x, β) it is then possible repeatedly to solve the equation adjusting β to conform with the value of S determined in the previous solution. Changes in slope profile can therefore be determined using a series of points along that profile.

Simulation models

Simulation models have been constructed by a number of workers including Young (1963), Ahnert (1976b), and Armstrong (1976). Because it is relatively simple the model of Armstrong will be described here.

The model has a land surface in three dimensions, represented as a matrix of unit cells, each of which has two important properties—a height and a soil depth. The matrix of heights represents the form of the basin at any one time, and the direction of mass transfer of material is determined by the gradient at any point, so that the form becomes an important variable which modifies the action of slope processes. The soil depth at any point represents the total amount of material which is potentially mobile. An initial form of the landscape is specified as a map of heights and depths.

The processes operating in the model are selected, their mode of operation specified as an equation, and their magnitude is assigned a value which represents a natural rate of operation. It is thus possible to specify that each iteration of the model represents a set period of time. Armstrong ran his model for 20 000 iterations with each iteration representing one year, so that he assumed a period of denudation of 20 000 years. By using a high-speed computer the evolution of the slopes over that period can be calculated in a matter of minutes.

The continuity equation for the slope system can be represented as a budget, so that over a unit of time (one iteration) can be calculated:

$$\delta H = I - O$$
$$\delta D = I - O + W,$$

where H is the height of the groundsurface above a datum;

where I is the inflow of material into the cell;

O is the outflow of material from the cell;

D is the soil depth;

W is the amount of weathering.

Thus in each iteration is computed the weathering component and the outflow from each cell (representing the magnitude of the transport process), which is then added to the inflow of the cell next downslope.

In the model only three processes are considered:

(1) *Weathering* is represented by:

$$W_a = W_p e^{-K_w D},$$

where W_a is the actual weathering rate;

W_p is the potential weathering rate at a bare rock surface;

K_w is a constant;

D is the soil depth;

e is the base of natural logarithms.

Values assigned are $W_p = 1 \times 10^{-4}$ m/year, $K_w = 2.0$, $D = 1$ m.

(2) *Slope transport* is represented in the form of soil creep at a rate calculated from:

$$C_s = K_s \sin\beta,$$

where C_s is the rate of soil movement;

K_s is a constant;

β is the slope angle.

Values assigned are $K_s = 10$ cm³/cm/year and slope angle was set by the initial landform and then modified according to the result of each iteration.

(3) *Fluvial transport* rate is given by:

$$C_r = K_r S Q E,$$

where C_r is the volumetric transport rate;

K_r is a constant;

S is the river slope;

Q is the discharge;

E is the river efficiency.

Values of S and Q are supplied by the model at each point, K_r is given a value of 4.13, and $E = 0.4$.

The results of the simulation are presented in Fig. 16.1. Starting from the initial form, block

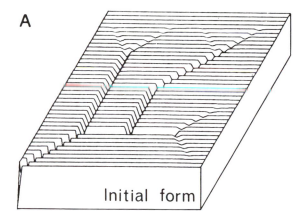

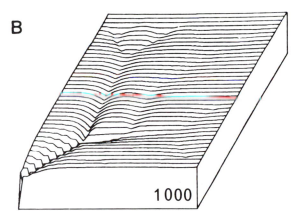

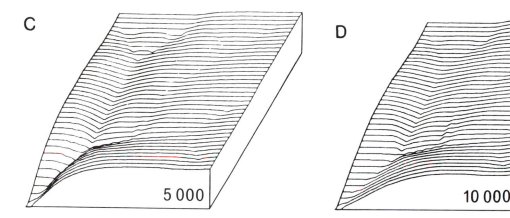

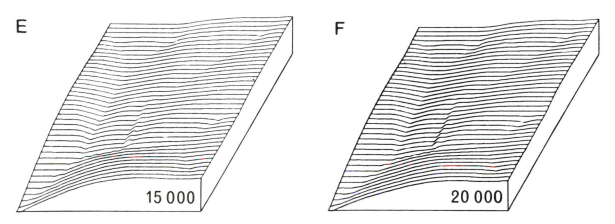

Fɪɢ. 16.1. Armstrong's (1976) computer-generated block drawings of a sequence of landforms. The numerals represent the number of iterations.

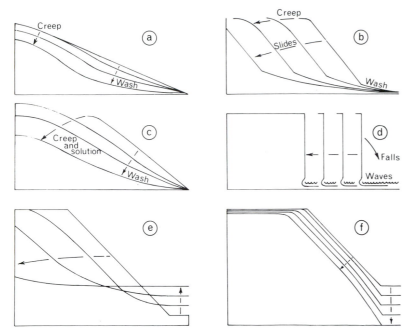

FIG. 16.2. Sequential slope profiles developed from mathematical models of slope change assuming given processes.

diagrams represent the result of the action of the three processes after the stated number of iterations. The main features of the landscape are the overall convexity of the landforms, their smoothness, and their stability. Individual slopes appear to be maintained in an equilibrium form, after first attaining convexity, with little change in shape, but a change in dimensions.

The simulation model of Ahnert (1976*b*) is more complex than that of Armstrong and includes weathering, structural effects, base-level change, waste transport by splash, overland flow and wash, plastic flow, viscous flow, and debris slides. As the capacity of computers grows we may expect to see further components and complexities being included in models.

Specific results of some published simulations which are well related to natural conditions suggest the following conclusions:

(1) processes involving downslope soil transportation tend to cause slope decline, and the slope is transport-limited;

(2) processes involving direct removal of material from the slope tend to cause parallel retreat, and the slope is weathering-limited;

(3) downslope transportation at a rate which

varies only with sinβ produces a smooth slope convexity;

(4) stream incision at a rate in excess of the rate of transportation on the slope produces steep basal slopes which may then fail by landsliding.

The simplest process–response models are those which assume that only one process is operating upon a slope. Creep processes produce an expanding upper convexity on a slope, wash produces an increasing lower concavity, uniform solution produces a parallel downwearing, and shallow landsliding on a transportational midslope produces parallel retreat of that slope unit and a lower concavity where deposition occurs.

Results of modelling selected processes with assumed modes of action are shown in Fig. 16.2. Six theoretical cases are depicted: (a) the gradual reduction of slopes to increasingly gentle inclinations as a result of creep on upper convexities with slope wash on middle and lower slopes; (b) the parallel recession of slopes which undergo uniform rates of weathering and transport across the main slope units, with basal wash slopes; (c) the gradual elimination of steep slope units and the joining of upper convexities and lower concavities; (d) the undercutting of a cliff by waves or streams; (e)

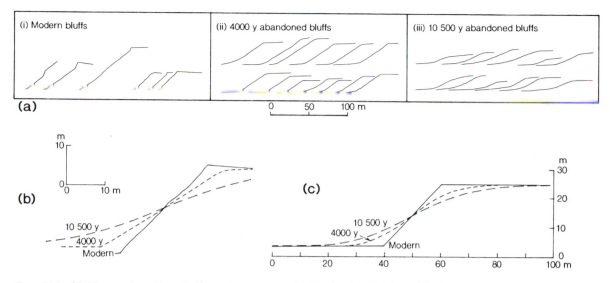

F IG. 16.3. (a) Measured profiles of: (i) modern wave-cut bluffs of Lake Michigan, (ii) 4000-year-old abandoned bluffs of Nipissing Great Lake, and (iii) 10 500-year-old degraded and abandoned bluff of Lake Algonquin. (b) Superimposed typical measured profiles, and (c) profiles modelled using the continuity equation. (After Nash 1980.)

accumulations at the base of a slope as a result of a rise in base-level; and (f) downcutting at the base of a slope as a result of a fall in base-level.

Verification of models

Tests of whether the results of modelling are in conformity with the forms, developed over time, of real hillslopes require a knowledge of the stages through which real hillslopes evolve. Such knowledge is not commonly available, but there are some conditions under which an initial and dated slope form may reasonably be assumed and subsequent forms also dated in a sequence. Fault scarps (Bucknam and Anderson 1979), riverside bluffs (Brunsden and Kesel 1973), bluffs of old lake shorelines (Nash 1980), spoilheaps of coal-mines (Haigh 1978), and marine terraces (Crittenden and Muhs 1986) present opportunities for studies of development of hillslopes in sediments and weak rock; more resistant rock of marine cliffs (Savigear 1952; Hutchinson 1973; Kirkby 1984), and high walls of open-pit mines (Vanarsdale *et al.* 1989) present opportunities to study rock slope evolution into soil-covered slopes. Such studies do not provide insights into the mechanics and rate of operation of individual processes, only of forms.

One of the clearest indications of evolution of slope forms has come from the work of Nash

(1980) who studied wave-cut bluffs of modern Lake Michigan and compared their forms with those of two abandoned bluffs formed by Glacial Lake Algonquin about 10 500 years ago and by Nipissing Lake about 4000 years ago. Both lakes occupied the Lake Michigan basin and cut bluffs in similar morainic materials to those forming the modern bluffs. The results (Fig. 16.3) show clearly a decrease, over time, in the inclination of the bluff mid-slope, with an increase in the extent of the convex crest and of the concave base of the bluff.

The results of modelling, using the continuity equation, show very clearly a close correspondence with the field conditions when an appropriate rate of downslope volumetric transfer of sediment is chosen. The overall conclusion is that the course of hillslope degradation observed in this study is similar to that occurring on other hillslopes underlain by cohesionless soil, covered with a thick growth of vegetation, in a temperate humid climate.

Modelling hillslope development by landsliding processes is more difficult than modelling development by creep, wash, and solution. Landsliding is not only episodic, but highly irregular in its distribution over the landscape. Kirkby (1987) has carried out an initial study of single slope profiles with the recognition of threshold slopes as a foundation.

Threshold hillslope inclinations and landsliding

In a landscape which has a full soil and vegetation cover, and the hillslopes are not being undercut by basal erosion, it is sometimes possible to recognize two characteristic gradients of hillslope units. (1) The steepest inclination is limited to that at which the soil would become unstable and landsliding would occur during adverse conditions. (2) The lower angles have reduced limits of steepness at which saturated landslide debris would come to rest and regain stability against the most adverse conditions.

The explanation for two characteristic inclinations of hillslope units has been set out by Carson (1976) and examined by several workers, including Carson and Petley (1970), Chandler (1982), Van Asch (1983), and Francis (1987). The theory, as set out by Carson, is based on the recognition that soils on most upland hillslopes are essentially noncohesive and the maximum angle at which the hillslope is stable against landsliding is equal to the angle of internal friction of the soil. The reasoning for this conclusion is as follows:

the forces acting parallel to the slope = $W\sin\beta$
the forces acting perpendicular to the slope
   = $W\cos\beta$,

where W is the weight of soil above a slide plane and β is the angle of slope, Fig. 13.8.

For conditions of equilibrium stability the forces promoting sliding

$$= \frac{\text{restraining forces}}{F}$$

thus $\quad W\sin\beta = \dfrac{W\cos\beta\tan\phi'}{F} \quad \left[\text{as } \dfrac{\sin\beta}{\cos\beta} = \tan\beta\right]$

then, $\quad F = \dfrac{\tan\phi'}{\tan\beta}$

when $\quad F = 1, \quad \tan\beta = \tan\phi'$

then, $\quad \beta = \phi'$.

This is true whether the slope is dry or submerged, and it may be of any height, provided that $c' = 0$ and there is no lateral seepage of water.

With soils on slopes, however, with a watertable at the groundsurface and seepage occurring parallel to the slope, at equilibrium ($m = 1$, $F = 1$) when shearing forces = resisting forces, from the infinite slope analysis:

$$\gamma z\sin\beta\cos\beta = (\gamma - \gamma_w)z\cos^2\beta\tan\phi'$$

which simplifies to:

$$\frac{\sin\beta}{\cos\beta} = \left(\frac{\gamma - \gamma_w}{\gamma}\right)\tan\phi'$$

as $\quad \dfrac{\sin\beta}{\cos\beta} = \tan\beta$

then, $\quad \tan\beta = \left(\dfrac{\gamma - \gamma_w}{\gamma}\right)\tan\phi'.$

As the unit weight (γ) of most granular soils is close to $20\,\text{kN/m}^3$ and the unit weight of water (γ_w) is $9.8\,\text{kN/m}^3$:

$$\tan\beta = \left(\frac{20 - 9.8}{20}\right)\tan\phi'$$

thus $\quad \tan\beta = \frac{1}{2}\tan\phi'$ (approximately).

For saturated cohesionless soils, with seepage parallel to the slope and the water-table at the surface, with $\phi' = 30°$, the maximum angle for slope stability is then $16°$, compared with $30°$ for the dry slope above it.

Many well-drained hillslopes, with frictional soils subject to planar, shallow translation landslides, have inclinations in the range $20\text{–}30°$. From back-analyses using the infinite slope equation, this implies values of m in the range $0.25\text{–}1.0$. In some conditions, it is clear that the soilwater seepage is not parallel to the hillslope surface and the porewater pressure ratio, r_u, is then relevant, consequently three characteristic maximum (i.e. limiting or threshold) angles for stability are recognized (Carson 1976):

(1) a frictional threshold slope angle when the soil is a dry rock rubble and $\beta = \phi$;
(2) a semi-frictional threshold slope angle for cohesionless soils when the water-table can rise to the surface and seepage downslope is parallel to it, then $\beta \approx \frac{1}{2}\phi'$;
(3) an artesian condition in which the piezometric surface is above the soil surface. This condition has yet to be analysed in enough situations for a general statement to be established, but β will be less than $\frac{1}{2}\phi'$.

It is implicit in this discussion that pore pressures, at their maximum, will be uniform downslope along a potential failure plane. In situations of shallow soils over bedrock of lower permeability this is probably common. In many areas, however, pore pressures will increase downslope from a ridge and this requires that the threshold angle must decrease downslope. This provides one hypothesis for the development of concave basal slopes.

The concept of threshold slopes applies primarily to straight slope segments between upper convexities and lower concavities and it relates only to those slopes which are subject to landsliding. In areas of rapid uplift and deep incision by streams, slope angles may be steeper than the threshold angles until landslides reduce the slope angle to the threshold angle. In high tectonically active mountains landslides are common. Once the rate of uplift is low enough for landsliding processes to adjust slopes to a threshold angle, this angle may be maintained for a while, but will eventually be reduced below the threshold value by creep, wash, and other processes. Threshold angles are thus temporary features of the landscape.

Across any hillslope the characteristic slope angles will vary because of variations in the strength of the soils and the pore-water conditions. Faces of spurs may be at frictional angles and hollows containing fine-grained colluvium may have high pore-water pressures and thus be close to semi-frictional angles. This may be the reason why many shallow landslides occur inside old landslide scars: for scars are collecting places for water and may be the heads of the drainage network.

Threshold slope angles are extremely varied. Examples in the literature include angles of 6–14° for soils from weathered clays and shales in the semi-frictional condition with the possibility of high water-tables; 19–28° for semi-frictional sandy soils in upland England and Wyoming, 21–42° in the Ardennes, and 33–55° for frictional soils in Colorado and California (e.g. Skempton 1948; Chandler 1970; Prior 1977; Carson 1976; Van Asch 1983).

Landslides will be able to adjust slopes towards threshold conditions only when trigger mechanisms set off movements—as during storms, earthquakes, or wet seasons. Longer-term changes such as deforestation cause a regrading of the slopes of entire uplands. In many parts of the temperate latitudes slopes may have evolved to angles appropriate to the climates of glacial or periglacial environments during glacial times. In areas like

northern New Zealand, where the uplands retained a full forest cover throughout the last glacial, slopes have been subjected to disturbance of equilibria in the period of human settlement and severe mass wasting is a characteristic of most uplands. Such processes are, of course, also promoted by the tectonic instability of the area.

In discussions of threshold slopes it is often assumed that soils are non-cohesive. This is only an approximation as most soils have some cohesion, even if it is only the apparent cohesion provided by plant roots. Measurements of friction angles are also approximate as the collection of large undisturbed soil samples from angular materials is difficult, and few very large shear boxes are in use. With these reservations the concept of threshold slopes does suggest why hillslope angles are related to the strength of rock and soils forming them, and why sets of characteristic slope angles occur in many landscapes.

Landscapes in dynamic equilibrium

It was proposed by Hack (1960) that the Appalachian region of the USA has a landscape with regoliths which are in a state of nearly stable equilibrium between the rates of weathering, rates of regolith removal, and the state of hillslope change so that the hillslope form is preserved but is progressively lowered. Such a concept implies that there is a steady rate of soil formation related to the bedrocks, so that deep regoliths will form on quartzofeldspathic rocks and thin regoliths on ultramafic rocks and on steep hillslope faces (see Fig. 9.4).

The fundamentals of Hack's proposition have been supported by Pavich et al. (1989) who have shown that the average rate of saprolite formation is of the order of 1 m/100 000 y and there is a steady state balance between rates of weathering and erosion. Hillslope forms are probably maintained by slow tectonic uplift and/or slow base-level lowering, because saprolite formation is most rapid under conditions of active groundwater circulation and dissolution of the most reactive minerals. Isovolumetric weathering creates the soil which can be readily removed by erosion and thick regoliths (15–20 m) can be preserved on interfluves for several million years.

Rates of uplift of 10–20 m/My, rates of saprolite formation of 10 m/My and regolith thickness on interfluves of 10–20 m have been preserved for the whole period of Pliocene to Pleistocene time. Most mid-slopes appear to be threshold slopes. It is not clear whether this is a unique situation or whether it applies more widely to uplands in humid environments.

Limitations of models

The power of mathematical models is restricted by two main limitations: the ability of the modeller to select correctly and represent the significant variables in the evolution of a landscape, and the accuracy with which an equation describes a particular process.

Selection of variables involves not only recognition of variables identified in the Universal Soil Loss Equation (Chapter 12) and the various forms of landslides (Chapters 13, 14, 15) but also the magnitude and frequency with which they operate (Chapter 18). Beyond this formidable array of possible processes are questions of scale, persistence, ecological processes of change, and climatic change. Furthermore, many processes are conditioned by feedback and thresholds (Douglas 1988).

Ecological processes involve continuing change of vegetation as trees age; suffer wind-throw, decay, disease, leaf-fall and replacement; grasslands put on growth after warm rains, mature, dry out and die, or are burned; and soil animals create burrows or build mounds. Each of these processes modifies interception, splash, runoff, and therefore erosion.

Climatic change occurs with well-recognized cyclicity with a variety of durations and effects on landforms (Rampino *et al.* 1987). The longer cycles have durations which exceed by far the instrumented record of climatic events, yet any realistic assessment of hillslope change which occurs over several hundred thousand years must involve several glacials and interglacials and even more stadials and interstadials, each of which is likely to have a different magnitude and frequency of major land-forming events. Studies of the Little Ice Age of the seventeenth to nineteenth centuries indicate the effects of relatively minor climatic shifts.

The question of scale has been overlooked in many models, yet scale effects are clearly recognized in the USLE where the importance of slope length, as an influence on surface runoff, is clear and incorporated into the design of many conservation works such as terraces and contour farming (Fig. 12.7, Plates 12.3, 12.14).

Feedback relationships are emphasized in Figs.

12.1, 12.2. Thick soils and dense vegetation cover check runoff and erosion; thin soils and sparse vegetation enhance runoff and erosion. Thresholds are evident in the required size of raindrops to produce splash, in the thickness of overland-flow water-films to cause entrainment, and in the hillslope inclination and pore-water pressure required to initiate landslides.

Persistence of paleosols, remnants of deposits formed under past environments and erosional forms which are no longer developing, create uncertainty about the condition of progressive change which is inherent in virtually all attempts at modelling.

The accuracy with which equations describe the functioning of a process is often unclear. For example, the equation used to describe the rate of soil creep is commonly of the form:

$$C = K\sin\beta.$$

There are good theoretical grounds for this assumption as the resultant of the force of gravity acting at the groundsurface varies in this way, but until far more long-term measurements are made of creep there can be no confidence that this equation is an adequate descriptor. If Young's (1978) measurements, indicating that creep is largely the result of solution and hence is largely an inwards directed process, are correct and universal, then the direct slope-angle function may be incorrect.

Improved accuracy in modelling is thus very dependent upon long-term and detailed field measurement of processes, and correct representation of process mechanisms in descriptive equations. There has been considerable progress in mathematical modelling of processes since 1960, especially of short-term processes (see, for example, papers in Anderson 1988), yet the complexity of natural systems cannot be represented in any models currently in use.

The overriding problem with all models of landscape evolution is the ability to test them against natural conditions. Landforms, even in small drainage basins, evolve over thousands, or even millions of years, in which the intensity and type of dominant processes may change in a direction which cannot be known. None the less, modelling is a method of testing hypotheses and is a much more objective way of envisaging the results of processes than conceptual models of the kind developed by W. M. Davis and those who followed his methods.

PLATE 16.1. The outer escarpment (background) and inner escarpment of the Fish River Canyon, Namibia. The outer escarpment has retreated some 10 km from the line of the river and its slopes are in strength equilibrium. The upper parts of the inner escarpment are also in strength equilibrium but the lower parts are undercut in some sections.

Evolution of rock slopes

Some of the most prominent rock slopes on Earth are the great escarpments which developed as Gondwanaland split into the modern southern continents. The cliffs forming the margins of the rifts are thought to have migrated inland by nearly parallel retreat occurring from the Late Cretaceous through Cenozoic time. The rocks in which these escarpments developed are mostly Precambrian to Paleozoic in age and resistant to erosion. Judging by the width of coastal zones between the base of the escarpments and the modern shoreline (70–200 km in many areas) rates of retreat have been about 1–2 km/My. Discussions of these great escarpments will be found in King (1962); in papers by Moon and Selby (1983); Ollier (1985); Ollier and Powar (1985); Ollier and Marker (1985); and Pain (1985).

Other great escarpments have formed by retreat of large slopes from incised canyons of such great rivers as the Fish River of southern Namibia and the Colorado River of southwestern USA (Plate 16.1). Schmidt (1989) has shown that escarpments formed on single units of thick, horizontally bedded sedimentary rocks have retreated from centres of uplift at rates of 0.5 to 6.7 km/My throughout the Cenozoic. The rate is controlled by the product of the thickness of the cap rock and the resistance of that rock (Table 16.1).

Large rock faces of many mountains are both more complex in their structure and the controls on their form than most of the great escarpments, as well as being far less accessible to most research workers.

Detailed study of rock slopes by geomorphologists has been limited, but engineers and geologists working in quarries, open-pit mines, cuttings, and tunnel portals have made considerable advances in understanding the properties of rock which control

TABLE 16.1. *Rates of scarp retreat and caprock thickness and resistance*

Cuesta scarp and caprock	Rate of scarp retreat (km/My)	Rank of retreat	Thickness of caprock (m)	Relative resistance	Thickness × resistance
Chocolate Cliffs Shinarump Conglomerate	6.7	1	20	7	140
Black Mesa Dakota Sandstone	4.5	2.5	25	8	200
Black Mesa Salt Wash Sandstone	4.5	2.5	10	5	50
Mesa Verde Mesaverde Sandstone	3.2	4	200	1.5	300
Black Mesa Mesaverde Sandstone	3	5.5	180	1.5	270
Cedar Mesa Cedar Mesa Sandstone	3	5.5	160	4	640
Pink Cliffs Wasatch Formation	2	7	300	3	900
Red House Cliffs Kayenta-Wingate Sandstone	1	8	150	6	900
Grand Canyon Kaibab Limestone	0.5	9	100	9	900

Source: Schmidt (1989).

the development of natural slopes. There is still much to be achieved by combining available knowledge and methods in attempts at developing a coherent synthesis.

Deterministic modelling of rock masses has been avoided because of the complexity and variability in the characteristics of joints within rock bodies and their effects on weathering and water flow within the rock. Where the rock mass is closely jointed, however, and therefore behaves as a granular material, and where the rock is being weathered so that grains and small fragments fall from free faces, modelling has been successfully carried out.

Deterministic models with talus formation

Cliff faces formed on rock masses retreat under the action of weathering processes and removal of debris. Uniform weathering and removal of thick layers of rock involves parallel retreat of the cliff face. The fallen debris accumulates at the base of the slope as a talus. It has been shown by Fisher (1866) and Lehmann (1933, 1934) that the rock

slope buried by the talus will theoretically develop a convex shape (Fig. 16.4). The actual curvature of the convexity will depend upon the ratio of the volume of rock removed from the cliff to the volume which accumulates at the cliff base. Where all falling debris accumulates, the volume of debris will exceed the volume of solid rock that is removed because of the higher void space in talus. Solution and removal by streams may, however, reduce the volume of the resulting talus to less than the volume of the intact rock which is removed.

Few exposures through talus into bedrock exist, but most reported profiles do not appear to support the theory that a convex rock-core slope will develop beneath talus below a retreating cliff, and the value of this model is still uncertain.

A special case of the Fisher–Lehmann theory has been stated by Bakker and Le Heux (1952). In this model all the talus is steadily moved by rolling and sliding and the basal rock slope evolves at the angle of rest of the talus material. Thus rock slopes with thin veneers of debris should be formed. It has been noted that such slopes are found

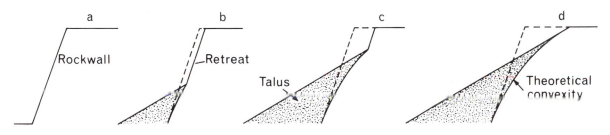

FIG. 16.4. A theoretical cliff, basal convex rock slope, and talus sequences in which it is assumed that there is no removal of talus from the base.

PLATE 16.2. Richter denudation slopes, with a thin veneer of talus, extending headwards into sandstone cliffs, Transantarctic Mountains.

quite widely in alpine and polar areas, and in the extreme environment of Antarctica they appear to be relatively common (Selby 1974*a*; Chardon, 1976) (Plate 16.2). The analysis of Bakker and Le Heux appears to be generally valid even though it assumes (falsely) that talus slopes are usually at the angle of rest of their materials (see Chapter 15).

We can visualize the development of this type of slope—which is known as a Richter slope—by imagining particles falling from a cliff on to the top of a talus. If the newly fallen material just covers a little of the base of the cliff the next fall will be over the new talus and hence the base of the cliff will now be higher, so the cliff will recede by a series

PLATE 16.3. Straight and even slopes cut across granite are interpreted as the final stages of Richter-slope formation, Koettlitz Valley, Antarctica.

of minute steps at the angle of the talus. These steps may then be removed by weathering or the abrasion of sliding talus—or they may remain. In either case an essentially straight rock slope (i.e. a rectilinear slope) will be formed below the cliff. Eventually the free face should be eliminated and a smooth rock slope of uniform angle will be produced (Plates 16.2, 16.3). Suggested successive stages in this development are shown in Fig. 16.5. Once the stage has been reached of either a uniform bare rock slope, or a uniform talus-covered slope, further evolution will depend upon the nature of the processes operating. Where uniform weathering occurs over a rock slope and the debris is blown away then it is likely that the rock slope will continue to get smaller but retain the same angle (Fig. 16.5); where wash processes occur the slope is likely to decline in angle as progressive weathering reduces the size of particles and basal regoliths thicken (Fig. 16.6).

Deterministic models are capable of using the fundamental conditions recognized in stability equations such as the Culmann wedge analysis (Fig. 13.22), and analysis for critical height for stability of a cliff face (Fig. 15.8) (see Kirkby 1987), however, in their simple form they assume uniformity of material properties through the rock mass and are of questionable validity for rock slopes.

Hillslopes controlled by their rock-mass strength

It has been shown that cliff gradients and forms may be controlled by the mass strength of their rocks (Selby 1980, 1982*a*,*b*,*c*; Moon and Selby 1983; Moon 1984*b*). Confirmation of the relationship is displayed in Fig. 6.13 in which, for all rock slope units which have gradients in conformity with the rock-mass strength, the plotted data points fall within the strength-equilibrium envelope. The displayed relationship has been established from

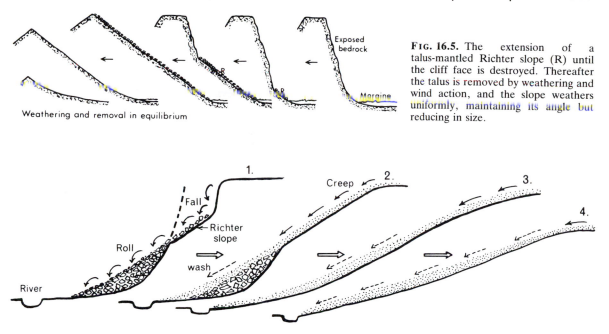

FIG. 16.5. The extension of a talus-mantled Richter slope (R) until the cliff face is destroyed. Thereafter the talus is removed by weathering and wind action, and the slope weathers uniformly, maintaining its angle but reducing in size.

FIG. 16.6. A model of slope evolution by the development of a talus and Richter slope. Weathering produces a veneer of fine-grained debris and wash processes reduce the angle of the slope.

study of more than 300 slope units with a variety of lithologies and in several climatic environments.

All available data indicate that strength-equilibrium hillslopes are common on escarpments and in mountains where slope undercutting, talus and soil cover, inherited slope forms, and structural controls are not dominant. For hillslopes to reach a state of strength equilibrium they must be able to retreat sufficiently for the effects of any extraneous control, such as faulting and under-cutting by rivers and glaciers, to be eliminated. Such retreat requires considerable periods of time for known rock-slope retreat rates are of the order of 20–6000 m/My, unless cliffs are being undercut by the sea when rates are much higher. Retreats of 10 m may be at the lower end of the range of retreat needed to establish strength equilibrium, so the time needed is probably in the range of 1500 y to 0.5 My with the longer period being required by many slopes on strong rock masses.

Sites where strength-equilibrium slopes have been identified include mountain slopes in Antarctica (Selby 1980), slopes on cliffs formed in relatively weak sedimentary rocks in humid areas of northern New Zealand (Selby 1980), inselbergs in the Namib desert (Selby 1982c), escarpments

around the margins of southern Africa (Moon and Selby 1983), slopes in the Cape Mountains (Moon 1984b, 1986), and limestone scarps in the Napier Range of northern Australia (Allison and Goudie 1990). With the exception of the New Zealand examples, all of the rocks studied, so far, have been well-indurated.

The recognition of strength-equilibrium slopes makes it possible to examine critically some established ideas about rock-slope development—especially to decide whether slopes decline in angle or retreat parallel to themselves as they evolve. The recognition that many rock slopes forming scarps and faces of inselbergs are in equilibrium implies that such slopes will maintain a constant angle as long as rock strength is constant; these slopes will retreat parallel to themselves. If strength increases or decreases into the outcrop then the hillslope angle will increase or decrease in conformity as the slope retreats. This theory has, so far, been tested on equilibrium slopes on sandstones, tillites, shales, dolomite, marble, dolerite, gneiss, schist, basalt, pegmatite, and some granites; there was much variety in bedding and joint patterns.

Application of the mass-strength classification

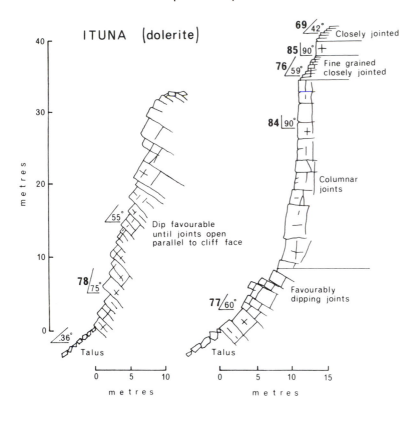

FIG. 16.7. Profiles of slopes in dolerite showing joint spacing and inclination and their effect upon slope angles, Ituna Valley, Antarctica.

shows that major changes in a slope profile may result from variability in just one of the parameters. In Fig. 16.7 it can be seen that, in a dolerite of uniform intact strength, changes in the dip of joint blocks from about 35° into the slope to vertical columnar jointing can change slope angles from 60° to 90°, and a change from columnar to closely spaced jointing can cause slope angles to decline from 90° to 59–42°. Similar changes are seen in the dolerites shown in Plate 16.4. The changes in jointing dip are minor but strongly influence slope angle, as does the incidence of closely spaced jointing at the large step in the profile.

Deterministic models of rock-slope development which incorporate variables such as the rate of downcutting at the base of a slope, rock-mass strength, and a rate of slope retreat which is proportional to the weathering rate, could be realistic. Areas with many layers of rock, as in the Grand Canyon in Arizona, would be ideal for this purpose. A study by Aronsson and Linde (1982) has used a rather similar approach, but their model did not simulate the conditions of weathering and

rock resistance that occur in nature. None the less they have illustrated that the method could be applied. A study by Schmidt (1987) demonstrates the lack of correspondence between models and natural forms in several pioneer studies.

Rock slopes which are not in strength equilibrium

Strength-equilibrium slopes exist only where the rock-mass properties are controlling factors of hillslope form. There are also other factors which may act as controls: undercutting, structure, internal stress, inheritance of form, solution, and talus or soil cover.

Undercutting of a hillslope is commonly by agencies such as rivers, glaciers, and waves. It results in maintenance of steep slopes or cliffs which may be at an inclination which, for very strong rock masses is in strength equilibrium but for weaker rock masses is too steep for equilibrium. In the latter case the upper cliff face will adjust towards an equilibrium inclination by rockfall and other processes but will not attain overall equilibrium as long as undercutting continues.

PLATE 16.4. Dolerite cliffs showing the effect upon slope profiles of subtle changes in joint orientation and spacing, Britannia Range, Antarctica.

PLATE 16.5. An amphitheatre head of a valley created by sapping and undercutting of Navajo Sandstone in the Colorado Plateau, Arizona.

A particular example of undercutting is spring sapping which works headwards towards the source of water (Plate 16.5). This situation has been recognized in many areas of soluble rocks, such as limestones, and in siliceous rocks such as the sedimentary rocks, with low angles of dip, of the Colorado Plateau. The rocks of the Plateau are sandstones and shales with interbeds of generally more soluble materials. The sandstones are quartzose and have relatively high porosity and permeability for such rocks. The upper surface of the Plateau has incised into it minor channels which carry little water, but the flanks of the Plateau are fretted by canyons with steep walls, amphitheatre-shaped headwalls, hanging tributary channels, long main valleys with short tributaries, irregular angles of channel junction, and valley widths which remain constant in a down-valley direction (Laity and Malin 1985; Howard *et al.* 1988). The features described are clearly not those usually associated with river erosion.

The features which are recognizably those derived from sapping are illustrated in Fig. 16.8 in which (a), (b), and (c) illustrate conditions where water can move through the porous Navajo Sandstone and concentrate above the less permeable Kayenta Formation. Sapping progresses towards the water source in much the same way that subsurface pipes in soils extend headwards and draw in progressively larger water volumes (Fig. 12.14). Progression is commonly enhanced by dominant joints creating a rectangular and highly asymmetrical valley pattern. In contrast, in Fig. 16.8 (d), where drainage arises on less permeable

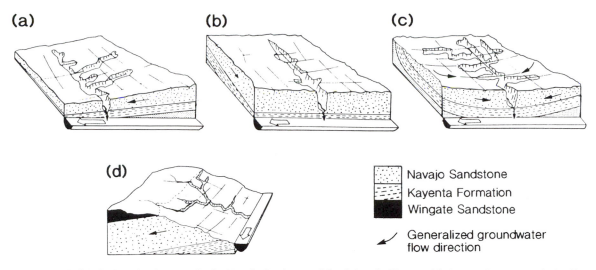

F IG. 16.8. Model of valley development in the Navajo Sandstone of the Colorado Plateau. (a) A strongly asymmetrical valley pattern has developed where headwards sapping can exploit an orthogonal joint set. The black arrow shows the direction of groundwater flow. (b) Groundwater flow is towards the main river and along a dominant joint. The valley pattern is symmetrical. (c) Flow follows the synclinal structure, and valley patterns are symmetrical. (c) Flow is down-dip away from the hillslope. Valley forms are therefore not influenced by sapping. (After Laity and Malin 1985.)

cover beds and the dip of the lower rock units carries groundwater away from the incising river, drainage patterns are more nearly dendritic and symmetrical, and valley heads are less incised and narrower than the channels further down-valley.

Sapping, then, results in undercutting headwalls by preferential wetting of rock along zones of permeable and more fractured rock above less permeable beds at a water-table (Nicholas and Dixon 1986). Head and side walls of valleys are kept steep where undercutting is active, but elsewhere evolve under the influence of a variety of processes including stress-release fracture, weathering by granular disintegration, exfoliation, rockfall, and slab failure (Oberlander 1977).

Structural controls on hillslope development are primarily expressed through the presence or absence of buttresses which prevent planar sliding. In the absence of buttresses sliding occurs of rock units along joints which dip out of the rock face at angles which are too steep for stability (Plates 16.6, 16.7).

The presence of buttresses in the form of rock units supporting the base of a rock body is seldom a permanent condition: the buttress may fail by buckling or be removed by erosion, or excavation machinery. The upper rock mass is then left without support but can only slide when either: (1)

there can be sufficient dilation along the critical joint for asperities along that joint to disengage; or (2) weathering reduces the strength of the asperities so that they fail by shearing through their bases; such shearing is most probable through small second-order asperities and less common through larger first-order asperities (Fig. 6.4). The critical angle for failure along a joint is ϕ_j (see Table 6.1). It is evident from the data in this table that there is a large range of inclinations over which rock slopes can be near critical states for stability and the nature of any infill in a joint may be a dominant control.

The example of slabs of sedimentary rock near the critical angle for stability in Plate 16.7 represents a very obvious condition. Many hillslopes are formed in rock masses in which dominant discontinuities do not follow bedding planes, faults, and other sharp discontinuities, but develop from linking of irregular, stepped, and segmented fractures. The result may be a slope failure along a zone of shattered and brecciated rock or a zone of plastic deformation. Such conditions are particularly common in closely bedded and laminated rocks, such as schists (Plate 16.8) in which the zone of weakness may be formed by aligned mineral grains such as those of micas, planes of separation between sedimentary bedforms, by lenses of

PLATE 16.6. Buttressed units of sedimentary rock forming triangular facets along the flank of an anticline in the Carr Boyd Range, Kimberley Plateau region, Australia. Although the rocks have dips of 50° or more they cannot develop planar slides.

PLATE 16.7. Small blocks on a dipping bedding plane at the critical angle for sliding.

weaker or more soluble materials, and by zones of preferential water movement.

In any rock mass, joints critical for stability have frictional properties that are related to the basic properties of the rock and the joint roughness, but these are not necessarily constant through a rock mass, especially in a rock mass which has very high boundary walls, such as large escarpments or deep valleys. Many escarpments have concave profiles especially where they are not being actively undercut. Such basal concavities may develop as a result of weathering, soil formation, and the action of

creep, rilling, and wash. Some such concavities may be controlled by properties of the joints in the rock.

Concave profiles of high cliffs may arise from the effect of overburden stresses acting across critical joints. This situation has been analysed by Selby (1982a) who has shown that the normal stress, due to the overburden, on a joint in a rock mass forming a high cliff, is likely to be insufficient to crush and shear the asperities along the joint where that joint is high in the cliff and therefore under a low overburden stress. At points progressively

PLATE 16.8. The margins and failure zone of a large landslide in schists. This failure is no longer active but could become so if the toe were undercut. Queenstown, New Zealand.

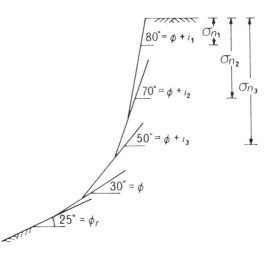

FIG. 16.9. A theoretical concave profile of a rock slope produced by frictional strength along joints. Slope angle decreases as normal stresses at greater depths shear asperities. (After Selby 1982*a*.)

lower in the cliff the asperities are more likely to be crushed by the overburden load. This results in the possibility that for a cliff, with random joint orientations, closed joints dipping out of the cliff at up to 80° may still be closed and resist planar sliding. In lower sections of the cliff the stable angle along the joint will become lower until at the base of a high cliff the stable joint dip will be at $\phi_{jr}°$ for the rock, or about 25°. As the cliff retreats, the unfavourably dipping joints will therefore control the cliff form which will be concave (Fig. 16.9). Such conditions have not been studied in the field, and for them to operate dilation must occur. The possibility of their existence is, however, a further warning that rock-slope stability may be a very difficult condition to assess. It is not surprising that unanticipated slope failures occur even from the

most carefully investigated sites in high mountains.

Residual stress in rock masses has been discussed already (Figs. 4.13, 4.14, 8.5, 15.4), it is necessary here only to reassert the importance of this condition in many rock bodies and especially in those which are massive, with few joints along which internal stresses can be isolated or dissipated.

Inheritance of form is a condition in which the controls of the form of a rock body have developed under the influence of processes which no longer operate. A very obvious example is that of glaciated valleys in which glacier retreat is so recent that rock slopes have not yet developed profiles adjusted to modern environments. In some cases, therefore, they are oversteepened with respect to the mass strength of their rocks.

A common condition in areas which once had deep weathering profiles formed above the rock, but under more recent environmental conditions the regoliths have been stripped, is the exposure of the weathering front. The form of the weathering front may be highly irregular and, on granitic rocks particularly, may leave tors and bornhardts standing above surrounding lowlands (see Chapter 9).

The tors and bornhardts will gradually develop joints through weathering and stress-release, and then strength-equilibrium forms. Similar development towards equilibrium forms occurs in bornhardts with domes created at original crystallization from igneous melts (Selby 1982*b*).

Solution processes may act across whole ground-surfaces or preferentially where water can collect, as at the bases of slopes. It may result, therefore, in a range of karstic features in both carbonate and siliceous rocks (Plate 9.7). Many features may be out of strength equilibrium but some escarpments are known to develop strength equilibrium forms (Allison and Goudie 1990).

Talus cover of bedrock is known to control the form of Richter denudation slopes, which may have inclinations at angles far lower than those that their mass strength could maintain (Plate 16.2). If basal convexities have developed in the way which is theoretically possible then removal of the talus would reveal an inherited form.

Under certain conditions recognized in deserts (Gerson and Grossman 1987; Schmidt 1989), remnants of talus become separated by rill erosion from the rock faces against which they formed. The remnants may preserve evidence of more humid environments and provide indications of rates of cliff retreat (Fig. 16.10).

Conclusions

(1) Recognition of the possible range of controls on rock-slope form is a necessary prelude to field analysis. It is then possible to map the areas under a particular influence (Fig. 16.11). Assessments of slope stability and geomorphological development can then be focused on specific sites.

(2) Because of the great intact strength of most rocks, compared with the limited power of most erosional processes, when these are considered at an instant of time, rock slopes usually develop forms which are controlled by the resistance of the rock.

(3) Cliffs, escarpments, and large rock faces which are in strength equilibrium are widespread. This suggests that, in the absence of other controls, rock-mass strength equilibrium is a common condition for rock slopes.

(4) As a result of the strength-equilibrium condition, many rock slopes will retreat to such a profile form that equilibrium is preserved; parallel retreat will occur in rock masses which have uniform mass strength; slope decline will occur in rock masses which have lower mass strength away from the free face; slope steepening will occur

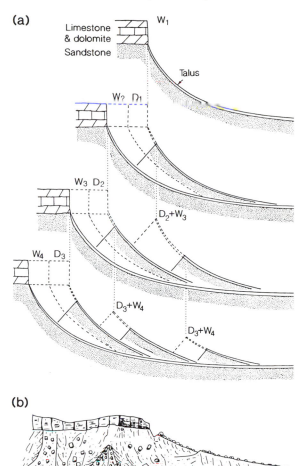

(a)

(b)

FIG. 16.10. (a) A generalized model of the effects of major environmental changes on escarpment retreat in which remnant wedges of sandstone, covered with a veneer of limestone talus, are left as residuals on the toe slope. Talus development occurs during a semiarid (relatively wet period, W), and scarp recession in following arid and wet phases as rills cut into the sandstone (b). (a, after Gerson and Grossman 1987; b, based on Schmidt 1989.)

where mass strength increases away from the face.

(5) Slopes formed on buttressed rock masses may be stable at any angle until cross joints dilate and then slopes develop towards inclinations which lie within a strength-equilibrium envelope.

(6) The stability of an unbuttressed rock slope, with joints dipping steeply out of the slope, is controlled by the available shear strength along the weakest continuous joint. This strength is a func-

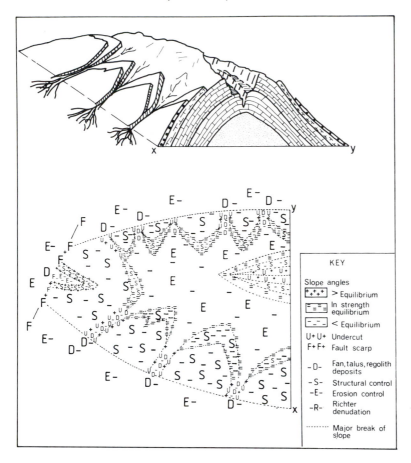

Fig. 16.11. A schematic landscape of the kind shown in Plate 16.6 and a map showing the controls on the form of its slope units. (After Selby 1987*a*.)

tion not only of frictional resistance to sliding of the rock material of the joint walls, but also of the normal stress acting across the joint, the uniaxial compressive strength of joint-wall rock and the roughness of the joint walls.

(7) If weathering, and dilation with movement, along critical joints increases during the development of rock slopes, then profile inclinations will decrease and may form concave profiles on high cliffs as normal stresses, increasing towards the base of the cliff, crush asperities of the critical joints.

(8) Modelling of the development of whole rock faces has not progressed as far as has modelling of soil slopes. This is mainly because of the geological complexities which have to be incorporated into realistic models, and also because many landscapes formed in rock have very long histories and retain elements of many controls and former environmental conditions.

(9) Modelling of soil slopes has demonstrated that, unless they are undercut, they are progressively lowered and retreat towards a concavo-convex profile.

(10) There are conditions, however, in which hillslope form is maintained over very long periods, by uniform downwearing, in a state of dynamic equilibrium.

(11) Threshold slopes belong to one category of forms of dynamic equilibrium which may be relatively stable for long periods but will eventually be replaced by another equilibrium state. Various names such as meta-stable and quasi-equilibrium have been attached to conditions in which there is a transition from one state of general stability to another state of general stability.

17 Landslide Hazards: Avoidance and Protection

For the seven-year period 1973–9, UNESCO records show that the average annual death rate from natural disasters was about 225 000 people; some of these deaths were due to landslides. Landslides and other forms of mass movement cause annual costs in excess of $1 billion (1 billion $= 10^9$) in USA, $1.1 billion in Italy, and $1.5 billion in Japan (Schuster and Krizek 1978; Arnould and Frey 1978), and the figures are rising steadily as the Earth's human population grows and people occupy more hazardous areas. It has become increasingly important, therefore, to plan land use so that hazards are avoided and so that construction projects can be designed to limit slope failures.

This brief chapter cannot be a detailed account of this topic; it is intended only as an introduction to the general principles and as an indication of where further information may be found. Useful elaborations of the subject will be found in Hansen (1984), Varnes (1984), Walker et al. (1987), and Cooke and Doornkamp (1990). The paper by Keefer et al. (1987) gives an account of a landslide warning system which is being developed for the San Francisco Bay region. A number of the papers in Culshaw et al. (1987) discuss various aspects of hazards and planning.

Hazard and risk

Natural phenomena are not hazards of themselves; a hazard only occurs where people are threatened with injury, damage, or loss. There is a need for clarity in the use of such words as hazard, risk, and vulnerability and the UNESCO terminology (Varnes 1984) is adopted here.

Natural hazard (H) means the probability of occurrence, within a specified period of time and within a given area, of a potentially damaging phenomenon.

Vulnerability (V) means the degree of loss to a given element or set of elements at risk (see below) resulting from the occurrence of a natural phenomenon of a given magnitude. It is expressed on a scale from 0 (no damage) to 1 (total loss).

Specific Risk (R_s) means the expected degree of loss due to a particular natural phenomenon. It may be expressed by the product of H times V.

Elements at Risk (E) means the population, properties, economic activities, including public services, etc. at risk in a given area.

Total Risk (R_t) means the expected number of lives lost, persons injured, damage to property, or disruption of economic activity due to a particular natural phenomenon, and is therefore the product of specific risk (R_s) and elements at risk (E). Thus:

$$R_t = (E)(R_s) = (E)(H \times V).$$

Regional investigations

Implicit in these definitions is a process of investigation and evaluation which will permit a statement of total risk. The necessary phases in a landslide-hazard investigation may be summarized.

1. Identify the localities which are of concern.
2. Define the extent of the area which requires study.
3. Identify areas of active and past landsliding.
4. Identify slope units which are potentially unstable.
5. Assess what buildings, property, facilities, and activities are at risk.
6. Determine the threat to lives.
7. Evaluate the potential economic, social, and environmental costs of landsliding.

8. Produce a landslide-hazard zoning map which shows areas of potential failures and the property, facilities, and activities which are vulnerable. The map will offer grades of hazard and grades of vulnerability.
9. Define the needs for further investigations especially at specific sites.
10. Indicate the need for avoidance, prevention, and correction of landslides.
11. Incorporate hazard zoning, and areas requiring specific investigation, in planning and legislation.

It will be seen from this list that items 3, 4, 8, and 9 are the province of earth scientists, and 10 is the province of engineers.

Areas of active and past landsliding are the first indicators of the potential for further landsliding because the geological, geomorphological, and hydrological conditions which have led to past and present failures are likely to be the conditions under which future landsliding will occur. The absence of failures in the past does not mean that failures will not occur in future: in Scandinavia, for example, some quick-clay lateral spreads have occurred on sites which have no previous history of slope failure. Study of past records, air-photos, and other available evidence can give an indication of the frequency of failures. Frequency indications may also be estimated from rainfall records where the frequency of landsliding is related to rainfalls of known intensity, duration, and return period. In the North Island of New Zealand, for example, it has been concluded that grassland slopes on greywacke ranges are subject to landsliding as a result of rainstorms with a recurrence interval of thirty years or more; for a similar degree of landsliding of soils with a forest cover much more intensive storms with a return period of a hundred years are required (Selby 1976a). A similar condition is recognized by Eyles *et al.* (1978).

For mapping purposes it is necessary to recognize and classify areas according to the severity of the hazard. Crozier (1984) has proposed one such classification with a six-point scale.

Class I Slopes with active landslides. Movement may be continuous or seasonal.
Class II Slopes frequently subject to new or renewed landslide activity. Triggering of landslides results from events with recurrence intervals of up to five years.
Class III Slopes infrequently subject to new or

renewed landslide activity. Recurrence intervals greater than five years.
Class IV Slopes with evidence of previous landslide activity but which have not undergone movement in the preceding hundred years.
Subclass IVa: erosional forms still evident.
Subclass IVb: erosional forms no longer evident—previous activity indicated by deposits.
Class V Slopes showing no evidence of landslide activity but which are considered likely to develop landslides in the future. Landslide potential indicated by comparison with other slopes, detailed stability analyses, changes in land use, drainage, or undercutting of the slope.
Class VI Slopes show no evidence of previous activity and there are no other indications of potential failures.

Classification of landslide processes is difficult if carried out from air-photos and from field-study of ancient scars and deposits. In such cases it may be useful to use simple indices such as Depth/Length (D/L) ratios and either use them to indicate probable processes and probable most susceptible areas (Table 17.1) or to map them as classification indices. It will be seen from the table that there is a wide range of values when data from many sources are put together. Lower limiting angles of slope are likely to be more narrowly defined for a single area or lithology.

Geology is traditionally mapped as an aid to understanding geological history; rocks are therefore mapped primarily on age and formation and only secondarily on lithology. A more useful map for geotechnical purposes is one based on lithology

TABLE 17.1. *Typical D/L ratios and limiting (threshold) slope inclinations for landslides in soils*

Landslide type	D/L (%)	Lower limit (°)
Debris slides, avalanches	5–10	22–38
Slumps	15–30	8–16
Flows	0.5–3.0	3–20

Sources: Skempton (1953a), Crozier (1973), Selby (1967a), Sidle *et al.* (1985).

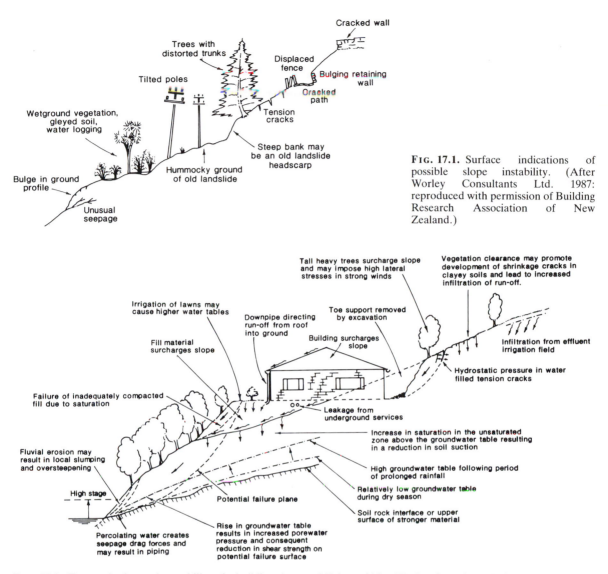

Fig. 17.1. Surface indications of possible slope instability. (After Worley Consultants Ltd. 1987: reproduced with permission of Building Research Association of New Zealand.)

Fig. 17.2. Factors detrimental to stability of a building site on a hillslope. (After Worley Consultants Ltd. 1987: reproduced with permission of Building Research Association of New Zealand.)

with particular attention being paid to late Tertiary and Quaternary deposits, for these are most commonly involved in slope failures.

For studies of rock bodies the Rock Mass Strength rating, with a classification of whether the rock mass is below, within, or above the equilibrium-strength envelope, is a first step in consideration of stability. The second step is detailed consideration of any rock units with joints unfavourable for stability. This approach has been adopted by Barisone and Bottino (1990) for a study in the European Alps. The unfavourable dips can be classified on a scale which takes into account joint dip, joint roughness, and clay interbeds as well as the possibility of sliding, wedge failure, and toppling. Further work is needed on these topics.

Soils can be assessed on their vane shear strength and Atterberg limits. This is a useful method when it is applied to clay soils. A classification has been

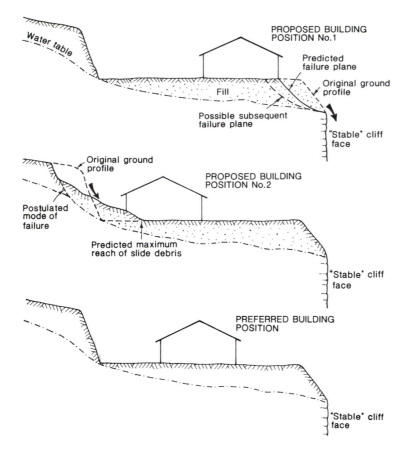

FIG. 17.3. Avoidance solution to potential instability hazards on a hillslope. (After Worley Consultants Ltd. 1987: reproduced with permission of Building Research Association of New Zealand.)

used by Stevenson (1977) which is worthy of examination for wider use. Its scoring method is as follows:

P. Clay factor. Use range of available values of plasticity index for the soil units involved in sliding:

PI in the lower third of the range	1
PI in the middle third of the range	2
PI in the upper third of the range	3

W. Water factor. Highest position annually of piezometric surface relative to failure plane:

Below the plane	1
Between plane and mid-depth of slide	2
Above mid-depth of slide	3

S. Slope angle. Use range of angles appropriate to local geology:

Within lower third of the range	1
Within mid-third of the range	2
Within upper third of the range	3

C. Slope complexity.

Simple slope	1
Old failure, now partly obliterated by erosion	2
New failure, stable but not eroded	3

U. Land use.

Woodland	1
Cleared of trees or built on with special precautions	1.25
Built on without special precautions	1.5

The factors are taken in pairs: clay behaviour is modified by water; and slope angle and complexity are taken together. The final risk, R, is defined as the product of three terms:

$$R = (P + 2W)(S + 2C)(U).$$

Experience indicates that a score above 50 should be treated as a warning of possible instability, and values over 60 are associated with failures. In the equation, $(P + SW)$ $(S + 2C)$ is analagous to hazard and U is a form of measure of vulnerability.

Hydrological factors are included in the scheme described above but heights of water-tables are often not known. Hydrological mapping may, however, indicate relevant conditions. High drainage densities are a sign of impervious strata, highly erodible soil, high rainfall, limited vegetation

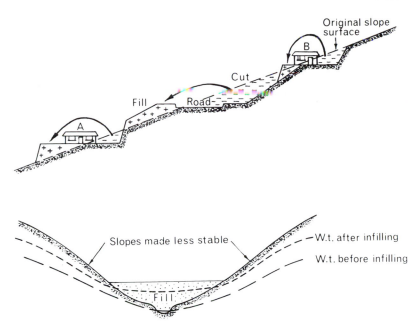

Fig. 17.4. Cutting and filling sites for first a road and then houses on a slope. House site A was formed by cutting into old, poorly designed fill, house site B was formed by cutting into a slope already steepened by cutting for the road. (After Taylor *et al.* 1977.)

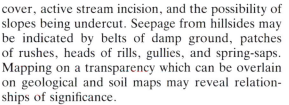

Fig. 17.5. Deposition of fill in a depression causes a rise in the water-table at the base of surrounding slopes.

cover, active stream incision, and the possibility of slopes being undercut. Seepage from hillsides may be indicated by belts of damp ground, patches of rushes, heads of rills, gullies, and spring-saps. Mapping on a transparency which can be overlain on geological and soil maps may reveal relationships of significance.

Geomorphological maps record not only the location of landslides and their classification, but also characteristics of slopes which are related to failure; such as tension cracks, steep slope units, drainage lines, and indicators of hydrological conditions.

Data assembly and mapping

In theory it is possible to assemble data of the kind discussed above and record it for each terrain unit, or grid square, or cell on a map. This has been done for certain sites as a trial of the method (Bernknopf *et al.* 1986); the Cincinnati, Ohio, area was divided into 14 255 square cells of land, 100 m on each side. For each cell, the probability of a landslide occurring was estimated on the basis of statistical relationships among attributes of the cell, such as mean and maximum ground slope, strength of the soil, and presence of potential triggering activities such as new road construction. The distribution of residential property values in each cell was obtained from the 1980 census and relating this to the probability of slope failure

yielded an annual expectation of economic loss. The results were mapped and used as a basis for planning mitigation measures on a cost-effectiveness basis.

For many areas the cell method is far too costly, but by selecting only those areas which have a high degree of risk and mapping them in detail it is possible to concentrate measures on sites which are selected according to criteria which suit the wealth or values of the local community. Examples of completed maps and proforma data collection sheets are given in Varnes (1984) and Hansen (1984). Examples of geomorphological mapping specifically for recognition of hazards encountered during road building in remote and steep mountains are given in Fookes *et al.* (1985). Ives and Bovis (1978) have produced maps of snow-avalanche hazards for a mountain area in which development may occur. Carrara *et al.* (1991) have incorporated relevant geological and geomorphological factors into a probability model and mapped landslide susceptibility, using Geographical Information System technology (GIS), in a drainage basin in Central Italy.

Site investigations

The objective of site investigations in an area with a potential risk of slope failure is the acquisition of

PLATE 17.1. In (a) can be seen the deeply gullied slide-flow complex landscapes formed on mudrocks of the Mangatu Formation (see also Plate 13.16, p. 295). Note the deep infilling of the valley floor with debris. In (b) the established forest can be seen to have stabilized the slopes and fans (New Zealand Forest Service, Crown Copyright reserved).

information, on site conditions, which will permit analyses of all influences upon hillslope stability and the formulation of recommendations for hazard avoidance and remedial measures to reduce risk to acceptable levels. The examples given below assume that the project is a medium-scale construction.

Investigations should always involve:

(1) consideration of influences on the site of interest from the surrounding area, e.g. the possibility of rockfall from a cliff which is above the boundary of the site;

(2) detailed site investigation before construction;

(3) recognition of factors which might reasonably be expected to influence the site after con-

struction, e.g. planting or removal of trees, with a high demand for water, from sites with clay soils which will dry out or swell, and hence shrink, crack, or expand with moisture change;

(4) continue site investigations during construction.

Four phases of investigation are commonly recognized:

(1) *initial studies* to collect available data on the site and its locality, and to identify potential problems so that a specific site investigation plan can be developed;

(2) *field investigations of the site conditions* and potential conditions which may influence hillslope stability;

PLATE 17.1(b)

(3) *laboratory testing* of samples to determine the strength and behavioural properties which will be used in stability analyses and as the basis of recommendations;

(4) *presentation of a report* in the form of a model of the site conditions clearly presented with plans, sections, and tabulated data; interpretation and recommendations should be clearly distinct from the data and the quality of the data should be evaluated. Presentation should be related to the needs of the client, the complexity and cost of the project, and to the needs of engineers who will design remedial works.

Summary of contents of investigations:

(1) *Initial studies* may include: (a) collection of published reports on local geology, geomorphol-ogy, and soils; (b) local climatic data; (c) earlier reports on site conditions in the locality and at the specific site; (d) maps and data on topography and drainage; (e) records of local slope instability with climatic, seismic, and observational data; (f) interpretation of aerial photographs; (g) pre-liminary reconnaissance of the locality and site to note areas worthy of detailed investigation, and influences beyond the site boundary; (h) prepara-tion of plan of site investigation.

(2) *Field investigations* may include: (a) topo-graphic survey; (b) geomorphological mapping; (c) geological mapping with emphasis on rocks and deposits which may influence hydrology and slope stability; (d) geophysical surveys; (e) test pit and trench observations and collection of specimens; (f) use of shafts or adits to gain access to potential

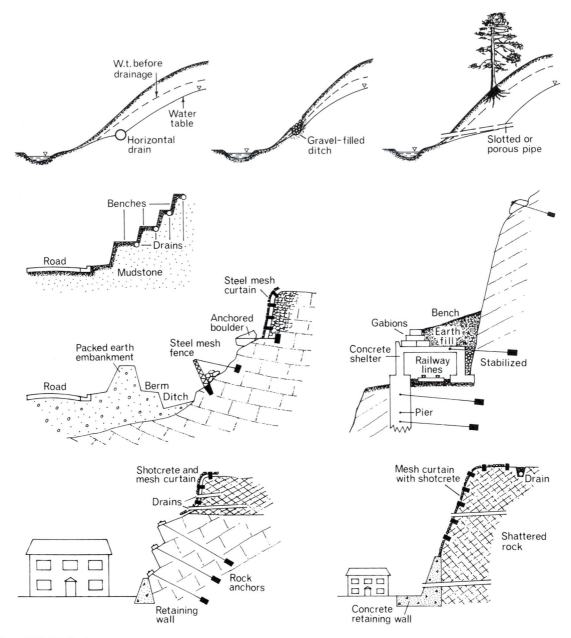

FIG. 17.6. Methods of stabilizing hillslopes. (After Selby 1987*b*.)

failure surfaces and to find critically weak beds or units; (g) drill holes, and logging of cores; (h) samples for laboratory analyses; (i) *in situ* strength testing; (j) instrumentation and logging of critical areas.

(3) *Laboratory tests* usually involve tests of shear strength and behaviour for soil classification purposes but, depending on the project and the site, they may include tests for such characteristics as soil dispersion and sensitivity.

(4) *Presentation of a report* is as important as any other phase of the investigation. It is not

always easy for people who do not know a site to understand the possible influences on hillslope stability. Diagrams, such as those in Figs. 17.1, 17.2, and 17.3, are particularly helpful ways of presenting information. These three figures assume that the project under discussion is consideration of alternative sites for location of a house, and of the problems of potential hillslope stability.

Detailed information on field and laboratory investigations is available in many texts. Reference should always be made to the relevant national standards (see Chapters 5 and 6) but international standards are being developed and works such as Bell (1987), Brown (1981), Matula (1981*a,b*), and Walker *et al.* (1987) are especially useful.

Protection against landslide hazards

The methods which may be adopted to reduce or control the effects of landslides may be broadly grouped into four categories:

(1) avoid, remove, or divert the problem;
(2) reduce the forces tending to cause movement;
(3) increase the forces resisting the movement;
(4) provide warning systems of imminent danger.

Details of the available methods are presented in Zaruba and Mencl (1976), Schuster and Krizek (1978), and Veder (1981). Gray and Leiser (1982) discuss biotechnical methods.

(1) *Avoidance* of danger by relocating an activity or construction away from a threatened area can be the cheapest solution to a problem, but is not always feasible. Roads through mountains, for example, are usually constructed along the floors of valleys which connect through low passes or ridges narrow enough to be breached by tunnels or cuttings. It is seldom possible to route the road around the end of the mountains. Along the valley-floor, however, recognition of the most hazardous zones for runout of debris flows, sturzstrom, and snow avalanches may make it possible to avoid some hazards (see Fookes *et al.* 1985).

Avoidance also involves good design and not creating a problem. In Figs. 17.4 and 17.5 are shown two situations, common in development of urban areas in hilly terrain, involving cutting of roads and platforms for buildings. Commonly this results in the undercutting of the toe of the slope above the road, and the increase of load on the slope below if spoil is tipped down the slope. Debris is also frequently deposited in valley floors, thus raising the local water-tables. The sealing of surfaces, water from roofs, and irrigation of lawns can all cause increases of runoff (Fig. 17.2).

Diversion of a causative factor, such as a river which is undercutting a slope, or a drainage channel which is supplying water to the head of an ancient landslide which is periodically reactivated, is sometimes both feasible and economic. In the broadest sense, the practice of afforesting catchments to reduce runoff and limit the supply of debris to channels is a diversionary tactic, in that the runoff is diverted to evapotranspiration and groundwater systems (Plate 17.1). It may not directly control movement of deep-seated failures but it may have other beneficial effects.

Removal of the problem may be a small or large endeavour. Stripping a cliff of potentially unstable boulders may eliminate a rockfall problem for many years. Removing a small landslide may be feasible, but for a large one removal is usually too expensive. Removal may also reduce the stability of materials higher up a slope.

(2) *Reducing the forces* tending to cause movement is a common strategy and often effective if high water-levels are the problem (Fig. 17.6). Drainage may be carried out by any combination of drains and ditches but, again, the face of a slope should not be left unsupported if ditches are used. Ditches may be filled with gravel to provide lateral support and the porosity of the gravel protected by lining the trench with a geotextile which will prevent clay and silt being washed into the gravel.

A second general cause of failure is a greater magnitude of driving forces than resisting forces. The head of a slope may be treated by removing soil or rock, effective lengths of slopes may be decreased by terracing, and infiltration of water reduced by coatings such as shotcrete.

(3) *Increasing the resisting forces* is accomplished by loading the toe and by retaining walls of which there are many designs. One of the cheapest and most effective devices, for low cost and remote area projects, is the use of 'gabions'—wire baskets filled with rocks. Shotcrete and mesh curtains reduce weathering and bind ravelling surfaces; rock anchors increase the normal stresses across joints and prevent dilation of the rock mass. Shear strength of some clays may be increased chem-

ically, as by the addition of hydrated lime to Na-montmorillonite, which can raise ϕ_r from 4° to 10° for the newly formed Ca-montmorillonite. Use of dowels or piles can also increase total soil resistance.

Many resistance devices are essentially protection walls or nets designed to stop and retain soil and rock materials from impacting sites needing protection.

(4) *Warning systems* are used where either the cost of protective works is too high, or where the threatened area is too poorly defined or too large to be protected. Warning devices may be placed across tension cracks at the head of landslides which are too large for other measures. Warnings of regional risk of landslides are being developed which are based upon recognition of duration, amount, and intensity of rainfall upon urban areas (Keefer *et al.* 1987).

Analyses of the effects of any attempts to stabilize hillslopes are an essential prerequisite to implementation of protective measures. Examples of analyses of the effects of loading and unloading landslides with non-circular and circular failure planes have been given by Hutchinson (1984) and reference should be made to that paper for full details.

It is a common procedure to load (fill) the toe and/or unload (cut) the head of a landslide to increase stability. The location, shape, and weight of the earthworks has usually been decided, in the past, by a combination of experience and a succession of trial stability analyses. Hutchinson's (1977, 1984) analysis calculates the influence of an added or reduced load at successive locations down a landslide by determining the changed factor of safety as the influence of the load falls on different slices. A major effect on the overall stability is the pore-water pressure in the slice.

Consider a slope which is represented in two dimensions by a longitudinal section $Y-Y$ (Fig. 17.7a). If an arbitrary influence load, ΔW, is assumed to act in turn at successive positions between the head and toe of the landslide, an influence line for the resulting effect on the overall factor of safety can be derived, with the ratio of the new factor of safety (F_1) produced by the influence load with the original factor of safety (F_0) being taken as a measure of this effect:

$$\frac{F_1 \ (= F_0 \pm \Delta F)}{F_0} \quad \text{or} \quad \frac{\text{overall } F \text{ with } \Delta F \text{ acting}}{\text{original value of } F}.$$

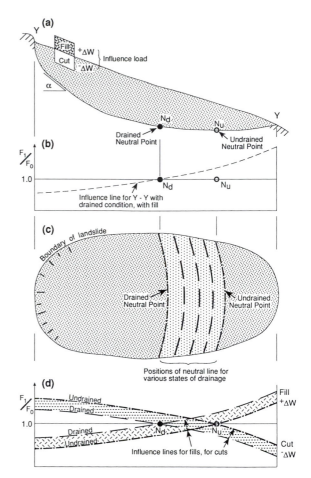

FIG. 17.7. (a) Section of a landslide with an influence load; this may have a positive or negative influence on slope stability, depending upon the position of the affected slice. Location of drained and undrained neutral points are shown. (b) A plot of an influence line through a drained neutral point for an influence load created by fill at the head of the landslide. (c) A plan of the landslide under investigation. The neutral lines have locations ranging from that of the drained to the undrained condition. (d) Plots of influence lines for fills and cuts with drained and undrained soil conditions. (After Hutchinson 1984.)

An influence load or fill, with ΔW having a positive value, will tend to decrease the existing factor of safety when the load acts at the head of the slide, and will increase the factor of safety when it acts near the toe. Of particular interest is the point at which $\Delta F = 0$, or $F_1/F_0 = 1.0$, this point is termed the neutral point. The neutral point indicates where an influence load will have no effect on F_0. The trace of the neutral points in plan is termed the neutral line (Fig. 17.7c). The neutral line therefore forms the boundary between that area of the landslide on which a fill, or cut, would increase the stability, and that area on which a fill, or cut, would decrease the stability. Neutral lines can be drawn for various pore-water pressures for each cut and fill and such lines will then indicate the effects of drainage on stability.

For an analysis using the Conventional Method of Slices:

$$F_0 = \frac{\Sigma \text{ available resisting forces}}{\Sigma \text{ driving forces}} = \frac{\Sigma R_0}{\Sigma D_0},$$

the new factor of safety will then be:

$$F_1 = \frac{\Delta \sigma'_n \tan\phi' + \Sigma R_0}{\Delta W \sin\alpha + \Sigma D_0},$$

where α is the inclination of the base of the slice under consideration and $\Delta\sigma'_n$ is the change in effective stress produced on the base of that slice by the influence load ΔW, given by (see Fig. 13.12):

$$\Delta\sigma'_n = \Delta W(\cos\alpha - \Delta ul).$$

Magnitudes and Frequencies of Erosional Events

The geomorphic importance of an erosional event is governed by the magnitude of the energy it expends upon the landscape, by the frequency with which it recurs, and by the work performed by processes operating in the period between severe erosional events. The greater the magnitude of an event the lower is the probability of its recurrence.

Recurrence intervals are expressed as a probability that an event will occur in a stated number of years (Table 18.1) A ten-year return period event has a 10 per cent chance of occurring in any one year, and a hundred-year event a one per cent chance. An event of such magnitude that it has a probability of returning every ten years will not necessarily occur in every ten-year period, but it has a 99.9 per cent chance of occurring in every fifty-year period. A statement of a return period is, consequently, not a forecast.

For reliable calculation of probabilities of occurrence the length of record should be at least as long as the recurrence interval. Calculation is from the relationship:

$$\text{return period} = \frac{N + 1}{M},$$

where N is the number of years of record and M is the rank of an individual event in an array of annual maximum events of a similar class, such as floods or rainstorms (Dalrymple 1960). The data for an individual site are usually plotted on logarithmic probability paper and a straight-line relationship describes the recurrence interval of events of given magnitudes (Fig. 18.1).

Records for longer than a few tens of years are not available for many parts of the world, and it is not possible to estimate the magnitude of very long return period events with confidence. A second source of uncertainty occurs with climatic events for, as the record gets longer, it becomes increasingly subject to climatic changes, so that the calculated recurrence interval may no longer represent the probability of return of an event under modern conditions. Estimates of the magnitude of past events may be obtained from geological deposits, such as alluvial terraces which are known to have formed in a short period of time and may be related to single extreme storms, landslide deposits, or suddenly raised beaches which may be related to seismic events (Plate 18.1). Even where these deposits can be dated, however, they can seldom be fitted into a data array but only used as indicators.

The magnitude of an event usually influences the area it will effect. Very severe storms or earthquakes are usually experienced over large areas, but the area of the greatest intensity, and hence return period, is confined and lesser intensities occur away from the centre. This can be seen particularly well in the isoseismals for earthquakes (Fig. 18.2). Earthquakes which cause severe modification of landforms have intensities, that is degrees of shaking, of VII or greater on the Modified Mercalli Scale. At the centre of the earthquake shaking may be severe, but it declines away from the source. Where earthquakes are common, isolines for the return period of events of given intensity can be mapped and seismically active and quiet zones delineated. It will be evident from Fig. 18.2 that there has been considerable extrapolation from recorded data by using geological evidence as a guide to earthquake occurrence and intensity (Adams 1980).

Some storms have some of their effect through their prolonged duration (Fig. 18.3) compared with low-magnitude events, but this is not usually the case with earthquakes as, no matter what their intensity, the latter are all of similar duration.

TABLE 18.1. *Probabilities of events occurring in set periods*

One hundred years	Fifty years	Twenty-five years	Ten years	Any one year	Return period years
					Percent chance of getting one or more such or bigger floods in this many years
				50	2
				40	
				30	
				25	
				20	5
		99	80	15	
	99.9	94	65	10	10
	90.5	71	40	5	20
86	61	40	18	2	50
64	39	22	9.6	1	100
40	22	12	5	0.5	200
18	9.5	5	2	0.2	500
10	4.8	2.5	1	0.1	1000
5	2.3	1.2	0.5	0.05	2000
2	1.0	0.5	0.2	0.02	5000
1	0.5	0.25	0.1	0.01	10 000

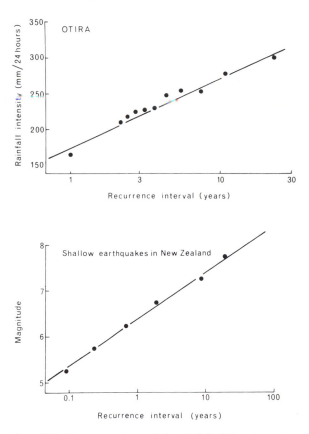

FIG. 18.1. Recurrence intervals for rainfall of given intensity for Otira Township in the New Zealand Alps, and for all shallow earthquakes of given magnitudes in New Zealand. (Data compiled by J. Adams.)

Equilibrium

The concepts of magnitude and frequency of landscape-forming events implies that energy inputs and change are not steadily progressive, unless viewed from the very long-term scale of the geological record in which indications of all discrete events are eradicated, or subsumed in the end-product of landform evolution.

In the shorter time scale of a few hundreds, or thousands, of years there appears to be an approximate relationship between landforms and the processes acting to modify them, such that hillslope processes which are active frequently, and are normally effective every year, do not disturb the approximate balance which usually exists be-

tween the rate of weathering and the removal of regolith. Except in deserts, therefore, there is an approximate equilibrium between weathering, soil development, the vegetation cover, and the rate of erosion on hillslopes. Extreme events destroy this equilibrium, breaking the vegetation cover, stripping away regolith along the track of gullies and landslides, and producing large influxes of sediment into valley floors. As it is uncommon for several extreme events to occur in a short interval, the scars and the deposits may then be slowly modified by lesser intensity processes such as creep, solution, and wash, while weathering, soil formation, and vegetation gradually re-establish a surface which is in approximate equilibrium with the energy of the processes usually acting on the slope. After a period the equilibrium may be broken by another extreme storm. This type of

PLATE 18.1. Boulder lines in a colluvial deposit indicating at least two periods of mobilization of large slope debris separated by a period of lower energy soil deposition. Inland Karroo, South Africa.

episodic development of the land surface can be visualized as occurring in a step-like manner (Fig. 18.4) in which storm events are followed by gradual periods of adjustment to the normally active processes and then, when an approximate equilibrium is re-established, by a period of relative stability.

Step-like functions are also evident from studies of seismic energy release (Fig. 18.5) (Robinson 1979). There are clearly periods of seismic quiescence followed by the release of accumulated strain energy. At no time does the rate of energy release appear to have a value close to the long-term average for any length of time. It is still uncertain if the pattern of energy release can be extrapolated to provide a reliable prediction of earthquake occurrence. Even if it could, the earthquakes would not necessarily cause landform change, as their effect depends upon their location and depth as well as their intensity.

The term 'dynamic equilibrium' has been defined in many ways, but is frequently used to describe the balanced fluctuations about a constantly changing condition which is characterized by a sequence of unrepeated average states through time. A hillslope in a humid climate, undergoing modification only by creep and solution processes, appears to be in a steady state but is, presumably, experiencing some fluctuation in the intensity of those processes. These minor fluctuations appearing to be in a steady state define the dynamic equilibrium for that hillslope.

A second hillslope, also in a humid climate, but with higher relief and steeper slopes, may evolve episodically by landsliding followed by a period of adjustment to the landsliding event. This adjustment may involve gradual modification of the landslide debris and revegetating of the scar and debris lobe. For such a hillslope the dynamic equilibrium includes the severe events. Some writers (e.g. Chorley and Kennedy 1971) have used the term 'dynamic metastable equilibrium' to describe this condition, but the distinction becomes arbitrary, and unhelpful, as the return period of the dominant event becomes shorter.

Thresholds

Any change in the landscape depends upon a threshold being passed in which the strength of a rock or soil material is exceeded by an applied stress. For a sand grain to be moved by surface wash the stress applied by the moving water to initiate transport need be only very low ($< 1 \, N/m^2$), but the stress required to initiate a landslide is thousands, or millions, of times larger.

The exceeding of a threshold stress causes a step-like change in landforms, but the nature of the threshold may be one of three kinds.

(1) An increase in external stress produces a sudden change—as when a storm causes an increase in flow velocity and depth over bare soil. If the soil is vegetated, however, the soil particles may remain immobile until a shear stress large enough to cut through the ground cover is applied. Then, a threshold will be crossed suddenly and rill or gully erosion of the underlying soil will be rapid and, perhaps, severe.

(2) A reduction in internal resistance by pro-

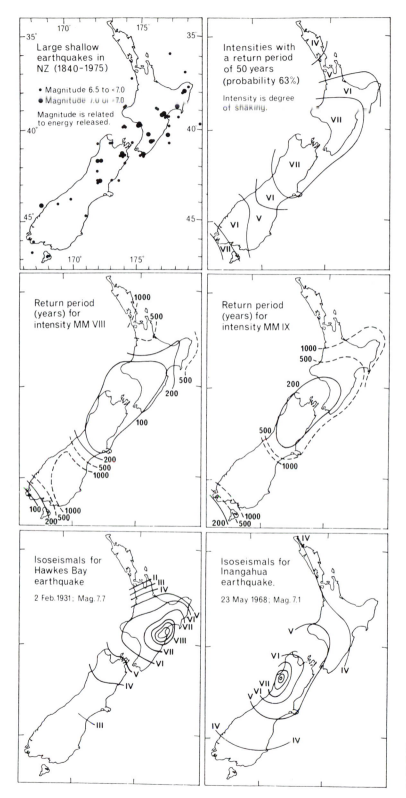

FIG. 18.2. Magnitudes, intensities, return periods, and isoseismals of New Zealand earthquakes. (Data from Smith 1978*a*, *b*.)

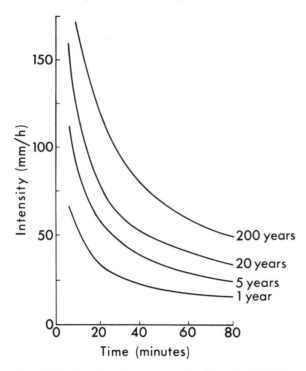

Fig. 18.3. Intensity–duration relationship of rainfall for various periods. (After Wiltshire 1960.)

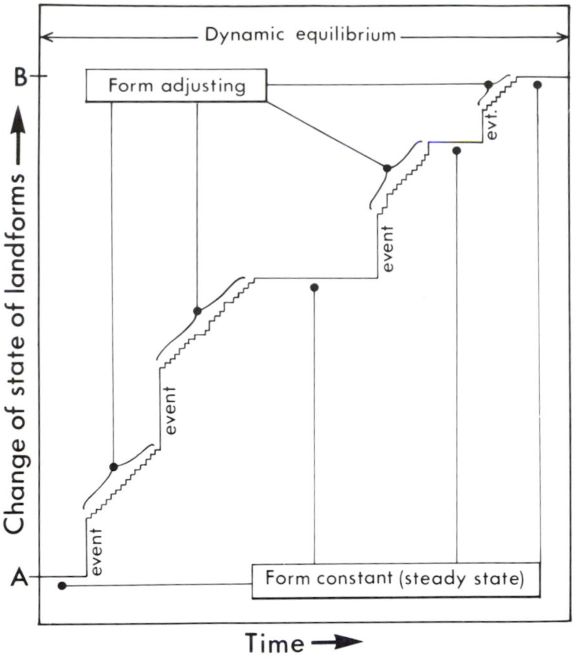

Fig. 18.4. Within the term dynamic equilibrium there are subsumed the three states of: (i) a landforming event; (ii) the adjustment of form which follows after that event; and (iii) a period of steady state in which there is virtually no adjustment of form. The curve which represents the change of landforms with time may, therefore, rise very steeply, gradually, or hardly at all, depending upon the magnitude and frequency of the dominant process. (After Selby 1974b.)

gressive weathering may operate less obviously by lowering the shear stress that is required to initiate instability. Internal changes can consequently give rise to threshold conditions without the operation of large external stresses.

(3) Another type of threshold is that resulting from gradual landform changes until a condition of potential instability is reached. Where a clay stratum is overlain by a resistant sandstone stratum progressive weakening of the clay by seeping water may cause it to fail, and the sandstone will collapse once a critical support has been removed.

Periods of form adjustment

It has been shown that the energy required to reach a threshold stress is related to gradient, lithology, soil cover, vegetation, and climatic or seismic events. Similarly the period of recovery from such an event is controlled not only by the magnitude of the event but by many external and internal factors. A severe storm producing say 300 mm of rain in 24 hours may cause landsliding on steep, but forested, hillslopes in a humid climatic

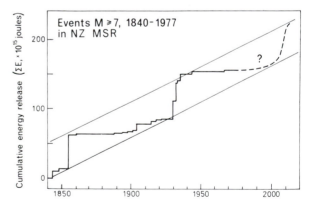

Fig. 18.5. Cumulative energy release for the main seismic region of New Zealand for earthquakes of magnitude greater or equal to 7. The step-like pattern is evident and a possible extrapolation is shown by the pecked line. (After Robinson 1979.)

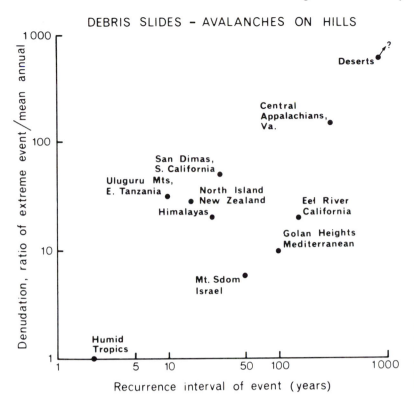

FIG. 18.6. Denudation of hillslopes, as indicated by volume of debris removed, during storm events in various parts of the world. The data points indicate the importance of extreme, compared with common, events in different environments. (Modified from the compilation by Wolman and Gerson 1978.)

zone. Within five years the scars may be covered by grasses and herbs, and fifty years later by maturing trees: virtually no trace of the storm may then survive in the landscape. The same storm in a semiarid zone may cause so much erosion, by wash, gullying, and debris flows, that its effect will still be noticeable hundreds of years later, because of debris flow levees on the hillslopes and alluvial infills in the headwater channels.

The period of adjustment after an event is consequently a guide to the work done by that event in modifying the landscape. Its *effectiveness* is, therefore, best understood by scaling it as a ratio to the mean annual erosion (Wolman and Gerson 1978). Recovery times from one event of a certain magnitude may thus vary from one to hundreds of years depending upon the local environment (Fig. 18.6). For such data to be accumulated for different parts of the world it is essential that all events be reported with a statement of the recurrence interval and recovery period from the event.

It must also be recognized that the dominant geomorphic event on hillslopes may vary with the slope angle so that, as Simonett (1970) reported in

New Guinea, the threshold angles of slope for various processes may be: mudflows—2°, rotational slumps—8°, debris slides and complex landslides—15°, debris avalanches—25 to 30°.

Extreme events in slope evolution

The importance of extreme events in hillslope change has been a neglected topic until recently because of the lack of data, and inability to date past large-magnitude events. The occurrence of a number of extreme events in the twentieth century has focused attention on extremes (e.g. Starkel 1979).

Mountains are particularly subject to the influences of extreme events because of high precipitation at high altitudes caused by extreme windiness and large water-vapour flux; their high relief energy; high drainage density; glacially oversteepened or deeply incised river valleys; the presence of unconsolidated glacial and colluvial debris; the presence of ice and snow; the lack of vegetation at high altitudes; and the occurrence of earthquakes in young fold mountains. As a result the gradual processes of solution and creep may be of far less relative significance in mountains than upon the lowlands of the world.

Large rockfall, rock avalanche, and rock slide events may leave their imprint on the landscape for thousands of years. In parts of the Rocky Mountains cliff falls involving 1 to 100 Mm³ of rock have occurred with a frequency of slightly less than once in a thousand years, since deglaciation, with greatest frequency occurring during and immediately after deglaciation (Gardner 1977). Rock avalanches of the magnitude of the Huascarán event are likely to occur in the Andes only once in a thousand years or more, and in individual valleys catastrophic events, which do enormous amounts of work, may have an even lower frequency. For example, a rainstorm, in the basin of the Guil River in the southern Alps of France, lasting only 48 hours, removed more talus from the mountain catchment than had been moved in the preceding 10 000 years (Tricart 1962).

Individual sites in alpine areas may be affected by mudflows perhaps once in ten or once in a hundred years, but cliff faces may experience repeated or nearly continuous minor rockfalls in the course of one year.

As a result of nearly continuous processes cliff faces may recede at rates of 0.1 to 2.5 mm/year with common values being around 0.7 mm/year (Caine 1974; Barsch 1977). This is probably 10 to 1000 times the rate at which broad interfluves are lowered. The great difficulty is to assess the significance for total denudation of the rare events. It has been suggested by Whalley (1974) that very infrequent events can still have a major effect upon total erosion. For example, the Rhine from its Alpine reaches is depositing about 1 Mm³/year of sediment in Lake Constance. The volume of the Flims rock slide was about 12 Gm³: if such a slide occurs only once in 1000 years it will still be making a large contribution to denudation in the Alps and may exceed the significance in a large catchment of all minor processes together. Its debris, of course, still has to be removed by other processes.

In any part of the world mountains are far more subject to extreme events than lowlands, but even in mountains the effects of extreme denudation are frequently limited to quite small localities. Thus in a mountain range catastrophic events might occur nearly every year, but each time hitting a different area so that overall the frequency is much less than

one event a year. Similarly the sediment produced by a storm may range from a few tonnes/km² to hundreds of thousands of tonnes/km² and the average downwearing of the affected area from 0.01 to 200 mm. This range occurs because the proportion of the ground, in one small valley, suffering extreme erosion during a storm may range from less than 1 per cent to more than 50 per cent.

Regions of extreme climatic events

The importance of extreme events in different regions of the world is very variable. Starkel (1976) has reviewed studies of extreme events and concluded that four classes of region may be distinguished.

(1) Regions with a frequency of extreme events of 5–10 per century and with events of such magnitude that in each the denudation greatly exceeds the denudation produced by all low-intensity processes during a hundred years. Such regions are most common in tropical monsoon and Mediterranean climates, and in steep uplands and farming lands of the temperate zone where human action has removed or changed the vegetation cover.

(2) Regions in which extreme events are rare and in which such events do not exceed the total denudation of a hundred years of low-intensity processes. In these areas intense rainfalls or snowmelts occur each year. Such regions are common in the semiarid zone.

(3) Regions with very rare extreme events. Because of normally very low precipitation a storm may achieve much more denudation than usually occurs in a hundred years. Arid zones and some parts of the boreal zone are in this category.

(4) Stable regions show little variation from normal denudation rates. Many Arctic, Antarctic, continental-boreal, and lowland zones of the temperate regions are in this group.

Data on the frequency of extreme events are not available for many parts of the world, but some comparisons are presented in Fig. 18.7. In the hills and lowlands of Western Europe periods of landsliding appear to be related to intervals of wet climate since the last glacial (Fig. 18.8). These periods have frequencies of about 1 in 2000 years, although each wet period lasted several hundred years and may have contained many extreme storms. At the other end of the scale the frontal ranges of the Himalayas north of the Ganges delta

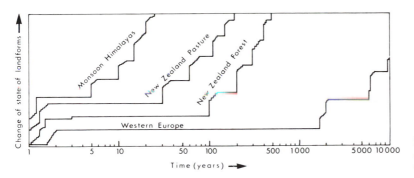

Fig. 18.7. The magnitude and frequency of landsliding in different environments.

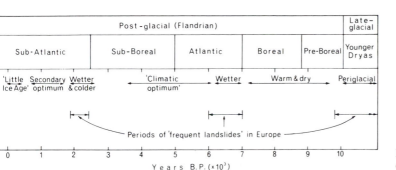

Fig. 18.8. Starkel's (1966) periods of frequent landslides in Europe.

receive prolonged and intense rainfalls nearly every year and extreme landforming events probably have a frequency of about once in five years or less.

Less frequent are the storms causing landslides in the North Island of New Zealand. Under original forest the hills (steeper than 20°) are affected by shallow landsliding about once every hundred years, but in the last century many uplands have been cleared of forest and pastures now cover the hills. These are less protective than the forests, and storms of low intensity, which recur about once in thirty years, now cause slope instability (Selby 1967a,b, 1976a; Pain 1969; Starkel 1972a,b; Jones 1973).

Effect on valley floors

The debris from a destructive storm may be carried directly into a large river and removed, but in the small catchments of many uplands it is stored, at least temporarily. The debris may come to rest as fans, talus slopes, or debris cones, but frequently it becomes incorporated in a debris flow which causes rapid infilling of valley floors. The infill may later be dissected by streams to leave a terrace (Plates 18.2 and 18.3) (Starkel 1987).

Periodic infilling of valley floors has been dated in North Island, New Zealand, by the oldest trees on a terrace and from the known age of volcanic ash deposits which have fallen on each terrace surface—the basic assumption being that the terrace cannot be younger than the trees or ashes on it. In the last 1800 years there have been eight major periods of erosion and alluvial sedimentation in the eastern North Island, New Zealand (Grant 1985). The more recent terraces suggest that very large catastrophic events may occur every 200 years or so, but it is certain that more frequent severe storms also cause much erosion. The debris of lesser events, however, may be largely removed from the catchment either in a series of pulses as large floods carry waves of debris down the channels, or by more gradual processes occurring every few months (Grant 1965). In forested areas fallen trees temporarily block stream channels creating infilled floodplains behind the dam. When the dam breaks the infill may again be incised and a low terrace formed (Pain 1968). This type of infilling and terrace formation is clearly a temporary phase in mountain valley development, for the long-term trend is towards stream incision and debris removal. It is, however, very common in

PLATE 18.2. A large alluvial fan formed by deposition of landslide debris during intense storms, Northern Ruahine Range, New Zealand. Subsequent stream incision has created large terraces in less than a year.

PLATE 18.3. Deposition of gravels, originally derived from landslides, has buried the lower terraces on which trees were growing and left a record of the storm event. Northern Ruahine Range, New Zealand.

upland areas (e.g. Machida 1966) and emphasizes the great importance of extreme events, not only for slope evolution but also for valley floor changes (Baker 1988; Ohmori and Hirano 1988; Brunsden 1990).

Accelerated, induced, and normal erosion

Enough has been said to demonstrate that the rate of change of most landscapes is extremely variable, yet terms such as 'accelerated' and 'normal' erosion are in common use. Without very long-term records of climate and landform change it is not possible to say confidently what is a 'normal' rate, and once the frequency of landform-changing events is greater than a few hundred years the effect of climatic change has to be considered. It is probable therefore that 'normal' rates of erosion are inherently variable and that periods of faster or 'accelerated' erosion are part of the common sequence of events.

Landform change induced by human interference, especially with the vegetation cover, has often accelerated erosion rates by 10 to 1000 times those of normal erosion beneath natural vegetation. To avoid confusion with climatically controlled accelerated erosion it is useful to call that resulting from human interference 'induced' erosion.

Conclusion

Extreme or catastrophic events are of greater significance, in the long-term development of hill-slopes, than has been commonly accepted. Their significance can, however, be recognized only when their magnitude and frequency are known, and when the length of the recovery time is known, so that the effect of one event can be evaluated against the rate of change caused by slower, less obvious processes which modify the landscape.

It has been contended that in rivers most of the work of transportation is carried out by floods which are of such magnitude that they recur with a frequency of at least once in five years (Wolman and Miller 1960). On many hillslopes subject to landsliding the dominant process of change may have a much lower frequency, but greater magnitude, and the valley floors of low order headwater channels in uplands may be strongly influenced by the landslide debris, so that the frequency of dominant events is far less than once in five years. More exact statements must await the accumulation of data over long periods, of at least fifty years, from a much greater range of environments.

19 Rates of Denudation and Their Implications

Denudation is the result of weathering and the stripping of weathering products from the surface of the Earth by the processes of erosion.

Four types of evidence are commonly used for estimating the rate of terrestrial denudation:

(1) The most common method, and the one most suitable for large-area studies, is to measure sediment and dissolved load discharged by rivers and then to convert this information to a rate of downwearing of the landscape.

(2) For small areas the sediment accumulated in reservoirs in a given period of time is a valuable, and sometimes very accurate, indicator of erosion and can be used to estimate the results of human interference with the landscape.

(3) Processes occurring on slopes may be measured directly; the rates of soil creep, surface wash, and landslides can be computed and, in a few cases, their contribution to total denudation assessed.

(4) Accurately dated land surfaces and the landform changes which have occurred on them can give indications of total areal denudation.

Pioneer studies of denudation rates fall into two main classes. (1) The first reliable estimate of the overall rate of ground loss was made from a summary of river sediment loads in the United States by Dole and Stabler (1909). (2) The first attempts to record quantitatively all the slope processes within a stream catchment were those of Jäckli (1957) in Switzerland, Rapp (1960a) in Kärkevagge, Sweden, and Iveronova (1969) in the Kirghiz SSR. The available data on rates of erosion by individual processes, and on catchment yield, have been summarized by Young (1974a), Saunders and Young (1983), and Young and Saunders (1986).

Methods of reporting data

Many different ways of reporting data on erosion rates are in use; consequently it is often difficult to compare information from different areas. A major problem is that most data are derived from discharge of material by streams and consequently are expressed as the mass, or volume of sediment, transported out of a catchment (in kg, t, m^3, or parts per million (ppm), with $1 g/m^3 = 1 ppm$ for sediment in suspension). This method has the advantage that masses or volumes may be averaged for a unit area of the catchment in a unit of time, to give a mean rate of lowering of the land surface. Data on transport rates of soil on hillslopes, however, are usually defined by the velocity of tracer material (m/year) or by the discharge of superficial soil through a unit contour length (m^3/m/year). The two classes of data are not readily converted and the results may be difficult to visualize.

A convenient measure is the Bubnoff Unit (B) with $1 B = 1 mm/1000$ years, which is equivalent to $1 m/M$ years. Average rates of ground lowering may be converted to volumes or masses of material removed, by the relationship: 1 mm of ground lowering = removal of $1000 m^3/km^2$, which may be converted to masses by multiplying by the bulk density of soil or rock. Average densities (Mg/m^3) are:

| silts | 1.25 | silty gravel | 1.8 |
| gravel | 2.1 | bedrock | 2.65 |

Hence 1 mm/year of bedrock lowering removes $2650 t/km^2$ per year, so $1 B = 2.65 t/km^2$ per year for removal of bedrock. The Bubnoff Unit for rate of ground lowering is convenient as it can be easily visualized either for rates of channel incision, surface lowering, or cliff retreat. It indicates a rate

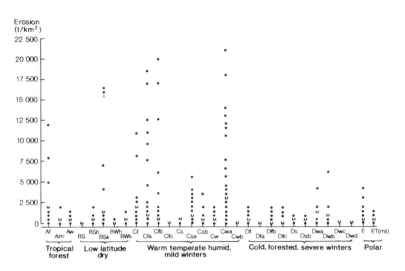

FIG. 19.1. Sediment yields from climatic zones as classified by the Köppen system. M is mean sediment yield. (After Jansson 1988.)

of change of landforms. It has the disadvantage of being an average which suggests uniformity through time and space; such a condition is fundamentally unrealistic.

Denudation rates and regional climate

River sediment discharges

In addition to the three pioneer studies of Corbel (1964), Fournier (1960), and Strakhov (1967), there have been attempts at producing maps of world sediment yield by Milliman and Meade (1983), Walling and Webb (1983a), Dedkov and Mozzherin (1984), and Jansson (1988). In her critique, Jansson pointed out that: (1) there are notable differences among the world maps and conclusions of the authors; (2) there is considerable variation in the quality and representativeness of the data used; (3) the data show a large range in annual yield (expressed in t/km²) for several climatic regions, especially where relief, lithology, and deforestation have major influences on erosion rates; (4) there is no clear theoretical or statistical relationship between annual runoff and sediment yield; and, (5) the use of mean values for sediment yield from major climatic zones obscures the extremes and the magnitude of the range of yields (Fig. 19.1).

In Fig. 19.1, a relatively small number of countries account for data giving yields which are notably higher than the mean values: for Af

climates—Java; for Bsk, Dwa, and Dwb—China; for Cf—highland Papua New Guinea; for Cfa and Cwa—Taiwan; for Cfb—New Zealand; and for ET—Alaska. These areas of high values have one or more features which may account for the high values: mountainous relief, highly erodible soils and weak rocks, dense human populations, and active volcanism.

The studies used by Young and Saunders (1986) are mostly of erosion rates in small catchments or on experimental plots. This is in contrast with the river-sediment yield data used in most other studies. The small catchment data are therefore more specific but, because of the paucity of data from many climatic zones, are less representative than data from river discharges. However, there is considerable agreement between the broad conclusions drawn by Jansson and by Young and Saunders (Fig. 19.2). Both studies recognize high sediment yields in the humid tropical regions (Af, Am) with moderate to steep relief, in semiarid zones (BS), and in the humid monsoonal subtropics (Cwa); low sediment yields and denudation are characteristic of the temperate zones (Cfb, Df), with minima in continental subarctic areas (Dwc) and in deserts (BW).

The effect of human activity on sediment yields is made very clear by the work of Milliman *et al.* (1987). They point out that rivers draining from southern Asia and the high-standing islands of Oceania, together contribute about 70 per cent of the fluvial sediment reaching the oceans (Milliman and Meade 1983). These areas have high relief,

(a)

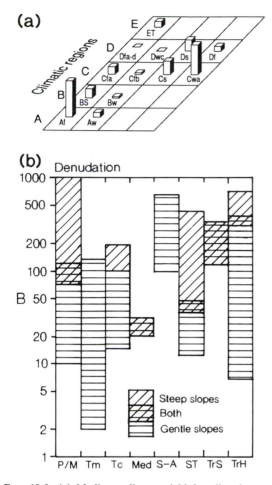

(b)

FIG. 19.2. (a) Median sediment yield for climatic groups arranged by climatic regions. (After Jansson 1988.) (b) Generalized ranges of observed rates of denudation for climatic types. P/M—polar/montane; Tm—temperate maritime; Tc—temperate continental; Med—Mediterranean; S-A—semiarid; ST—subtropical humid; TrS—tropical subhumid; TrH—tropical humid. (After Young and Saunders 1986.) The relationship between the two climatic classifications is set out in Table 19.1.

easily erodible sediments and rocks, seasonally heavy rains, especially in monsoon and typhoon regions, and poor soil conservation in many rice-farming areas. Taiwan, for example, discharges nearly as much sediment to the ocean as all rivers draining contiguous USA. The contribution of one river, the Huanghe (Yellow River) of China, is perhaps the most extreme example of the effect of human interference with a fragile environment.

Most of the sediment in the Huanghe is derived from the easily eroded loess plateau of northern China. Upstream of the plateau the river water is relatively clear, but downstream of the plateau sediment concentrations range from 15 to 500 g/l, 2 to 3 orders of magnitude greater than the world average. Before about 200 BC, the loess plateau was largely an unfarmed, wooded steppe, the sediment load of the river was low with an estimated annual load of 0.1 Gt ($0.1 km^3/y$) of sediment, 10 per cent or less of the modern annual load of 1.2 Gt ($1.2 km^3/y$)—assuming a dry weight density for sediment of $1.0 t/m^3$.

Rates of surface processes

Soil creep measurements have been discussed in some detail in Chapter 13. The analysis of Young and Saunders (Fig. 19.3) confirms that movement near the surface in temperate maritime climates is typically 0.5–2 mm/y. In areas where frost action is severe in winter, as in temperate continental climates, it ranges up to 15 mm/y. Rates for solifluction have notably higher maxima.

A conclusion from many studies is that creep rates are higher on moist sites and may be particularly high in clays which are montmorillonitic; rates on sandy soils are generally low. These conclusions are much as might be expected. Their implication is that creep can rarely be a major process of modification of landscape because it operates entirely by transfer of material downslope. Ground loss occurs therefore only at the crest of a slope, which may help explain the origin of upper slope convexities, as Gilbert and Davis reasoned, but the rate of slope lowering or retreat is necessarily very small. The most obvious effect of creep is probably the general smoothing of the groundsurface which is a feature of many grass-covered hillsides in humid environments.

Solifluction rates are generally an order of magnitude higher than those of soil creep. The availability of moisture is the major control, and slope angle, soil texture, and vegetation have secondary importance, but, where other things are equal, solifluction is faster on steeper slopes and on silty soils.

Modern rates of solifluction may be substantially higher than those of earlier, colder, periods and the quoted rates may be unrepresentative of hillslopes because the data are heavily biased towards measurement of the most active features, such as lobes. The relatively high rates, compared with soil creep, suggest that solifluction can account for

TABLE 19.1. *Ranges of denudation rates associated with selected climatic zones*

Climate	Abbreviations used in figs.	Approximate Köppen equivalent	Relief	Denudation rates (B)
Glacial	—	—	Gentle (ice sheets)	50–200
			Steep (valleys)	1000–5000
Polar/Montane	P/M	E	Mostly steep	10–1000
Temperate maritime	Tm	Cfb, Cfc	Mostly gentle	5–100
Temperate continental	Tc	Df	Gentle	10–100
			Steep	100–200+
Mediterranean	Med	Cs	—	10–?
Semiarid	S-A	BS	Gentle	100–1000
Arid	Arid	Bw	—	10–?
Subtropical humid	ST	Cfa	—	10–1000?
Tropical subhumid	TrS	Aw	—	100–500
Tropical humid	TrH	Af, Am	Gentle	10–100
			Steep	100–1000
Any climate badlands	—	—		1000–10^6

Source: Young and Saunders (1986).

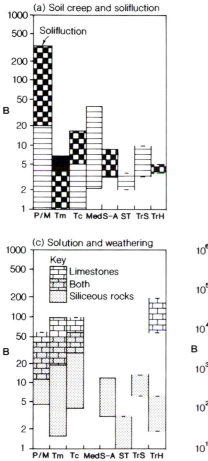

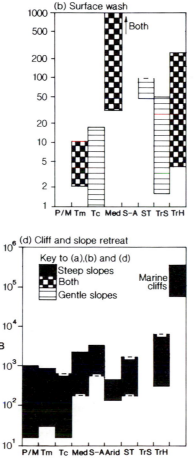

FIG. 19.3. Generalized ranges of observed rates of surface processes and slope retreat, grouped according to climate and slope angle. Key to climatic zones is that set out in the caption of Fig. 19.2. Pecked upper and lower boundaries are based on only one record. (After Young and Saunders 1986.)

a substantial proportion of total slope retreat—perhaps as much as half.

Surface wash includes raindrop impact and surface flow. The data indicate that rates of soil removal by the combined detachment and wash processes are extremely variable and are in the range from 1 B in some woodlands to 1000–10 000 B on badlands. The more common range is 1–200 B. The reasons for variability are set out in Chapter 12, and include rainfall intensity, vegetation cover, soil erodibility, slope inclination, slope length, and, above all, variations in land use.

Surface wash is almost certainly the major process of denudation in semiarid and subhumid climates, possibly in deserts and probably under rainforests. In temperate and cool climates with a forest cover, and therefore with a thick humic surface horizon, wash is probably of negligible effectiveness.

Solution data have been converted to average ground loss in Fig. 19.3c. Rates shown here are never very high, even from limestone terrains, which have ranges of 10–100 B. Siliceous rocks have ranges of 2–60 B.

The data show, however, that solution is a very important process of denudation because it operates by direct removal rather than the downslope transport of all other processes. Furthermore, solutes are seldom stored in a catchment, and usually leave it very quickly. Solution at any rate means slope retreat at that rate—as also from flat surfaces.

Landsliding is seldom reported with data from which average rates of denudation can be derived. The best estimates indicate a very large range in all climatic zones. The major point is, of course, that they occur only on moderate to steep hillslopes, most being steeper than about 25°. They are clearly activated by water and earthquakes but primarily in areas of modest to high available relief. Vast areas of continental surfaces have few or no landslides. It may be that landsliding is the dominant process of most major mountain systems, but the debris is stored on the land surfaces. An example from Japan (Aniya 1985) indicates average denudation of 1.5 kB (1.5 mm/y) by landsliding in one river basin.

Cliff retreat has been studied most commonly in mountains and marine cliffs. The mean values are around 100 B but the range of rates is extreme with the range from very strong rocks to those of medium strength being unmeasurably low to about 6000 B (Chapter 16).

Many other processes are either very specific to a few environments, have very local effects, or have not been studied in sufficient detail to provide usable data: snow avalanching; action of burrowing rabbits and gophers; gullying, piping, and throughflow provide examples in these three classes.

Sources of error

Data collection problems are the main cause of the different values, obtained in the studies quoted, for world-wide erosion rates. A secondary contributing factor is the different methods of calculation.

Rates of denudation are derived from the rates of removal of earth material by a river from its catchment. Thus the annual rate of transport of earth material past a point in a stream, whose catchment area above that point is known, gives a rate in t/km^2 per year.

The areas of catchment can be measured accurately but the mass of material moved by the river is difficult to measure. Rivers transport material as bed load, in suspension, and in solution. Most early estimates of denudation were based on suspended load and solution load only. It is still impossible to measure bed-load accurately and estimates of the proportion of bed-load to total load vary from nil in the lower reaches of streams to 55 per cent or more in mountain reaches.

The accuracy of estimates of suspended and solute load transport depends upon the frequency of the measurements and the length of the period over which the measurements are made. The annual suspended load may vary by as much as a factor of five in successive years. Dole and Stabler (1909), with only one year's record of suspended sediment transport, estimated that the rate of denudation for the whole of the United States of America is 33 B, but Judson and Ritter (1964) with a longer record estimated the rate at 61 B.

For the few catchments in which daily suspended sediment and solute samplings have been carried out over an extended period, it is possible to construct sediment and solute hydrographs and annual discharges of load can then be calculated. Where the measurements of load are less complete, load and water discharges are correlated and total load is then estimated from the water-discharge hydrograph. When sufficient correlations have been made the flow duration, or annual frequencies of discharge, can be converted to a suspended sediment or solute curve. With corrections for bed load this can be made into a denudation curve. Suspended sediment loads vary even with equal discharges so the reliability of the correlations depends upon the size of the sample of measurements (Walling and Webb 1983b).

Inaccuracies may result from inadequate measurement of water discharge; inaccurate laboratory analyses of sediment and solutes; poor sample collection in the field; and inadequate sampling times. From studies carried out on rivers in Devon, Walling (1978) has suggested that absolute errors associated with suspended sediments can be as high as +60 per cent for annual loads and between +400 and −80 per cent for monthly loads. For solution loads errors may be up to ±60 per cent for monthly loads.

Storage of sediment is one of the major characteristics of drainage basins. Many slope processes involve only transfers of material within a drainage basin and not its export through the channel. Slope debris is commonly stored in fans, talus slopes, on terraces, or on flood plains, so that even though hillslopes may be changing rapidly this change is not necessarily expressed in channel transport. The disparity is particularly noticeable with increasing area of the catchment, as hillslope debris in small catchments is commonly delivered directly into the channel, but in large catchments is held in a store. In a study of ten large river basins in southeastern USA Trimble (1977) found that while upland erosion was proceeding at about 9.5 B, sediment yields at the mouths of catchments were only 0.53 B. The delivery ratio was thus only 6 per cent, and the difference is stored in valleys and channels. Storage times in some mountain valleys may reach thousands of years where late Pleistocene glacial and talus deposits are not being removed from the catchment.

The opposite situation may also occur with little erosion on fully vegetated slopes but channel bank erosion producing large quantities of debris. In such areas the denudation rate as measured at stream outlets is far greater than that from the slopes. Similarly if a river suddenly erodes unconsolidated glacial or talus deposits its load will increase far above the rate of slope denudation.

Human interference with natural vegetation may be the major source of error in estimates of denudation. Most stream load data collected by national agencies are obtained not for the purposes of geomorphological research but for water quality, river training, flood control, or some other purpose. Consequently they are most readily available from drainage basins with intense land use and often with considerable population. Deforestation and wasteful farming practices have been characteristic features of the opening up of North America and other new lands for settlement. The increase in runoff and acceleration of soil erosion which has resulted still continues in spite of soil conservation measures. Silted stream beds, buried floodplains, infilled reservoirs and estuaries are all witness to the results of human interference. It has been estimated that the conversion of forest to cropland in the middle Atlantic seaboard states of USA has increased sediment yield up to tenfold (Meade 1969). It has also been estimated that accelerated erosion has stripped some three to four Gm^3 of sediment from the upland of the Piedmont in southeastern USA since AD 1700. Over 90 per cent of this sediment remains stored on hillslopes, in stream bottoms, and reservoirs. Only a small proportion of the sediment has so far reached the sea (Meade 1976). Upland soil profiles have lost 150 mm. Since the decline of agricultural land use in the early 1900s small upland tributaries have adjusted to decreased sediment loads by entrenchment into, and erosion of, sediment deposited since the initiation of colonial agriculture (Costa 1975).

The general pattern of relationships between land use, erosion rates, and channel conditions was demonstrated by Wolman (1967) for the area around Washington, DC (Fig. 19.4). Such changes are severe but locally they can be even more catastrophic. Pearce (1976) has shown that the destruction of vegetation around Sudbury, Ontario, over an area of 125 km^2 has increased local denudation rates by two orders of magnitude to about 37 000 B. It has to be remembered that such rates cannot continue for long as they are caused by erosion of the soil and regolith which may have taken thousands of years to form. Denudation rates will fall when the regolith has been severely depleted.

Coal-mining, urbanization, and highway construction have all increased sediment yields. By the end of the nineteenth century some of the rivers of Pennsylvania became so choked with anthracite coal debris that their bottom sediments could be

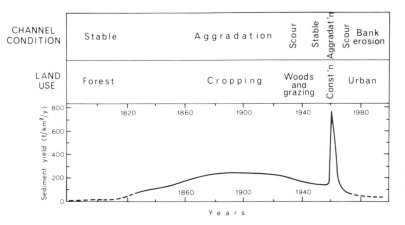

FIG. 19.4. The effect of land use changes on sediment yields near Washington, DC. (After Wolman 1967.)

dredged profitably for the coal they contained. Anthracite debris from the Susquehanna River basin is found in the modern bottom sediments of Chesapeake Bay as far as 40 km beyond the mouth of the river. Even though the Pennsylvania anthracite industry has declined since 1917, 10 per cent of the suspended matter measured in the Susquehanna River in the spring flood of 1960 was coal. It has been shown that similar increases of loads of rivers in populated areas occur in many other parts of USA, in Australia, Malaysia, and Europe (e.g. Douglas 1967; Judson 1968).

As a result of his studies of river discharge data for streams draining to the Atlantic seaboard of the USA, Meade (1969) has concluded that

... as the present dissolved load of the Atlantic streams is nearly equivalent to the detrital load ... the errors in estimating each type of load are roughly equivalent. Considering the distribution and effects of different land uses in the Atlantic states, I estimate that the present sediment loads are four to five times what they were before the European settler arrived. On the basis of studies made in North Carolina and New Hampshire I estimate that about one-fourth of the dissolved loads of the streams represents material contributed by the atmosphere. Another one-tenth of the dissolved load may represent material added directly to the streams by human activity. Previous estimates of the natural rate of denudation of the Atlantic states therefore have probably been too large by at least a factor of two.

Dissolved loads are particularly difficult to assess for many natural sources of solute may contribute to the total solute load. Janda (1971) has pointed out that dissolved gases, cyclic salt (i.e. salt from oceans or land areas, carried into the basin by wind), soil dust, volcanic ash, connate water, and soil organic matter may all greatly increase the solute load and give erroneously high estimates of denudation (Fig. 12.15). Janda has estimated that computations of chemical denudation rates which use total dissolved load—as most published computations have—exaggerate those rates by 1.4 to 2.4 times the real significance of this process. In areas of close human population solutes from sewerage, pesticides, fertilizers, and industrial wastes can also greatly increase solute loads. It has been pointed out that in 1958 the discharge of chloride ion alone into the Rhine River at Rees-Lobith was about 9.2 Mt (Durum *et al.* 1960).

Not only human influences make calculations of denudation difficult. Natural sources of contamination are also a problem. Rainfall and snow carry low concentrations of soluble salts, some of which may find their way into the drainage waters. Coastal areas are particularly subject to precipitation containing marine salts and the decline in chloride content of rain away from the coast is often very distinct. In continental areas salts from endoreic drainage basins, soil dust, and, in a few places, dust derived from decaying vegetation are added to drainage waters.

An estimate (Anon. 1972 in *Nature*, 240: 320–1) of the amounts of particulate matter taken up each year by the atmosphere, over the northern hemisphere, assesses the quantity of natural particles as a steady 690 Mt (110 Mt from oceans and 580 Mt from land surfaces). Added to this are contributions derived from human activities, put at 120 Mt in 1880, 480 Mt in 1970, and a predicted 760 Mt by AD 2000.

The timing of stream sampling is important. The composition of dissolved load varies seasonally, and between rising and falling stages of stream discharge. A 'spring' burst of released solutes is a well-recognized phenomenon in many catchments. The non-seasonal release of geothermal waters into rivers also has a marked effect upon dissolved loads.

An as yet unassessed source of error in measurements of dissolved load is the quantity of solute which is transported as a gel and that which is adsorbed on to colloids. It may be expected that in drainage basins yielding sediment with colloids of high base-exchange capacity—such as montmorillonitic clays—a considerable proportion of the cations may be adsorbed. It has been estimated by Pitty (1971) that in some catchments, during periods of high sediment concentration, cations adsorbed on suspended sediment may approach or even exceed the cations carried in solution. By ignoring this condition it is possible that the geomorphic importance of mechanical weathering in relation to chemical weathering could be greatly exaggerated.

It is clear that estimates of rates of natural denudation are liable to serious error. It is also clear that human activities have major influences on sediment yields. However, the human influence has increased greatly in the last 200 years and was negligible before about 3000 years ago. Modern rates of denudation should be corrected to eliminate the human influence, if they are to be used to estimate rates of landform change.

Small site and catchment studies have their own set of problems and sources of error, as well as those also associated with large catchment studies.

The primary problem is that of the representativeness of a small slope or first-order basin study. Does the study site represent a lithological, landform, soil, or vegetation regional complex? Do the locations within the site represent the slope units characteristic of the region? Are the processes which are being monitored characteristic of a wider area? Particularly, are the obviously active processes being accorded special attention while the less obvious processes, such as solution, are being ignored or inadequately monitored?

A second problem is that virtually all small site investigations are carried out over relatively brief periods, often of a year or two, much better for ten years (e.g. Rapp 1960a), but rarely for longer than that. Inevitably, therefore, the longer return period events may be missed, yet we now recognize that in many areas it is such events which have a major influence upon landform development.

Denudation and relief

In spite of all the reservations mentioned above, it is abundantly clear that relief has a very large influence upon denudation rates (Phillips 1990), as is very obvious in the results from those few mountain ranges from which the data permit a reliable estimation of sediment yield from small catchments which can be taken to represent sections of the range. The data from the New Zealand Alps (Whitehouse 1988) has permitted a clear representation of the relationship between relief, precipitation, and sediment yield (Figs. 19.5, 19.6), but it cannot take into account the effect of individual earthquakes which are major features of a tectonic-plate boundary.

It has been shown by Adams (1980), Pearce and O'Loughlin (1985), and Pearce and Watson (1986) that a single earthquake, the Murchison, 1929, quake with a magnitude of 7.7, has delivered up to $400\,000\,m^3/km^2$ of sediment into stream channels. First- to third-order channels were buried to depths of as much as 10 m with sedimentary rock debris. At least 50 per cent to 75 per cent of granitic debris is retained in fourth-order catchments after fifty years. This suggests that earthquake-induced landsliding is the principal sediment supply mechanism in the region.

The frequency with which major earthquakes occur has been estimated (Figs. 18.2, 18.5). It seems probable that they are of sufficient frequency to have major effects in at least the most seismically active parts of the Southern Alps. It is reasonable to expect similar effects in other seismically active regions. If this conclusion is valid, how useful is a small site study, carried out over 2–10 years, in assessing the major denudation processes of a region in which the dominant event has a return period of 100–300 years?

The last question should not inspire a sense of futility, rather it should inform questions about the objectives of process studies. The content of earlier chapters demonstrates the practical value of understanding the mechanisms and consequences of natural processes; its significance for erosion control, prediction, avoidance, and amelioration of hazards; and its value for recognizing the

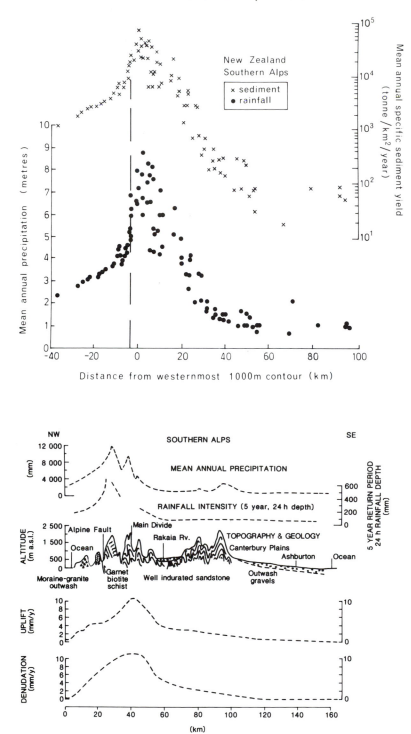

Fig. 19.5. The relationship between mean annual precipitation and sediment yield for a transect across the New Zealand Alps. (Supplied by M. McSaveney.)

Fig. 19.6. The relationship between precipitation, relief, uplift, and denudation along a cross-section of the central South Island, New Zealand. (After Whitehouse 1988.)

influences on sites for buildings, open-pit mines, and transport routes are evident. Where the objective is to understand the relative importance of individual processes, well-designed observational and monitoring programmes are still essential and, in areas where the energy of long-return period processes is inadequate to dominate the environment, studies of modest duration may reasonably be expected to yield data which is relevant to understanding landform change over periods of a few hundreds to a few thousand years.

With clear recognition of the uncertainties inherent in the estimates, it is appropriate to consider the implications of our best estimates of the operation of major processes (Table 19.1).

The general validity of these conclusions is supported by a number of studies which have used geological phenomena to derive long-term estimates of change and of denudation.

Differences in the elevation of Cenozoic erosion surfaces, in Kenya, suggest a rate of 8.4 B as characteristic of late Cenozoic time—a rate similar to that from Kenyan forest areas with an annual precipitation of about 750 mm at the present, and which is consistent with rates of downwearing by solution alone. By contrast, the estimates of soil erosion in Kenya, on overgrazed land, are as high as 10 kB (Dunne *et al*. 1978).

Tectonic uplift and denudation

The existence of large mountain chains for prolonged periods is clearly dependent upon continuing uplift. A mountain 4000 m high could, theoretically, be levelled in 0.5 My at the rate of denudation recorded for the central Taiwan Alps of 5.5 kB (Li 1976). Rates of sediment transport in rivers in Taiwan, however, suggest that the denudation rate approximately equals the uplift rate. This is also generally true for the Southern Alps of New Zealand (Fig. 19.6) where rates are approximately 10 kB for the zone of highest uplift and denudation. It should also be noted that denudation rates decline eastwards as rain-shadow effects reduce rainfall, with the result that high relief can survive in an area of relatively low uplift rates (Whitehouse 1988).

Where a land surface can be dated accurately the modification of that surface since a known date can be expressed in quantitative terms. Since radiometric dating has been available several measure-

ments have been made of the dissection of lava flows. Ruxton and McDougall (1967) working in the Hydrographers Range, Papua, on lavas ranging in age from 650 000 to 700 000 years BP (by potassium-argon dating) have found that calculated denudation rates are directly related to relief and range from 80 B at 60 m above sea level to 800 B at 760 m above sea-level. These rates are similar to those calculated by other methods for hot moist hilly areas.

In a study of denudation rates in the White Mountains, of eastern California, Marchand (1971) used the base of a basalt lava flow, dated at 10.8 million years, as his datum level and calculated the volume of material removed from below this level. Present rates of chemical denudation, corrected for contributions from the atmosphere and biosphere, he estimated as ranging from 1.4 to 21.0 B.

Denudation rates in the European Alps have been estimated by Clark and Jäger (1969) at between 400 and 1000 B. They have used dates of mineral assemblages derived from Rb/Sr and K/Ar ratios and assumed that the dates indicate the age at which the minerals were crystallized in metamorphism during Alpine folding. The metamorphic mineral assemblage indicates the range of temperatures to which the rock was subjected, and these temperatures must have occurred at pressures, or depths, that are compatible with the observed mineralogy. The rate of denudation can be deduced, since a rock collected at the surface at the present time was at a depth corresponding to the critical temperature of the mineral (in this case biotite) at the time in the past given by the age of that mineral. This rate contrasts with that of 40 B derived from an estimate of rock eroded from the Pelvoux Massif in the last 70 My (Montjuvent 1973) but Montjuvent's study could not include the unknown general lowering of mountain peaks.

High general rates of tectonic uplift are about 8 kB, and the highest rates of post-glacial isostatic rebound are about 11 kB (for the Gulf of Bothnia (Walcott 1972)). No known rates of denudation for large areas equal the maximum known rates of uplift. For small areas where rates are known it is possible to determine the actual increase in relief in a given time. In the Transverse Ranges of southern California, for example, local uplift rates reach 7.6 kB but the maximum rate of denudation, although high, is only 2.3 kB (Scott and Williams 1978). The relationship between the two rates for the Japanese Alps has been demonstrated by

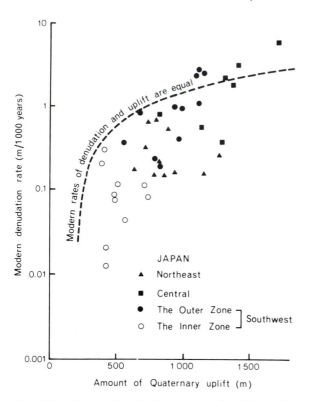

FIG. 19.7. The relationship between total uplift in the Quaternary and modern denudation rates for Japan. (After Yoshikawa 1974.)

Yoshikawa (1974) who has found that denudation rates may exceed uplift rates in the Outer Zone of southwest Japan and in the central mountains, but elsewhere modern rates of uplift exceed rates of denudation (Fig. 19.7). The western Caucasus also appears to be increasing in altitude with uplift rates apparently reaching 20–25 kB but denudation rates being close to 200 B (Gabrielyan 1971).

The wide variations in rates of uplift and downwearing make it difficult to generalize on how long it would take to erode a mountain chain, but through the operation of isostatic adjustments, as denudation occurs, it appears to be closer to a 100 My, for a high chain, than to 10 My. Oscillating rates appear to preclude the possibility of any long-term steady state relief (Ahnert 1970).

Conclusions

The relationship between climate and rates of denudation appears to be close but is not necessarily uniform within one climatic zone. It operates primarily through the completeness of vegetation cover on potentially mobile soil and regolith materials, and is very strongly influenced by intensely seasonal precipitation.

The relationship between denudation and lithology is strong but not easy to quantify. Resistant, often ancient, massive rock has relatively low rates of denudation. Where rocks have a thin regolith even human interference may not greatly increase rates of denudation over large areas. It has been shown by Gordon (1979), for example, that in central New England, where soils are generally thin, most of the sediment removed has been trapped in Long Island Sound, and that over the last 8000 years the rate has been nearly uniform and low at around 50 B. Most of the sediment has come from river-bank collapse and not from soil erosion. Similar results were obtained by Pearce and Elson (1973) for areas in Quebec.

There appears to be rather broad agreement in the available data that for lowland areas in many parts of the world mean rates of 10 to 100 B are common, but there is also evidence that rates of downwearing of landscapes with soils upon ancient hard rocks, such as granite, may be limited by the rate of weathering to <5 B (e.g. Owens and Watson 1979). For mountainous areas, and even lowland areas with highly erodible rocks such as bare shales, rates are commonly two to ten times higher than mean rates for lowlands. There is still considerable doubt about how generally applicable are higher rates, in excess of 100 B. Slaymaker (1977), for example, has found that for the Canadian Cordillera rates of 60 to 120 B are common. Young and Saunders (1986) by contrast, suggest that mountain lands more usually have rates in the range 100–1000 B. There is no doubt that differences in relief and slope angle are of paramount importance in controlling denudation rates, but whether the available data are biased by excessive sampling in rapidly eroding areas is not clear, and must await more measurements taken in a statistically acceptable manner.

Another major problem lies in uncertainty about how far modern rates can be extrapolated to explain rates of change of landforms. This is clearly a contentious area, for relict landforms, such as bornhardts and duricrusts, may survive in the landscape for millions of years (Twidale 1976), but with what degree of alteration of form and volume is unknown: valley terraces and some slope deposits may last for many thousands, or even hundreds of thousands, of years. It is still not always possible to determine whether this survival is the result of slow process rates, or of selective preservation. The question of the degree to which landforms are relict from past environments must be answered, in the absence of direct dating, by a better knowledge of the variation of denudation rates in small areas and by better estimates of the variation in process intensities.

If measured rates are extrapolated they indicate that in one million years a mountain range is reduced in altitude by 1000 m (ignoring uplift) and a gently sloping landscape by 10 to 100 m. A possible estimate for the lowering of a major erosion surface is 5 to 20 m in one million years.

Such extrapolations ignore the effects of human interference with erosion which appears to exceed normal geological rates by at least a factor of 2 to 10, but such accelerated rates cannot be sustained as the soil store is limited. The extrapolations also ignore the effects of climatic change.

The resulting time-scale for geomorphological change suggests that slopes are unlikely to have been greatly modified during the last 10 000 years, so studies of slope forms must take into account the results of late glacial and Holocene climatic changes. Minor valleys 20–50 m deep are reasonably attributed to the late Pleistocene but intermediate-sized features have to be considered in relation to the whole of Pleistocene time.

Appendix

The International System (SI) of Units

(1) The SI is based on six primary units

Quantity	Unit	Symbol
length	metre	m
mass	kilogram	kg
time	second	s
electric current	ampere	A
temperature	degree Kelvin	°K
luminous intensity	candela	cd

(Note: the spellings 'gramme' and 'meter' are used in some countries.)

(2) The following are some of the supplementary and derived units used in the system:

Quantity	Unit	Symbol
plane angle	radian	rad
area	square metre	m^2
volume	cubic metre	m^3
velocity	metre per second	m/s
acceleration	metre per second squared	m/s^2
density	kilogram per cubic metre	kg/m^3
force	newton	$N = kg \, m/s^2$
moment of force	newton metre	N m
pressure, stress	{ newton per square metre {=pascal	N/m^2 Pa
viscosity: kinematic	metre squared per second	m^2/s
dynamic	newton second per metre squared	$N \, s/m^2$
work, energy, quantity of heat	joule	$J = N \, m$
power, heat flow rate	watt	$W = J/s$
temperature (customary unit)	degree Celsius	°C
density of energy flow rate	watt per square metre	W/m^2

(3) Prefixes used to extend the magnitude of the basic units are in steps of one thousand, and are:

Prefix	Symbol	Magnitude
giga	G	10^9
mega	M	10^6
kilo	k	10^3
milli	m	10^{-3}
micro	μ	10^{-6}
nano	n	10^{-9}
pico	p	10^{-12}

(4) *Writing SI units*

(a) The abbreviations do not take an 's' in the plural: thus km (NOT kms). Symbols are NOT followed by a full stop.

(b) Only one multiplying prefix is applied at one time to a given unit: thus two megametres squared $(2\,\text{Mm}^2)$ (NOT two kilo-kilometres squared).

(c) Multiplying prefixes are printed immediately adjacent to the unit symbols with which they are associated: thus km, mW.

(d) The product of two or more units is indicated by a space between unit symbols: N m (NOT Nm).

(e) The quotient of two units may be expressed by use of the solidus or of negative exponents: m/s or $\text{m\,s}^{-1}$; $\text{kg\,m}^2/(\text{s}^3\,\text{A})$ or $\text{kg\,m}^2\,\text{s}^{-3}\,\text{A}^{-1}$. The solidus must not be repeated on the same line of one expression unless ambiguity is avoided by the use of parentheses: m/s^2 or (m/s)/s, but NOT m/s/s.

(f) SI symbols and numerals should be spaced from one another: thus 13 g (NOT 13g).

(g) To facilitate reading of numbers of more than four digits, the digits should be spaced in groups of three on either side of the decimal point: 10 479.014. The comma is NOT used for spacing because in many countries it is used for the decimal point.

(h) The use of the prefixes hecta (10^2), deca (10), deci (10^{-1}), and centi (10^{-2}) is discouraged.

(5) *Notes*

(a) There is a basic coherence in the SI, for example, 1 watt = 1 joule/second exactly = 1 newton $\times$ 1 metre/1 second exactly.

(b) The kilogram is a unit of mass (i.e. quantity of matter), not of weight. Weight varies slightly with gravity at different places on Earth (and more in space) and is the force applied by a unit of mass. A force is a product of mass (kg) and acceleration (m/s^2). Thus force is $\text{kg\,m/s}^2$ and $1\,\text{kg\,m/s}^2 = 1\,\text{N}$. The acceleration due to gravity on Earth is approximately $9.81\,\text{m/s}^2$. Thus the weight of a half kilogram mass is $0.5 \times 9.81 = 4.9\,\text{N}$.

(c) The tonne (t) is commonly used for one thousand kilograms $(1000\,\text{kg} = 1\,\text{t} = 1\,\text{Mg})$.

(d) The term 'unit weight' expresses the gravitational force per unit volume (N/m^3).
   The term 'density' refers to mass per unit volume (kg/m^3).

(e) The litre per second is sometimes used for small discharges: $1\,\text{m}^3/\text{s} = 1000$ litres per second.

References

Abrahams, A. D. and Parsons, A. J. (1987), 'Identification of strength equilibrium rock slopes: Further statistical considerations', *Earth Surface Processes and Landforms*, 12: 631–5.

—— —— Cook, R. U., and Reeves, R. W. (1984), 'Stone movement on hillslopes in the Mojave Desert, California: A 16-year record', *Earth Surface Processes and Landforms*, 9: 365–70.

Adams, J. (1980), 'Contemporary uplift and erosion of the Southern Alps, New Zealand', *Geological Society of America Bulletin*, 91, pt. II: 1–114.

—— (1982), 'Stress-relief buckles in the McFarland quarry, Ottawa', *Canadian Journal of Earth Sciences*, 19: 1883–7.

Ahnert, F. (1964), 'Quantitative models of slope development as a function of waste cover thickness', *Abstracts of papers, 20th IGU Congress (London)*.

—— (1970), 'Functional relationships between denudation, relief, and uplift in large mid-latitude drainage basins', *American Journal of Science*, 268: 243–63.

—— (1976a), 'Darstellung des Strukture-influsses auf die Oberflächenformen im theoretischen Modell', *Zeitschrift für Geomorphologie, Supplementband*, 24: 11–22.

—— (1976b), 'Brief description of a comprehensive three-dimensional process–response model of landform development', *Zeitschrift für Geomorphologie, Supplementband*, 25: 29–49.

—— (1988), 'Modelling landform change', in M. G. Anderson (ed.), *Modelling Geomorphological Systems* (Chichester: Wiley), 375–400.

Aires-Barros, L., Graça, R. C., and Velez, A. (1975), 'Dry and wet laboratory tests and thermal fatigue of rocks', *Engineering Geology*, 9: 249–65.

Akerman, H. J. (1984), 'Notes on talus morphology and processes in Spitsbergen', *Geografiska Annaler*, 66A: 267–84.

Aldridge, R. and Jackson, R. J. (1968), 'Interception of rainfall by Manuka (*Leptospermum scoparium*) at Taita, New Zealand', *New Zealand Journal of Science*, 11: 301–17.

Allbrook, R. F. (1980), 'The drop-cone penetrometer method for determining Atterberg limits', *New Zealand Journal of Science*, 23: 93–7.

Allison, R. J. (1988), 'A non-destructive method of determining rock strength', *Earth Surface Processes and Landforms*, 13: 729–36.

—— and Goudie, A. S. (1990). 'The form of rock slopes in tropical limestone and their associations with rock mass strength', *Zeitschrift für Geomorphologie*, 34: 129–48.

Anderson, M. G. (ed.) (1988), *Modelling Geomorphological Systems* (Chichester: Wiley).

—— and Burt, T. P. (1978), 'The role of topography in controlling throughflow generation', *Earth Surface Processes*, 3: 331–44.

—— —— (1982), 'Throughflow and pipe monitoring in the humid temperate environment', in R. Bryan and A. Yair (eds.), *Badland Geomorphology and Piping* (Norwich: Geo Books), 337–53.

—— and Kemp, M. J. (1987), 'Suction-controlled triaxial testing: Laboratory procedures in relation to resistance envelope methods', *Earth Surface Processes and Landforms*, 12: 649–54.

—— —— and Lloyd, D. M. (1990), 'Design and installation of a combined automatic tensiometer-piezometer system', *Earth Surface Processes and Landforms*, 15: 63–71.

—— —— and Shen, J. M. (1987), 'On the use of resistance envelopes to identify the controls on slope stability in the tropics', *Earth Surface Processes and Landforms*, 12: 637–48.

Aniya, M. (1985), 'Contemporary erosion rate by landsliding in the Amahata River basin', *Zeitschrift für Geomorphologie*, 29: 301–14.

Arguden, A. T. and Rodolfo, K. S. (1990), 'Sedimentologic and dynamic differences between hot and cold laharic debris flows of Mayon Volcano, Philippines', *Geological Society of America Bulletin*, 102: 865–76.

Armstrong, A. (1976), 'A three-dimensional simulation of slope forms', *Zeitschrift für Geomorphologie, Supplementband*, 25: 20–8.

Arnould, M. and Frey, P. (1978), 'Analyse des responses

a une Enquête Internationale de l'UNESCO sur les glissements de terrain', *Bulletin of the International Association of Engineering Geology*, 17: 114–18.

Aronsson, G. and Linde, K. (1982), 'Grand Canyon: A quantitative approach to the erosion and weathering of a stratified bedrock', *Earth Surface Processes and Landforms*, 7: 589–99.

Atkinson, B. K. (1984), 'Subcritical crack growth in geological materials', *Journal of Geophysical Research*, 89(B6): 4077–114.

Atkinson, T. C. (1978), 'Techniques for measuring sub-surface flow on hillslopes', in M. J. Kirkby (ed.), *Hillslope Hydrology* (Chichester: Wiley), 73–120.

Atterberg, A. (1911), 'Über die Physicalische Bodenuntersuchung and über die Plastimatat der Tone, *Internationale Mitteilungen für Bodenkunde*, 1: 5–10.

Attewell, P. B. and Farmer, I. W. (1976), *Principles of Engineering Geology* (London: Chapman and Hall).

Augusthitis, S. S. and Otteman, J. (1966), 'On diffusion rings and spheroidal weathering', *Chemical Geology*, 1: 201–9.

Augustinus, P. C. (1988), 'The influence of the lithological and geotechnical properties of rocks on the morphology of glacial valleys', D.Phil. thesis, University of Waikato.

—— and Selby, M. J. (1990), 'Rock slope development in McMurdo Oasis, Antarctica, and implications for interpretations of glacial history', *Geografiska Annaler*, 72A: 55–62.

Aydan, O. and Kawamoto, T. (1990), 'Discontinuities and their effect on rock mass', in N. Barton and O. Stephansson (eds.), *Rock Joints* (Rotterdam: Balkema), 149–56.

Bagnold, R. A. (1954), 'Experiments on a gravity-free dispersion of large solid spheres in a Newtonian fluid under shear', *Proceedings, Royal Society, London*, ser. A, 225: 49–63.

Baker, V. R. (1988), 'Cataclysmic processes in geomorphological systems', *Zeitschrift für Geomorphologie Supplementband*, 67: 25–32.

Bakker, J. P. and Le Heux, J. W. N. (1946), 'Projective geometric treatment of O. Lehmann's theory of the transformation of steep mountain slopes', *Koninklijke Nederlandsche Akademie van Wetenschappen*, B49: 533–47.

—— —— (1952), 'A remarkable new geomorphological law', *Koninklijke Nederlandsche Akademie van Wetenschappen*, B55: 399–410 and 554–71.

Balk, R. (1937), 'Structural behaviour of igneous rocks', *Bulletin of the Geological Society of America, Memoir*, 5.

Ballantyne, C. K. and Kirkbride, M. P. (1987), 'Rockfall activity in upland Britain during the Loch Lomond Stadial', *Geographical Journal*, 153: 86–92.

Bandis, S., Lumsden, A. C., and Barton, N. R. (1981),

'Experimental studies of scale effects on the shear behaviour of rock joints', *International Journal of Rock Mechanics and Mining Sciences and Geomechanics Abstracts*, 8: 1–21.

—— —— —— (1986), 'Experimental studies of scale effects on the shear behaviour of rock joints', *Norwegian Geotechnical Institute Publication*, N152 (Oslo: Norwegian Geotechnical Institute), 1–21.

Barden, L., McGown, A., and Collins, K. (1973), 'The collapse mechanism in partly saturated soil', *Engineering Geology*, 7: 49–60.

Barisone, G. and Bottino, G. (1990), 'A practical approach for hazard evaluation of rock slopes in mountainous areas', *Proceedings, 6th International Association for Engineering Geology Congress*, 3 (Rotterdam: Balkema), 1509–15.

Barsch, D. (1977), 'Eine Abschätzung von Schuttproduktion und Schutttransport im Bereich aktiver blockgletscher der Schweizer Alpen', *Zeitschrift für Geomorphologie, Supplementband*, 28: 148–60.

Barshad, I. (1966), 'The effect of variation in precipitation on the nature of clay mineral formation in soils from acid and basic igneous rocks', *Proceedings, International Clay Conference (Jerusalem)*, 167–73.

Barton, D. C. (1916), 'Notes on the disintegration of granite in Egypt', *Journal of Geology*, 24: 382–93.

Barton, M. E. (1988), 'The sedimentological control of bedding plane shear surfaces', *Proceedings, 5th International Symposium on Landslides (Lausanne)*, (Rotterdam: Balkema), 1: 73–6.

Barton, N. (1973), 'Review of a new shear-strength criterion for rock joints', *Engineering Geology*, 7: 287–332.

—— and Choubey, V. (1977), 'The shear strength of rock joints in theory and practice', *Rock Mechanics*, 10: 1–54.

—— and Kjaernsli, B. (1981), 'Shear strength of rockfill', *Journal of the Geotechnical Engineering Division, Proceedings, American Society of Civil Engineers*, 107, GT7: 873–91.

Basher, L. R., Tonkin, P. J., and McSaveney, M. J. (1988), 'Geomorphic history of a rapidly uplifting area on a compressional plate boundary: Cropp River, New Zealand', *Zeitschrift für Geomorphologie Supplementband*, 69: 117–31.

Bates, T. (1962), 'Halloysite and gibbsite formation in Hawaii', *Clays and Clay Minerals*, 9: 315–28.

Bauer, R. A. (1987), 'The effects of valleys on the strength of rock materials at depth', *Proceedings, 28th US Symposium on Rock Mechanics (Tucson)*, 345–9.

Baynes, F. J. and Dearman, W. R. (1978), 'The microfabric of a chemically weathered granite', *Bulletin of the International Association of Engineering Geology*, 18: 91–100.

—— —— and Irfan, T. Y. (1978), 'Practical assessment of grade in a weathered granite', *Bulletin of the Inter-*

national Association of Engineering Geology, 18: 101–9.

Beattie, A. G. (1990), 'Petrological controls on the geomechanical behaviour of coal-measure soft rocks, Waikato, New Zealand', M.Sc. thesis, University of Waikato.

Beckett, P. M. T. and Webster, R. (1971), 'Soil variability: a review', Soils and Fertilizers, 34: 1–15.

Bedzyk, M. J., Bommarito, G. M., Caffrey, M., and Penner, T. L. (1990), 'Diffuse-double layer at a membrane–aqueous interface measured with X-ray standing waves', Science, 248: 52–6.

Bell, D. H. and Pettinga, J. R. (1988), 'Bedding-controlled landslides in New Zealand soft rock terrain', Proceedings, 5th International Symposium on Landslides (Lausanne) (Rotterdam: Balkema), 1: 77–83.

Bell, F. G. (ed.) (1987), Ground Engineer's Reference Book (London: Butterworth).

Bennett, P. and Siegel, D. L. (1987), 'Increased solubility of quartz in water due to complexing by organic compounds', Nature, 326: 684–6.

Bentley, S. P. (1979), 'Viscometric assessment of remoulded sensitive clays', Canadian Geotechnical Journal, 16: 414–19.

Bernknopf, R. L., Campbell, R. H., and Shapiro, C. D. (1986), 'A quantitative approach to mapping landslide risk and its economic implications', US Geological Survey Yearbook for 1985: 88–92.

Betson, R. P. (1964), 'What is watershed runoff?' Journal of Geophysical Research, 69: 1541–51.

Beven, K. (1978), 'The hydrological response of headwater and sideslope areas', Hydrological Sciences Bulletin, 23: 419–37.

Beverage, J. P. and Culbertson, J. K. (1964), 'Hyperconcentrations of suspended sediment', Journal of the Hydraulics Division, American Society of Civil Engineers, 90, HY6: 117–26.

Bieniawski, Z. T. (1973), 'Engineering classification of jointed rock masses' Transactions, South African Institution of Civil Engineers, 15: 335–44.

—— (1975), 'The point-load test in geotechnical practice', Engineering Geology, 9: 1–11.

—— (1989), Engineering Rock Mass Classifications (New York: Wiley).

Birkeland, P. W. (1984), Soils and Geomorphology (New York: Oxford University Press).

—— (1990), 'Soil-geomorphic research: A selective overview', Geomorphology, 3: 207–24.

Bisdom, E. B. A. (1967), 'The role of microcrack systems in the spheroidal weathering of an intrusive granite in Galicia (N.W. Spain)', Geologie en Mijnbouw, 46: 333–40.

—— Stoops, G., Delvigne, P., and Altemuller, H.-J. (1982), 'Micromorphology of weathering biotite and its secondary products', Pedologie, 32: 225–52.

Bishop, A. W. (1955), 'The use of the slip circle in the stability analysis of earth slopes', Géotechnique, 5: 7–17.

—— (1971), 'The influence of progressive failure on the method of stability analysis', Géotechnique, 21: 168–72.

—— and Bjerrum, L. (1960), 'The relevance of the triaxial test to the solution of stability problems', Proceedings, American Society of Civil Engineers Research Conference on Shear Strength of Cohesive Soils, 437–501.

—— Green, G. E., Garga, V. K., Andresen, A., and Brown, J. D. (1971), 'A new ring shear apparatus and its application to the measurement of residual strength', Géotechnique, 21: 273–328.

Bishop, D. M. and Stevens, M. E. (1964), 'Landslides in logged areas in S. E. Alaska', US Department of Agriculture, Forest Service, Research Paper, NOR-1: 1–18.

Bjerrum, L. (1967), 'Progressive failure in slopes of overconsolidated plastic clay and clay shales', Journal of Soil Mechanics and Foundation Engineering Division, American Society of Civil Engineers, 93: 1–49.

—— and Jørstad, F. (1957), 'Rockfalls in Norway', Norwegian Geotechnical Institute Publication, 79: 1–11.

—— —— (1968), 'The stability of rock slopes in Norway', Norwegian Geotechnical Institute, Publication, 79 (Oslo: Norwegian Geotechnical Institute), 1–11.

Blackwelder, E. (1925), 'Exfoliation as a phase of rock weathering', Journal of Geology, 33: 793–806.

—— (1933), 'The insolation hypothesis of rock weathering', American Journal of Science, 26: 97–113.

Blong, R. J. (1970), 'The development of discontinuous gullies in a pumice catchment', American Journal of Science, 268: 369–83.

—— (1973), 'A numerical classification of selected landslides of the debris slide-avalanche-flow type', Engineering Geology, 7: 99–114.

—— (1974), 'Landslide form and hillslope morphometry: An example from New Zealand', Australian Geographer, 12: 425–38.

—— (1975), 'Hillslope morphometry and classification: A New Zealand example', Zeitschrift für Geomorphologie, 19: 405–29.

Bock, H. (1979), 'Experimental determination of the residual stress field in a basaltic column', Proceedings, International Congress on Rock Mechanics (Montreux) (Rotterdam: Balkema), i. 45–9; iii. 136–7.

Bock, H. and Wallace, K. (1978), 'Testing intact rock specimen', in H. Bock (ed.), An Introduction to Rock Mechanics (Townsville: James Cook University of North Queensland), 131–53.

Bonnard, C. (ed.) (1988), Landslides, Proceedings, 5th International Symposium on Landslides (Lausanne), 3 vols. (Rotterdam: Balkema).

Bouchard, M. (1985), 'Weathering and weathering residuals on the Canadian Shield', *Fennia*, 163: 327–32.

Bouyoucos, G. J. (1935), 'The clay ratio as a criterion of susceptibility of soils to erosion', *Journal of the American Society of Agronomists*, 27: 738–41.

Bovis, M. J. (1982), 'Uphill-facing (antislope) scarps in the Coast Mountains southwest British Columbia', *Geological Society of America Bulletin*, 93: 804–12.

—— (1985), 'Earthflows in the interior plateau, southwest British Columbia', *Canadian Geotechnical Journal*, 22: 313–34.

—— and Dagg, B. R. (1988), 'A model for debris accumulation and mobilization in steep mountain streams', *Hydrological Sciences Journal*, 33: 589–604.

—— —— and Kaye, D. (1985), 'Debris flows and debris torrents in the southern Canadian Cordillera: Discussion', *Canadian Geotechnical Journal*, 22: 608.

Boyce, J. R. (1985), 'Some observations on the residual strength of tropical soils', *Proceedings, 1st International Conference on Geomechanics in Tropical Lateritic and Saprolitic Soils (Brasilia)*, 1: 229–37.

Brabb, E. E. and Harrod, B. L. (eds.) (1989), *Landslides: Extent and Economic Significance* (Rotterdam: Balkema).

Bradley, W. C. (1963), 'Large-scale exfoliation in massive sandstones of the Colorado Plateau', *Geological Society of America Bulletin*, 74: 519–28.

Brady, B. H. G. and Brown, E. T. (1985), *Rock Mechanics For Underground Mining* (Hemel Hempstead: George Allen & Unwin).

Brady, N. C. (1990), *The Nature and Properties of Soils*, 10th edn. (New York: Macmillan).

Brandl, H. (1988), 'Stabilization of deep cuts in unstable slopes', *Proceedings, 5th International Symposium on Landslides (Lausanne)*, 2 (Rotterdam: Balkema), 867–72.

Brazier, V. and Ballantyne, C. K. (1989), 'Late Holocene debris cone evolution in Glen Feshie, western Cairngorm Mountains, Scotland', *Transactions, Royal Society of Edinburgh: Earth Sciences*, 80: 17–24.

—— Whittington, G., and Ballantyne, C. K. (1988), 'Holocene debris cone evolution in Glen Etive, western Grampian Highlands, Scotland', *Earth Surface Processes and Landforms*, 13: 525–31.

Brice, J. C. (1966), 'Erosion and deposition in the loess-mantled Great Plains, Medicine Creek drainage basin, Nebraska', *US Geological Survey, Professional Paper*, 352-H: 255–339.

Briceño, H. O. and Schubert, C. (1990), 'Geomorphology of the Gran Sabana, Guyana shield, southeastern Venezuela', *Geomorphology*, 3: 125–41.

Bridgman, P. W. (1911), 'Water, in the liquid and five solid forms, under pressure', *American Academy of Arts and Science Proceedings (Daedalus)*, 47: 439–558.

Brighenti, G. (1979), 'Mechanical behaviour of rocks under fatigue', *Proceedings, International Congress on Rock Mechanics (Montreux)* (*International Society for Rock Mechanics*), 1: 67–70.

Brindley, G. W. and Brown, G. (1980), *Crystal Structures of Clay Minerals and their X-ray Identification* (London: Mineralogical Society).

Brink, A. B. A. (1979–83). *Engineering Geology of Southern Africa*, (Pretoria: Building Publications) (i. 1979; ii. 1981; iii. 1983).

Broch, E. (1974), 'The influence of water on some rock properties', *Proceedings, 3rd International Congress of the Society for Rock Mechanics (Denver)*, 2, pt. A: 33–8.

—— (1979), 'Changes in rock strength caused by water', *Proceedings, 4th Congress of the International Society for Rock Mechanics*, 1: 71–5.

—— and Franklin, J. A. (1972), 'The point-load strength test', *International Journal of Rock Mechanics and Mining Science*, 9: 669–97.

Bromhead, E. N. (1986), *The Stability of Slopes* (London: Surrey University Press).

Brown, A. (1982), 'Toppling induced movements in large, relatively flat rock slopes', in R. E. Goodman and F. E. Heuze (eds.), *Issues in Rock Mechanics* (New York: Society of Mining Engineers of the American Institute of Mining, Metallurgical and Petroleum Engineers), 1035–47.

Brown, C. B. and Sheu, M. S. (1975), 'Effects of deforestation on slopes', *Journal of the Geotechnical Engineering Division, American Society of Civil Engineers*, GT2: 147–65.

Brown, E. T. (ed.) (1981), *Rock Characterization and Monitoring: ISRM Suggested Methods* (Oxford: Pergamon).

—— and Hoek, E. (1978), 'Trends in relationships between measured in situ stresses and depth', *International Journal of Rock Mechanics and Mining Sciences and Geomechanics Abstracts*, 15: 211–15.

Brown, G., Newman, A. C. D., Rayner, J. H., and Weir, A. H. (1978), 'The structures and chemistry of soil clay minerals', in D. J. Greenland and M. H. B. Hayes (eds.), *The Chemistry of Soil Constituents* (Chichester: Wiley), 29–178.

Brown, K. W. (1977), 'Shrinking and swelling of clay, clay strength and other properties of clay soils and clays', in J. B. Dixon and S. B. Weed (eds.), *Minerals in Soil Environments* (Madison: Soil Science Society of America, Inc.), ch. 19.

Browning, J. M. (1973), 'Catastrophic rock slide, Mount Huascarán, north-central Peru, May 31, 1970', *American Association of Petroleum Geologists Bulletin*, 57: 1335–41.

Brunori, F., Penzo, M. C., and Torri, D. (1989), 'Soil shear strength: Its measurement and soil detachability', *Catena*, 16: 59–71.

Brunsden, D. (1984), 'Mudslides', in D. Brunsden and

D. B. Prior (eds.), *Slope Instability* (Chichester: Wiley), 363–418.

—— (1990), 'Tablets of stone: Toward the Ten Commandments of geomorphology', *Zeitschrift für Geomorphologie Supplementband*, 79: 1–37.

—— and Jones, D. K. C. (1976), 'The evolution of landslide slopes in Dorset', *Philosophical Transactions, Royal Society London*, A283: 605–31.

—— and Kesel, R. H. (1973), 'Slope development on a Mississippi River bluff in historic time', *Journal of Geology*, 81: 576–97.

—— and Prior, D. B. (eds.) (1984), *Slope Instability* (Chichester: Wiley).

—— Jones, D. K. C., and Goudie, A. S. (1984), 'Particle size distribution on the debris slopes of the Hunza Valley', in K. J. Miller (ed.), *The International Karakoram Project*, ii. (Cambridge: Cambridge University Press), 536–80.

—— Doornkamp, J. C., Fookes, P. G., Jones, D. K. C., and Kelley, J. M. H. (1975), 'Geomorphological mapping techniques in highway engineering', *Journal of the Institution of Highway Engineers*, 22: 35–41.

Bryan, R. B. (1987), 'Processes and significance of rill development', *Catena Supplement*, 8: 1–15.

—— and Poesen, J. (1989), 'Laboratory experiments on the influence of slope length on runoff, percolation and rill development', *Earth Surface Processes and Landforms*, 14: 211–31.

—— and Yair, A. (eds.) (1982), *Badland Geomorphology and Piping* (Norwich: Geo Books).

—— and Hodges, W. K. (1978), 'Factors controlling the initiation of runoff and piping in Dinosaur Provincial Park badlands, Alberta, Canada', *Zeitschrift für Geomorphologie Supplementband*, 29: 151–68.

Bucknam, R. C. and Anderson, R. E. (1979), 'Estimation of fault-scarp ages from a scarp-height–slope-angle relationship', *Geology*, 7: 11–14.

Büdel, J. (1957), 'Die "Doppelten Einebnungs-flächen" in den feuchten Tropen', *Zeitschrift für Geomorphologie*, 1: 201–88.

—— (1982), *Climatic Geomorphology*, trans. L. Fischer and D. Busche (Princeton: Princeton University Press).

Burdine, N. T. (1963), 'Rock failure under dynamic loading conditions', *Society of Petroleum Engineers' Journal*, Mar. 1963, 1–8.

Burt, T. P. (1989), 'Storm runoff generation in small catchments in relation to the flood response of large basins', in K. Beven and P. Carling (eds.), *Floods: Hydrological, Sedimentological and Geomorphological Implications* (Chichester: Wiley), 11–35.

Butler, B. E. (1959), 'Periodic phenomena in landscapes as a basis for soil studies', *CSIRO Australia, Soil Publication*, 14.

—— (1967), 'Soil periodicity in relation to landform development in southeastern Australia', in J. N. Jennings and J. A. Mabbutt (eds.), *Landform Studies*

from Australia and New Guinea (Canberra: ANU Press), 231–55.

Butler, D. R. (1989), 'Subalpine snow avalanche slopes', *Canadian Geographer*, 33: 269–73.

Cabrera, J. G. (1986), 'Buckling and shearing of basalt flows beneath deep valleys', *Proceedings, 5th International Association of Engineering Geology Congress (Buenos Aires)*, 1: 589–94.

—— and Smalley, I. J. (1973), 'Quick clays as products of glacial action: a new approach to their nature, geology, distribution and geotechnical properties', *Engineering Geology*, 7: 115–33.

Caine, N. (1969), 'A model for alpine talus slope development by slush avalanching', *Journal of Geology*, 77: 92–100.

—— (1974), 'The geomorphic processes of the alpine environment, in J. D. Ives and R. G. Barry (eds.), *Arctic and Alpine Environments* (London: Methuen), 721–48.

—— (1980), 'The rainfall intensity-duration control of shallow landslides and debris flows', *Geografiska Annaler*, 62A: 23–7.

—— (1982), 'Toppling failures from alpine cliffs on Ben Lomond, Tasmania', *Earth Surface Processes and Landforms*, 7: 133–52.

—— and Mool, P. K. (1982), 'Landslides in the Kolpu Khola drainage, middle mountains, Nepal', *Mountain Research and Development*, 2: 157–73.

Calkin, P. and Cailleux, A. (1962), 'A quantitative study of cavernous weathering (taffonis) and its application to glacial chronology in Victoria Valley, Antarctica', *Zeitschrift für Geomorphologie*, 6: 317–24.

Campbell, B. L. (1983), 'Application of environmental caesium-137 for the determination of sedimentation rates in reservoirs and lakes and related catchment studies in developing countries', in *Radioisotopes in Sediment Studies*, IAEA-TECDOC, 298: 7–29.

Campbell, C. S. (1989), 'Self-lubrication of long runout landslides', *Journal of Geology*, 97: 653–65.

Campbell, D. J. (1976), 'Plastic limit determination using a drop cone penetrometer', *Journal of Soil Science*, 27: 295–301.

Campbell, R. H. (1975), 'Soil slips, debris flows, and rainstorms in the Santa Monica Mountains and vicinity, southern California', *US Geological Survey, Professional Paper*, 851: 1–51.

Carrara, A., Cardinali, M., Detti, R., Guzzetti, F., Pasqui, V., and Reichenbach, P. (1991), 'GIS techniques and statistical models in evaluating landslide hazard', *Earth Surface Processes and Landforms*, 16: 427–45.

Carroll, R. D. (1969), 'The determination of the acoustic parameters of volcanic rocks from compression velocity measurements', *International Journal of Rock Mechanics and Mining Sciences*, 6: 557–80.

Carson, M. A. (1976), 'Mass-wasting, slope development and climate', in E. Derbyshire, (ed.), *Geomorphology and Climate* (London: Wiley), 101–36.

Carson, M. A. (1977), 'Angles of repose, angles of shearing resistance and angles of talus slopes', *Earth Surface Processes*, 2: 363–80.

—— and Kirkby, M. J. (1972), *Hillslope Form and Process* (London: Cambridge University Press).

—— and Petley, D. J (1970), 'The existence of threshold slopes in the denudation of the landscape', *Transactions, Institute of British Geographers*, 49: 71–92.

Casagrande, A. (1948), 'Classification and identification of soils', *Transactions, American Society of Civil Engineers*, 113: 901.

Chalcraft, D. and Pye, K. (1984), 'Humid tropical weathering of quartzite in southeastern Venezuela', *Zeitschrift für Geomorphologie*, 28: 321–32.

Chamley, H. (1989), *Clay Sedimentology* (Heidelberg: Springer-Verlag).

Chandler, M. P., Parker, D. C., and Selby, M. J. (1981), 'An open-sided field direct shear box', *British Geomorphological Research Group, Technical Bulletin*, 27.

Chandler, R. J. (1970), 'A shallow slab slide in the Lias Clay near Uppingham, Rutland' *Géotechnique*, 20: 253–60.

—— (1971), 'Landsliding on the Jurassic escarpment near Rockingham, Northamptonshire', *Institute of British Geographers, Special Publication*, 3: 111–28.

—— (1972), 'Lias Clay: weathering processes and their effect on shear strength', *Géotechnique*, 22: 403–31.

—— (1973), 'The inclination of talus: Arctic talus terraces and other slopes composed of granular material', *Journal of Geology*, 81: 1–14.

—— (1982), 'Lias clay slope sections and their implications for the prediction of limiting or threshold slope angles', *Earth Surface Processes and Landforms*, 7: 427–38.

—— Kellaway, G. A., Skempton, A. W., and Wyatt, R. J. (1976), 'Valley slope sections in Jurassic strata near Bath, Somerset', *Philosophical Transactions, Royal Society, London*, A 283: 527–56.

Chardon, M. (1976), 'Observations sur la formation des versants regularisés ou versants de Richter', *Actes du symposium sur les versants en pays méditerranéens, à Aix-en-Provence, France, 28–30 avril 1975, Centre d'Etudes Géographiques et de Recherches Méditerranéenes*, 5: 25–7.

Cheng, J. D. (1988), 'Subsurface stormflows in the highly permeable forested watersheds of southwestern British Columbia', *Journal of Contaminant Hydrology*, 3: 171–91.

Chigira, M. (1990), 'A mechanism of chemical weathering of mudstone in a mountainous area', *Engineering Geology*, 29: 119–38.

Chinn, T. J. H. (1981), 'Use of rock weathering-rind thickness for Holocene absolute age-dating in New Zealand', *Arctic and Alpine Research*, 13: 33–45.

Chorley, R. J. and Kennedy, B. A. (1971), *Physical Geography: A Systems Approach* (London: Prentice-Hall International).

Churchill, R. R. (1982), 'Aspect-induced differences in hillslope processes', *Earth Surface Processes and Landforms*, 7: 171–82.

Ciolkosz, E. J., Carter, B. F., Hoover, M. T., Cronce, R. C., Waltman, W. J., and Dobos, R. R. (1990), 'Genesis of soils and landscapes in the Ridge and Valley province of central Pennsylvania', *Geomorphology*, 3: 245–61.

Claridge, G. G. C. (1960), 'Clay minerals, accelerated erosion, and sedimentation in the Waipaoa River catchment', *New Zealand Journal of Geology and Geophysics*, 3: 184–91.

—— (1970), 'Studies in element balances in a small catchment at Taita, New Zealand', *International Association of Scientific Hydrology, Publication* 96: 523–40.

Clark, S. P. and Jäger, E. (1969), 'Denudation rate in the Alps from geochronologic and heat flow data', *American Journal of Science*, 267: 1143–60.

Clemence, S. P. and Finbarr, A. O. (1981), 'Design considerations for collapsible soils', *Journal of the Geotechnical Engineering Division, Proceedings of the American Society of Civil Engineers*, 107: 305–17.

Clemency, C. V. (1975), 'Simultaneous weathering of a granitic gneiss and an intrusive amphibolite dike near São Paulo Brazil, and the origin of clay minerals', *Proceedings, International Clay Conference (Mexico)*, 157–72.

Cloos, H. (1936), *Einfuhrung in die Geologie* (Berlin: Reimer).

Close, U. and McCormick, E. (1922), 'Where the mountains walked', *National Geographic Magazine*, 445–64.

Coles, N. and Trudgill, S. T. (1985), 'The movement of nitrate fertilizer from the soil surface to drainage waters by preferential flow in weakly structured soils, Slapton, south Devon', *Agriculture, Ecosystems and Environment*, 13: 241–59.

Colman, S. M. (1981), 'Rock-weathering rates as functions of time', *Quaternary Research*, 15: 250–64.

—— (1982), 'Chemical weathering of basalts and andesites: Evidence from weathering rinds', *US Geological Survey, Professional Paper*, 1246.

—— and Dethier, D. P. (eds.) (1986), *Rates of Chemical Weathering of Rocks and Minerals* (New York: Academic Press).

Conacher, A. J. (1988), 'The geomorphic significance of process measurements in an ancient landscape', *Catena Supplement*, 13: 147–64.

—— and Dalrymple, J. B. (1977), 'The nine unit landsurface model: an approach to pedogeomorphic research', *Geoderma*, 18: 1–154.

Conca, J. L. and Astor, A. M. (1987), 'Capillary moisture flow and the origin of cavernous weathering in dolerites of Bull Pass, Antarctica', *Geology*, 15: 151–4.

—— and Cubba, R. (1986), 'Abrasion resistance hardness testing of rock materials', *International Journal of Rock Mechanics and Mining Science and Geomechanics Abstracts*, 23: 141–9.

—— and Rossman, G. R. (1982), 'Case-hardening of sandstone', *Geology*: 10: 520–3.

—— —— (1985), 'Core softening in cavernously weathered tonalite', *Journal of Geology*, 93: 59–73.

Conway, B. E. (1977), 'The state of water and hydrated ions at interfaces', *Advances in Colloid Interface Science*, 8: 91–211.

Cooke, R. U. (1981), 'Salt weathering in deserts', *Proceedings of the Geologists' Association*, 92: 1–16.

—— and Doornkamp, J. C. (1990), *Geomorphology in Environmental Management* (Oxford: Clarendon Press).

—— and Reeves, R. W. (1976), *Arroyos and Environmental Change in the American South-West* (Oxford: Clarendon Press).

—— and Smalley, I. J. (1968), 'Salt weathering in deserts', *Nature*, 220: 1226–7.

—— and Warren, A. (1973), *Geomorphology in Deserts* (London: Batsford).

Coombs, D. S. and Norris, R. J. (1981), 'The East Abbotsford, Dunedin, New Zealand, landslide of August 8, 1971, and interim report', *Bulletin de Liaison des Laboratoires des Ponts et Chaussées (Paris)*, Special X: 27–34.

Corbel, J. (1959), 'Vitesse de l'érosion', *Zeitschrift für Geomorphologie*, 3: 1–28.

—— (1964), 'L'érosion terrestre, étude quantitative (Méthodes–techniques–résultats)', *Annales de Géographie*, 73: 385–412.

Cordery, I. and Pilgrim, D. H. (1983), 'On the lack of dependence of losses from flood runoff on soil and cover characteristics', *IAHS Publication*, 140: 187–95.

Costa, J. E. (1975), 'Effects of agriculture on erosion and sedimentation in the Piedmont province, Maryland', *Geological Society of America Bulletin*, 86: 1281–6.

—— (1984), 'Physical geomorphology of debris flows', in J. E. Costa and P. J. Fleischer (eds.), *Developments and Applications of Geomorphology* (Berlin: Springer-Verlag), 268–317.

Cotecchia, V. (1987), 'Earthquake-prone environments', in M. G. Anderson and K. S. Richards (eds.), *Slope Stability* (Chichester: Wiley), 287–330.

—— and Melidoro, G. (1974), 'Some principal geological aspects of the landslides of southern Italy', *Bulletin of the International Association of Engineering Geology*, 9: 23–32.

Cotterill, R. M. J. (1985), *The Cambridge Guide to the Natural World* (Cambridge: Cambridge University Press).

Cotton, C. A. and Te Punga, M. T. (1955), 'Solifluxion and periglacially modified landforms of Wellington, New Zealand', *Transactions, Royal Society of New Zealand*, 82: 1001–31.

Coulomb, C. A. (1776), 'Essais sur une application des règles des maximis et minimis à quelques problems de statique relatifs à l'architecture', *Mémoirs présentées par divers Savants* (Paris: Academie des Sciences).

Crandell, D. R. (1989), 'Gigantic debris avalanche of Pleistocene age from ancestral Mount Shasta volcano, California, and debris-avalanche hazard zonation', *US Geological Survey Bulletin*, 1861: 1–32.

—— Miller, C. D., Glicken, H. X., Christiansen, R. L., and Newhall, C. G. (1984), 'Catastrophic debris avalanche from ancestral Mount Shasta volcano, California', *Geology*, 12: 143–6.

Cripps, J. C. and Taylor, R. K. (1981), 'The engineering properties of mudrocks', *Quarterly Journal of Engineering Geology*, 14: 325–46.

Crittenden, R. and Muhs, D. R. (1986), 'Cliff height and slope–angle relationships in a chronosequence of Quaternary marine terraces, San Clemente Island, California', *Zeitschrift für Geomorphologie*, 30: 291–301.

Crossley, R. (1986), 'Sedimentation by termites in the Malawi Valley', in Frostick, L. E. *et al.* (eds.), *Sedimentation in the African Rifts*, Geological Society Special Publication, 25 (London: Geological Society), 191–9.

Crouch, R. J. (1976), 'Field tunnel erosion—a review', *Journal of the Soil Conservation Service of New South Wales*, 32: 98–111.

—— and Blong, R. J. (1989), 'Gully sidewall classification: Methods and applications', *Zeitschrift für Geomorphologie*, 33: 291–305.

—— and Novruzi, T. (1989), 'Threshold conditions for rill initiation on a vertisol, Gunnedah, N.S.W., Australia', *Catena*, 16: 101–10.

Crozier, M. J. (1973), 'Techniques for the morphometric analysis of landslips', *Zeitschrift für Geomorphologie*, 17: 78–101.

—— (1984), 'Field assessment of slope instability', in D. Brunsden and D. B. Prior (eds.), *Slope Instability* (Chichester: Wiley), 103–42.

—— (1986), *Landslides: Causes, Consequences and Environment* (London: Croom Helm).

—— Vaughan, E. E., and Tippett, J. M. (1990), 'Relative instability of colluvium-filled bedrock depressions', *Earth Surface Processes and Landforms*, 15: 329–39.

—— Eyles, R. J., Marx, S. L., McConchie, J. A., and Owen, R. C. (1980), 'Distribution of landslips in the Wairarapa hill country', *New Zealand Journal of Geology and Geophysics*, 23: 575–86.

Cruden, D. M. (1988), 'Thresholds for catastrophic instabilities in sedimentary rock slopes: Some examples from the Canadian Rockies', *Zeitschrift für Geomorphologie Supplementband*, 67: 67–76.

—— (1989), 'Limits to common toppling', *Canadian*

Geotechnical Journal, 26: 737–42.

Cruden, D. M. and Krahn, J. (1973), 'A re-examination of the geology of the Frank Slide', *Canadian Geotechnical Journal*, 10: 581–91.

Cruse, R. M. and Larson, W. E. (1977), 'Effect of soil shear strength on soil detachment due to raindrop impact', *Soil Science Society of America, Journal*, 41: 777–81.

Culling, W. E. (1963), 'Soil creep and the development of hillside slopes', *Journal of Geology*, 71: 127–61.

—— (1983), 'Rate–process theory in geomorphic soil creep', *Catena Supplement*, 4: 191–214.

—— (1986), 'Highly erratic spatial variability of soil-pH on Iping Common, West Sussex', *Catena*, 13: 81–98.

—— (1988), 'Mudflows as a rate process', *Catena*, 15: 249–67.

Culmann, C. (1866), *Graphische Statik* (Zurich).

Culshaw, M. G., Bell, F. G., Cripps, J. C., and O'Hara, M. (eds.) (1987), *Planning and Engineering Geology*, *Engineering Geology Special Publication*, 4 (London: Geological Society).

Curtis, C. D. (1976), 'Chemistry of rock weathering: Fundamental reactions and controls', in E. Derbyshire (ed.), *Geomorphology and Climate* (London: Wiley), 25–57.

Dackombe, R. V. and Gardiner, V. (1983), *Geomorphological Field Manual* (London: George Allen & Unwin).

Dale, T. N. (1923), 'The commercial granites of New England', *US Geological Survey Bulletin*, 738.

Dalrymple, J. B., Blong, R. J., and Conacher, A. J. (1968), 'A hypothetical nine-unit landsurface model', *Zeitschrift für Geomorphologie*, 12: 60–76.

Dalrymple, T. (1960), 'Flood frequency analysis: Manual of Hydrology, Part 3, Flood flow techniques', *US Geological Survey, Water Supply Paper*, 1543A: 1–80.

D'Andrea, D. V., Fisher, R. L., and Fogelson, D. E. (1965), 'Prediction of compressive strength of rock from other rock properties', *US Bureau of Mines, Report of Investigations*, 6702.

Dapples, E. C. (1959), *Basic Geology for Science and Engineering* (New York: Wiley).

Dardis, G. F., Beckedahl, H. R., Bowyer-Bower, T. A. S., and Hanvey, P. M. (1988), 'Soil erosion forms in southern Africa', in G. F. Dardis and B. P. Moon (eds.), *Geomorphological Studies in Southern Africa* (Rotterdam: Balkema), 187–213.

Davidson, D. A. (1983), 'Problems in the determination of plastic and liquid limits of remoulded soils using a drop-cone penetrometer', *Earth Surface Processes and Landforms*, 8: 171–5.

Davidson, M. R. (1985), 'Numerical calculation of saturated-unsaturated infiltration in a cracked soil', *Water Resources Research*, 21: 709–14.

Davis, W. M. (1892), 'The convex profile of bad-land divides', *Science*, 20: 245.

—— (1899), 'The geographical cycle', *Geographical Journal*, 14: 481–504.

—— (1909), *Geographical Essays* (repr. 1954) (New York: Dover).

—— and Snyder, W. H. (1898), *Physical Geography* (Boston: Ginn).

Day, M. J. and Goudie, A. S. (1977), 'Field assessment of rock hardness using the Schmidt test hammer', *British Geomorphological Research Group, Technical Bulletin*, 18: 19–29.

—— Leigh, C., and Young, A. (1980), 'Weathering of rock discs in temperate and tropical soils', *Zeitschrift für Geomorphologie Supplementband*, 35: 11–15.

Dearman, W. R. (1974), 'Weathering classification in the characterisation of rock for engineering purposes in British practice', *Bulletin of the International Association of Engineering Geology*, 9: 33–42.

—— (1976), 'Weathering classification in the characterisation of rock: a revision', *Bulletin of the International Association of Engineering Geology*, 13: 123–7.

Dedkov, A. P. and Mozzherin, V. I. (1984), 'Eroziya i Stok Nanosov na Zemle', *Izdatelstvo Kazanskogo Universiteta*.

Deere, D. U. (1968), 'Geological considerations', in O. C. Zienkiewicz and D. Stagg (eds.), *Rock Mechanics in Engineering Practice* (New York: Wiley), 1–20.

—— and Miller, R. P. (1966), 'Engineering classification and index properties for intact rock', *Technical Report No. AFNL-TR-65-116*, (Air Force Weapons Laboratory, New Mexico).

De Graff, J. M. and Aydin, A. (1987), 'Surface morphology of columnar joints and its significance to mechanics and direction of joint growth', *Geological Society of America Bulletin*, 99: 605–17.

De Ploey, J. (1983), 'Runoff and rill generation on sandy and loamy topsoils', *Zeitschrift für Geomorphologie Supplementband*, 46: 15–23.

—— and Cruz, O. (1979), 'Landslides in the Serra do Mar, Brazil', *Catena*, 6: 111–22.

—— and Poesen, J. (1985), 'Aggregate stability, runoff generation and interrill erosion', in K. S. Richards, R. R. Arnett, and S. Ellis (eds.), *Geomorphology and Soils* (London: George Allen & Unwin), 99–120.

—— —— (1987), 'Some reflections on modelling hillslope processes', *Catena Supplement*, 10: 67–72.

Derski, W., Izbicki, R., Kisiel, I., and Miroz, Z. (1989), *Rock and Soil Mechanics* (Amsterdam: Elsevier).

De Sitter, L. U. (1956), *Structural Geology* (New York: McGraw-Hill).

Dietrich, W. E. and Dorn, R. (1984), 'Significance of thick deposits of colluvium on hillslopes: a case study involving the use of pollen analysis in the coastal mountains of northern California', *Journal of Geology*, 92: 147–58.

—— and Dunne, T. (1978), 'Sediment budget for a small

catchment in mountainous terrain', *Zeitschrift für Geomorphologie Supplementband*, 29: 191–206.

Dinsdale, A. and Moore, F. (1962), 'Viscosity and its measurement', *Institute of Physics and the Physical Society, Monographs for Students* (London: Chapman and Hall).

Dobereiner, L. and De Freitas, M. H. (1986), Geotechnical properties of weak sandstones', *Géotechnique*, 36: 79–94.

Dole, R. B. and Stabler, H. (1909), 'Denudation', in 'Papers in the conservation of water resources', *US Geological Survey, Water Supply Paper*, 234: 78–93.

Donohue, D. (1986), 'Review and analysis of slow mass movement mechanisms with reference to a Weardale catchment, N. England', *Zeitschrift für Geomorphologie Supplementband*, 60: 41–54.

Dorn, R. I. (1989), 'Cation-ratio dating of rock varnish: A geographic assessment', *Progress in Physical Geography*, 13: 559.

—— and Dragovich, D. (1990), 'Interpretation of rock varnish in Australia: Case studies from the arid zone', *Australian Geographer*, 21: 18–32.

—— and Oberlander, T. M. (1982), 'Rock varnish', *Progress in Physical Geography*, 6: 317–67.

Douglas, I. (1967), 'Man, vegetation and the sediment yields of rivers', *Nature*, 215: 925–8.

—— (1968), 'The effects of precipitation chemistry and catchment area lithology on the quality of river water in selected catchments in eastern Australia', *Earth Science Journal*, 2: 126–42.

Douglas, I. (1988), 'Restrictions of hillslope modelling', in M. G. Anderson (ed.), *Modelling Geomorphological Systems* (Chichester: Wiley), 401–20.

Dragovich, D. (1969), 'The origin of cavernous surfaces (tafoni) in granite rocks of southern Australia', *Zeitschrift für Geomorphologie*, 13: 163–81.

Drever, J. I. (1982), *The Geochemistry of Natural Waters* (Englewood-Cliffs: Prentice Hall).

Dunne, T. (1978a), 'Rates of chemical denudation of silicate rocks in tropical catchments', *Nature*, 274: 244–6.

—— (1978b), 'Field studies of hillslope flow processes', in M. J. Kirkby, (ed.), *Hillslope Hydrology* (Chichester: Wiley), 227–93.

—— and Aubry, B. F. (1986), 'Evaluation of Horton's theory of sheetwash and rill erosion on the basis of field experiments', in A. D. Abrahams (ed.), *Hillslope Processes* (Boston: Allen and Unwin), 31–53.

—— and Black, R. G. (1970a), 'An experimental investigation of runoff production in permeable soils', *Water Resources Research*, 6: 478–90.

—— and Black, R. G. (1970b), 'Partial area contributions to storm runoff in a small New England watershed', *Water Resources Research*, 6: 1296–311.

—— and Dietrich, W. E. (1980), 'Experimental studies of Horton overland flow on tropical hillslopes: 1. Soil conditions, infiltration and frequency of runoff',

Zeitschrift für Geomorphologie Supplementband, 35: 40–59.

—— —— and Brunengo, M. J. (1978), 'Recent and past erosion rates in semiarid Kenya', *Zeitschrift für Geomorphologie, Supplementband*, 29: 130–40.

Durum, W. H., Heidel, S. G., and Tison, L. J. (1960), 'World-wide runoff of dissolved solids', *International Association of Scientific Hydrology*, 51: 618–28.

Dury, G. H. (1969), 'Rational descriptive classification of duricrusts', *Earth Science Journal*, 3: 77–86.

Eberl, D. D. (1984), 'Clay mineral formation and transformation in rocks and soils', *Philosophical Transactions, Royal Society, London*, A311: 241–57.

Eichler, H. (1981), 'Kleinformen der hocharktischen Verwitterung in Bereich der Oobloya Bay, N-Ellesmere Island, NWT, Kanada: Formengenese und Prozesse', *Heidelberger Geographische Arbeiten*, 69: 465–86.

Einstein, H. H. and Dershowitz, W. S. (1990), 'Tensile and shear fracturing in predominantly compressive stress fields: a review', *Engineering Geology*, 29: 149–72.

Einstein, H. J., Bruhn, R. W., and Hirschfield, R. C. (1970), 'Mechanics of jointed rock. Experimental and theoretical studies', *Department of Civil Engineering Report* (Cambridge, Mass.: MIT).

Eissa, E. A. and Kazi, A. (1988), 'Relation between static and dynamic Young's moduli of rocks', *International Journal of Rock Mechanics and Mining Sciences and Geomechanics Abstracts*, 25: 479–82.

Ellen, S. D. and Wieczorek, G. F. (eds.) (1988), 'Landslides, floods and marine effects of the storm of January 3–5, 1982, in the San Francisco Bay Region, California', *US Geological Survey, Professional Paper*, 1434: 1–310.

Ellison, W. D. (1947), 'Soil erosion studies', *Agricultural Engineering*, 28: 145–6, 197–201, 245–8, 297–300, 349–51, 402–5, 442–50.

Elton, D. J. and Hadj-Hamou, T. (1990), 'Liquefaction potential map for Charleston, South Carolina', *Journal of Geotechnical Engineering*, 116: 244–65.

Elwell, H. A. (1977), 'Soil loss estimation system for southern Africa', *Dept. of Conservation and Extension, Research Bulletin*, 22 (Salisbury, Rhodesia).

Emmett, W. W. (1978), 'Overland flow', in M. J. Kirkby (ed.), *Hillslope Hydrology* (Chichester: Wiley), 145–76.

Ericksen, G. E., Plafker, G., and Fernandez, J. C. (1970), 'Preliminary report on the geological events associated with the May 31, 1970, Peru earthquake', *US Geological Survey, Circular*, 639: 1–25.

Erismann, T. H. (1979), 'Mechanisms of large landslides', *Rock Mechanics*, 12: 15–46.

Evans, I. S. (1970), 'Salt crystallization and rock

weathering: a review', *Revue de Géomorphologie Dynamique*, 19: 153–77.

Evans, R. S. (1981), 'An analysis of secondary toppling rock failures: The stress redistribution method', *Quarterly Journal of Engineering Geology, London*, 14: 77–86.

Evans, S. F. (1989), 'Rock avalanche run-up record', *Nature*, 340: 271.

Evans, S. G., Claque, J. J., Woodsworth, G. J., and Hungr, O. (1989), 'The Pandemonium Creek rock avalanche, British Columbia', *Canadian Geotechnical Journal*, 26: 427–46.

Everett, A. G. (1979), 'Secondary permeability as a possible factor in the origin of debris avalanches associated with heavy rainfall', *Journal of Hydrology*, 43: 347–54.

Exon, N. F., Langford-Smith, T., and McDougall, I. (1970), 'The age and geomorphic correlations of deep weathering profiles, silcrete and basalt in the Roma-Amby District, Queensland', *Journal of the Geological Society of Australia*, 17: 21–30.

Eyles, R. J., Crozier, M. J., and Wheeler, R. H. (1978), 'Landslips in Wellington City', *New Zealand Geographer*, 34: 58–74.

Fahey, B. D. (1983), 'Frost action and hydration as rock weathering mechanisms on schist: a laboratory study', *Earth Surface Processes and Landforms*, 8: 535–45.

—— (1986*a*), 'A comparative laboratory study of salt crystallization and salt hydration as potential weathering agents in deserts', *Geografiska Annaler*, 68A: 107–11.

—— (1986*b*), 'Weathering pit development in the central Otago mountains of southern New Zealand', *Arctic and Alpine Research*, 18: 337–48.

—— and Dagesse, D. F. (1984), 'An experimental study of the effect of humidity and temperature variations on the granular disintegration of argillaceous carbonate rocks in cold climates', *Arctic and Alpine Research*, 16: 291–8.

Fair, T. J. (1947), 'Slope form and development in the interior of Natal', *Geological Society of South Africa Transactions*, 50: 105–20.

—— (1948*a*), 'Slope form and development in the coastal hinterland of Natal', *Geological Society of South Africa, Transactions*, 51: 37–53.

—— (1948*b*), 'Hillslopes and pediments of the semiarid Karoo', *South African Geographical Journal*, 30: 71–9.

Fairbridge, R. W. (1988), 'Cyclical patterns of exposure, weathering and burial of cratonic surfaces, with some examples from North America and Australia', *Geografiska Annaler*, 70A: 277–83.

Falconer, J. D. (1911), *The Geology and Geography of Northern Nigeria* (London: Macmillan).

FAO (1965), *Soil Erosion by Water: Some Measures for its Control on Cultivated Lands* (Rome: FAO/UNESCO).

Farmer, I. W. (1968), *Engineering Properties of Rocks* (London: Butler and Tanner).

—— (1983), *Engineering Behaviour of Rocks*, 2nd edn., (London: Chapman and Hall).

Farres, P. (1978), 'The role of time and aggregate size in the crusting process', *Earth Surface Processes*, 3: 243–54.

—— and Cousen, S. M. (1985), 'An improved method of aggregate stability measurement', *Earth Surface Processes and Landforms*, 10: 321–9.

Feda, J. (1966), 'Structural stability of subsident loess soil from Praha-Dejvice', *Engineering Geology*, 1: 201–19.

Fellenius, W. (1936), 'Calculation of the stability of earth dams', *Transactions, 2nd Congress on Large Dams (Washington, DC)*, 4: 445–65.

Ferguson, H. F. and Hamel, J. V. (1981), 'Valley stress relief in flat-lying sedimentary rocks', *Proceedings of the International Symposium on Weak Rock (Tokyo)*, 1235–40.

Fieldes, M. (1968), 'Clay mineralogy', *New Zealand Soil Bureau Bulletin*, 26, pt. 2: 22–39.

—— and Swindale, L. D. (1954), 'Chemical weathering of silicates in soil formation', *New Zealand Journal of Science and Technology*, 36B: 140–54.

Finlayson, B. L. (1985), 'Soil creep: A formidable fossil of misconception', in K. S. Richards, R. R. Arnett, and S. Ellis (eds.), *Geomorphology and Soils* (London: George Allen & Unwin), 141–58.

Fischer, R. V. and Schminke, H. U. (1984), *Pyroclastic Rocks* (Berlin: Springer-Verlag).

Fisher, O. (1866), 'On the disintegration of a chalk cliff', *Geological Magazine*, 3: 354–6.

Fitzharris, B. B. and Owens, I. F. (1984), 'Avalanche tarns', *Journal of Glaciology*, 30: 308–12.

FitzPatrick, E. A. (1983), *Soils: Their Formation, Classification and Distribution* (Harlow: Longman).

Fleming, R. W. and Johnson, A. M. (1975), 'Rates of seasonal creep of silty clay soil', *Quarterly Journal of Engineering Geology*, 8: 1–29.

Folk, R. L. and Patton, E. B. (1982), 'Buttressed expansion of granite and development of grus in central Texas', *Zeitschrift für Geomorphologie*, 26: 17–32.

Fookes, P. G., Sweeney, H., Manby, C. N. D., and Martin, R. I. P. (1985), 'Geological and geotechnical engineering aspects of low-cost roads in mountainous terrain', *Engineering Geology*, 21: 1–152.

Ford, D. C. and Williams, P. W. (1989), *Karst Geomorphology and Hydrology* (London: Unwin Hyman).

Foster, G. R. and Meyer, L. D. (1975), 'Mathematical simulation of upland erosion by fundamental erosion mechanics. Present and prospective technology for predicting sediment yields and sources', *US Dept. of Agriculture, Agricultural Research Services Report*, ARS-S-40: 190–207.

—— —— and Onstad, C. A. (1977*a*), 'A runoff erosivity

factor and variable slope length exponents for soil loss estimates', *American Society of Agricultural Engineers, Transactions*, 20: 683–7.

—— —— —— (1977*b*), 'An erosion equation derived from basic erosion principles', *American Society of Agricultural Engineers, Transactions*, 20: 678–82.

Fournier, F. (1960), *Climat et Erosion: la relation entre l'érosion du sol par l'eau et les précipitations atmosphériques* (Paris: Presses Universitaire de France), 1–201.

Francis, S. C. (1987), 'Slope development through the threshold concept', in M. G. Anderson and K. S. Richards (eds.), *Slope Stability* (Wiley: Chichester), 601–24.

Francou, B. (1988*a*), 'Talus formation in high mountain environments, Alps and tropical Andes' (abstract) (Caen: Centre de Géomorphologie du CNRS).

—— (1988*b*), 'Éboulis statifiés dans les Hautes Andes Centrales du Pérou', *Zeitschrift für Geomorphologie*, 32: 47–76.

Franklin, J. A. and Chandra, R. (1972), 'The slake durability test', *International Journal for Rock Mechanics and Mining Sciences*, 9: 325–41.

Franzle, O. (1971), 'Die Opferkessel im quartzitischen Sandstein von Fontainebleau', *Zeitschrift für Geomorphologie*, 15: 212–35.

Fredlund, D. G. and Krahn, J. (1977), 'Comparison of slope stability methods of analysis', *Canadian Geotechnical Journal*, 14: 429–39.

Freeze, R. A. (1972), 'Role of subsurface flow in generating surface runoff. 2. Upstream source areas', *Water Resources Research*, 8: 1271–83.

French, H. M. (1976), *The Periglacial Environment* (London: Longman).

Friedmann, E. I. (1982), 'Endolithic microorganisms in the Antarctic cold desert', *Science*, 215: 1045–53.

Freitas, M. H. De and Watters, R. J. (1973), 'Some field examples of toppling failures', *Géotechnique*, 23: 495–514.

Gabrielyan, H. (1971), 'Quantitative characteristics of the denudation of the Caucasus', *Acta Geographia Debrecina*, 10: 37–40.

Gage, M. (1966), 'Franz Josef Glacier', *Ice*, 20: 26–7.

Galan, C. and Lagarde, J. (1988), 'Morphologie et evolution des cavernes et formes superficielles dans les quartzites du Roraima (Venezuela)', *Karstologie*, 11–12: 49–60.

Gardner, J. S. (1977), 'High magnitude rockfall-rockslide: frequency and geomorphic significance in the Highwood Pass area, Alberta', *Great Plains–Rocky Mountain Geographical Journal*, 6: 228–39.

—— (1983*a*), 'Accretion rates on some debris slopes in the Mt. Rae area, Canadian Rocky Mountains', *Earth Surface Processes and Landforms*, 8: 347–55.

—— (1983*b*), 'Rockfall frequency and distribution in the Highwood Pass area, Canadian Rocky Mountains', *Zeitschrift für Geomorphologie*, 27: 311–24.

Geikie, A. (1880), 'Rockweathering, as illustrated in Edinburgh churchyards', *Proceedings, Royal Society of Edinburgh*, 10: 518–32.

Geological Society Engineering Group Working Party Report (1990), 'Tropical Residual Soils', *Quarterly Journal of Engineering Geology*, 23: 1–101.

Gerber, E. (1980), 'Geomorphological problems in the Alps', *Rock Mechanics Supplement*, 9: 93–107.

—— and Scheidegger, A. E. (1969), 'Stress-induced weathering of rock masses', *Ecologae Geologicae Helveticae*, 62: 401–15.

—— —— (1973), 'Erosional and stress-induced features on steep slopes', *Zeitschrift für Geomorphologie Supplementband*, 18: 38–49.

—— —— (1975), 'Geomorphological evidence for the geophysical stress-field in mountain massifs', *Revista Italiana Di Geofisica*, 2: 47–52.

Gerits, J., Imeson, A. C., Verstraten, J. M., and Bryan, R. B. (1987), 'Rill development and badland regolith properties', *Catena Supplement*, 8: 141–60.

Gerrard, A. J. (1985), 'Soil erosion and landscape stability in southern Iceland: a tephrochronological approach', in K. S. Richards, R. R. Arnett, and S. Ellis (eds.), *Geomorphology and Soils* (London: George Allen & Unwin), 78–94.

—— (1990), 'Soil variations on hillslopes in humid temperate climates', *Geomorphology*, 3: 225–44.

Gerson, R. and Grossman, S. (1987), 'Geomorphic activity on escarpments and associated fluvial systems in hot deserts', in M. R. Rampino, J. E. Sanders, W. S. Newman, and L. K. Königsson (eds.), *Climate, History, Periodicity, and Predictability* (New York: Van Nostrand Reinhold), 300–22.

Gibbs, H. J. and Bara, J. P. (1962), 'Predicting surface subsidence from basic soil tests', *American Society for Testing Materials, Special Technical Publication*, No. 322: 231–47.

Gibbs, H. S. (1945), 'Tunnel-gully erosion on the Wither Hills, Marlborough', *New Zealand Journal of Science and Technology*, A27: 135–46.

Gilbert, G. K. (1904), 'Domes and dome structure of the High Sierra', *Geological Society of America Bulletin*, 15: 29–36.

—— (1909), 'The convexity of hilltops', *Journal of Geology*, 17: 344–50.

Gile, L. H., Hawley, J. W., and Grossman, R. B. (1981), 'Soils and geomorphology in the Basin and Range area of southern New Mexico', *Guidebook to the Desert Project, New Mexico Bureau of Mines and Mineral Resources Memoir*, 39.

Gillott, J. E., Penner, E., and Eden, W. J. (1974), 'Microstructure of Billing Shale and biochemical alteration products', *Canadian Geotechnical Journal*, 11: 482–9.

Giraud, A., Rochet, L., and Antoine, P. (1990), 'Processes of slope failure in crystallophyllian formations', *Engineering Geology*, 29: 241–53.

Goguel, J. (1978), 'Scale-dependent rockslide mechanisms, with emphasis on the role of pore fluid vaporization', in B. Voight (ed.) *Rockslides and Avalanches*, i. (Amsterdam: Elsevier), 693–705.

Goldich, S. S. (1938), 'A study on rock weathering', *Journal of Geology*, 46: 17–58.

Goldstein, M. and Ter-Stepanian, G. (1957), 'The long-term strength of clays and deep creep of slopes', *Proceedings, 4th International Conference of Soil Mechanics and Foundation Engineering*, 2: 311–14.

Goodman, R. E. (1970), 'The deformability of joints' in *Determination of the in situ modulus of deformation of rock*, Special Publication, 477 (American Society for Testing and Materials), 174–96.

—— (1980), *Introduction to Rock Mechanics* (New York: Wiley).

—— and Bray, J. W. (1976), 'Toppling of rock slopes', in *Rock Engineering for Foundations and Slopes* (New York: American Society of Civil Engineers).

Gordon, J. E. (1988), *The Science of Structures and Materials* (New York: Scientific American Library, Freeman).

Gordon, R. B. (1979), 'Denudation rate of central New England determined from estuarine sedimentation', *American Journal of Science*, 279: 632–42.

Gossman, H. (1976), 'Slope modelling with changing boundary conditions: Effects of climate and lithology', *Zeitschrift für Geomorphologie, Supplementband*, 25: 72–88.

Goudie, A. S. (1973), *Duricrusts in Tropical and Subtropical Landscapes* (Oxford: Clarendon Press).

—— (1974), 'Further experimental investigation of rock weathering by salt and other mechanical processes', *Zeitschrift für Geomorphologie Supplementband*, 21: 1–12.

—— (ed.) (1981), *Geomorphological Techniques* (London: George Allen & Unwin).

—— (1985), 'Duricrusts and landforms', in K. S. Richards, R. R. Arnett, and S. Ellis (eds.), *Geomorphology and Soils* (London: George Allen & Unwin), 37–57.

—— and Watson, A. (1984), 'Rock block monitoring of rapid salt weathering in southern Tunisia', *Earth Surface Processes and Landforms*, 9: 95–8.

Grabowska-Olszewska, B., Osipov, V., and Sokolov, V. (1984), *Atlas of the Microstructure of Clay Soils* (Warsaw: PWN).

Grant, P. J. (1963), 'Forests and recent climatic history of the Huiarau Range, Urewera region, North Island', *Transactions, Royal Society of New Zealand, Botany*, 2: 144–72.

—— (1965), 'Major regime changes of the Tukituki River, Hawke's Bay, since about 1650 AD', *Journal of Hydrology (N.Z.)*, 4: 17–30.

—— (1985), 'Major periods of erosion and alluvial sedimentation in New Zealand during the late Holocene', *Journal of the Royal Society of New Zealand*, 15: 67–121.

Grant-Taylor, T. L. (1964), 'Stable angles in Wellington greywacke', *New Zealand Engineering*, 19: 129–30.

Gray, D. H. (1970), 'Effects of forest clear-cutting on the stability of natural slopes', *Bulletin of the Association of Engineering Geologists*, 7: 45–66.

—— and Leiser, A. J. (1982), *Biotechnical Slope Protection and Erosion Control* (New York: Van Nostrand Reinhold).

Greenway, D. R. (1987), 'Vegetation and slope stability', in M. G. Anderson and K. S. Richards (eds.), *Slope Stability* (Chichester: Wiley), 187–230.

Gregory, K. J. and Walling, D. E. (1973), *Drainage Basin Form and Process* (London: Arnold).

Griffith, A. A. (1921), 'The phenomenon of rupture and flow in solids', *Philosophical Transactions, Royal Society, London*, A221: 163–98.

Griffiths, A. and McSaveney, M. J. (1983), 'Hydrology of a basin with extreme rainfalls: Cropp River, New Zealand', *New Zealand Journal of Science*, 26: 293–306.

Griggs, D. T. (1936), 'The factor of fatigue in rock exfoliation', *Journal of Geology*, 44: 783–96.

Grim, R. E. (1968), *Clay Mineralogy* (New York: McGraw-Hill).

Grimstad, E. and Nesdal, S. (1990), 'The Loen rockslides: A historical review', in N. Barton and O. Stephansson (eds.), *Rock Joints* (Rotterdam: Balkema), 3–8.

Grove, J. M. (1972), 'The incidence of landslides, avalanches and floods in western Norway during the Little Ice Age', *Arctic and Alpine Research*, 4: 131–8.

Gryta, J. J. and Bartholomew, M. J. (1989), 'Factors influencing the distribution of debris avalanches associated with the 1969 Hurricane Camille in Nelson County, Virginia', *Geological Society of America, Special Paper*, 236: 15–28.

Gunn, R. and Kinzer, G. D. (1949), 'The terminal velocity of fall from water droplets in stagnant air', *Journal of Meteorology*, 6: 243–8.

Gutiérrez, M., Benito, G., and Rodriguez, J. (1988), 'Piping in badland areas of the middle Ebro Basin, Spain', *Catena Supplement*, 13: 49–60.

Habermehl, M. A. (1980), 'The Great Artesian Basin, Australia', *BMR Journal of Geology and Geophysics*, 5: 9–38.

Habib, P. (1975), 'Production of gaseous pore pressure during rock slides', *Rock Mechanics*, 7: 193–7.

Hack, J. T. (1960), 'Interpretation of erosional topography in humid temperate regions', *American Journal of Science*, 258-A: 80–97.

—— (1979), 'Rock control and tectonism: Their importance in shaping the Appalachian highlands', *US Geological Survey, Professional Paper*, 1126-B.

Haefli, R. (1948), 'The stability of slopes acted upon by parallel seepage', *Proceedings, 2nd International*

Conference on Soil Mechanics and Foundation Engineering, 1: 134–48.

—— (1965), 'Creep and progressive failure in snow, soil, rock and ice', Proceedings, 6th International Conference on Soil Mechanics and Foundation Engineering, 3: 134–48.

Haigh, M. J. (1978), 'Evolution of slopes on artificial landforms: Blaenavon, U.K.', University of Chicago Department of Geography, Research Paper, 183.

Haimson, B. C. (1974), 'Mechanical behaviour of rock under cyclic loading', Proceedings, 3rd Congress of the International Society for Rock Mechanics (Denver), 2A: 373–8.

Hall, A. M. (1986), 'Deep weathering patterns in north-east Scotland and their geomorphological significance', Zeitschrift für Geomorphologie, 30: 407–22.

—— (1988), 'The characteristics and significance of deep weathering in the Gaick area, Grampian Highlands, Scotland', Geografiska Annaler, 70A: 309–14.

Hall, K. (1986), 'Rock moisture content in the field and the laboratory and its relationship to mechanical weathering studies', Earth Surface Processes and Landforms, 11: 131–42.

—— (1987), 'The physical properties of quartz-micaschist and their application to freeze–thaw weathering studies in the maritime Antarctic', Earth Surface Processes and Landforms, 12: 137–49.

Hallbauer, D. K., Wager, H., and Cook, N. G. W. (1973), 'Some observations concerning the microscopic and mechanical behaviour of quartzite specimens in stiff, triaxial compression tests', International Journal of Rock Mechanics and Mining Sciences, 10: 713–26.

Hallet, B. (1983), 'The breakdown of rock due to freezing: A theoretical model', Proceedings, 4th International Conference on Permafrost (Fairbanks, Alaska). 433–8.

Hampton, M. A. (1972), 'The role of subaqueous debris flow in generating turbidity currents', Journal of Sedimentary Petrology, 42: 775–93.

—— (1979), 'Buoyancy in debris flows', Journal of Sedimentary Petrology, 49: 753–8.

Hamrol, A. A. (1961), 'A quantitative classification of the weathering and weatherability of rocks', Proceedings, 5th International Conference on Soil Mechanics and Foundation Engineering, 2: 771–4.

Handy, R. L. (1976), 'Measurement of in situ shear strength', Proceedings, Conference on In Situ Measurement of Soil Properties, American Society of Civil Engineers, 2: 143–9.

Hansbo, S. (1979), 'Mechanical behaviour of clay explained in microstructural terms', in K. E. Easterling (ed.), Mechanisms of Deformation and Fracture (Oxford: Pergamon), 321–7.

Hansen, A. (1984), 'Landslide hazard analysis', in D. Brunsden and D. B. Prior (eds.), Slope Instability (Chichester: Wiley), 523–602.

Hansen, W. R. (1965), 'Effects of the earthquake of March 27, 1964, at Anchorage, Alaska', US Geological Survey, Professional Paper, 542-A: 1–68.

Harden, J. W. (1990), 'Soil development on stable landforms and implications for landscape studies', Geomorphology, 3: 391–8.

—— and Taylor, E. M. (1983), 'A quantitative comparison of soil development in four climatic regimes', Quaternary Research, 20: 342–59.

Hardy, H. R. and Chugh, Y. P. (1971), 'Failure of geologic materials under low-cycle fatigue', Proceedings, 6th Canadian Rock Mechanics Symposium (Montreal) (Ottawa: Mines Branch, Dept of Energy Mines and Resources), 33–47.

Harp, E. L., Wells, W. G., and Sarmiento, J. G. (1990), 'Pore pressure response during failure in soils', Geological Society of America Bulletin, 102: 428–38.

—— Wilson, R. C., and Wieczorek, G. F. (1981), 'Landslides from the February 4, 1976, Guatemala earthquake', US Geological Survey, Professional Paper, 1204-A: 1–35.

Harris, C. (1987), 'Mechanisms of mass movement in periglacial environments', in M. G. Anderson and K. S. Richards (eds.), Slope Stability (Chichester: Wiley), 531–59.

Harrison, J. V. and Falcon, N. L. (1937), 'The Saidmarreh landslip, southwest Iran', Geographical Journal, 89: 42–7.

Haughey, A. (1979), 'Man's effect on water quality', Soil and Water, 15: 9–12.

Hawkes, I. and Mellor, M. (1970), 'Uniaxial testing in rock mechanics laboratories', Engineering Geology, 4: 177–285.

Hawkins, A. B., Lawrence, M. S., and Privett, K. D. (1988), 'Implications of weathering on the engineering properties of the Fuller's Earth formation', Géotechnique, 38: 517–32.

Haxby, W. F. and Turcotte, D. L. (1976), 'Stresses induced by the addition or removal of overburden and associated thermal effects', Geology, 4: 181–4.

Head, K. H. (1980–6), Manual of Soil Laboratory Testing, 3 vols. (London: Pentech Press).

—— (1989), Soil Technicians' Handbook (London: Pentech Press).

Hedges, J. (1969), 'Opferkessel', Zeitschrift für Geomorphologie, 13: 22–55.

Heede, B. H. (1971), 'Characteristics and processes of soil piping in gullies', USDA Forest Service, Research Paper, RM-68 (Fort Collins, Colorado: Rocky Mountain Forest and Range Experiment Station).

—— (1976), 'Gully development and control', USDA Forest Service, Research Paper, RM-169: 1–42.

—— (1977), 'Case study of a watershed rehabilitation project: Alkali Creek, Colorado', US Forest Service, Research Paper, RM189.

Heim, A. (1882), 'Der Bergsturz von Elm', Zeitschrift der deutschen geologischen Gesellschaft, 34: 74–115.

Heim, A. (1933), *Bergsturz und Menschenleben* (Zurich: Fretz und Wasmuth).

Hencher, S. R. (1987), 'The implications of joints and structures for slope stability', in M. G. Anderson and K. S. Richards (eds.), *Slope Stability* (Chichester: Wiley), 145–86.

—— and Richards, L. R. (1989), 'Laboratory direct shear testing of rock discontinuities', *Ground Engineering*, 22: 24–31.

Henderson, P. (1982), *Inorganic Geochemistry* (Oxford: Pergamon).

Hendron, A. J. (1968), 'Mechanical properties of rock', in K. G. Stagg and O. C. Zienkiewitz (eds.), *Rock Mechanics in Engineering Practice* (New York: Wiley), 21–53.

—— and Patton, F. D. (1985), 'The Vaiont slide, a geotechnical analysis based on new geologic observations of the failure surface', *U.S. Army Corps of Engineers, Technical Report*, GL-85-5 (Washington, DC).

Henkel, D. J. (1982), 'Geology, geomorphology and geotechnics', *Géotechnique*, 32: 175–94.

Herwitz, S. R. (1986), 'Infiltration-excess caused by stemflow in a cyclone-prone tropical forest', *Earth Surface Processes and Landforms*, 11: 401–12.

Heuberger, H., Masch, L., Preuss, E., and Schrocker, A. (1984), 'Quaternary landslides and rock fusion in central Nepal and in the Tyrolean Alps', *Mountain Research and Development*, 4: 345–62.

Hewitt, K. (1988), 'Catastrophic landslide deposits in the Karakoram Himalaya', *Science*, 242: 64–7.

Hewlett, J. D. (1961), *Watershed Management, Report for 1961* (Asheville, NC: South Eastern Forest Experimental Station, U.S. Forest Service).

—— and Hibbert, A. R. (1967), 'Factors affecting the response of small watersheds to precipitation in humid areas', in *International Symposium on Forest Hydrology* (Oxford: Pergamon), 275–90.

Higgins, C. G. (1982), 'Grazing-step terracettes and their significance', *Zeitschrift für Geomorphologie*, 26: 459–72.

Hillel, D. (1971), *Soil and Water: Physical Principles and Processes* (New York: Academic Press).

Hills, R. C. (1970), 'The determination of the infiltration capacity of field soils using the cylinder infiltrometer', *British Geomorphological Research Group, Technical Bulletin*, 3.

Hockmann, A. and Kessler, D. W. (1950), 'Thermal moisture expansion studies of some domestic granites', *Journal of Research, US Bureau of Standards*, 44: 395–410.

Hodder, A. P. W. (1984), 'Thermodynamic interpretation of weathering indices and its application to engineering properties of rocks', *Engineering Geology*, 20: 241–51.

—— and Hetherington, J. R. (1992), 'A quantitative study of the weathering of greywacke', *Engineering Geology*, 31: 353–68.

—— Green, B. E., and Lowe, D. J. (1990), 'A two-stage model for the formation of clay minerals from tephra-derived volcanic glass', *Clay Minerals*, 25: 313–27.

Hodges, W. K. (1982), 'Hydrologic characteristics of a badland pseudopediment slope system during simulated rainstorm experiments', in R. B. Bryan and A. Yair (eds.), *Badland Geomorphology and Piping*, (Norwich: Geo Books), 127–52.

—— and Bryan, R. B. (1982), 'The influence of material behaviour on runoff initiation in the Dinosaur Badlands, Canada', in R. B. Bryan and A. Yair (eds.), *Badland Geomorphology and Piping* (Norwich: Geo Books), 13–46.

Hoek, E. (1968), 'The brittle failure of rock', in O. C. Zienkiewicz and D. Stagg (eds.), *Rock Mechanics in Engineering Practice* (New York: Wiley), 99–124.

—— (1990), 'Estimating Mohr-Coulomb friction and cohesion values from the Hoek–Brown failure Criterion', *International Journal of Rock Mechanics and Mining Sciences and Geomechanics Abstracts*, 27: 227–9.

—— and Bray, J. (1977), *Rock Slope Engineering*, 2nd edn. (London: Institution of Mining and Metallurgy).

—— and Brown, E. T. (1980), 'Empirical strength criterion for rock masses', *Journal of the Geotechnical Engineering Division, Proceedings of the American Society of Civil Engineers*, 106 (GT9): 1013–35.

—— and Franklin, J. A. (1968), 'A simple triaxial cell for field and laboratory testing of rock', *Transactions, Institution of Mining and Metallurgy*, 77: A22–6.

Hohberger, K. and Einsele, G. (1979), 'Die Bedeutung des Lösungsabtrags verschiedener Gesteine für die Landschaftsentwicklung in Mitteleuropa', *Zeitschrift für Geomorphologie*, 23: 361–82.

Hollingworth, S. E., Taylor, J. H., and Kellaway, G. A. (1944), 'Large-scale superficial structures in the Northampton ironstone field', *Quarterly Journal of the Geological Society*, 100: 1–44.

Holzhausen, G. R. (1989), 'Origin of sheet structure, 1. Morphology and boundary conditions', *Engineering Geology*, 27: 225–78.

Hoogmoed, W. B. and Stroosnijder, L. (1984), 'Crust formation on sandy soils in the Sahel: 1. Rainfall and infiltration', *Soil and Tillage Research*, 4: 5–24.

Horn, H. M. and Deere, D. U. (1962), 'Frictional characteristics of minerals', *Géotechnique*, 12: 319–35.

Horswill, O. and Horton, A. (1976), 'Cambering and valley bulging in the Gwash Valley at Empingham, Rutland', *Philosophical Transactions, Royal Society, London*, A283: 427–62.

Horton, R. E. (1933), 'The role of infiltration in the hydrological cycle', *American Geophysical Union, Transactions*, 14: 446–60.

—— (1945), 'Erosional development of streams and their drainage basins: hydrophysical approach to quantitative morphology', *Geological Society of*

America Bulletin, 56: 275–370.

Hoskins, E. R., Jaeger, J. C., and Rosengren, K. J. (1967), 'A medium-scale direct friction experiment', *International Journal of Rock Mechanics and Mining Sciences*, 5: 143–54.

Howard, A. D., Kochel, R. C., and Holt, H. E. (1988), *Sapping Features of the Colorado Plateau* (Washington, DC: NASA).

Howard, J. K. and Higgins, C. G. (1987), 'Dimensions of grazing-step terracettes and their significance', in V. Gardiner (ed.), *International Geomorphology 1986, Proceedings of the 1st Conference* (Chichester: Wiley), 545–68.

Hsi, G. and Nath, J. H. (1970), 'Wind drag within a simulated forest', *Journal of Applied Meteorology*, 9: 592–602.

Hsü, K. J. (1975), 'Catastrophic debris streams (Sturzstroms) generated by rockfalls', *Geological Society of America Bulletin*, 86: 129–40.

—— (1978), 'Albert Heim: Observations on landslides and relevance to modern interpretations', in B. Voight (ed.), *Rockslides and Avalanches* (Amsterdam: Elsevier), 71–93.

—— (1989), *Physical Principles of Sedimentology* (Berlin: Springer-Verlag).

Hu, J. and Wan, Y. (1987), 'The origins and classification of earthquake landforms in the alluvial plains of eastern China', *Zeitschrift für Geomorphologie Supplementband*, 63: 167–71.

Huang, C., Bradford, J. M., and Cushman, J. H. (1982), 'A numerical study of raindrop impact phenomena: the rigid case', *Soil Science Society of America Journal*, 46: 14–19.

Huang, W. H. and Kiang, W. C. (1972), 'Laboratory dissolution of plagioclase feldspars in water and organic acids at room temperature', *American Mineralogist*, 57: 1849–59.

Hudec, P. P. and Sitar, N. (1975), 'Effect of water sorbtion on carbonate rock expansivity', *Canadian Geotechnical Journal*, 12: 179–86.

Huder, J. (1976), 'Creep in Büdner Schist', in N. Janbu, F. Jørstad, and B. Kjaernsli (eds.), *Laurits Bjerrum Memorial Volume* (Oslo: Norwegian Geotechnical Institute), 125–53.

Hudson, J. A., Brown, E. T., and Fairhurst, C. (1971), 'Shape of the complete stress-strain curve for rock', *Proceedings, 13th US Rock Mechanics Symposium* (*University of Illinois, Urbana*), 773–95.

—— and Cooling, C. M. (1988), 'In situ rock stresses and their measurement in the U.K.: pt. 1, the current state of knowledge', *International Journal of Rock Mechanics and Mining Sciences*, 25: 363–70.

Hudson, N. W. (1961), 'An introduction to the mechanics of soil erosion under conditions of subtropical rainfall', *Rhodesia Science Association, Proceedings*, 49: 14–25.

—— (1971), *Soil Conservation* (London: Batsford).

—— and Jackson, D. C. (1959), 'Results achieved in the measurement of erosion and runoff in Southern Rhodesia', Paper presented to the *Third Inter-African Soil Conference* (*Dalaba*), 1959.

Huggett, R. J. (1985), *Earth Surface Systems* (Berlin: Springer-Verlag).

Hungr, O., Morgan, G. C., and Kellerhals, R. (1984), 'Quantitative analysis of debris torrent hazards for design of remedial measures', *Canadian Geotechnical Journal*, 21: 663–77.

Huppert, F. (1986), 'Petrology of Soft Tertiary Sedimentary Rocks and its Relationship to Geomechanical Behaviour, Central North Island, New Zealand', Ph.D. thesis, University of Auckland.

—— (1988), 'Influence of microfabric on geomechanical behaviour of Tertiary fine-grained sedimentary rocks from central North Island, New Zealand', *Bulletin of the International Association of Engineering Geology*, 38: 83–9.

Hurault, J. (1971), 'La significance morphologique des lavaka', *Revue de Géomorphologie Dynamique*, 19: 121–8.

Hutchinson, J. N. (1968), 'Mass movement', in R. W. Fairbridge (ed.), *The Encyclopedia of Geomorphology* (New York: Reinhold), 688–95.

—— (1973), 'The response of London Clay to differing rates of toe erosion', *Geologica Applicata E Idrologia*, 8: 221–39.

—— (1977), 'Assessment of the effectiveness of corrective measures in relation to geological conditions and types of slope movement', *Bulletin of the International Association of Engineering Geology*, 16: 131–55.

—— (1984), 'An influence line approach to the stabilization of slopes by cuts and fills', *Canadian Geotechnical Journal*, 21: 363–70.

—— (1986), 'A sliding-consolidation model for flow slides', *Canadian Geotechnical Journal*, 23: 115–26.

—— (1988), 'General report: Morphological and geotechnical parameters of landslides in relation to geology and hydrogeology', *Proceedings, 5th International Symposium on Landslides (Lausanne), 1* (Rotterdam: Balkema), 3–35.

—— and Bhandari, R. K. (1971), 'Undrained loading, a fundamental mechanism of mudflows and other mass movements', *Géotechnique*, 21: 353–8.

Idnurm, M. and Senior, B. R. (1978), 'Palaeomagnetic ages of Late Cretaceous and Tertiary weathered profiles in the Eromanga Basin, Queensland', *Palaeogeography, Palaeoclimatology, Palaeoecology*, 24: 263–77.

Ilier, I. G. (1966), 'Attempt to estimate the degree of weathering of intrusive rocks from their physicomechanical properties', *Proceedings, 1st Congress of the International Society for Rock Mechanics*, 2: 109–19.

Imazu, M. (1986), 'Data base system and evaluation of mechanical properties of rock', *Proceedings, International Symposium on Engineering in Complex Rock Formations* (Beijing: Science Press), 111–20.

Imeson, A. C. and Verstraten, J. M. (1988), 'Rills on badland slopes: a physico-chemically controlled phenomenon', *Catena Supplement*, 12: 139–50.

International Society for Rock Mechanics (1985), 'Suggested method for determining point load strength', *International Journal of Rock Mechanics and Mining Sciences and Geomechanics Abstracts*, 22: 51–60.

Irfan, T. Y. and Dearman, W. R. (1978), 'Engineering classification and index properties of a weathered granite, *Bulletin of the International Association of Engineering Geology*, 17: 79–90.

—— and Woods, N. W. (1988), 'The influence of relict discontinuities on slope stability in saprolitic soils', *Proceedings, 2nd International Conference on Geomechanics in Tropical Soils* (Singapore), 267–76.

Ishihara, K. (1985), 'Stability of natural deposits during earthquakes', *Proceedings, 11th International Conference on Soil Mechanics and Foundation Engineering*, 1: 321–76.

—— (1989), 'Liquefaction-induced landslide and debris flow in Tajikistan, USSR', *Landslide News*: 3, 6–7.

Iveronova, M. I. (1969), 'An attempt at the quantitative analyses of contemporary denudation processes', *National Lending Library Translation*, RTS7436 (from Russian language original).

Iverson, R. M. and LaHusen, R. G. (1989), 'Dynamic pore-pressure fluctuations in rapidly shearing granular materials', *Science*, 246: 796–9.

Ives, J. D. and Bovis, M. J. (1978), 'Natural hazard maps for land-use planning, San Juan Mountains, Colorado, USA', *Arctic and Alpine Research*, 10: 185–212.

Iwasaki, T., Tokida, K., and Tatsuoka, F. (1981), 'Soil liquefaction potential evaluation, with use of simplified procedure', *Proceedings, International Conference on Recent Advances in Geotechnical Earthquake Engineering and Soil Dynamics (St Louis, Mo.)*.

Jäckli, H. (1957), 'Gegenwartsgeologie des bundnerischen Rheingebietes-ein Beitrag zur exogenen Dynamik Alpiner Gebirgslandschaften', *Beiträge zur Geologie der Schweiz, Geotechnische Serie*, 36: 1–126.

Jackson, M. L. and Sherman, G. D. (1953), 'Chemical weathering of minerals in soils', *Advances in Agronomy*, 5: 219–318.

Jacobson, R. B., Cron, E. D., and McGeehin, J. P. (1989), 'Slope movements triggered by heavy rainfall, November 3–5, 1985, in Virginia and West Virginia, USA', *Geological Society of America, Special Paper*, 236: 1–13.

Jacquet, D. (1990), 'Sensitivity to remoulding of some volcanic ash soils in New Zealand', *Engineering Geology*, 28: 1–25.

Jaeger, J. C. (1971), 'Friction of rocks and stability of rock slopes', *Géotechnique*, 21: 97–134.

—— and Cook, N. G. W. (1976), *Fundamentals of Rock Mechanics* (London: Chapman and Hall).

Jahn, A. (1981), 'Some regularities of soil movement on the slope as exemplified by the observations in Sudety Mts.', *Transactions, Japanese Geomorphological Union*, 2: 321–8.

—— (1989), 'The soil creep of slopes in different altitudinal and ecological zones of Sudeten Mountains', *Geografiska Annaler*, 71A: 161–70.

James, P. M. (1971), 'The role of progressive failure in clay slopes', *Proceedings, First Australia–New Zealand Conference on Geomechanics*, 1: 344–8.

Janbu, N., Bjerrum, L., and Kjaernsli, B. (1956), 'Soil mechanics applied to some engineering problems' [in Norwegian with English summary], *Norwegian Geotechnical Institute Publication*, 16 (Oslo: Norwegian Geotechnical Institute).

Janda, R. J. (1971), 'An evaluation of procedures used in computing chemical denudation rates', *Geological Society of America Bulletin*, 82: 67–80.

Jansson, M. B. (1988), 'A global survey of sediment yield', *Geografiska Annaler*, 70A: 81–98.

Jennings, J. E. (1971), 'A mathematical theory for the calculation of the stability of slopes in opencast mines', *Proceedings, Open-Pit Mining Symposium, South African Institute of Mining and Metallurgy*, 6: 87–102.

Jennings, J. M. (1985), *Karst Geomorphology*, 2nd edn. (Oxford: Blackwell).

Jenny, H. (1941), *Factors of Soil Formation* (New York: McGraw-Hill).

Jerwood, L. C., Robinson, D. A., and Williams, R. B. G. (1990), 'Experimental frost and salt weathering of chalk–II', *Earth Surface Processes and Landforms*, 15: 699–708.

Jibson, R. W. (1989), 'Debris flows in southern Puerto Rico', *Geological Society of America, Special Paper*, 236: 29–55.

Jocelyn, J. (1972), 'Stress patterns and spheroidal weathering', *Nature Physical Science*, 240: 39–40.

Johnson, A. M. (1970), *Physical Processes in Geology* (San Francisco: Freeman, Cooper & Co.).

—— and Rodine, J. R. (1984), 'Debris flow', in D. Brunsden and D. B. Prior (eds.), *Slope Instability* (Chichester: Wiley), 257–361.

Johnson, D. L. (1990), 'Biomantle evolution and the redistribution of earth materials and artifacts', *Soil Science*, 149: 84–101.

—— Keller, E. A., and Rockwell, T. K. (1990), 'Dynamic pedogenesis: New views on some key soil concepts, and a model for interpreting Quaternary soils', *Quaternary Research*, 33: 306–19.

—— and Watson-Stegner, D. (1987), 'Evolution model of pedogenesis', *Soil Science*, 143: 349–66.

Johnson, P. G. (1984), 'Paraglacial conditions of instability and mass movement: A discussion', *Zeitschrift für Geomorphologie*, 28: 235–50.

Johnston, I. W. and Chiu, H. K. (1984), 'Strength of weathered Melbourne Mudstone', *Journal of Geotechnical Engineering*, 110: 875–98.

Jones, F. O. (1973), 'Landslides of Rio de Janeiro and the Serra das Araras escarpment, Brazil', *US Geological Survey, Professional Paper*, 697: 1–42.

Jones, J. A. A. (1987a), 'Soil piping and stream channel initiation', *Water Resources Research*, 7: 602–10.

—— (1987b), 'The effects of soil piping on contributing areas and erosion patterns', *Earth Surface Processes and Landforms*, 12: 229–48.

Judson, S. (1968), 'Erosion rates near Rome, Italy', *Science*, 160: 1444–6.

—— and Ritter, D. F. (1964), 'Rates of regional denudation in the United States', *Journal of Geophysical Research*, 69: 3395–401.

Kaiser, P. K. and Simmons, J. V. (1990), 'A reassessment of transport mechanisms of some rock avalanches in the Mackenzie Mountains, Yukon and Northwest Territories, Canada', *Canadian Geotechnical Journal*, 27: 129–44.

Keefer, D. K. (1984), 'Landslides caused by earthquakes', *Geological Society of America Bulletin*, 95: 406–21.

—— and Johnson, A. M. (1983), 'Earth flows: Morphology, mobilization, and movement', *US Geological Survey, Professional Paper*, 1264: 1–56.

—— et al. (1987), 'Real-time landslide warning during heavy rainfall', *Science*, 238: 921–5.

Kemp, E. M. (1981), 'Tertiary palaeogeography and the evolution of Australian climate', in A. Keast (ed.), *Ecological Biogeography of Australia*, (The Hague: W. Junk), 33–49.

Kent, P. E. (1966), 'The transport mechanism in catastrophic rock falls', *Journal of Geology*, 74: 79–83.

Kerr, A., Smith, B. J., Whalley, W. B., and McGreevy, J. P. (1984), 'Rock temperatures from southeast Morocco and their significance for experimental rock-weathering studies', *Geology*, 12: 306–9.

Keys, J. R. and Williams, K. (1981), 'Origin of crystalline, cold desert salts in the McMurdo region, Antarctica', *Geochimica et Cosmochimica Acta*, 45: 2299–309.

Kiersch, G. A. (1965), 'Vaiont reservoir disaster', *Geotimes*, 9: 9–12.

Kieslinger, A. (1960), 'Residual stress and relaxation in rocks', *International Geological Congress, Copenhagen, Session 21*: 270–6.

King, L. C. (1953), 'Canons of landscape evolution', *Geological Society of America Bulletin*, 64: 721–51.

—— (1962 and 1967), *The Morphology of the Earth* (Edinburgh: Oliver and Boyd).

King, M. S. (1983), 'Static and dynamic elastic properties of rocks from the Canadian Shield', *International Journal of Rock Mechanics and Mining Sciences and Geomechanics Abstracts*, 20: 237–41.

Kingdon-Ward, J. (1952), *My Hill So Strong* (London: Jonathan Cape).

—— (1955), 'Aftermath of the great Assam earthquake of 1950', *Geographical Journal*, 121: 290–303.

Kinnell, P. I. A. (1990), 'The mechanics of raindrop-induced flow transport', *Australian Journal of Soil Research*, 28: 497–516.

Kirby, S. H. (1984), 'Introduction and digest to the special issue on chemical effects of water on the deformation and strengths of rocks', *Journal of Geophysical Research*, 89(B6): 3991–5.

Kirkby, A. V. T. and Kirkby, M. J. (1974), 'Surface wash at the semiarid break in slope', *Zeitschrift für Geomorphologie Supplementband*, 21: 151–76.

Kirkby, M. J. (1971), 'Hillslope process–response models based on the continuity equation', *Institute of British Geographers, Special Publication*, 3: 15–30.

—— (ed.) (1978), *Hillslope Hydrology* (Chichester: Wiley).

—— (1984), 'Modelling cliff development in South Wales: Savigear rereviewed', *Zeitschrift für Geomorphologie*, 28: 405–26.

—— (1985), 'A model for the evolution of regolith-mantled slopes', in M. J. Woldenberg (ed.), *Models in Geomorphology* (Boston: Allen and Unwin), 213–37.

—— (1987), 'General models of long-term slope evolution through mass movement', in M. G. Anderson and K. S. Richards (eds.), *Slope Stability* (Chichester: Wiley), 359–79.

—— (1988), 'Hillslope runoff processes and models', *Journal of Hydrology*, 100: 315–39.

—— and Statham, I. (1975), 'Surface stone movement and scree formation', *Journal of Geology*, 83: 349–62.

Kohlbeck, F., Scheidegger, A. E., and Sturgul, J. R. (1979), 'Geomechanical model of an alpine valley', *Rock Mechanics*, 12: 1–14.

Kojan, E. and Hutchinson, J. N. (1978), 'Mayunmarca rockslide and debris flow, Peru', in B. Voight (ed.), *Rockslides and Avalanches* (Amsterdam: Elsevier), 315–61.

Komura, S. (1976), 'Hydraulics of slope erosion by overland flow', *Journal of the Hydraulics Division ASCE*, 102: 1573–86.

Konishchev, V. N. (1982), 'Characteristics of cryogenic weathering in the permafrost zone of the European USSR', *Arctic and Alpine Research*, 14: 261–5.

Koo, Y. C. (1982), 'The mass strength of jointed residual soils', *Canadian Geotechnical Journal*, 19: 225–31.

Kotarba, A. and Strömquist, L. (1984), 'Transport, sorting and deposition processes of alpine debris slope deposits in the Polish Tatra Mountains', *Geografiska Annaler*, 66A: 285–94.

Krahn, J., Fredlund, D. G., and Klassen, M. J. (1989), 'Effect of soil suction on slope stability at Notch Hill',

Canadian Geotechnical Journal, 26: 269–78.

Krinsley, D. H. and Smalley, I. J. (1972), 'Sand', *American Scientist*, 60: 286–91.

Kuenen, P. H. (1951), 'Properties of turbidity currents of high density', *Society for Economic Paleontology and Mineralogy, Special Publication*, 2: 14–33.

Kukal, Z. (1990), 'The rate of geological processes', *Earth-Science Reviews*, 28: 1–284.

Kutter, J. K. and Rautenberg, A. (1979), 'The residual shear strength of filled joints in rock', *Proceedings, 4th Congress of the International Society for Rock Mechanics*, 1: 221–7.

Laffan, M. D. and Sutherland, R. D. (1988), 'Treatment of tunnel-gully erosion in loess colluvium on the Wither Hills, New Zealand', in D. N. Eden and R. J. Furkert (eds.), *Loess: Its Distribution, Geology and Soils* (Rotterdam: Balkema), 83–90.

Laity, J. E. and Malin, M. C. (1985), 'Sapping processes and the development of theatre-headed valley networks on the Colorado Plateau', *Geological Society of America Bulletin*, 96: 203–17.

Lajtai, E. Z. and Alison, J. R. (1979), 'A study of residual stress effects in sandstone', *Canadian Journal of Earth Sciences*, 16: 1547–57.

Lama, R. D. and Vutukuri, V. S. (1978), *Handbook on Mechanical Properties of Rock*, ii–iv (Clausthal: Trans Tech).

Lambe, T. W. and Whitman, R. V. (1979), *Soil Mechanics: SI Version* (Chichester: Wiley).

Lasaga, A. C. (1984), 'Chemical kinetics of water-rock interactions', *Journal of Geophysical Research*, 89: 4009–25.

Laws, J. O. (1941), 'Measurements of fall-velocity of water-drops and raindrops', *Transactions of the American Geophysical Union*, 22: 709.

—— and Parsons, D. A. (1943), 'The relation of raindrop size to intensity', *Transactions of the American Geophysical Union*, 24: 452.

Leaf, C. F. and Martinelli, M. (1977), 'Avalanche Dynamics', *US Department of Agriculture, Forest Service, Research Paper*, RM-183: 1–51.

Lee, C. F. (1978), 'Stress relief and cliff stability at a power station near Niagara Falls', *Engineering Geology*, 12: 193–204.

Lee, F. T. (1989), 'Slope movements in the Cheshire Quartzite, southwestern Vermont', *Geological Society of America, Special Paper*, 236: 89–102.

Lees, G., Abdelkater, M. O., and Hamdani, S. K. (1982), 'Effect of the clay fraction on some mechanical properties of lime–soil mixtures', *Journal of the Institution of Highway Engineers*, 29: 33–9.

Lehmann, O. (1933), 'Morphologische Theorie der Verwitterung von steinschlag wänden', *Vierteljahrsschrift der Naturforschende Gesellschaft in Zurich*, 87: 83–126.

—— (1934), 'Ueber die morphologischen Folgen der Wandwitterung', *Annals of Geomorphology*, 8: 93–9.

Lembo Fazio, A., Tommasi, P., and Ribacchi, R. (1990), 'Sheared bedding joints in rock engineering: Two case histories in Italy, in N. Barton and O. Stephansson (eds.), *Rock Joints* (Rotterdam: Balkema), 83–90.

Leopold, L. B., Emmett, W. W., and Myrick, R. M. (1966), 'Channel and hillslope processes in a semiarid area, New Mexico', *US Geological Survey, Professional Paper*, 352-G: 193–252.

—— Wolman, M. G., and Miller, J. P. (1964), *Fluvial Processes in Geomorphology* (San Francisco: W. H. Freeman).

Leslie, D. M. (1973), 'Quaternary deposits and surfaces in a volcanic landscape on Otago Peninsula', *New Zealand Journal of Geology and Geophysics*, 16: 557–66.

Lessard, G. and Mitchell, J. K. (1985), 'The causes and effects of aging in quick clays', *Canadian Geotechnical Journal*, 22: 335–46.

Li, Y.-H. (1976), 'Denudation of Taiwan Island since the Pliocene Epoch', *Geology*, 4: 105–7.

Li Jian and Luo Defu (1981), 'The formation and characteristics of mudflow and flood in the mountain area of the Dachao River and its prevention', *Zeitschrift für Geomorphologie*, 25: 470–84.

—— and Wang Jingrung (1986), 'The mudflows of Xiaojiang Basin', *Zeitschrift für Geomorphologie*, 58: 155–64.

Lin, Z. and Liang, W. (1982), 'Engineering properties and zoning of loess and loess-like soils in China', *Canadian Geotechnical Journal*, 19: 76–91.

Lindner, E. (1976), 'Swelling rock: a review', in *Rock Engineering for Foundations and Slopes*, 1 (New York: American Society of Civil Engineers), 141–81.

Lo, K. Y. and Lee, C. F. (1975), 'Stress distributions in rock slopes under high *in situ* stresses', in *Mass Wasting, 4th Guelph Symposium on Geomorphology, 1975* (Norwich: Geo Abstracts) 35–55.

Loague, K. and Gander, G. A. (1990), 'R-5 revisited: 1. spatial variability of infiltration on a small rangeland catchment', *Water Resources Research*, 26: 957–71.

Lohnes, R. A. and Handy, R. L. (1968), 'Slope angles in friable loess', *Journal of Geology*, 76: 247–58.

López-Bermúdez, F. and Romero-Diaz, M. A. (1989), 'Piping erosion and badland development in south-east Spain', *Catena Supplement*, 14: 59–73.

Loughran, R. J., Campbell, B. L., and Walling, D. E. (1987), 'Soil erosion and sedimentation indicated by caesium-137: Jackmoor Brook catchment Devon, England', *Catena*, 14: 201–12.

—— —— Elliot, G. L., and Shelley, D. J. (1990), 'Determination of the rate of sheet erosion on grazing land using caesium-137', *Applied Geography*, 10: 125–33.

—— —— —— Cummings, D., and Shelley, D. J. (1989), 'A caesium-137–sediment hillslope model with tests from south-eastern Australia', *Zeitschrift für Geomorphologie*, 2: 235–50.

Lowe, D. J. (1986), 'Controls on the rates of weathering and clay mineral genesis in airfall tephras: A review and New Zealand case study', in S. M. Coleman and D. P. Dethier (eds.), *Rates of Chemical Weathering of Rocks and Minerals* (New York: Academic Press), 265–330.

Luckman, B. H. (1977), 'The geomorphic activity of avalanches', *Geografiska Annaler*, 59A: 31–48.

—— (1978), 'Geomorphic work of snow avalanches in the Canadian Rocky Mountains', *Arctic and Alpine Research*, 10: 261–76.

Lull, H. W. (1964), 'Ecological and silvicultural aspects', sect. 6 in Ven Te Chow (ed.), *Handbook of Applied Hydrology* (New York: McGraw-Hill).

Lupini, J. F., Skinner, A. E., and Vaughan, P. R. (1981), 'The drained residual strength of cohesive soils', *Géotechnique*, 31: 181–213.

Lutenegger, A. J. (1982), 'Engineering properties and zoning of loess and loess-like soils in China: discussion', *Canadian Geotechnical Journal*, 20: 192–3.

McCaig, M. (1983), 'Contributions to storm quickflow in a small headwater catchment—the role of natural pipes and soil macropores', *Earth Surface Processes and Landforms*, 8: 239–52.

McConnell, R. G. and Brock, R. W. (1904), 'The great landslide at Frank, Alberta', *Canada Department of the Interior Annual Report 1902–1903*, pt. 8., app.

McDowell, P. W. and Millett, N. (1984), 'Surface ultrasonic measurement of longitudinal and transverse wave velocities through rock samples', *International Journal of Rock Mechanics and Mining Sciences and Geomechanics Abstracts*, 21: 223–7.

McFadden, L. D. and Knuepfer, P. J. K. (1990), 'Soil geomorphology: the linkage of pedology and surficial processes', *Geomorphology*, 3: 197–205.

McFarlane, M. J. (1976), *Laterite and Landscape* (London: Academic Press).

McGarr, A. and Gay, N. C. (1978), 'State of stress in the earth's crust', *Annual Review of Earth and Planetary Sciences*, 6: 405–36.

McGreal, W. S. (1979), 'Factors promoting coastal slope instability in southeast County Down, N. Ireland', *Zeitschrift für Geomorphologie*, 23: 76–90.

McGreevy, J. P. (1982), '"Frost and salt" weathering: Further experimental results', *Earth Surface Processes and Landforms*, 7: 475–88.

—— (1985), 'Thermal properties as controls on rock surface temperature maxima, and possible implications for rock weathering', *Earth Surface Processes and Landforms*, 10: 125–36.

—— and Smith, B. J. (1982), 'Salt weathering in hot deserts: Observations on the design of simulation experiments', *Geografiska Annaler*, 64A: 161–70.

McGregor, G. R. (1989), 'Snow avalanche terrain of the Craigieburn Range, central Canterbury, New Zealand', *New Zealand Journal of Geology and Geophysics*, 32: 401–9.

Machida, H. (1966), 'Rapid erosional development of mountain slopes and valleys caused by large landslides in Japan', *Geographical Reports of Tokyo Metropolitan University*, 1: 55–78.

McRoberts, E. C. and Morgenstern, N. R. (1974), 'The stability of thawing slopes', *Canadian Geotechnical Journal*, 11: 447–69.

McTigue, D. F. and Mei, C. C. (1981), 'Gravity-induced stresses near topography of small slope', *Journal of Geophysical Research*, 86, B10: 9268–78.

Maekado, A. (1990), 'Critical length of ledge developed on an artificially cut slope: an example', *Transactions, Japanese Geomorphological Union*, 11: 363–8.

Mahr, T. (1977), 'Deep-reaching gravitational deformation of high mountain slopes', *Bulletin of the International Association of Engineering Geology*, 16: 121–7.

—— and Nemčok, A. (1977), 'Deep-seated creep deformations in the crystalline caves of the Tatry Mountains', *Bulletin of the International Association of Engineering Geology*, 16: 104–6.

Maignien, R. (1966), *A Review of Research on Laterite* (UNESCO, Natural Resources Research, 4).

Mainguet, M. (1972), *Le Modelé Des Grès*, 2 vols. (Paris: Institut Géographique National).

Marchand, D. E. (1971), 'Rates and modes of denudation, White Mountains, eastern California', *American Journal of Science*, 270: 109–35.

Marron, D. C. (1985), 'Colluvium in bedrock hollows on steep slopes, Redwood Creek drainage basin, northwestern California', *Catena Supplement*, 6: 59–68.

Marshall, T. J. and Holmes, J. W. (1988), *Soil Physics*, 2nd edn. (Cambridge: Cambridge University Press).

Martin, R. P. (1986), 'Use of index tests for engineering assessment of weathered rocks', *Proceedings, 5th International Association for Engineering Geology Congress (Buenos Aires)*, 2: 433–50.

Massey, J. B. and Pang, P. L. R. (1988), 'General report: Stability of slopes and excavations in tropical soils', *Proceedings, 2nd International Conference on Geomechanics in Tropical Soils (Singapore)*, 2 (Rotterdam: Balkema), 551–70.

Matheson, D. S. and Thomson, S. (1973), 'Geological implications of valley rebound', *Canadian Journal of Earth Sciences*, 10: 961–78.

Matsukura, Y. (1988), 'Cliff instability in pumice flow deposits due to notch formation on the Asama mountain slope, Japan', *Zeitschrift für Geomorphologie*, 32: 129–41.

—— and Maekado, A. (1984), 'Slope stability analysis for "Murose" debris-slide triggered by the 1949 Imaichi earthquake', *Annual Report of the Institute of Geoscience, University of Tsukuba, Japan*, 10: 63–5.

—— Hayashida, S., and Maekado, A. (1984), 'Angles

of valley-side slope made of 'Shirasu' ignimbrite in South Kyushu, Japan', *Zeitschrift für Geomorphologie*, 28: 179–91.

Matthews, J. A. (1985), 'Radiocarbon dating of surface and buried soils: Principles, problems and prospects', in K. S. Richards, R. R. Arnett, and S. Ellis (eds.), *Geomorphology and Soils* (London: George Allen & Unwin), 269–88.

Matula, M. (1981*a*), 'Recommended symbols for engineering geological mapping, report by the IAEG Commission on Engineering Geological Mapping', *Bulletin of the International Association of Engineering Geology*, 24: 227–34.

—— (1981*b*), 'Rock and soil description and classification for engineering geological mapping, report by the IAEG Commission, on Engineering Geological Mapping', *Bulletin of the International Association of Engineering Geology*, 24: 235–74.

Meade, R. H. (1969), 'Errors in using modern stream-load data to estimate natural rates of denudation', *Geological Society of American Bulletin*, 80: 1265–74.

—— (1976), 'Sediment problems in the Savannah river basin', in B. L. Dillman and J. M. Stepp (eds.), *The Future of the Savannah River, Proceedings of a Symposium held at Hickory Knob State Park, SC, 14–15 October 1975* (Water Resources Research Institute, Clemson University, SC), 105–29.

Mellett, J. S. (1990), 'Ground-penetrating radar enhances knowledge of Earth's surface layer', *Geotimes*, 35(9): 12–14.

Mellor, M. (1978), 'Dynamics of snow avalanches', in B. Voight (ed.), *Rockslides and Avalanches* (Amsterdam: Elsevier), 753–92.

Melosh, H. J. (1979), 'Acoustic fluidization; a new geologic process?', *Journal of Geophysical Research*, 84: 7513–20.

—— (1987), 'The mechanics of large rock avalanches', *Geological Society of America, Reviews in Engineering Geology*, 7: 41–9.

—— (1990), 'Giant rock avalanches', *Nature*, 348: 483–4.

Menges, C. M. (1990), 'Soils and geomorphic evolution of bedrock facets on a tectonically active mountain front, western Sangre de Cristo Mountains, New Mexico', *Geomorphology*, 3: 301–32.

Menzies, B. K. (1988), 'A computer-controlled hydraulic triaxial testing system', *American Society for Testing and Materials, Special Technical Publication*, 977, ASTM STP 977.

Meyer, L. D., Foster, G. R., and Romkens, P. (1975), 'Source of soil eroded by water from upland slopes', *USDA Agricultural Research Service Report*, ARS-S-40: 177–89.

Meyerhof, G. G. (1969), 'Safety factors in soil mechanics', *Proceedings, 7th International Conference on Soil Mechanics and Foundation Engineering (Mexico)*, 479–81.

Michalske, T. A. and Freiman, S. W. (1983), 'A molecular mechanism for stress corrosion in vitreous silica', *Journal of the American Ceramic Society*, 66: 284–8.

Middleton, H. E. (1930), 'Properties of soils which influence soil erosion', *US Department of Agriculture, Technical Bulletin*, 178.

Mietton, M. (1988), 'Mesures continués des temperatures dans le socle granitique en région soudanienne (Février 1982–Juin 1983, Ouagadougou, Burkina Faso)', *Catena Supplement*, 12: 77–93.

Milliman, J. D. and Meade, R. H. (1983), 'World-wide delivery of river sediment to the oceans', *Journal of Geology*, 91: 1–21.

—— Yun-Shan, Q., Mei-E, R., and Saito, Y. (1987), 'Man's influence on the erosion and transport of sediment by Asian rivers: The Yellow River (Huanghe) example', *Journal of Geology*, 95: 751–62.

Millot, G. (1970), *Geology of Clays* (New York: Springer-Verlag).

Mills, H. H. (1981), 'Boulder deposits and the retreat of mountain slopes, or "gully gravure" revisited', *Journal of Geology*, 89: 649–60.

—— (1987), 'Variation in sedimentary properties of colluvium as a function of topographic setting, Valley and Ridge province, Virginia', *Zeitschrift für Geomorphologie*, 31: 277–92.

—— (1990), 'Thickness and character of regolith on mountain slopes in the vicinity of Mountain Lake, Virginia, as indicated by seismic refraction, and implications for hillslope evaluation', *Geomorphology*, 3: 143–57.

Milne, G. (1935), 'Some suggested units for classification and mapping, particularly for East African soils', *Soil Research, Berlin*, 4: 183–98.

Mining Research and Development Establishment (1979), 'NCB Cone Indentor', *MRDE Handbook*, no. 5.

Mitchell, J. K. (1986), 'Practical problems from surprising soil behaviour', *Journal of Geotechnical Engineering*, 112: 259–89.

Miura, K. (1973), 'Weathering in plutonic rocks (pt. 1): Weathering during the late Pliocene of Götsu plutonic rocks', *Journal of the Society for Engineering Geology of Japan*, 14(3).

Mochinaga, R. (1941), 'Earthwork on the Aso volcanic ash soil: A case of Kyushu Expressway', *Japan Road Association, Annual Report on Roads*, 42–7.

Moeyersons, J. (1983), 'Measurements of splash-saltation fluxes under oblique rain', *Catena Supplement*, 4: 19–31.

—— (1988), 'The complex nature of creep movements on steeply sloping ground in southern Rwanda', *Earth Surface Processes and Landforms*, 13: 511–24.

Mohr, E. C. and Baren, van F. A. (1954), *Tropical Soils* (New York: Interscience).

Mollard, J. D. (1977), 'Some regional landslide types in Canada', in D. R. Coates (ed.), *Landslides, Reviews in Engineering Geology*, iii (Geological Society of America), 29–56.

Montgomery, D. R. and Dietrich, W. E. (1988), 'Where do channels begin?', *Nature*, 336: 232–4.

—— —— (1989), 'Source areas, drainage density, and channel initiation', *Water Resources Research*, 25: 1907–18.

Montjuvent, G. (1973), 'L'érosion sur les Alpes françaises d'après l'exemple du massif de Pelvoux', *Revue Géographie Alpine*, 61: 107–20.

Moon, B. P. (1982), 'Rock mass strength and the morphology of rock slopes in the Cape Mountains', Ph.D. thesis, University of the Witwatersrand, Johannesburg.

—— (1984a), 'Refinement of a technique for determining rock mass strength for geomorphological purposes', *Earth Surface Processes and Landforms*, 9: 189–93.

—— (1984b), 'The forms of rock slopes in the Cape Fold Mountains', *South African Geographical Journal*, 66: 16–31.

—— (1986), 'Controls on the form and development of rock slopes in fold terrane', in A. D. Abrahams (ed.), *Hillslope Processes* (Boston: Allen and Unwin), 225–43.

—— and Selby, M. J. (1983), 'Rock mass strength and scarp-forms in southern Africa', *Geografiska Annaler*, 65A: 135–45.

Moore, R. (1991), 'The chemical and mineralogical controls upon the residual strength of pure and natural clays', *Géotechnique*, 41: 35–47.

Morgan, R. P. C. (1977), 'Soil erosion in the United Kingdom: Field studies in the Silsoe area, 1973–5', *National College of Agricultural Engineering, Silsoe, Occasional Paper*, 4.

—— (1980), 'Field studies of sediment transport by overland flow', *Earth Surface Processes*, 5: 307–16.

—— (1986), *Soil Erosion and Conservation* (Harlow: Longman).

—— Morgan, D. D. V., and Finney, H. J. (1984), 'A predictive model for the assessment of soil erosion risk', *Journal of Agricultural Engineering Research*, 30: 245–53.

Morgenstern, N. R. and Nixon, J. F. (1971), 'One-dimensional consolidation of thawing soils', *Canadian Geotechnical Journal*, 8: 558–65.

Morin, J., Benyamini, Y., and Michaeli, A. (1981), 'The effect of raindrop impact on the dynamics of soil surface crusting and water movement in the profile', *Journal of Hydrology*, 52: 321–36.

Morisawa, M. (1968), *Streams: Their dynamics and morphology* (New York: McGraw-Hill).

Moss, A. J. and Green, P. (1983), 'Movement of solids in air and water by raindrop impact. Effects of drop-size and water-depth variations', *Australian Journal of Soil Research*, 21: 257–69.

—— and Walker, P. H. (1978), 'Particle transport by continental water flows in relation to erosion, deposition, soils and human activities', *Sedimentary Geology*, 20: 81–139.

—— , Green, P., and Hutka, J. (1981), 'Static breakage of granitic detritus by ice and water in comparison with breakage by flowing water', *Sedimentology*, 28: 261–72.

—— —— —— (1982), 'Small channels: Their formation, nature and significance', *Earth Surface Processes and Landforms*, 7: 401–15.

Moye, D. E. (1955), 'Engineering Geology for the Snowy Mountains scheme', *Journal of the Institution of Engineers, Australia*, 27: 281–99.

Mugridge, S.-J. and Young, H. R. (1983), 'Disintegration of shale by cyclic wetting and drying and frost action', *Canadian Journal of Earth Sciences*, 20: 568–76.

Muhunthan, B. (1991), 'Liquid limit and surface area of clays', *Géotechnique*, 41: 135–8.

Müller, L. (1958), 'Geomechanische Auswertung gefügekundlicher Details', *Geologie und Bauwesen*, 24: 4–21.

—— (1963), *Der Felsbau* (Stuttgart: Enke).

—— (1964), 'The rock slide in the Vajont Valley', *Rock Mechanics and Engineering Geology, Vienna*, 2: 148–212.

Musgrave, G. W. (1947), 'Quantitative evaluation of factors in water-erosion: A first approximation', *Journal of Soil and Water Conservation*, 2: 133–8.

Myrvang, A. M. and Grimstad, E. (1984), 'Coping with the problem of rockbursts in hard rock tunnelling', *Tunnels and Tunnelling*, 16: 13–15.

Nagaraj, T. S. and Srinivasa Murthy, B. R. (1986), 'A critical reappraisal of compression index equations', *Géotechnique*, 36: 27–32.

Nahon, D. and Trompette, R. (1982), 'Origin of siltstones: Glacial grinding versus weathering', *Sedimentology*, 29: 25–35.

Nankano, R. (1967), 'On weathering and change of properties of Tertiary mudstone related to landslides', *Soil and Foundations (Tokyo)*, 7: 1–14.

Nash, D. (1980), 'Forms of bluffs degraded for different lengths of time in Emmet County, Michigan, USA', *Earth Surface Processes and Landforms*, 5: 331–45.

—— (1987), 'A comparative review of limit equilibrium methods of stability analysis', in M. G. Anderson and K. S. Richards (eds.), *Slope Stability* (Chichester: Wiley), 11–75.

—— Brunsden, D. K., Hughes, R. E., Jones, D. K. C., and Whalley, B. F. (1985), 'A catastrophic debris flow near Gupis, Northern areas, Pakistan', *Proceedings, 11th International Conference on Soil Mechanics and Foundation Engineering*, 3: 1163–6.

Neall, V. F. (1977), 'Genesis and weathering of andosols in Taranaki, New Zealand', *Soil Science*, 123: 400–8.

Neary, D. G. and Swift, L. W. (1987), 'Rainfall

thresholds for triggering a debris avalanching event in the southern Appalachian Mountains', *Geological Society of America, Reviews in Engineering Geology*, 7: 81–92.

Nemčŏk, A. (1977), 'Geological/tectonic structures: An essential condition for genesis and evolution of slope movement', *Bulletin of the Association of Engineering Geology*, 16: 127–30.

—— Pašek, J., and Rybar, J. (1972), 'Classification of landslides and other mass movements', *Rock Mechanics*, 4: 71–8.

Nesbitt, H. W. and Young, G. M. (1989), 'Formation and diagenesis of weathering profiles', *Journal of Geology*, 97: 129–47.

Netterberg, F. (1967), 'Some roadmaking properties of South African calcretes', *Proceedings, 4th Regional Conference for Africa on Soil Mechanics and Foundation Engineering (Cape Town)*, 1: 77–81.

Nguyen, V. U. and Chowdhury, R. N. (1984), 'Probabilistic study of spoil pile stability in strip coal mines: Two techniques compared', *International Journal of Rock Mechanics and Mining Science and Geomechanics Abstracts*, 21: 303–12.

Nicholas, R. M. and Dixon, J. C. (1986), 'Sandstone scarp form and retreat in the Land of Standing Rocks, Canyonlands National Park, Utah', *Zeitschrift für Geomorphologie*, 30: 167–87.

Norton, S. A. (1973), 'Laterite and bauxite formation', *Economic Geology*, 68: 353–61.

Nyberg, R. (1989), 'Observations of slushflows and their geomorphological effects in the Swedish Mountain area', *Geografiska Annaler*, 71A: 185–98.

Oberlander, T. M. (1977), 'Origin of segmented cliffs in massive sandstones of southeastern Utah', in D. O. Doehring (ed.), *Geomorphology in Arid Regions* (Boston: Allen and Unwin), 79–114.

Obermeier, S. F. (1989), 'The New Madrid earthquakes: An engineering–geologic interpretation of relict liquefaction features', *US Geological Survery, Professional Paper*, 1336-B: 1–114.

Obert, L. and Duvall, W. I. (1967), *Rock Mechanics and the Design of Structures in Rock* (New York: Wiley).

Oen, I. S. (1965), 'Sheeting and exfoliation in the granites of Sermersoq, South Greenland', *Meddelelser om Grønland*, 176, no. 6: 1–40.

Ogrosky, H. O. and Mockus, V. (1964), 'Hydrology of agricultural lands', in V. T. Chow (ed.), *Handbook of Applied Hydrology* (New York: McGraw-Hill), sect. 21.

Ohmori, H. and Hirano, M. (1988), 'Magnitude, frequency and geomorphological significance of rocky mud flows, land-creep and the collapse of steep slopes', *Zeitschrift für Geomorphologie Supplementband*, 67: 55–65.

Olivier, H. J. (1979), 'A new engineering: Geological

rock durability classification', *Engineering Geology*, 14: 255–79.

Ollier, C. D. (1978), 'Inselbergs of the Namib Desert, processes and history', *Zeitschrift für Geomorphologie, Supplementband*, 31: 161–76.

—— (1984), *Weathering*, 2nd edn. (Harlow: Longman).

—— (1985), 'Morphotectonics of passive continental margins: introduction', *Zeitschrift für Geomorphologie Supplementband*, 54: 1–9.

—— (1988), 'Deep weathering, groundwater and climate', *Geografiska Annaler*, 70A: 285–90.

—— and Ash, J. E. (1983), 'Fire and rock breakdown', *Zeitschrift für Geomorphologie*, 27: 363–74.

—— and Galloway, R. W. (1990), 'The laterite profile, ferricrete and unconformity', *Catena*, 17: 97–109.

—— and Marker, M. E. (1985), 'The Great Escarpment of southern Africa', *Zeitschrift für Geomorphologie Supplementband*, 54: 37–56

—— and Powar, K. B. (1985), 'The Western Ghats and the morphotectonics of peninsular India', *Zeitschrift für Geomorphologie Supplementband*, 54: 57–69.

—— and Tuddenham, W. G. (1962), 'Inselbergs of Central Australia', *Zeitschrift für Geomorphologie*, 5: 257–76.

O'Loughlin, C. (1974), 'The effect of timber removal on the stability of forest soils', *Journal of Hydrology (N.Z.)*, 13: 121–34.

Olsson, W. A. and Peng, S. S. (1976), 'Microcrack nucleation in marble', *International Journal of Rock Mechanics and Mining Sciences*, 13: 53–9.

Olyphant, G. A. (1983), 'Analysis of the factors controlling cliff burial by talus within Blanca Massif, southern Colorado, USA', *Arctic and Alpine Research*, 15: 65–75.

Onalp, A. (1988), 'Slope stability problems on the southeastern coast of Black Sea', *Proceedings, 5th International Symposium on Landslides (Lausanne)* 1 (Rotterdam: Balkema), 275–8.

Orr, C. M. (1974), 'The geological description of *in situ* rock masses as input data for engineering design', *C.S.I.R. South Africa, Report*, ser. MEG/344, no. ME1274.

Osipov, V. I. (1975), 'Structural bonds and the properties of clays', *Bulletin of the International Association of Engineering Geology*, 12: 13–20.

—— (1978), 'Structural bonds as the basis of the engineering-geological classification of clayey soils', *International Association of Engineering Geology, 3rd International Congress (Madrid)*, sect. II, 2: 160–5.

—— Nikolaeva, S. K., and Sokolov, V. N. (1984), 'Microstructural changes associated with thixotropic phenomena in clay soils', *Géotechnique*, 34: 293–303.

Osterkamp, W. R., Hupp, C. R., and Blodgett, J. C. (1986), 'Magnitude and frequency of debris flows, and areas of hazard on Mount Shasta, northern California', *US Geological Survey, Professional Paper*, 1396-C: C1–C21.

Owens, L. B. and Watson, J. P. (1979), 'Landscape reduction by weathering in small Rhodesian watersheds', *Geology*, 7: 281–4.

Pacher, F. (1958), 'Kennziffern des Flächengefüges', *Geologie und Bauwesen*, 24: 223–7.

Pain, C. F. (1968), 'Geomorphic effects of floods in the Orere River catchment, eastern Hunua Ranges', *Journal of Hydrology (N.Z.)*, 7: 62–74.

—— (1969), 'The effect of some environmental factors on rapid mass movement in the Hunua Ranges, New Zealand', *Earth Science Journal*, 3: 101–7.

—— (1971), 'Rapid mass movement under forest and grass in the Hunua Ranges, New Zealand', *Australian Geographical Studies*, 9: 77–84.

—— (1975), 'The Kaugel diamicton: A late Quatenary mudflow deposit in the Kaugel Valley, Papua New Guinea', *Zeitschrift für Geomorphologie*, 19: 430–42.

—— (1985), 'Morphotectonics of the continental margins of Australia', *Zeitschift für Geomorphologie Supplementband*, 54:23–35.

—— and Bowler, J. M. (1973), 'Denudation following the November 1970 earthquake at Madang, Papua New Guinea', *Zeitschift für Geomorphologie, Supplementband*, 18: 92–104.

Palmer, B. A. and Neall, V. E. (1989), 'The Murimotu Formation: 9500 year old deposits of debris avalanche and associated lahars, Mount Ruapehu, North Island, New Zealand', *New Zealand Journal of Geology and Geophysics*, 32: 477–89.

Palmer, R. S. (1965), 'Waterdrop impact forces', *American Society of Agricultural Engineers*, 8: 70–2.

Parker, A. (1970), 'An index of weathering for silicate rocks', *Geological Magazine*, 107: 501–4.

Parker, G. G. and Jenne, E. A. (1967), 'Structural failure of western highways caused by piping', *Highway Research Record*, 203: 57–76.

Parsons, A. J. (1978), 'A technique for the classification of hillslope forms', *Institute of British Geographers Transactions, New Series*, 3: 432–43.

—— (1988), *Hillslope Form* (London: Routledge).

Pathak, P. C., Pandey, A. N., and Singh, J. S. (1985), 'Apportionment of rainfall in central Himalayan forests', *Journal of Hydrology*, 76: 319–32.

Patton, F. D. (1966), 'Multiple modes of shear failure in rock and related materials', Ph.D. thesis, University of Illinois (University Microfilms no. 667786).

—— and Deere, D. U. (1971), 'Geologic factors controlling slope stability in open-pit mines', in C. O. Brawner and V. Milligan (eds.), *Stability in Open-Pit Mining* (New York: Society of Mining Engineers, American Institute of Mining, Metallurgical and Petroleum Engineers), 23–48.

Paulding, B. W. (1970), 'Coefficient of friction of natural rock surfaces', *Proceedings, American Society of Civil Engineers, Journal of the Soil Mechanics and Foundations Division*, 96(SM2): 385–94.

Pavich, M. J. (1989), 'Regolith residence time and the concept of surface age of the Piedmont "Peneplain"', *Geomorphology*, 2: 181–96.

—— Leo, G. W., Obermeier, S. F., and Estabrook, J. R. (1989), 'Investigations of the characteristics, origin, and residence time of the upland residual mantle of the Piedmont of Fairfax County, Virginia', *US Geological Survey, Professional Paper*, 1352: 1–51.

Pearce, A. J. (1976), 'Geomorphic and hydrologic consequences of vegetation destruction, Sudbury, Ontario', *Canadian Journal of Earth Sciences*, 13: 1358–73.

—— and Elson, J. A. (1973), 'Postglacial rates of denudation by soil movement, free face retreat, and fluvial erosion, Mont St. Hilaire, Quebec', *Canadian Journal of Earth Sciencs*, 10: 91–101.

—— and O'Loughlin, C. L. (1985), 'Landsliding during a M7.7 earthquake: Influence of geology and topography', *Geology*, 13: 855–8.

—— and Watson, A. J. (1986), 'Effects of earthquake-induced landslides on sediment budget and transport over a 50-yr period', *Geology*, 14: 52–5.

—— Black, R. D., and Nelson, C. S. (1981), 'Lithologic and weathering influences on slope form and process, eastern Raukumara Range, New Zealand', *IAHS Publication*, 132: 95–122.

Peel, R. F. (1974), 'Insolation weathering: Some measurements of diurnal temperature changes in exposed rocks in the Tibesti region, central Sahara', *Zeitschrift für Geomorphologie Supplementband*, 21: 19–28.

Penck, W. (1924), *Die Morphologische Analyse: Ein Kapitel der Physikalischen Geologie*, Geographische Abhandlungen 2, Reihe, Heft, 2, Stuttgart.

Penner, E. (1963), 'Sensitivity in Leda Clay', *Nature*, 197: 347–8.

Pennock, D. J., Zebarth, B. J., and De Jong, E. (1987), 'Landform classification and soil distrubution in hummocky terrain, Saskatchewan, Canada', *Geoderma*, 40: 297–315.

Perla, R. I. and Martinelli, M. (1976), *Avalanche Handbook, US Department of Agriculture Forest Service, Agriculture Handbook*, 489.

Persons, B. S. (1970), *Laterite Genesis, Location, Use* (New York: Plenum Press).

Pettinga, J. R. (1987), 'Ponui landslide: A deep-seated wedge failure in Tertiary weak-rock flysh, southern Hawke's Bay, New Zealand', *New Zealand Journal of Geology and Geophysics*, 30: 415–30.

Philip, J. R. (1957–8), 'The theory of infiltration', *Soil Science*, 83: 345–57, 438–48; 84: 163–77, 257–64, 329–39; 85: 278–86, 333–7.

Phillips, J. D. (1990), 'Relative importance of factors influencing fluvial soil loss at the global scale', *American Journal of Science*, 290: 547–68.

Pierson, T. C. (1980), 'Erosion and deposition by debris flows at Mt Thomas, North Canterbury, New

Zealand', *Earth Surface Processes*, 5: 227–47.

Pierson, T. C. (1982), 'Classification and hydrological characteristics of scree slope deposits in the northern Craigieburn Range, New Zealand', *Journal of Hydrology (N.Z.)*, 21: 34–60.

—— and Costa, J. E. (1987), 'A rheological classification of subaerial sediment-water flows', *Geological Society of America, Reviews in Engineering Geology*, 7: 1–12.

Pilgrim, D. H. and Huff, D. D. (1983), 'Suspended sediment in rapid subsurface stormflow on a large field plot', *Earth Surface Processes and Landforms*, 8: 451–63.

Piteau, D. R. (1971), 'Geological factors significant to the stability of slopes cut in rock', *Proceedings, Open-Pit Mining Symposium, South African Institute of Mining and Metallurgy*, 33–53.

—— (1973), 'Characterizing and extrapolating rock joint properties in engineering practice', *Rock Mechanics, Supplement*, 2: 5–31.

Pitty, A. F. (1971), *Introduction to Geomorphology* (London: Methuen).

Plafker, G. and Ericksen, G. E. (1978), 'Nevados Huascarán avalanches, Peru', in B. Voight (ed.), *Rockslides and Avalanches* (Amsterdam: Elsevier), 277–314.

Poesen, J. (1985), 'An improved splash transport model', *Zeitschrift für Geomorphologie*, 29: 193–211.

Polynov, B. B. (1973), *Cycle of Weathering*, trans. A. Muir (London: Murby).

Popescu, M. E. (1986), 'A comparison between the behaviour of swelling and of collapsing soils', *Engineering Geology*, 23: 145–63.

Pouyllau, M. and Seurin, M. (1985), 'Pseudo-karst dans les roches greso-quartzitiques de la formation Roraima', *Karstologia*, 5: 45–52.

Price, A. M. (1975), 'The effects of confining pressure on the post-yield deformation characteristics of rocks', Ph.D. thesis, University of Newcastle-upon-Tyne.

Priest, S. D. and Brown, E. T. (1983), 'Probabilistic stability analysis of variable rock slopes', *Transactions, Institution of Mining and Metallurgy* (sect. A), 92: A1–A12.

Prior, D. B. (1977), 'Coastal mudslide morphology and processes on Eocene clays in Denmark', *Geografisk Tidsskrift*, 77: 14–33.

—— and Suhayda, J. N. (1979), 'Application of infinite slope analysis to subaqueous sediment instability, Mississippi Delta', *Engineering Geology*, 14: 1–10.

Protod'yakonov, M. M. (1960), 'New methods of determining mechanical properties of rock', *Proceedings, International Conference on Strata Control (Paris)*, Paper C2: 187–95.

—— and Koifman, M. I. (1969), *Mechanical Properties of Rocks* (Jerusalem: Israel Program for Scientific Translations).

Pusch, R. (1979), 'Cohesion in fine-grained soils', in K.

E. Easterling (ed.), *Mechanics of Deformation and Fracture* (Oxford: Pergamon), 137–44.

Pye, K. (1983), 'Formation of quartz silt during humid tropical weathering of dune sands', *Sedimentary Geology*, 34: 267–82.

—— (1987), *Aeolian Dust and Dust Deposits* (London: Academic Press).

Quansah, K. (1985), 'The effect of soil type, slope, flow rate and their interactions on detachment by overland flow with and without rain', *Catena Supplement*, 6: 19–28.

Radbruch-Hall, D. H. (1978), 'Gravitational creep of rock masses on slopes', in B. Voight (ed.), *Rockslides and Avalanches* (Amsterdam: Elsevier), 607–57.

Rahn, P. H. (1986), *Engineering Geology* (New York: Elsevier).

Rampino, M. R., Sanders, J. E., Newman, W. S., and Königsson, L. K. (eds.) (1987), *Climate, History, Periodicity, and Predictability* (New York: Van Nostrand Reinhold).

Rapp, A. (1959), 'Avalanche boulder tongues in Lappland: A description of little known landforms of periglacial debris accumulation', *Geografiska Annaler*, 41: 34–48.

—— (1960a), 'Recent development of mountain slopes in Kärkevagge and surroundings, northern Scandinavia', *Geografiska Annaler*, 42: 73–200.

—— (1960b), 'Talus slopes and mountain walls at Tempelfjorden, Spitsbergen', *Norsk Polarinstitutt Skrifter*, 119: 1–96.

—— (1975), 'Studies of mass wasting in the Arctic and in the tropics', in *Mass Wasting* (Norwich: Geo Abstracts), 79–103.

Rashidian, K. (1986), 'A new technique for field measurement of soil creep displacement profiles', *Zeitschrift für Geomorphologie Supplementband*, 60: 93–103.

Rengers, N. (1971), 'Unebenheit und Reibungswiderstand von Gesteinstrennflächen' (Roughness and friction properties of separation planes in rock), thesis, Technische Hochschule Fredericiana, Karlsruhe.

Report by a Working Party under the auspices of the Geological Society (1982), 'Land Surface Evaluation for Engineering Practice', *Quarterly Journal of Engineering Geology, London*, 15: 265–316.

Retallack, G. J. (1988), 'Field recognition of paleosols', *Geological Society of America, Special Paper*, 216: 1–20.

Riquier, J. (1958), 'Le "lavaka" de Madagascar', *Bulletin de Société Geographique, Marseilles*, 69: 181–91.

Ritchie, J. C. (1987), 'Literature relevant to the use of radioactive fallout caesium-137 to measure soil erosion and sediment deposition', *USDA-ARS Hydrology Laboratory Technical Report*, HL-9.

Roberts, A. (1977), *Geotechnology* (Oxford: Pergamon).

Robertson, A. MacG. (1971), 'The interpretation of geological factors for use in slope theory', *Proceedings, Open-Pit Mining Symposium, South African Institute of Mining and Metallurgy*, 55–71.

Robertson, E. C. (1982), 'Continuous formation of gouge and breccia during fault displacement', in R. E. Goodman and F. E. Heuze (eds.), *Issues in Rock Mechanics* (New York: Society of Mining Engineers of the American Institute of Mining, Metallurgical and Petroleum Engineers), 397–403.

Robinson, D. A. and Williams, R. B. G. (1989), 'Polygonal cracking of sandstone at Fontainebleau, France', *Zeitschrift für Geomorphologie*, 33: 59–72.

Robinson, R. (1979), 'Variation of energy release, rate of occurrence and b-value of earthquakes in the main seismic region, New Zealand', *Physics of the Earth and Planetary Interiors*, 18: 209–20.

Rodine, J. D. (1974), 'Analysis of mobilization of debris flows', Ph.D. thesis, Stanford University, California.

Rogers, N. W. and Selby, M. J. (1980), 'Mechanisms of shallow translational landsliding during summer rainstorms: North Island, New Zealand', *Geografiska Annaler*, 62A: 11–21.

Roš, M. and Eichinger, A. (1928), 'Versuche zur Klärung der Frage der Bruchgefar. II. Nichtmetallische Stoffe Eidgenössische Materialprufungsanstalt der Eidgen', Zürich.

Rosenqvist, I. T. (1953), 'Considerations on the sensitivity of Norwegian quick clays', *Géotechnique*, 3: 195–200.

—— (1984), 'Colloidal physics as bases for quick clay properties', *Striae*, 19: 5–11.

Roth, E. S. (1965), 'Temperature and water content as factors in desert weathering', *Journal of Geology*, 73: 454–68.

Rouse, W. W. and Farhan, Y. I. (1976), 'Threshold slopes in South Wales', *Quarterly Journal of Engineering Geology*, 9: 327–38.

Rowe, P. W. (1972), 'The relevance of soil fabric to site investigation practice', *Géotechnique*, 22: 195–300.

Ruhe, R. V. (1975), *Geomorphology* (Boston: Houghton Mifflin).

Ruxton, B. P. (1988), 'Towards a weathering model of Mount Lamington Ash, New Guinea', *Earth-Science Reviews*, 25: 387–97.

—— and Berry, L. (1957), 'The weathering of granite and associated erosional features in Hong Kong', *Bulletin of the Geological Society of America*, 68: 1263–92.

—— and McDougall, I. (1967), 'Denudation rates in northeast Papua from potassium-argon dating of lavas', *American Journal of Science*, 265: 545–61.

Saito, M. and Uezawa, H. (1961), 'Failure of soil due to creep', *Proceedings, 5th International Conference on Soil Mechanics and Foundation Engineering*, 1: 315–18.

Salt, G. A. (1985), 'Aspects of landslide mobility', *Proceedings, 11th International Conference on Soil Mechanics and Foundation Engineering (San Francisco)*, 3: 1167–72.

Salter, R. T., Crippen, T. F., and Noble, K. E. (1981), 'Storm damage assessment of Thames-Te Aroha area following the storm of April 1981: Final report', *Water and Soil Science Centre, Aokautere, N.Z. Ministry of Works and Development, Internal Report*, no. 44.

Sancho, C. and Benito, G. (1990), 'Factors controlling tafoni weathering in the Ebro Basin (NE Spain)', *Zeitschrift für Geomorphologie*, 34: 165–77.

Sassa, K. (1985), 'The mechanisms of debris flow', *Proceedings, 11th International Conference on Soil Mechanics and Foundation Engineering*, 3: 1173–6.

Sassa, K. (1989), 'Geotechnical classification of landslides', *Landslide News*, 3: 21–4.

Saunders, I. and Young, A. (1983), 'Rate of surface processes on slopes, slope retreat and denudation', *Earth Surface Processes and Landforms*, 8: 473–501.

Savage, W. Z., Swolfs, H. S., and Powers, P. S. (1985), 'Gravitational stresses in long symmetric ridges and valleys', *International Journal of Rock Mechanics and Mining Sciences*, 22: 291–302.

—— and Varnes, D. J. (1987), 'Mechanics of gravitational spreading of steep-sided ridges (sackung)', *Bulletin of the International Association of Engineering Geology*, 35: 31–6.

Savigear, R. A. G. (1952), 'Some observations on slope development in South Wales', *Transactions, Institute of British Geographers*, 18: 31–51.

Schaetzl, R. J. and Follmer, L. R. (1990), 'Longevity of tree-throw microtopography: implications for mass wasting', *Geomorphology*, 3: 113–23.

Schattner, I. (1961), 'Weathering phenomena in the crystalline of the Sinai in the light of current notions', *Bulletin of the Research Council of Israel*, 10G: 247–66.

Scheidegger, A. E. (1963), 'On the tectonic stresses in the vicinity of a valley and mountain range', *Proceedings, Royal Society of Victoria*, 76: 141–5.

—— (1980), 'Alpine joints and valleys in the light of the neotectonic stress field', *Rock Mechanics Supplement*, 9: 109–24.

—— and Ai, N. S. (1986), 'Tectonic processes and geomorphological design', *Tectonophysics*, 136: 285–300.

Schmidt, K.-H. (1987), 'Factors influencing structural landform dynamics on the Colorado Plateau about the necessity of calibrating theoretical models by empirical data', *Catena Supplement*, 10: 51–66.

—— (1989), 'The significance of scarp retreat for Cenozoic landform evolution on the Colorado Plateau, USA', *Earth Surface Processes and Landforms*, 14: 93–105.

Schmidt, K.-H. (1989), 'Stufenhangabtragung und geomorphologische Entwicklung der Hamada de Meski, Südostmarokko', *Zeitschrift für Geomorphologie Supplementband*, 74: 33–44.

Scholz, C. H. and Engelder, J. T. (1976), 'The role of asperity indentation and ploughing in rock friction', *International Journal of Rock Mechanics, and Mining Science, and Geomechanics Abstracts*, 13: 149–63.

Schouten, C. J. and Rang, M. C. (1984), 'Measurement of gully erosion and the effects of soil conservation techniques in Puketurua Experimental Basin (New Zealand)', *Zeitschrift für Geomorphologie Supplementband*, 49: 151–64.

Schumm, S. A. (1956), 'Evolution of drainage systems and slopes in badlands at Perth Amboy, New Jersey', *Bulletin of the Geological Society of America*, 67: 597–646.

—— (1967), 'Rates of surficial creep on hillslopes in western Colorado', *Science*, 155: 560–1.

—— and Chorley, R. J. (1964), 'The fall of threatening rock', *American Journal of Science*, 262: 1041–54.

—— and Hadley, R. F. (1957), 'Arroyos and the semiarid cycle of erosion', *American Journal of Science*, 225: 161–74.

Schuster, R. L. (1979), 'Reservoir-induced landslides', *Bulletin of the International Association of Engineering Geology*, 20: 8–15.

—— and Krizek, J. (eds.) (1978), *Landslides: Analysis and Control* (Washington, DC: Transportation Research Board).

Scott, K. M. and Williams, R. P. (1978), 'Erosion and sediment yields in the Transverse Ranges, Southern California', *US Geological Survey, Professional Paper*, 1030: 1–38.

Seed, H. B. (1968), 'Landslides during earthquakes due to soil liquefaction', *Journal of the Soil Mechanics and Foundations Division, Proceedings, the American Society of Civil Engineers*, 94: 1053–122.

—— and Wilson, S. D. (1967), 'The Turnagain Heights Landslide, Anchorage, Alaska', *Journal of the Soil Mechanics and Foundations Division, Proceedings, American Society of Civil Engineers*, 93, SM4: 325–53.

—— and Idriss, I. M. (1971), 'Simplified procedure for evaluating soil liquefaction potential', *Journal of the Soil Mechanics and Foundations Division, Proceedings, American Society of Civil Engineers*, 97: 1249–73.

—— —— and Arango, I. (1983), 'Evaluation of liquefaction potential using field performance data', *Journal of the Geotechnical Engineering Division, Proceedings, American Society of Civil Engineers*, 109: 458–82.

Selby, M. J. (1967a), 'Aspects of the geomorphology of the greywacke ranges bordering the lower and middle Waikato Basins', *Earth Science Journal*, 1: 37–58.

—— (1967b), 'Erosion by high intensity rainstorms in the lower Waikato Basin', *Earth Science Journal*, 1: 153–6.

—— (1970), 'Design of a hand-portable rainfall-simulating infiltrometer, with trial results from the Otutira catchment', *Journal of Hydrology (N.Z.)*, 9: 117–32.

—— (1971a), 'Salt-weathering of landforms, and an Antarctic example', *Proceedings, 6th Geography Conference (New Zealand)*, 30–5.

—— (1971b), 'Slopes and their development in an ice-free arid area of Antarctica', *Geografiska Annaler*, 53A: 235–45.

—— (1974a), 'Slope evolution in an Antarctic oasis', *New Zealand Geographer*, 30: 18–34.

—— (1974b), 'Dominant geomorphic events in landform evolution', *Bulletin of the International Association of Engineering Geology*, 9: 85–9.

—— (1974c), 'Rates of creep in pumiceous soils and deposits, central North Island, New Zealand', *New Zealand Journal of Science*, 17: 47–8.

—— (1976a), 'Slope erosion due to extreme rainfall: a case study from New Zealand', *Geografiska Annaler*, 58A: 131–8.

—— (1976b), 'Loess', *New Zealand Journal of Geography*, 61: 1–18.

—— (1977a), 'Bornhardts of the Namib Desert', *Zeitschrift für Geomorphologie*, 21: 1–13.

—— (1977b), 'On the origin of sheeting and laminae in granitic rocks: evidence from Antarctica, the Namib Desert and the Central Sahara', *Madoqua*, 171–9.

—— (1980), 'A rock-mass strength classification for geomorphic purposes: With tests from Antarctica and New Zealand', *Zeitschrift für Geomorphologie*, 24: 31–51.

—— (1982a), 'Controls on the stability and inclinations of hillslopes formed on hard rock', *Earth Surface Processes and Landforms*, 7: 449–67.

—— (1982b), 'Form and origin of some bornhardts of the Namib Desert', *Zeitschrift für Geomorphologie*, 26: 1–15.

—— (1982c), 'Rock mass strength and the form of some inselbergs in the central Namib Desert', *Earth Surface Processes and Landforms*, 7: 489–97.

—— (1985), *Earth's Changing Surface* (Oxford: Clarendon Press).

—— (1987a), 'Rock slopes', in M. G. Anderson and K. S. Richards (eds.), *Slope Stability* (Chichester: Wiley), 475–504.

—— (1987b), 'Slopes and weathering', in K. J. Gregory and D. E. Walling (eds.), *Human Activity and Environmental Processes* (Chichester: Wiley), 183–205.

—— (1988), 'Landforms and denudation of the High Himalaya of Nepal: Results of continental collision', *Zeitschrift für Geomorphologie Supplementband*, 69: 133–52.

—— and Hosking, P. J. (1971), 'Causes of infiltration into yellow-brown pumice soils', *Journal of Hydrology (N.Z.)*, 19: 113–19.

—— Augustinus, P., Moon, V. G., and Stevenson, R. J.

(1988), 'Slopes on strong rock masses: Modelling and influences of stress distributions and geomechanical properties', in M. G. Anderson (ed.), *Modelling Geomorphological Systems* (Chichester: Wiley), 341–74.

Sergeyev, Y. M., Grabowska-Olszewska, B., Osipov, V. I., and Kolomenski, Y. N. (1980), 'The classification of microstructures of clay soils', *Journal of Microscopy*, 120: 237–60.

Shachak, M., Jones, C. G., and Granot, Y. (1987), 'Herbivory in rocks and the weathering of a desert', *Science*, 236: 1098–9.

Shanmuganathan, R. T. and Oades, J. M. (1983), 'Modification of soil physical properties by addition of calcium compounds', *Australian Journal of Soil Research*, 21: 285–300.

Sharpe, C. F. S. (1938), *Landslides and Related Phenomena* (New Jersey: Pageant).

Shehata, W. M. and Eissa, E. S. A. (1985), 'On the validity of Schmidt hammer in estimating the joint wall compressive strength', *Proceedings, International Symposium on Fundamentals of Rock Joints (Björkliden)*, 227–31.

Sherard, J. L., Dunnigan, L. P., and Decker, R. S. (1976a), 'Identification and nature of dispersive soils', *Journal of the Geotechnical Engineering Division, Proceedings, American Society of Civil Engineers*, 102: 287–301.

—— —— —— (1976b), 'Pinhole test for identifying dispersive soils', *Journal of the Geotechnical Engineering Division, Proceedings, American Society of Civil Engineers*, 102: 69–85.

Sherwood, P. T. and Ryley, M. D. (1970), 'An investigation of a cone-penetrometer method for the determination of the liquid limit', *Géotechnique*, 20: 203–8.

Shimuzu, F., Togo, M., and Matsuda, T. (1980), 'Origin of scarplets around Mt Noguchigoro-Dake in the Japan Alps, central Japan', *Geographical Review of Japan*, 53: 531–41.

Shreve, R. L. (1968), 'Leakage and fluidization in air-layer lubricated landslides', *Geological Society of America Bulletin*, 79: 653–8.

Sidle, R. C., Pearce, A. J., and O'Loughlin, C. L. (1985), 'Hillslope stability and land use', *American Geophysical Union, Water Resources Monograph Series*, 11.

Siever, R. (1959), 'Petrology and geochemistry of silica cementation in some Pennsylvanian sandstones', in H. A. Ireland (ed.), *Silica in sediments: A symposium, Society of Economic Paleontologists and Mineralogists, Special Publication*, 7: 56–76.

—— (1988), *Sand* (New York: Scientific American Library, Freeman).

Silvestri, T. (1961), 'Determinazione sperimentale de resistenza meccania del materiale constituente il corpo di una diga del tipo "Rockfill"', *Geotechnica*, 8: 186–91.

Silvestri, V. (1980), 'The long-term stability of a cutting slope in an overconsolidated sensitive clay', *Canadian Geotechnical Journal*, 17: 337–51.

Simon, A., Larsen, M. C., and Hupp, C. R. (1990), 'The role of soil processes in determining mechanisms of slope failure and hillslope development in a humid tropical forest, eastern Puerto Rico', *Geomorphology*, 3: 263–86.

Simonett, D. S. (1967), 'Landslide distribution and earthquakes in the Bewani and Torricelli Mountains, New Guinea', in J. N. Jennings and J. A. Mabbutt (eds.), *Landform Studies from Australia and New Guinea* (Canberra: A.N.U. Press), 64–84.

—— (1970), 'The role of landslides in slope development in the high rainfall tropics', *Final Report, Office of Naval Research, Geographical Branch*, Contract Number, 583(11), Task No., 389-133: 1–23.

Sitar, N. and Clough, G. W. (1983), 'Seismic response of steep slopes in cemented soils', *Journal of Geotechnical Engineering*, 109: 210–27.

Skempton, A. W. (1948), 'The rate of softening of stiff, fissured clays', *Proceedings, 2nd International Conference on Soil Mechanics and Foundation Engineering (Rotterdam)*, 2: 50–3.

—— (1953a), 'Soil mechanics in relation to geology', *Proceedings, Yorkshire Geological Society*, 29: 33–62.

—— (1953b), 'The colloidal "activity" of clays', *Proceedings, 3rd International Conference of Soil Mechanics and Foundation Engineering (Switzerland)*, 1: 57–61.

—— (1960), 'Terzaghi's discovery of effective stress', in L. Bjerrum, A. Casagrande, R. B. Peck, and A. W. Skempton (eds.), *From Theory to Practice in Soil Mechanics* (New York: Wiley), 42–53.

—— (1964), 'Long-term stability of clay slopes', *Géotechnique*, 14: 77–101.

—— (1970a), 'First-time slides in overconsolidated clays', *Géotechnique*, 20: 320–4.

—— (1970b), 'The consolidation of clays by gravitational compaction', *Quarterly Journal of the Geological Society, London*, 125: 373–411.

—— (1977), 'Slope stability of cuttings in brown London clay', *Proceedings, 9th International Conference on Soil Mechanics and Foundation Engineering (Tokyo)*, 3: 261–70.

—— (1985), 'Residual strength of clays in landslides, folded strata and the laboratory', *Géotechnique*, 35: 3–18.

—— and De Lory, F. A. (1957), 'Stability of natural slopes in London Clay', *Proceedings, 4th International Conference on Soil Mechanics and Foundation Engineering (London)*, 2: 378–81.

—— and Northey, R. D. (1952), 'The sensitivity of clays', *Géotechnique*, 3: 30–53.

—— and Weeks, A. G. (1976), 'The Quaternary history of the Lower Greensand escarpment and Weald Clay vale near Seven Oaks, Kent', *Philosophical Transactions, Royal Society, London*, A283: 493–526.

—— Leadbetter, A. D., and Chandler, R. J. (1989),

'The Mam Tor landslide, North Derbyshire', *Philosophical Transactions, Royal Society, London*, A329: 503–47.

Slaymaker, O. (1977), 'An overview of geomorphic processes in the Canadian Cordillera', *Zeitschrift für Geomorphologie*, 21: 169–86.

—— (1988), 'The distinctive attributes of debris torrents', *Hydrological Sciences Journal*, 33: 567–73.

Smalley, I. (1976), 'Factors relating to the landslide process in Canadian quickclays', *Earth Surface Processes*, 1: 163–72.

—— and Taylor, R. (1972), 'The quickclay enigma', *Scientific Era*, Dec.

Smettem, K. R. J., Chittleborough, D. J., Richards, B. G., and Leaney, F. W. (1991), 'The influence of macropores on runoff generation from a hillslope soil with contrasting textural class', *Journal of Hydrology*, 122: 235–52.

Smith, B. J. (1978), 'The origin and geomorphic implications of cliff foot recesses and tafoni on limestone hamadas in the northwest Sahara', *Zeitschrift für Geomorphologie*, 22: 21–43.

—— and McGreevy, J. P. (1983), 'A simulation study of salt weathering in hot deserts', *Geografiska Annaler*, 65A, 127–33.

Smith, W. D. (1978a), 'Earthquake risk in New Zealand: statistical estimates', *New Zealand Journal of Geology and Geophysics*, 21: 313–27.

—— (1978b), 'Spatial distribution of felt intensities for New Zealand earthquakes', *New Zealand Journal of Geology and Geophysics*, 21: 293–311.

Sommerhalder, E. (1965), 'Avalanche forces and the protection of objects', *US Forest Service Alta Avalanche Study Centre, Translation*, 6.

Sopper, W. E. and Lull, H. W. (eds.) (1967), *International Symposium on Forest Hydrology* (Oxford: Pergamon).

Spencer, E. W. (1969), *Introduction to the Structure of the Earth* (New York: McGraw-Hill).

Sperling, C. H. B. and Cooke, R. U. (1985), 'Laboratory simulation of rock weathering by salt crystallization and hydration processes in hot, arid environments', *Earth Surface Processes and Landforms*, 10: 541–55.

Sridharan, A. and Jayadeva, M. S. (1982), 'Double layer theory and compressibility of clays', *Géotechnique*, 32: 133–44.

—— and Venkatappa Rao, G. (1979), 'Shear strength behaviour of saturated clays and the role of the effective stress concept', *Géotechnique*, 29: 177–93.

Stacey, T. R. (1973), 'Stability of rock slopes in mining and civil engineering situations', *National Mechanical Engineering Institute, CSIR Report*, ME1202 (Pretoria).

Starkel, L. (1966), 'Post-glacial climate and the moulding of European relief', in *Proceedings of the International Symposium on World Climate 8000 to 0 B.C.* (London: Royal Meteorological Society), 15–32.

—— (1972a), 'The modelling of monsoon area of India as related to catastrophic rainfall', *Geographica Polonica*, 23: 153–73.

—— (1972b), 'The role of catastrophic rainfall in the shaping of the relief of the Lower Himalaya (Darjeeling Hills)', *Geographica Polonica*, 21: 103–47.

—— (1976), 'The role of extreme (catastrophic) meteorological events in contemporary evolution of slopes', in E. Derbyshire (ed.), *Geomorphology and Climate* (London: Wiley), 203–46.

—— (1979), 'The role of extreme meteorological events in the shaping of mountain relief', *Geographica Polonica*, 41: 13–20.

—— (1987), 'Long-term and short-term rhythmicity in terrestrial landforms and deposits', in M. R. Rampino, J. E. Sanders, W. S. Newman, and L. K. Königsson (eds.), *Climate, History, Periodicity, and Predictability* (New York: Van Nostrand Reinhold), 323–32.

Statham, I. (1973), 'Scree development under conditions of surface particle movement', *Transactions, Institute of British Geographers*, 59: 41–54.

Stehlík, O. (1975), 'Potenciálni eroze pudy proudící vodou na území ČSR', *Studia Geographica*, 42: Brno.

Steinbrenner, E. C. (1955), 'The effect of repeated tractor trips on the physical properties of forest soils', *Northwest Science*, 29: 155–9.

Stephens, C. G. (1971), 'Laterite and silcrete in Australia: A study of the genetic relationships of laterite and silcrete and their companion materials, and their collective significance in the formation of the weathered mantle, soils, relief and drainage of the Australian continent', *Geoderma*, 5: 5–52.

Stevenson, P. C. (1977), 'An empirical method for the evaluation of relative landslip risk', *Bulletin of the International Association of Engineering Geology*, 16: 69–72.

Stimpson, B. (1979), 'Simple equations for determining the factor of safety of a planar wedge under various groundwater conditions', *Quarterly Journal of Engineering Geology*, 12: 3–7.

Stocking, M. A. (1980), 'Examination of the factors controlling gully growth', in M. D. Boodt and D. Gabriels (eds.), *Assessment of Erosion* (Chichester: Wiley), 505–20.

Strakhov, N. M. (1967), *Principles of Lithogenesis*, i. (tran. from Russian edn. of 1964) (Edinburgh: Oliver and Boyd).

Sturgul, J. R. and Scheidegger, A. E. (1967), 'Tectonic stresses in the vicinity of a wall', *Rock Mechanics and Engineering Geology*, 5: 137–49.

—— and Grinshpan, Z. (1976), 'Finite-element model of a mountain massif', *Geology*, 4: 439–42.

Sugden, M. B., Van Wieringen, M., and Knight, K. (1977), 'Slip failures in bedded sediments', *Proceedings, 9th International Conference on Soil Mechanics and Foundation Engineering*, 2: 155–60.

Swanson, F. J. and Swanston, D. N. (1977), 'Complex

mass-movement terrains in the western Cascade Range, Oregon', *Geological Society of America, Reviews in Engineering Geology*, 3: 113–24.

Swanson, M. L., Kondolf, G. M., and Boison, P. J. (1989), 'An example of rapid gully initiation and extension by subsurface erosion: Coastal San Mateo County, California', *Geomorphology*, 2: 393–403.

Swanston, D. N. (1970), 'Mechanics of debris avalanching in shallow till soils of southeast Alaska', *USDA Forest Service, Research Paper*, PNW-103, 1–17.

Sweeting, M. M. (1972), *Karst Landforms* (London: Macmillan).

Swolfs, H. S., Handin, J., and Pratt, H. R. (1974), 'Field measurements of residual strain in granitic rock masses', *Proceedings, 3rd Congress of the International Society for Rock Mechanics*, 11A: 563–8.

Takahashi, T. (1978), 'Mechanical characteristics of debris flows', *Journal of the Hydraulics Division, American Society of Civil Engineers*, 104, HY8: 1153–69.

Talibudeen, O. (1981), 'Cation exchange in soils', in D. S. Greenland and M. H. B. Hayes (eds.), *Chemistry of Soil Processes* (Chichester: Wiley), 115–78.

Taylor, D. K., Hawley, J. E., and Riddolls, B. W. (1977), 'Slope stability in urban development', *Department of Scientific and Industrial Research (New Zealand), Information Series*, 122: 1–71.

Taylor, G. and Ruxton, B. P. (1987), 'A duricrust catena in South-east Australia', *Zeitschrift für Geomorphologie*, 31: 385–410.

Taylor, R. K. and Spears, D. A. (1970), 'The breakdown of British Coal Measure rocks', *International Journal of Rock Mechanics and Mining Science*, 7: 481–501.

—— —— (1981), 'Laboratory investigation of mudrocks', *Quarterly Journal of Engineering Geology*, 14: 291–309.

Taylor, R. M. (1987), 'Non-silicate oxides and hydroxides', in A. C. D. Newman (ed.), *Chemistry of Clays and Clay Minerals, Mineralogical Society Monograph*, 6 (London: Longman Scientific and Technical).

Tennett, R. M. (1971), *Science Data Book* (Edinburgh: Oliver and Boyd).

Ter-Stepanian, G. I. (1977), 'Deep-reaching gravitational deformation of mountain slopes', *Bulletin of the International Association of Engineering Geology*, 16: 87–94.

Terzaghi, K. (1936a). 'The shearing resistance of saturated soils', *Proceedings, First International Conference on Soil Mechanics*, 1: 54–6.

—— (1936b), 'Stability of slopes in natural clay', *Proceedings, 1st International Conference on Soil Mechanics and Foundation Engineering*, 1: 161–5.

—— (1950), 'Mechanism of Landslides', *Geological Society of America, Engineering Geology (Berkey) Volume*, 83–123.

—— (1953), 'Some miscellaneous notes on creep', *Proceedings, 3rd International Conference on Soil Mechanics and Foundation Engineering*, 3: 205–6.

—— (1962), 'Stability of steep slopes in hard unweathered rock', *Géotechnique*, 12: 251–70.

—— and Peck, R. B. (1948), *Soil Mechanics in Engineering Practice* (New York: Wiley).

Tharp, T. M. (1987), 'Conditions for crack propagation by frost wedging', *Geological Society of America Bulletin*, 99: 94–102.

Thomas, M. F. (1965), 'Some aspects of the geomorphology of domes and tors in Nigeria', *Zeitschrift für Geomorphologie*, 9: 63–81.

Thomas, M. F. (1989a), 'The role of etch processes in landform development: I. etching concepts and their applications', *Zeitschrift für Geomorphologie*, 33: 129–42.

—— (1989b), 'The role of etch processes in landform development: II. etching and the formation of relief', *Zeitschrift für Geomorphologie*, 33: 257–74.

Tobutt, D. C. (1982), 'Monte Carlo simulation methods for slope stability', *Computers and Geoscience*, 8: 199–208.

Toksoz, M. N., Cheng, G. H., and Timur, A. (1976), 'Velocities of seismic waves in porous rocks', *Geophysics*, 41: 621–45.

Tonkin, P. J. and Basher, L. R. (1990), 'Soil-stratigraphic techniques in the study of soil and landform evolution across the Southern Alps, New Zealand', *Geomorphology*, 3: 547–75.

Torrance, J. K. (1983), 'Towards a general model of quick-clay development', *Sedimentology*, 30: 547–55.

—— (1984), 'A comparison of marine clays from Ariake Bay, Japan and the South Nation River landslide site, Canada', *Soils and Foundations, Japanese Society of Soil Mechanics and Foundation Engineering*, 24: 75–81.

—— (1988), 'Mineralogy, pore-water chemistry and geotechnical behaviour of Champlain Sea and related sediments', *Geological Association of Canada, Special Paper*, 35: 259–75.

Towner, G. D. (1973), 'An examination of the fall-cone method for the determination of some strength properties of remoulded agricultural soils', *Journal of Soil Science*, 24: 470–80.

Trendall, A. F. (1962), 'The formation of "apparent peneplains" by a process of combined laterisation and surface wash', *Zeitschrift für Geomorphologie*, 6: 183–97.

Tricart, J. (1962), 'Mécanismes normaux et phénomènes catastrophiques dans l'évolution des versants du bassin du Guil (Hautes-Alpes, France)', *Zeitschrift für Geomorphologie*, 5: 277–301.

Trimble, S. W. (1977), 'The fallacy of stream equilibrium in contemporary denudation studies', *American Journal of Science*, 277: 876–87.

Trudgill, S. (1985), *Limestone Geomorphology* (Harlow: Longman).

Trustrum, N. A. and De Rose, R. D. (1988), 'Soil depth—age relationship of landslides on deforested hillslopes, Taranaki, New Zealand', *Geomorphology*, 1: 143–60.

Tse, R. and Cruden, D. M. (1979), 'Estimating joint roughness coefficients', *International Journal of Rock Mechanics and Mining Sciences and Geomechanics Abstracts*, 16: 303–7.

Tuncer, E. R., Ordemir, I. M., and Oud, M. (1989), 'Soil susceptibility to dispersion', *Proceedings, 12th International Conference on Soil Mechanics and Foundation Engineering (Rio de Janeiro)*, 3: 2115–18.

Turcotte, D. L. (1974), 'Are transform faults thermal contraction cracks?', *Journal of Geophysical Research*, 79: 2573–7.

Turk, N. and Dearman, W. R. (1985), 'Investigation of some rock joint properties: Roughness angle determination and joint closure', *Proceedings, International Symposium on Fundamentals of Rock Joints (Björkliden)*, 197–204.

Turner, M. J., Clough, R. W., Martin, G. C., and Topp, L. J. (1956), 'Stiffness and deflection analysis of complex structures', *Journal of Aeronautical Science*, 23: 805–24.

Tweedale, J. G. (1973), *Materials Technology*, vol. i, *The Nature of Materials* (London: Butterworth).

Twidale, C. R. (1976), 'On the survival of paleoforms', *American Journal of Science*, 276: 77–95.

—— (1982), *Granite Landforms* (Amsterdam: Elsevier).

—— (1990), 'The origin and implications of some erosional landforms', *Journal of Geology*, 98: 343–64.

—— and Bourne, J. A. (1975), 'The subsurface initiation of some minor granite landforms', *Journal of the Geological Society of Australia*, 22: 477–84.

Ui, T. (1983), 'Volcanic dry avalanche deposits: Identification and comparison with non-volcanic debris stream deposits', *Journal of Volcanology and Geothermal Research*, 18: 135–50.

—— Yamamoto, H., and Suzuki-Kamata, K. (1986), 'Characterization of debris avalanche deposits in Japan', *Journal of Volcanology and Geothermal Research*, 29: 231–43.

Updike, R. G., Egan, J. A., Moriwaki, Y., Idriss, I. M., and Moses, T. L. (1988*a*), 'A model for earthquake-induced translatory landslides in Quaternary sediments', *Geological Society of America Bulletin*, 100: 783–92.

—— Olsen, H. W., Schmoll, H. R., Kharaka, Y. K., and Stokoe, K. H. (1988*b*), 'Geologic and geotechnical conditions adjacent to the Turnagain Heights landslide, Anchorage, Alaska', *U.S. Geological Survey Bulletin*: 1817.

Ursic, S. J. and Dendy, F. E. (1965), 'Sediment yields from small watersheds under various land uses and forest covers', *Proceedings, Federal Inter-Agency Sedimentation Conference 1963, U.S. Department of Agriculture, Miscellaneous Publications*, 970: 47–52.

US Department of the Interior, Bureau of Reclamation (undated), *Engineering Geology Field Manual* (Washington, DC).

Valentine, K. W. G. and Dalrymple, J. B. (1975), 'The identification, lateral variation and chronology of two buried paleocatenas at Woodhall Spa and West Runton, England', *Quaternary Research*, 5: 551–90.

Valliapan, S. and Evans, R. S. (1980), 'Finite element of a slope at Illawarra Escarpment', *Proceedings, 3rd Australia–New Zealand Geomechanics Conference*, 2: 241–6.

Vanarsdale, R., Costello, P., and Marcelletti, N. (1989), 'Denudation of highwalls near Manchester, Kentucky', *Engineering Geology*, 26: 111–23.

Van Asch, Th. W. J. (1983), 'The stability of slopes in the Ardennes region', *Geologie en Mijnbouw*, 62: 683–8.

—— and Van Genuchten, P. M. B. (1990), 'A comparison between theoretical and measured creep profiles of landslides', *Geomorphology*, 3: 45–55.

—— Deimel, M. S., Haak, W. J. C., and Simon, J. (1989), 'The viscous creep component in shallow clayey soil and the influence of tree load on creep rates', *Earth Surface Processes and Landforms*, 14: 557–64.

Van de Graaff, W. J. E., Crowe, R. W. A., Bunting, J. A., and Jackson, M. J. (1977), 'Relict early Cainozoic drainages in arid Western Australia', *Zeitschrift für Geomorphologie*, 21: 379–400.

Van Dine, D. F. (1985), 'Debris flows and debris torrents in the southern Canadian Cordillera', *Canadian Geotechnical Journal*, 22: 44–68.

Van Olphen, H. (1977), *An Introduction to Clay Colloid Chemistry*, 2nd edn., (New York: Wiley).

Van Steijn, H., van Brederode, L. E., and Goedheer, G. J. (1984), 'Stratified slope deposits of the grèze-litée type in the Ardèche region in the south of France', *Geografiska Annaler*, 66A: 295–305.

Varnes, D. J. (1958), 'Landslide types and processes', *Highway Research Board, Special Report* (Washington, DC), 29: 20–47.

—— (1975), 'Slope movements in the Western United States', in *Mass Wasting* (Norwich: Geo Abstracts), 1–17.

—— (1984), *Landslide Hazard Zonation: A Review of Principles and Practice* (Paris: UNESCO).

Vaughan, P. R. (1988), 'Keynote paper: Characterising the mechanical properties of in situ residual soil', *Proceedings, 2nd International Conference on Geomechanics in Tropical Soils (Singapore)*, 2: 469–76.

Veder, C. (1981), *Landslides and Their Stabilization* (New York: Springer-Verlag).

Vickers, B. (1983), *Laboratory Work in Soil Mechanics*,

2nd edn., (London: Granada).

Vincent, P. J. and Clarke, J. V. (1976), 'The terracette enigma: A review', *Biuletyn Peryglacjalny*, 25: 65–77.

Voight, B. (1973a), 'Correlation between Atterberg plasticity limits and residual shear strength of natural soils', *Géotechnique*, 23: 265–7.

—— (1973b). 'The mechanics of retrogressive block-gliding, with emphasis on the evolution of the Turnagain Heights Landslide, Anchorage, Alaska', in K. A. DeJong and R. Scholten (eds.), *Gravity and Tectonics* (New York: Wiley), 97–121.

—— (ed.) (1978), *Rock Slides and Avalanches* (Amsterdam: Elsevier).

—— and Pariseau, W. G. (1978), 'Rockslides and avalanches: An introduction', in B. Voight (ed.), *Rockslides and Avalanches* (Amsterdam: Elsevier), 1–67.

—— and St. Pierre, B. H. P. (1974), 'Stress history and rock stress', *Proceedings, Third Congress of the International Society for Rock Mechanics (Denver)*, 2: 580–2.

—— Janda, R. J., Glicken, H., and Douglass, P. M. (1983), 'Nature and mechanics of the Mount St. Helens rockslide-avalanche of 18 May 1980', *Géotechnique*, 33: 243–73.

Vutukuri, V. S., Lama, R. D., and Saluja, S. S. (1974), *Handbook on Mechanical Properties of Rocks*, i. (Clausthal: Trans Tech Publications).

Vyalov, S. S. (1986), *Rheological Fundamentals of Soil Mechanics* (Amsterdam: Elsevier).

Wahlstrom, E. E. (1973), *Tunnelling in Rock* (Amsterdam: Elsevier).

Walcott, R. I. (1972), 'Late Quaternary vertical movements in eastern North America: Quantitative evidence of glacio-isostatic rebound', *Review of Geophysics and Space Physics*, 10: 849–84.

Walder, J. and Hallet, B. (1985), 'A theoretical model of the fracture of rock during freezing', *Geological Society of America Bulletin*, 96: 336–46.

—— —— (1986), 'The physical basis of frost weathering: Toward a more fundamental and unified perspective', *Arctic and Alpine Research*, 18: 27–32.

Waldron, L. J. (1977), 'The shear resistance of root-permeated homogeneous and stratified soil', *Soil Science Society of America Journal*, 41: 843–9.

Walker, B. F., Blong, R. J., and MacGregor, J. P. (1987), 'Landslide classification, geomorphology, and site investigations', in B. F. Walker and R. Fell (eds.), *Soil Slope Instability and Stabilisation* (Rotterdam: Balkema), 1–52.

Walker, P. H. (1963), 'Soil history and debris avalanche deposits along the Illawarra Scarpland', *Australian Journal of Soil Research*, 1: 223–30.

Walling, D. E. (1978), 'Reliability considerations in the evaluation and analysis of river loads', *Zeitschrift für Geomorphologie, Supplementband*, 29: 29–42.

—— and Bradley, S. B. (1990), 'Some applications of caesium-137 measurements in the study of erosion, transport and deposition', *IAHS Publication*, 189: 179–203.

—— and Webb, B. W. (1983a), 'Patterns of sediment yield', in K. J. Gregory (ed.) *Background to Paleohydrology* (Chichester: Wiley), 69–100.

—— —— (1983b), 'The dissolved loads of rivers: A global overview', *IAHS Publication*, 141: 3–20.

Waltham, A. C. (1989), *Ground Subsidence* (Glasgow: Blackie).

Ward, R. G. W. (1985), 'Geomorphological evidence of avalanche activity in Scotland', *Geografiska Annaler*, 67A: 247–56.

—— Langmuir, E. D. G., and Beattie, B. (1985), 'Snow profile and avalanche activity in the Cairngorm Mountains, Scotland', *Journal of Glaciology*, 31: 18–27.

Wasson, R. J. (1979), 'Stratified debris slope deposits in the Hindu Kush, Pakistan', *Zeitschrift für Geomorphologie*, 23: 301–20.

Wasti, Y. and Bezirci, M. H. (1986), 'Determination of the consistency limits of soils by the fall cone test', *Canadian Geotechnical Journal*, 23: 241–6.

Watkins, J. R. (1967), 'The relationship between climate and the development of landforms in Cainozoic rocks of Queensland', *Journal of the Geological Society of Australia*, 14: 153–68.

Wayland, E. J. (1933), 'Peneplains and some other erosional platforms', *Annual Report and Bulletin of the Protectorate of Uganda Geological Survey and Department of Mines*, note 1: 77–9.

Waylen, M. J. (1979), 'Chemical weathering in a drainage basin underlain by Old Red Sandstone', *Earth Surface Processes*, 4: 167–78.

Weaver, C. E. (1978), 'Mn-Fe coatings on saprolite fracture surfaces', *Journal of Sedimentary Petrology*, 48: 595–610.

Weissbach, G. and Kutter, H. K. (1978), 'The influence of stress and strain history on the shear strength of rock joints', *Proceedings of the International Association of Engineering Geology 3rd International Congress (Madrid)*, sect. II, 2: 88–92.

Wesley, L. D. (1977), 'Shear strength properties of halloysite and allophane clays in Java, Indonesia', *Géotechnique*, 27: 125–36.

Whalley, W. B. (1974), 'The mechanics of high magnitude, low-frequency rock failure', *Geographical Papers* (University of Reading), 27.

—— (1984), 'Rockfalls', in D. Brunsden and D. B. Prior (eds.), *Slope Instability* (Chichester: Wiley), 217–56.

—— Marshall, J. R., and Smith, B. J. (1982), 'Origin of desert loess from some experimental observations', *Nature*, 300: 433–5.

Whipkey, R. Z. and Kirkby, M. J. (1978), 'Flow within the soil', in M. J. Kirkby (ed.), *Hillslope Hydrology* (Chichester: Wiley), 121–44.

Whitaker, C. R. (1974), 'Split boulder', *Australian Geographer*, 12: 562–3.

White, W. B., Jefferson, G. L., and Haman, J. F. (1967), 'Quartzite karst in southeastern Venezuela', *International Journal of Speleology*, 2: 309–14.

Whitehouse, I. E. (1983), 'Distribution of large rock avalanche deposits in the central Southern Alps, New Zealand', *New Zealand Journal of Geology and Geophysics*, 26: 271–9.

—— (1988), 'Geomorphology of the central Southern Alps, New Zealand: The interaction of plate collision and atmospheric circulation', *Zeitschrift für Geomorphologie Supplementband*, 69: 105–16.

—— and Griffiths, G. A. (1983), 'Frequency and hazard of large rock avalanches in the central Southern Alps, New Zealand', *Geology*, 11: 331–4.

—— and McSaveney, M. J. (1983), 'Diachronous talus surfaces in the Southern Alps, New Zealand, and their implications to talus accumulation', *Arctic and Alpine Research*, 15: 53–64.

Whitlow, R. and Shakesby, R. A. (1988), 'Bornhardt micro-geomorphology: Form and origin of micro-valleys and rimmed gutters, Domboshava, Zimbabwe', *Zeitschrift für Geomorphologie*, 32: 179–94.

Whitman, R. V. and Bailey, W. A. (1967), 'Use of computers for slope stability analysis', *Proceedings, American Society of Civil Engineers*, 93 (SM4): 475–98.

Whyte, I. L. (1982), 'Soil plasticity and strength: A new approach using extrusion', *Ground Engineering*, 15: 16–24.

Wickham, G. E., Tiedemann, H. R., and Skinner, E. H. (1972), 'Support determinations based on geologic predictions', *Proceedings, First North American Rapid Excavation and Tunnelling Conference (New York)*, 1: 43–64.

Wild, A. (ed.) (1988), *Russell's Soil Conditions and Plant Growth*, 11th edn. (Harlow: Longman Scientific and Technical).

Wilhelmy, H. (1964), 'Cavernous rock surfaces (tafoni) in semi-arid and arid climates', *Pakistan Geographical Review (Lahore)*, 19: 9–13.

Williams, R. B. G. and Robinson, D. A. (1983), 'The effect of surface texture on the determination of the surface hardness of rock using the Schmidt hammer', *Earth Surface Processes and Landforms* 8: 289–92.

—— —— (1989), 'Origin and distribution of polygonal cracking of rock surfaces', *Geografiska Annaler*, 71A: 145–59.

Wiltshire, G. R. (1960), 'Rainfall intensities in New South Wales', *Journal of the Soil Conservation Service of New South Wales*, 16: 54–69.

Winkler, E. M. (1966), 'Important agents of weathering for building and monumental stone', *Engineering Geology*, 381–400.

—— (1970), 'The importance of air pollution in the corrosion of stone and metals', *Engineering Geology*, 4: 327–34.

—— (1977), 'Insolation of rock and stone, a hot item', *Geology*, 5: 188–9.

—— and Wilhelm, E. J. (1970), 'Salt burst by hydration pressures in architectural stone in urban atmosphere', *Geological Society of America Bulletin*, 81: 567–72.

Wischmeier, W. H. (1959), 'A rainfall erosion index for a universal soil-loss equation', *Proceedings, Soil Science Society of America*, 25: 246–9.

—— and Mannering, J. V. (1969), 'Relation of soil properties to its erodibility', *Proceedings, Soil Science Society of America*, 33: 131–7.

—— and Smith, D. D. (1965), 'Predicting rainfall-erosion losses from cropland east of the Rocky Mountains', *US Department of Agriculture Handbook*, 282: 1–47.

—— Johnson, C. B., and Cross, B. V. (1971), 'A soil erodibility nomograph for farmland and construction sites', *Journal of Soil and Water Conservation*, 26: 189–93.

—— Smith, D. D., and Uhland, R. E. (1958), 'Evaluation of factors in the soil-loss equation', *Agricultural Engineering*, 39: 458.

Wolfe, J. A. (1977), 'Large Holocene low-angle landslide, Somar Island, Philippines', in D. R. Coates (ed.), *Landslides, Reviews in Engineering Geology*, 3 (Boulder, Col.: Geological Society of America).

Wolle, C. M. and Hachich, W. (1989), 'Rain-induced landslides in southeastern Brazil', *Proceedings, 12th International Conference on Soil Mechanics and Foundation Engineering (Rio de Janeiro)*, 3 (Rotterdam: Balkema), 1639–42.

Wolman, M. G. (1967), 'A cycle of sedimentation and erosion in urban river channels', *Geografiska Annaler*, 49A: 385–95.

—— and Gerson, R. (1978), 'Relative scales of time and effectiveness of climate in watershed geomorphology', *Earth Surface Processes*, 3: 189–208.

—— and Miller, J. P. (1960), 'Magnitude and frequency of forces in geomorphic processes', *Journal of Geology*, 68: 54–74.

Wopfner, H. and Twidale, C. R. (1967), 'Geomorphological history of the Lake Eyre Basin' in J. N. Jennings and J. A. Mabbutt (eds.), *Landform Studies from Australia and New Guinea* (Canberra: ANU Press), 118–42.

Worley Consultants Ltd., (1987), *Slope Stability Assessment at Building Sites*, BRANZ Study Report, SR4 (Building Research Association of New Zealand).

Wroth, C. P. and Wood, D. M. (1978), 'The correlation of index properties with some basic engineering properties of soils', *Canadian Geotechnical Journal*, 15: 137–45.

Wu, T. H., Williams, R. L., Lynch, J. E., and Kulatilake, P. H. S. W. (1987), 'Stability of slopes in Red

Conemaugh Shale of Ohio', *Journal of Geotechnical Engineering*, 113: 248–64.

Wuerker, R. G. (1955), 'Annotated tables of strength and elastic properties of rock', *Transactions of the Australian Institute of Mining Engineers*, 202: 157.

Wyllie, D. C. (1980), 'Toppling rock slope failures, examples of analysis and stabilization', *Rock Mechanics*, 13: 89–98.

Xu, S. and De Freitas, M. H. (1990a), 'The complete shear stress-vs-shear displacement behaviour of clean and infilled rough joints', in N. Barton and O. Stephansson (eds.), *Rock Joints* (Rotterdam: Balkema), 341–8.

—— —— (1990b), 'Kinematic mechanisms of shear deformation and the validity of Barton's shear models', in N. Barton and O. Stephansson (eds.), *Rock Joints* (Rotterdam: Balkema), 767–74.

Xue-Cai, F. and An-Ning, G. (1986), 'The principal characteristics of earthquake landslides in China', *Geologia Applicata e Idrogeologia*, 21: 27–45.

Yaalon, D. H. (ed.) (1971), *Paleopedology: Origin, Nature and Dating of Paleosols* (Jerusalem: International Society of Soil Science and Israel Universities Press).

—— and Ganor, E. (1973), 'The influence of dust on soils during the Quaternary', *Soil Science*, 116: 146–55.

Yatsu, E. (1988), *The Nature of Weathering: An Introduction* (Tokyo: Sozosha).

Yee, C. S. and Harr, R. D. (1977), 'Influence of soil aggregation on slope stability in the Oregon Coast Ranges', *Environmental Geology*, 1: 367–77.

Yim, K. P., Heung, L. K., and Greenway, D. R. (1988), 'Effect of root reinforcement on the stability of three fill slopes in Hong Kong', *Proceedings, 2nd International Conference on Geomechanics in Tropical Soils (Singapore)*, 1: 293–9.

Yong, R. N. and Warkentin, B. P. (1966), *Introduction to Soil Behavior* (New York: Macmillan).

Yoshida, N., Morgenstern, N. R., and Chan, D. H. (1990), 'A failure criterion for stiff soils and rocks exhibiting softening', *Canadian Geotechnical Journal*, 27: 195–202.

Yoshikawa, T. (1974), 'Denudation and tectonic movement in contemporary Japan', *Bulletin of the Department of Geography, University of Tokyo*, 6: 1–14.

Yoshinaka, R. and Yamabe, T. (1981), 'Deformation behaviour of soft rocks', *Proceedings, International Symposium on Weak Rock (Tokyo)*, 87–92.

Young, A. (1961), 'Characteristic and limiting slope angles', *Zeitschrift für Geomorphologie*, 5: 126–31.

—— (1963), 'Deductive models of slope evolution', *Nachrichten der Akademie der Wissenschaften in Göttingen*, II. *MathematischPhysikalische Klasse*, 5: 45–66.

—— (1972), *Slopes* (Edinburgh: Oliver and Boyd).

—— (1974a), 'The rate of slope retreat', *Institute of British Geographers, Special Publication*, 7: 65–78.

—— (1974b), 'Slope profile survey', *British Geomorphological Research Group, Technical Bulletin*, 11.

—— (1978), 'A twelve-year record of soil movement on a slope', *Zeitschrift für Geomorphologie, Supplementband*, 29: 104–10.

—— and Saunders, I. (1986), 'Rates of surface processes and denudation', in A. D. Abrahams (ed.), *Hillslope Processes* (Boston: Allen and Unwin), 3–27.

Young, A. R. M. (1987), 'Salt as an agent in the development of cavernous weathering', *Geology*, 15: 962–6.

Young, R. W. (1985), 'Silcrete distribution in eastern Australia', *Zeitschrift für Geomorphologie*, 29: 21–36.

—— (1986), 'Tower karst in sandstone, Bungle Bungle massif, northwestern Australia', *Zeitschrift für Geomorphologie*, 30: 189–202.

—— (1987), 'Sandstone landforms of the tropical East Kimberley region, northwestern Australia, *Journal of Geology*, 95: 205–18.

—— (1988), 'Quartz etching and sandstone karst: Examples from the East Kimberleys, northwestern Australia', *Zeitschrift für Geomorphologie*, 32: 409–23.

Yu, Y. S. and Coates, D. F. (1970), 'Analysis of rock slopes using the finite element method', *Department of Energy, Mines and Resources Mines Branch, Mining Research Centre, Research Report*, R229 (Ottawa).

Zanbak, C. (1983), 'Design charts for rock slopes susceptible to toppling', *Journal of Geotechnical Engineering*, 109: 1039–62.

Zaruba, Q. and Mencl, V. (1969), *Landslides and their Control* (Amsterdam: Elsevier).

—— —— (1976), *Engineering Geology* (Amsterdam: Elsevier).

Zienkiewicz, O. C. (1971), *The Finite Element Method in Engineering Science* (New York: McGraw-Hill).

Zingg, A. W. (1940), 'Degree and length of land slope as it affects soil loss in runoff', *Agricultural Engineering*, 21: 59–64.

Index